A COAST TO EXPLORE

Coastal Geology and Ecology of Central California

Miles O. Hayes and Jacqueline Michel

Joseph M. Holmes

Illustrator

PANDION BOOKS

a Division of Research Planning, Inc.
Columbia, South Carolina

PANDION BOOKS
a division of Research Planning, Inc.
P.O. Box 328
Columbia, South Carolina USA 29202
Email: permissions@researchplanning.com

A CIP catalog record for this book has been applied for
from the Library of Congress.

ISBN 978-0-9816618-1-0

All photographs by Miles O. Hayes and Jacqueline Michel unless otherwise indicated.
Illustration and design by Joseph M. Holmes.

Excerpts from TO A GOD UNKNOWN by John Steinbeck, copyright 1933, renewed (c) 1961 by John Steinbeck. Used by
permission of Viking Penguin, a division of Penguin Group (USA) inc.

Excerpts from TORTILLA FLAT by John Steinbeck, copyright 1935, renewed (c) 1963 by John Steinbeck. Used by
permission of Viking Penguin, a division of Penguin Group (USA) inc.

Excerpts from CANNERY ROW by John Steinbeck, copyright 1945, renewed (c) 1973 by Elaine Steinbeck, John Steinbeck
IV and Thom Steinbeck. Used by permission of Viking Penguin, a division of Penguin Group (USA) inc.

Excerpts from EAST OF EDEN by John Steinbeck, copyright 1952, renewed (c) 1980 by Elaine Steinbeck, John Steinbeck
IV and Thom Steinbeck. Used by permission of Viking Penguin, a division of Penguin Group (USA) inc.

Excerpts from SWEET THURSDAY by John Steinbeck, copyright 1954, renewed (c) 1982 by Elaine Steinbeck, Thom
Steinbeck and John Steinbeck IV. Used by permission of Viking Penguin, a division of Penguin Group (USA) inc.

The paper used in this book meets the minimum requirements
of the American National Standard for Information
Sciences – Permanence of Paper for
Printed Library Materials,
ANSI/NISO Z39.48-1992

Printed in Korea by Tara TPS through
Four Colour Imports, Ltd., Louisville, KY

Front Cover: Near Avila Beach (San Luis Obispo County) on 5 December 2008.
Back Cover: Beach near mouth of Santa Ynez River (Santa Barbara County) on 5 December 2008.

DEDICATION

We were in the Elkhorn Slough on Sunday, 19 January 1992, preparing for a field training course for the Marine Spill Response Corporation that was scheduled to start the next day in Monterey. As we drove up to the visitor's center, we noticed that the flags were at half-mast. We learned when we got to Monterey that, on the previous Saturday, 11 January, two employees of the Oil Spill Prevention and Response unit of the California Department of Fish and Game had died while responding to a potential oil spill in the Carquinez Strait in San Francisco Bay. The helicopter they were riding in struck a power line and crashed. They were Greg Cook and Sonia Hamilton.

When we got back to Columbia, S.C. the next Sunday, Hayes wrote the following in his diary: *Turns out they were in our course in November (1991). Sonia seemed to be the most inspired of the biologists. I could tell by the look in her eyes that she felt this was the ultimate thing for her to do. Greg was a game warden who was so glad to be doing this. Guess they were so excited to be on this new team that they volunteered to work on a Saturday. I tried not to think about it too much, as that cloud hung low over the whole course.*

This book is dedicated to the memory of SONIA and GREG, and to all the others like them, be they field scientists or U.S. Coast Guard responders and strike-team members, who work the long hours for not much pay to do what little they can do to protect the coastal environment.

TABLE OF CONTENTS

SECTION II: COASTAL ECOLOGY OVERVIEW

SECTION III: MAJOR COMPARTMENTS

LIST OF FIGURES

PREFACE

One might rightfully wonder why a duo of scientists native to and permanent residents of the Carolinas would write a book on the coastal geology and ecology of the Central California Coast. It all started in March 1967, when a supertanker named the *Torrey Canyon* was shipwrecked off the west coast of Cornwall County, England, giving scientists their first taste of trying to respond to a major tanker oil spill in the western world. A couple of years later, in January 1969, an offshore rig in the Santa Barbara Channel suffered a blow-out, eventually spilling between 3,360,000 to 4,200,000 gallons of crude oil that oiled the coastline from Goleta to Rincon, and all four of the northern Channel Islands (Wikipedia Encyclopedia). Thus, the scientific study of and response to oil spills was begun in a big way in the U.S.

Meanwhile, co-author Hayes, who was a Professor in the Geology Department at the University of South Carolina at that time, was happily directing the research of an army of graduate students on the coasts of Alaska, South Carolina, and elsewhere, and co-author Michel was working on her Ph.D. degree in geology at the University of South Carolina, neither with any thoughts of consequence about oil spills.

But in the summer of 1975, soon after arriving back in South Carolina from a couple of months of research on the Alaska shore, Hayes was solicited by the National Science Foundation to check out an oil spill by the tanker *Metula* in the Strait of Magellan, Chile, to see if any information could be gained that might be applied to the Alaska area in the possible event of a major oil spill there some time in the future. This eventually evolved into a full-blown study of the spill in Chile, which Michel joined in. The formation of an oil-spill study group at the University of South Carolina evolved from that project.

Lessons learned in Chile, and later at the *Urquiola* oil spill in Spain in the spring of 1976, allowed us to devise a coastline mapping system, now called the Environmental Sensitivity Index (ESI), that provides planners a systematic database outlining the most critical areas and resources along their shoreline that should be protected in the event of an oil spill. This system was first devised by us after a field study in Lower Cook Inlet, Alaska in the late summer of 1976.

By the time of the *Exxon Valdez* oil spill in Alaska in March 1989, we had been working under contract for the National Oceanic and Atmospheric Administration (NOAA), providing scientific support to the U.S. Coast Guard for eight years, during which time we assisted in the response to most of the largest oil spills of that era (e.g., *Amoco Cadiz* in France and *Ixtoc I* in Mexico and South Texas). We had also formed our own private company, Research Planning, Inc. (RPI), which the NOAA contract was assigned to. Hayes had resigned from the University and worked full time for the company, and Michel, who would spend many months of field work at the *Exxon Valdez* oil spill site, was in charge of the oil-spill response work in the company. During that time frame, we conducted coastline sensitivity mapping projects in many coastal states, including California.

This experience allowed us to offer oil-spill training courses related to oil-spill response, coastal clean-up strategies, and sensitivity mapping for many groups around the country, especially the U.S. Coast Guard. Many private industries, such as Shell Oil, Chevron, and Amoco, sent participants to these courses. We have conducted a number of these courses on the California coast under the sponsorship of NOAA, the Oil Spill Prevention and Response (OSPR) unit of the California Department of Fish and Game, the U.S. Coast Guard, the Marine Spill Response Corporation (MSRC), and Chevron Oil Company. Daily field trips to the different coastal areas were a part of these courses. Several of these were taught in Monterey.

The coastal mapping projects by RPI have been in full sway in California (under sponsorship of NOAA and OSPR) since the early 1980s. For example, the first maps we created for the Southern California region were published in 1980, an updated digital version was published in 1995, and we are currently, in 2009, working on the latest update. The latest set of digital maps for the Northern California Coast was published (by NOAA and OSPR) in 2008, and the latest one for the Central California Coast was published in 2006. The project on the Central California Coast provided our primary inspiration for the publication of this book. Both of us participated in the fieldwork in Central California, personally conducting the overflights and coastal geomorphology mapping for the project.

Another key project that allowed us to examine in the field the entire California coast was making detailed strategies for the protection of, and prevention of oil passing through, the 170 tidal inlets in the state. This project, which was carried out in 1992, was a joint venture between MSRC and OSPR.

And last, but not least, our company, RPI, led by co-author Michel, has continued to be a part of the scientific teams responding to spills in California, most notably the following ones:

1) *Puerto Rican*, 1984, offshore of San Francisco – 1.74 million gallons of different types of oils

2) Shell Martinez tank failure in San Francisco Bay, 1988 – 400,000 gallons of crude oil

3) *T/V American Trader*, Huntington Beach, 1990 – 300,000 gallons of crude oil

4) Cantara Loop - Sacramento River, 1991 – 19,000 gallons of the herbicide metam sodium

5) Guadalupe oil field contamination, 1994 – 12 million gallons of diluent

6) *SS Cape Mohican*, 1996, San Francisco Bay – 6,000 gallons of Intermediate Fuel Oil

7) *M/V Cosco Busan*, 2007, San Francisco Bay – 58,000 gallons of Intermediate Fuel Oil

Whereas we won't be taking you on any boat rides, beach walks, or overflights in a small airplane, which is usually a part of our field seminars and mapping projects, we hope the descriptions and illustrations in this book will convey a little of the excitement that we always feel when we are out in the field along the unsurpassed Central California Coast.

Miles O. Hayes and Jacqueline Michel, Columbia, S.C., 10 February 2009.

ACKNOWLEDGEMENTS

First and foremost, we would like to thank Dr. Ed Clifton, Emeritus Scientist of the U.S. Geological Survey in Menlo Park, and fellow coastal geomorphologist, for his detailed scrutiny of our first draft of this book. Heeding his suggestions for changes throughout the book has greatly improved both its flow and content. His many years of research on the coast has enabled him to notably enlighten us on topics such as the origin of Monterey Bay, surf zone dynamics, and the fascinating rock suites at Pebble Beach and Point Lobos. What a break for us that Ed is now semi-retired and living in Monterey, where he continues his efforts to solve the mysteries of this remarkable coastline.

We would also like to thank Dr. Doris Sloan, Adjunct Professor in the Department of Earth and Planetary Science at UC Berkeley, who also reviewed our first draft, setting us straight on many aspects of the detailed and complex geological history of the area. We were very pleased that she found the time to review our first attempts to unravel the intricacies of this famous and challenging geological setting.

However, we should add that we take full responsibility for any misinterpretations we might have made in our conclusions, with respect to both the suggestions of Ed and Doris, and our interpretation of the impressive volume of scientific literature that now exists for the Central California Coast (on a variety of ecological and geological topics).

We would also like to thank our friends, associates, and sponsors with the National Oceanic and Atmospheric Administration (NOAA) and the Office of Spill Prevention and Response, California Department of Fish and Game (CDF&G), who have shared with us the experience of learning about this coast. At NOAA, John Robinson, Robert Pavia, and Jim Morris have supported our research along the California coast for years, including funding for the Environmental Sensitivity Index (ESI) Atlases that we used extensively for the ecology sections. Jan Roletto at the Gulf of the Farallones and Cordell Bank National Marine Sanctuaries provided data on oiled birds monitoring from the Beach Watch Program. At CDF&G, Kim McCleneghan, Don Lollock, John Tarpley, Jim Hardwick, Randy Imai, and Joe Lesch joined us in the field and have been close colleagues on the ESI projects and during spill responses.

Kenneth and Gabrielle Adelman generously allowed us to use their oblique aerial photographs taken over the years as part of the California Coastal Records Project, an aerial photographic survey of the California coastline that includes photographs as early as 1972. With these photographs, the reader can view the shoreline features from offshore, a perspective that is both unique and informative. We encourage you to visit their website (http://www.californiacoastline.org/); you will be amazed at the quality of the photographs and the value of being able to observe the changes over time.

At Research Planning, Inc., Mark White, Lee Diveley, and Katy Beckham used the ESI data to produce the figures showing shoreline habitats and pinniped haulouts. Todd Montello participated in the tidal inlet studies and gathered data for the discussion of leading edge coasts. Chris Boring (the senior ecologist on the ESI Atlases for Northern and Central California) and David Betenbaugh (the geologist who classified the shoreline for Southern California) provided good technical reviews on the draft. Wendy Early assisted in preparation of the manuscript and getting permissions for the figures and quotes. Pandion Books, a Division of Research Planning, Inc., provided funding as well as staff support, particularly by Joe Holmes, our illustrator/cartographer, and Jack Moore, who kept the process of producing the book running smoothly.

1 INTRODUCTION

The authors of this book have been studying coastlines for many years, with co-author Hayes' analysis of the coastlines of the world for the U.S. Navy in 1960-65 being a starting point (Hayes, 1964; 1967a). Since that time, the authors have worked on the shorelines of over 50 countries, including almost all of the shoreline of North America (with numerous projects on the California coast), the Middle East, parts of Central and South America, and West Africa. Hayes also supervised Masters and Doctoral research of 72 graduate students, most of which had to do with coastlines, while a Professor at the Universities of Massachusetts and South Carolina. Michel's specialty of oil-spill response has enabled her to carry out research in a wide variety of shoreline settings.

However, of all of the coasts we have studied, this one has been the most challenging to write about because of its almost unparalleled beauty, fame, and notoriety as recognized by millions of people. There are places on the other shorelines we have studied, for example in Alaska, southern Chile, and Oman, that are just as beautiful and unique, if not more so, but this is Carmel-by-the-Sea, the Monterey Aquarium, Steinbeck country, and the home of the winter/summer beach, a concept we had debunked so many times over the years.

Our goal for this book is to introduce you to the natural setting of the shoreline of Central California, without a doubt the most beautiful in all the "lower forty eight" states, and one you can enjoy exploring during all seasons of the year (see satellite image of the entire coastline under discussion in Figure 1). Before you look further, there is one significant omission from this discussion—San Francisco Bay. The Bay is such a large and diverse area that an adequate coverage of it would require a separate book; therefore, we have chosen to concentrate on the outer shoreline in this presentation.

The overall approach and underlying theme of this book are based first on a description of the **physical characteristics** of the coast, because that physical framework is the basic skeleton upon which other

There are two ways you may want to consider using this book:
1) If you wish to gain an introductory understanding of the coastal processes and land forms, and general ecological trends, you should spend some time right away reading Section I, and II.
2) On the other hand, if you are visiting the coast and you want to learn about a specific area, you can look up that area in Section III. Then, if you have questions about some aspect of that particular area, such as why the beach is eroding or what principal bird groups are present, you can read up on those topics in Sections I and II.

relevant features, such as coastal wetlands, major sand beaches, and bay/lagoon complexes rest. For example, certain fundamental questions, such as why do some segments of the shoreline erode dramatically while others do not, cannot be answered without a firm understanding of the physical processes that cause the erosion and the critical geological makeup of the landscape being eroded. However, none of the major coastal habitats, including their biological components, are neglected as we explore the coastline from Bodega Bay in the north to Point Conception in the south.

Before getting further into this discussion, we need to introduce the term **geomorphology** (*geo* = earth; *morph* = shape, form; *ology* = to study), a scientific discipline devoted to understanding the origin and three-dimensional shape of the landforms of the earth. The discipline of **coastal geomorphology** is focused on the understanding of how the different coastlines of the world originated, as well as their resulting forms and organization.

Both of the authors of this book are coastal geomorphologists. When we started our research in coastal geomorphology, understanding how the coast was made was a young field with but a few practitioners, such as Francis P. Shepard, Willard Bascom, and Douglas Inman, who produced many of the first ideas on the workings of the California Coast.

With regard to our early work, hypotheses on how barrier islands evolve, how the tides and waves shape the coast, and how coastal storms change things were there for the alert observer to deduce, assuming enough examples had been seen and that the global processes had been properly accounted for. We were using what Comet (1996) referred to as "one of Aristotle's two forms of logical inference" – namely inductive reasoning from the observation of a generalized pattern or distribution in order to develop a principle or law. In other words, we started with a large number of more or less random observations, not with a detailed data set on a specific topic. However, we did eventually try to figure out how to collect meaningful data sets to verify or disprove those deduced principles or laws.

Although we call ourselves coastal geomorphologists (or sometimes coastal geologists), our interests and professional activities have provided for us a broad experience in other aspects of **coastal ecology**. We are ardent birders and the work of our company, Research Planning, Inc. (RPI) of Columbia, South Carolina, in areas such as beach erosion, oil-spill response, and natural resources mapping, has broadened our understanding of coastal ecosystems. In fact, we assisted the geologists and biologists in our company as they compiled a summary of the coastal ecosystems of the shoreline under discussion in this book in 2006 under contract to the National Oceanic and Atmospheric Administration (NOAA) that included 41 detailed maps (NOAA/RPI, 2006). The data gathered during that project provided necessary information on many of the sites discussed in Section III.

This book begins in Section I with a brief comparison of this coastline with other shorelines around the world, with emphasis on those shorelines on the western boundaries of North and South America. This treatment is followed by a general discussion of the origin of the California shoreline, emphasizing the newest theories on its geological evolution. The rest of Section I includes a description of the dynamic physical processes that shape the present shoreline, namely waves, tides, major storms, and tsunamis. The major landforms of the Central California Coast are also discussed, specifically rocky coasts, gravel beaches, littoral cells, crenulate bays, tidal inlets, rocky headlands, coastal dune fields, natural groin/sand beach systems, and river deltas. Central California has an abundance of sand beaches, thus there is a detailed discussion of the beach cycle and a review

of the erosional impacts of **major coastal storms** on the beaches.

Section II is a discussion of the key ecological resources of Central California.

In Section III, we discuss and identify opportunities for naturalists to learn about and enjoy the diverse ecosystems that exist within the **seven major geo-morphological compartments** (as defined by us) of the coastal zone of Central California. The description of each compartment includes notes on its geological framework, beach morphology and sediments, and general ecology. Also, areas with significant beach erosion or landslide problems (if present within the compartment) are pointed out. A list and discussion of the **key places to visit** concludes the coverage of each of the seven compartments, which are labeled on Figure 1.

There is a glossary at the end for most of the technical terms we use. And an index is provided to help you find topics and locations.

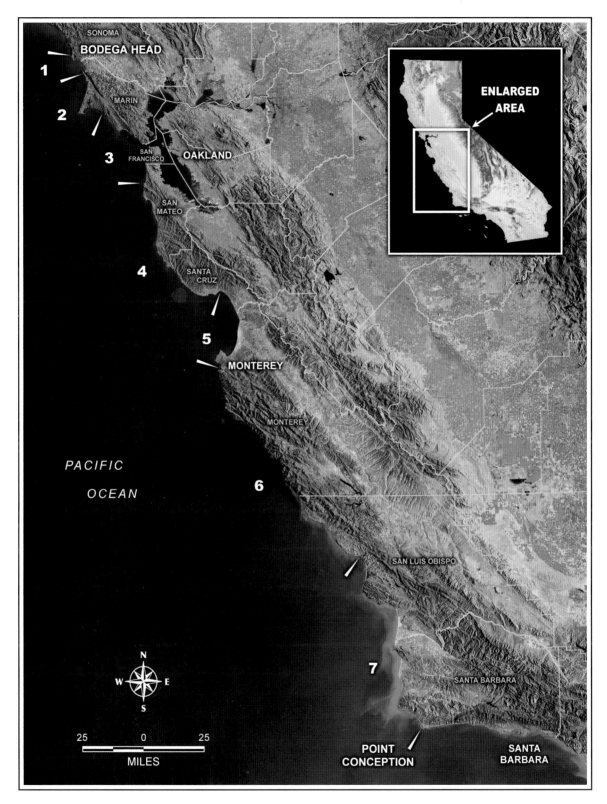

FIGURE 1. Satellite image in color of the Central California Coast. The boundaries of the seven geomorphological compartments discussed in the text are shown: 1) San Andreas Fault Valley; 2) Point Reyes; 3) San Francisco's Outer Shore; 4) Central Headland; 5) Monterey Bay and Vicinity; 6) Big Sur; and 7) The South Bays. The borders and names of the eight counties that occur along this shoreline are also shown. Image derived from National Elevation Dataset and Landsat 7 data acquired from 1999-2002. Courtesy of the California Spatial Information Library (CaSIL).

SECTION I
Coastal Processes and Geomorphology

2 GENERAL STRUCTURE AND ORIGIN

INTRODUCTION

"The rivers and the rocks, the seas and the continents, have changed in all their parts; but the laws which direct those changes and the rules to which they are subject, have remained invariably the same." With these impeccable words, Playfair (1802) conveyed Hutton's doctrine of uniformitarianism (normally expressed as "the present is the key to the past;" Hutton, 1785), and it is upon the strength of this concept that all geologists that study modern shorelines with the intent of understanding their history have built their science. With this in mind, we now begin our daunting task of relating the story of how the Central California Coast has evolved to its present form.

All coastlines have a rather complicated evolution, but the Central California Coast has been among the most difficult ones for geologists to interpret, because of its relatively unstable position along the Pacific Ocean's "ring of fire," as well as the impact the San Andreas Fault system has had in its development. When co-author Hayes was a graduate student at the University of Texas in the 1960s, very few of the professors in the elite, Ivy League universities "believed in" **continental drift**, although the idea had been seriously proposed and discussed by scientists since almost the beginning of the 20th century. The problem for the "non-believers" was that no credible mechanism had been proposed to explain why the continents had drifted apart. Now, of course, the concept of **plate tectonics** provides the basis for almost all current thinking relative to most topics in geology, including the mechanism for continental drift. Geophysical evidence, such as increasing age of parallel magnetic bands of volcanic rocks on the sea floor away from the mid-ocean ridges, provided indisputable evidence that the continents are indeed drifting apart. The exact mechanism causing this movement is more in doubt. The favored theory at this time is that huge individual slabs of rock, called **plates**, are slowly moved apart by convection cells of hot, softened mantle material

below the rigid plates. There are other more complex theories. One fine day all of this will be figured out. Meanwhile, let us get back to the indisputable fact that the continents are moving.

Based on a study of the world's inner continental shelves that was sponsored by the U.S. Navy through the Defense Research Laboratory at the University of Texas, co-author Hayes (1964; 1965) proposed the following preliminary coastal classification based to a large extent on the concept of plate tectonics.

Class	Tectonism
A. Tectonic coasts	rapid uplift
1. Young mountain range coasts	
2. Glacial rebound coasts	
B. Plateau-shield coasts	stable
C. Depositional coasts	rapid downwarp

Tectonism is defined as the forces involved in or producing deformation of the Earth's crust, such as folding or faulting of the rocks, with plate motion usually being the driving mechanism for this deformation. In this case, the term **rapid** is used in a geological sense (measured in millimeters or centimeters per year). A similar concept that related shoreline types to the plate tectonic postulate was presented in considerably more detail by Inman and Nordstrom in 1971. They used the term **collision coasts**, a category into which most young mountain coasts would fit, the term **trailing-edge coasts**, into which many depositional coasts would fit, and so on.

The cross-section given in Figure 2 shows one of the basic concepts of plate tectonics, in which a plate made up of a continental crust (e.g., the North American Plate, which is relatively light because of the abundance of aluminum, silica, sodium, etc. in the rocks) is riding over a sinking oceanic plate (e.g., the Pacific Plate,

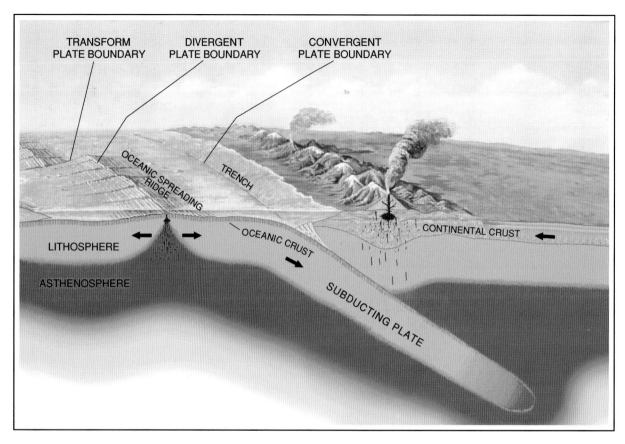

FIGURE 2. Illustration of many of the elements involved in the process of the migration (and collision) of tectonic plates and seafloor spreading. Two key elements in the ocean basin are: 1) the oceanic spreading zone, where new material is added to the sea floor as the two adjacent oceanic plates move away from each other; and 2) the area where the continental and oceanic plates collide, resulting in the creation of an oceanic trench and the subduction of the oceanic plate underneath the continental plate. Modified after José F. Vigil, U.S. Geological Survey (1999).

which is relatively heavy because of the abundance of iron, magnesium, etc. in the rocks). If we look at two of the simpler examples, the North and South American Plates, a striking contrast can be seen between the western shorelines of these continents and their eastern shorelines. Along much of the western shores, where the continental and oceanic plates **collide**, or the **leading edge** of the continental plates, a series of young mountain ranges flank the coast. This comment applies to such areas as the southwest shoreline of Alaska and the entire west coast of South America. The California Coast south of Cape Mendocino is an exception to this general rule, because the Coast Ranges have a distinctly different genesis related to activity along the San Andreas Fault system (discussed in detail later). In contrast, on the eastern shore, or the **trailing edge** of the continental plate, which is moving westward at a rate of around 4 inches/year (Monroe and Wicander, 1998), broad deltaic and coastal plains are present in places. To use this simple example, the western shorelines of a significant

amount of the North American continent and most of South American are dominated by **young mountain range coasts** (collision coasts as defined by Inman and Nordstrom, 1971) characterized by: 1) high mountains caused by relatively recent orogenic (mountain building) activity; 2) bedrock of variable age but dominated by relatively young sedimentary and volcanic rocks; 3) active tectonic uplift (e.g., with old beach lines lifted up the sides of the mountains thousands of feet in some places); and 4) mostly short, steep rivers emptying into the sea. This is a presentation of the leading-edge concept in its simplest terms. The eastern side of the North American Plate is dominated by **depositional coasts** (Amero-trailing edge coasts as defined by Inman and Nordstrom, 1971) characterized by: 1) coastal zone made up chiefly of broad coastal and deltaic plains; 2) bedrock composed generally of Paleogene/Neogene and Quaternary sediments (young geologically speaking; see geological time scale in Table 1); 3) tectonically subsiding areas (e.g., thousands of feet of sediments have accumulated in some areas); and 4) many large and long rivers emptying into the sea. A more elaborate discussion of a coastline of this type is presented by the authors in their recently published book - *A Coast For All Seasons: A Naturalist's Guide to the Coast of South Carolina* (Hayes and Michel, 2008).

For the record, the shorelines termed **plateau-shield coasts** by Hayes fall mostly in Inman and Nordstrom's **Afro-trailing edge coasts**. Hayes (1965) noted that the coastal zones of plateau-shield coasts typically are made up of plateaus and moderate mountain ranges related to pre-Cenozoic orogenic (mountain building) activity. The bedrock along the shoreline is typically composed of ancient basement complexes of granite and gneiss (in shield areas) and Paleozoic and Mesozoic sedimentary rocks (in plateau areas). They are tectonically stable and some moderately long rivers may be present, depending upon the climate. The geological time scale is given in Table 1.

Inman and Nordstrom also defined a major category called **marginal sea coasts** (e.g., the shoreline of the Gulf of Mexico and the Arctic islands of Alaska). Most of these shorelines would be classed as depositional coasts in the Hayes scheme.

CHARACTERISTICS OF DEPOSITIONAL COASTS

Clearly, the Central California Coast, the focus of this book, occurs along the leading edge of the North American Plate, although the impact of the San Andreas Fault system needs to be taken into account (discussed in detail later). Before beginning a more detailed treatment of leading-edge coasts, a brief consideration of the relevant aspects of depositional coasts is in order to clarify the contrasts between these two opposing types of shorelines. Another purpose for this approach is to define some important concepts that have evolved from the study of depositional coasts, which have received somewhat more attention from geologists than leading-edge coasts, particularly those wishing to construct depositional models to be used in oil-and-gas exploration (an activity co-author Hayes and his students were actively engaged in for many years). As already noted, the nature of global tectonic crustal movements are the fundamental cause of the major differences between depositional coasts (commonly present on the trailing edges of continental plates) and young mountain range coasts (commonly present on the leading edge of continental plates). Further research by Hayes (1965; 1976) concluded

TABLE 1. Geologic time. Time intervals are listed in millions of years before the present [from a chart published by the International Commission on Stratigraphy (2008)].

Era/Period/Epoch			Time (Million yr ago)
Cenozoic era "Recent Life"	Quarternary period	Holocene epoch	0.012-0
		Pleistocene epoch	2.6-0.012
	Neogene period	Pliocene epoch	5.3-2.6
		Miocene epoch	23.0-5.3
	Paleogene period	Oligocene epoch	33.9-23.0
		Eocene epoch	55.8-33.9
		Paleocene epoch	65.5-55.8
Mesozoic era	Cretaceous period		145.5-65.5
	Jurassic period		199.6-145.5
	Triassic period		251.0-199.6
Paleozoic era	Permian period		299.0-251.0
	Carboniferous (Mississippian/Pennsylvanian) period		359.2-299.0
	Devonian period		416.0-359.2
	Silurian period		443.7-416.0
	Ordovician period		488.3-443.7
	Cambrian period		542.0-488.3
Precambrian	Proterozoic era		2500-542
	Archaeozoic (Archean) era		4600-2500

that the following hierarchy of factors determine the geomorphological makeup of **depositional coasts** – in decreasing order of importance: 1) hydrodynamic regime; 2) climate; 3) sediment supply and sources; and 4) local geological history and sea-level change. There are some significant differences in the tectonic impacts along trailing-edge and other types of depositional coasts; for example, some areas are sinking more rapidly than others because of heavy sediment loads (e.g., at the mouth of the Mississippi River). However, these differences do not appear to have a significant enough imprint on the regional geomorphology of depositional coasts to warrant further discussion here. Accordingly, we conclude that the factor that exerts the most control on the geomorphological nature of a depositional coastline is the interaction of water in motion with the coastal sediments – typically called the **hydrodynamic regime** (*hydro* = water; *dynamics* = kinetic energy of the water). For purposes of interpreting the morphology of coastal features, the hydrodynamic regime is commonly expressed as the ratio of wave energy to tidal energy (i.e., how big the waves are versus how large the tidal range is). Generally speaking, on trailing edge or epicontinental sea coastlines the tidal range is of utmost importance. On leading edge coastlines, wave action commonly plays a dominant role. During Hayes' early study of the world's shoreline for the U.S. Navy, he observed a striking correlation between tidal range (vertical distance

between high and low tide) and shoreline characteristics of depositional coasts. Eventually in 1976, he classified depositional shoreline types based on this observation, differentiating among **microtidal** (tidal range less than 6 feet), **mesotidal** (tidal range = 6-12 feet), and **macrotidal** (tidal range greater than 12 feet) shoreline types.

Curiously, Hayes (1965) also noted a relatively equal distribution of the three originally defined classes of tidal range around the world's shoreline, which was a surprise to him, because at that time, most of his field experience had been limited to the microtidal shorelines of North Carolina and Texas. Consequently, upon leaving Texas, much of his early research was focused on the mesotidal shorelines of New England and South Carolina. Eventually, macrotidal shorelines with tidal ranges greater than 30 feet were also studied in Alaska, France, and Chile. In addition, based partly on publications by the original guru of coastal geomorphology, W. Armstrong Price, he publicized two additional shoreline types (on depositional coasts), **wave-dominated coasts** and **tide-dominated coasts** (Hayes, 1965; 1976).

As a generalization, most **microtidal depositional coasts** are wave-dominated and characterized by such features as deltas with smooth outer margins and abundant long barrier islands (see Figure 3A). In North America, wave-dominated, depositional coasts occur on much of the shoreline of Texas, the Outer Banks of North Carolina, and in Kotzebue Sound, Alaska.

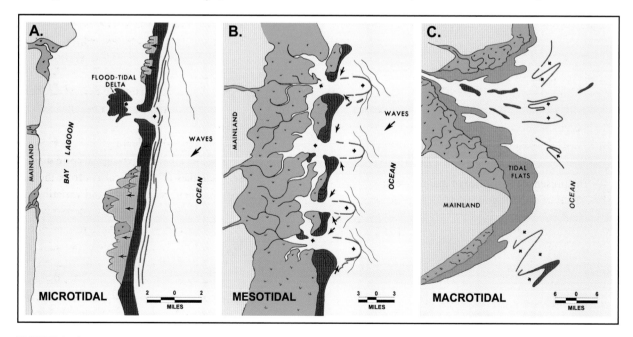

FIGURE 3. Generalized models of the three basic types of depositional coasts (excluding major deltas): (A) Microtidal (tidal range less than 6 feet, usually wave-dominated); (B) Mesotidal (tidal range = 6-12 feet, usually mixed energy); and (C) Macrotidal (tidal range greater than 12 feet, usually tide-dominated). The red lines in A represent nearshore, subtidal wave-formed sand bars, and those in B and C indicate intertidal and shallow subtidal sand deposits.

On the other hand, most **macrotidal depositional coasts** are tide-dominated and characterized by such features as open mouthed, multi-lobate river deltas, abundant large estuarine complexes, and extensive tidal flats and salt marshes (Figure 3C). In North America, tide-dominated depositional coasts occur in Bristol Bay and Cook Inlet, Alaska and in the northern reaches of the Gulf of California.

To complicate matters somewhat, most **mesotidal depositional coasts** have intermediate-sized tides and waves, thus another class of coast, **mixed-energy coasts**, has been recognized by us. Most mixed-energy depositional coasts are characterized by short, drumstick-shaped barrier islands with complex tidal flats and coastal wetlands on their landward flanks (see Figure 3B). In North America, mixed-energy, depositional coasts occur in Georgia and South Carolina, as well as along the Copper River Delta shoreline of Alaska.

A second major factor that determines certain key characteristics of depositional coasts is the regional **climate**. Wind patterns along a coast and offshore storms determine the average size of the waves, which have a major impact on the nature of depositional coasts. Storms, such as the hurricanes that strike the Texas and South Carolina coasts, typically generate the largest waves that occur in those areas. In certain microtidal areas, such as the south Texas coast, wind-driven tides control the geomorphological and ecological conditions of the coastal lagoons. The type of sediments on tidal flats and beaches is strongly related to the regional climate (Hayes, 1967a). For example, with regard to beach sediments, in tropical areas they are commonly composed of coral and algal fragments, carbonate precipitates, and shell. In temperate regions, quartz sand is usually the dominant sediment type, with rock fragments and feldspar being abundant in sand near river mouths and along coasts with eroding bedrock. Other obvious climatic influences include:

1) The presence or absence of major rivers;
2) Effects of glaciers and freeze/thaw processes;
3) Types of weathering (chemical or mechanical) that produce the sediments that make up the coastal systems; and
4) Presence of beachrock in the tropics and permafrost in polar regions.
[NOTE: Permafrost is permanently frozen subsoil. Beachrock is formed by rapid solidification of beach sand through the precipitation of calcium carbonate from sea water into the pore spaces between the sand grains, a process more likely to occur where the water is warm.]

Although tides and waves determine the general framework of a depositional coastline, the climate determines the type of vegetation that occurs on the surface of the coastal habitats. Intertidal zones sheltered from waves are commonly populated by mangroves in the humid tropics, salt marshes in more temperate regions, and algal mats and sediments called evaporates, that have an abundance of precipitated minerals in them, such as halite (NaCl, or salt) and gypsum [$CaSO_4 \cdot (2H_2O)$], in the arid tropics.

Another control on the nature of depositional coasts that is somewhat overrated in our opinion is the **source and supply of sediments**. Without such a supply, there would be no depositional coasts. On depositional coasts, most of the sediment along the shore is derived from the large rivers that empty into coastal waters. Much of the sand on the beaches on the Central California Coast is also delivered by streams, but wave erosion of coastal cliffs is an important contributor in some areas. On both depositional and young mountain range coasts, some beaches are artificially nourished with sand from a variety of sources.

As shown in Figure 4, there are three major classes of sediments found on shorelines – gravel (particles >2 millimeters in diameter, which consists of four separate classes, culminating with the largest - boulders), sand (particles between 0.0625 and 2 millimeters in diameter, ranging from very fine to very coarse grained), and mud (particles <0.0625 millimeters in diameter and consisting primarily of silt and clay). Finer-grained sediments (sand and mud) usually dominate on depositional coasts, but on the Central California Coast, there is a wide range in sediment grain size, from large boulders at the base of many of the eroding cliffs to mud on the tidal flats in coastal water bodies.

The last two factors affecting the nature of depositional shorelines are **local geological controls** and **sea-level rise**. Examples of geological controls include such factors as glacial effects, faulting, and variations in shoreline bedrock type. With regard to sea-level rise, sea level on a global scale has been relatively constant within the past 5,000 years. A key question at the moment is will sea level continue to rise globally as it has been in the past few decades (at the rate of about one foot/century on tectonically stable coastlines like those in North and South Carolina) and, if so, how much? This issue is discussed in more detail later.

The factors that determine the geomorphological characteristics of leading-edge (or collision) shorelines are discussed next.

General Class	Wentworth Scale (Size Description)		Grain Diameter d (mm)
GRAVEL	Boulder		256.0
	Cobble		64.0
	Pebble		4.0
	Granule		2.0
SAND	Sand	Very Coarse	1.0
		Coarse	0.5
		Medium	0.25
		Fine	0.125
		Very Fine	0.0625
MUD	Silt		0.00391
	Clay		0.00024
	Colloid		

FIGURE 4. Definition of sediment types based on size in millimeters (mm). From Wentworth (1922).

CHARACTERISTICS OF LEADING-EDGE COASTS

The coastline under discussion in this book is situated on the leading edge of the North American Plate. However, whereas the classic features of collision coasts, such as young mountain ranges rising adjacent to a major tectonic trench under the influence of a subducting oceanic plate, have been present in the area now known as California throughout the past 300 plus million years or so, that is not the case at the present time. The last collision coast configuration ended 25-30 million years ago, because the present Central California sliding plate boundary is not a collision boundary in the classic sense (more discussion to follow in the next section – Origin of the Central California Coast). However, even if the present shoreline doesn't fit perfectly, it is more like the leading edge model than the trailing edge one. Also, as noted by Ed Clifton (pers. comm.), the "dominant factor in the specific character of a leading-edge coast is **rapid tectonic uplift**." With regard to the Central California Coast, he noted that "most of the coast is rising, and

where it is not, important depositional basins or major inflections in the coastline, like San Francisco and Monterey Bays, develop."

Detailed discussions of the geomorphological characteristics of leading-edge shorelines beyond Inman and Nordstrom's (1971) general description of leading-edge, collision coasts and those outlined by Hayes (1965; 1976; briefly discussed earlier) are not available in the geological literature. In order to elaborate in somewhat more detail upon the characteristics of such shorelines, Hayes and associates at RPI (Todd Montello and David Betenbaugh) carried out a study of the entire length of the leading-edge shorelines of North and South America (a straight-line distance of almost 16,000 miles), based on previous field work and imagery from Google Earth.

Generally speaking, four attributes of these leading-edge shorelines readily distinguish them from trailing-edge, depositional coasts: 1) the abundance of rocky shores; 2) lack of major river deltas; 3) less extensive development of coastal wetlands; and 4) relative lack of barrier islands. Based on our study of the imagery available, the shoreline of western South America south from Chincha Alta, Peru to San Antonio, Chile, a straight-line distance of over 1,600 miles, has up to **50% rocky shores** (mostly wave-cut rock cliffs and/or wave-cut rock platforms). Rocky shorelines are also abundant in other areas where the mountains rising under the influence of the converging plates are located near the shoreline. For example, the Pacific shoreline of the Aleutian Islands chain, which has active volcanoes along a strong subduction zone, contains 25% rocky coasts. Subduction zones are defined as regions where portions of the Earth's tectonic plates are forced beneath other plates (shown in Figure 2). In the case of the Aleutians, the Pacific Plate is moving beneath the North American Plate. On the other hand, the shoreline on the trailing edge of the North American Plate from Yucatan, Mexico to Long Island, New York contains only a minimum number of rocky shores. The shorelines of New England and eastern Canada do contain some spectacular rocky areas, but they are, generally speaking, along coasts which continue to rise even to this day as a result of the removal of the vast ice sheets associated with the last Pleistocene glaciation (known as **glacial rebound** coasts).

Only two significant major river deltas occur along the outer shorelines of the leading edges of the North and South American Plates – the Copper River Delta in Alaska and the Colorado River Delta, Mexico. The delta at the head of San Francisco Bay is by far the largest one on the Central California Coast. The Copper River Delta is there because of an overwhelming amount of sediment bought to the shoreline by the river considering its relatively small size (annual sediment load of 100 x 10^6 metric tons - Reimnitz 1966; Hayes and Ruby, 1994). Much of this sediment is derived from glacial sources. The Colorado River has a rather unique positioning related to the opening of the Gulf of California as a result of an amazing lateral (south to north) motion along the trace of the San Andres Fault system (discussed in more detail later). In comparison, there are many river deltas, such as the Santee/Pee Dee Delta in South Carolina, the Mississippi River Delta in Louisiana, Rio Grande Delta in Texas/Mexico, the Orinoco River Delta in Venezuela, and the delta of the Amazon River in Brazil, where lengthy rivers have made their way to the edges of the wide coastal plains on the trailing edges of the continents.

Some fairly extensive coastal wetlands do occur in the widely separated estuarine systems of the leading-edge coasts of North and South America. Examples include those in San Francisco Bay, California; Laguna Ojo de Liebre, Baja California; Golfo de Fonseca, El Salvador/Nicaragua; and Golfo de Guayaquil, Ecuador, but these cannot match in scale the extensive wetlands of the South Carolina/Georgia coast, the Mississippi Delta region, the mouths of the Amazon and Orinoco Rivers, and other locations on the trailing edges of the continents of North and South America.

Barrier islands are elongate, shore-parallel accumulations of unconsolidated sediment, some parts of which are situated above the high-tide line (supratidal) most of the time, except during major storms. They are separated from the mainland by bays, lagoons, estuaries, or wetland complexes and are typically intersected by deep tidal channels called tidal inlets (see Figures 3A and 3B). In order for barrier islands to exist, there must be a supply of sand in the longshore sediment transport system from which the islands can be formed, as well as waves large enough to build them. Also, the tidal range must be less than about 12 feet (i.e., microtidal or mesotidal), because the strong tidal currents on macrotidal shorelines transport the available sand offshore. Waves need to be focused within a relatively limited vertical zone within the intertidal zone in order to construct the islands. Finally, barrier islands can only form on a relatively flat coastal zone that provides the sediments space to accumulate. Therefore, barrier islands never occur on steep rocky shores.

[NOTE: There are three very relevant terms regarding segments of shorelines influenced by tides. Supratidal areas are those only very rarely covered with water, such as during significant storm surges. Intertidal areas are confined to within the zone of the highest and lowest astronomical tides. Subtidal areas are almost always covered with water. These distinctions are particularly important on rocky shorelines like the Central California Coast.]

There are six areas along the leading edges of the North and South American plates that contain significant barrier islands, with one of the most interesting areas being along the Pacific coast of Colombia, where a continuous chain of 62 barrier islands occurs along 415 miles of the shoreline. The sand that composes these barrier islands is delivered to the coast by the abundant runoff of eight rivers that is generated by 200 inches of rainfall per year (Pilkey and Fraser, 2003).

There are altogether approximately 1,182 miles of barrier islands along the leading-edge shorelines of North and South America, a somewhat surprising number considering the general models of the two contrasting shoreline types that have been published to date. However, this number pales in comparison to the fact that barrier islands make up at least 80% of the trailing-edge shoreline of the North American Plate between Campeche, Mexico and Long Island, New York, a straight-line distance of almost 4,000 miles. A recent study of satellite imagery of barrier islands along the world's shoreline by Matthew Stutz and Orin Pilkey of Duke University revealed that 63% of the barrier islands occur on trailing-edge coasts, 16% on leading-edge coasts, and 21% on marginal seas. Going one step further, barrier islands compose only 1/16th of the total leading-edge shorelines of the two continents that we studied (North and South America). However, this unexpected occurrence of over 1,000 miles of barrier islands on these two leading-edge shorelines is an interesting puzzle that begs for explanation.

In the earlier discussion of the determining characteristics of depositional coasts, the effects of climate and sediment supply were considered to be less important than hydrodynamic regime as determined by the ratio of wave energy to tidal energy. On leading-edge coasts, however, the tidal range *per se* is not as important as a general rule. Of course, for leading-edge coasts on open oceans with an almost limitless fetch (on the windward side of the continents), the waves are usually quite large (a relative constant). Therefore, after the impacts of tectonic uplift and the consistent influence of the waves are taken into account, the climate and sediment supply assume the next most important functions with regard to the overall geomorphological makeup of leading-edge coastlines. Two examples of this is the importance climate had in creating the large sand supplies for the barrier islands of the Copper River Delta, Alaska (glacier-aided huge sediment load of the Copper River), and those of the Colombia coastline (200 inches of rainfall per year).

The general geological cross-section normally used to illustrate the collision of continental and oceanic plates is shown in Figure 2. At the collision point, the continental plate rides over the oceanic plate, creating a high, young mountain range. Material being transported along the ocean floor, which includes the oceanic crust, of the Pacific Ocean in the case under discussion, is subducted downward to depths that allow melting of this material.

[NOTE: The term **subduction** is defined as the action or process of the edge of one crustal plate descending below the edge of another, with crust being defined as the solid material (i.e., rocks, minerals, and sediments) that form the outer rind of the Earth's interior.]

Some of this subducted solid crust that has been melted (called **magma**), which is a result of its reaching depths where temperatures exceed 1,200 degrees C (2,192 degrees F), then rises to the surface of the Earth in the form of volcanoes (Figure 2; discussed in more detail later). This chain of volcanoes typically forms above the sinking plate in a line parallel to the trench into which its parent material sank at a distance of 50 to 200 miles from the shore (Alt and Hyndman, 2000). According to this general, classic model, the seaward edge of the continental plate is rising and the shoreline is dominated by eroding rocky shores. Excellent examples of coastal areas that fit this model closely occur along the "lost coast" of northern California, the southwestern Pacific shoreline of mainland Mexico, the south coast of Peru, and the northern coast of Chile.

However, close inspection of the two leading-edge shorelines of North and South America reveals some significant variations along the collision zone. For example, the edges of the continental plates are not rising everywhere at the present time. The outer shore of the Kenai Peninsula, Alaska contains some irregular islands composed of sinking mountaintops. It is well documented that this area sunk as much as 3 feet during the Good Friday earthquake of 1964 (Plafker, 1969). In the long term, these mountains have risen in response to the major subducting system, but the type of variation in the general trend on the outer shore of the modern Kenai Peninsula (i.e., recent sinking) makes for some unexpected shoreline types. Similar recent shoreline depression is evident along the southern tip of Chile.

And last, but certainly not least, is the effect of the San Andreas Fault system, which has greatly modified the shoreline in Southern and Central California. Details on how this came about are given in the next major section, which explores the geological evolution of the Central California Coast.

ORIGIN OF THE CENTRAL CALIFORNIA COAST

This history is designed now and ever to keep the sneers from the lips of sour scholars.

John Steinbeck – TORTILLA FLAT

Introduction

Unraveling the long history of how the present California coastline originated has fascinated geologists for many decades, and they often travel from afar to study these famous rocks (e.g., the granitic rocks in the Sierra Nevada and the sedimentary and metamorphic rocks of the Coast Ranges).

[NOTE: The three major classes of rocks – igneous, sedimentary, and metamorphic – were first defined by the pioneering Scottish geologist James Hutton in the mid 1700s. He recognized a class of rocks called **igneous rocks** that crystallized from an extremely hot molten mass of material (magma) that was formed as if by fire (*ignis* = fire). One class of igneous rocks, which includes granite and diorite, develops by slow cooling of the molten mass at great depths within the Earth's crust. As a result of the slow cooling, large crystals of individual minerals form that can be as much as an inch or so in diameter. These minerals, typically including feldspar and quartz, eventually coalesce to form a rock mass, which may later be pushed up to the Earth's surface by mountain-building processes. A second class of igneous rocks, called volcanic rocks (e.g., basalt and rhyolite), crystallize rapidly as a result of molten lava flowing suddenly onto the Earth's surface. The crystals of these minerals form so rapidly that they are, for the most part, too small to be seen by the naked eye. On the other hand, **sedimentary rocks** are most commonly formed by weathering and erosion of pre-existing rocks on the Earth's surface. This process creates sediment that can be transported by water, or wind in some cases, to be deposited in large masses on river deltas, beaches, offshore on the continental shelf, in deep ocean depths, and so on. Once deposited, these sediments may become buried where they are consolidated by chemical cementation and other processes. The sedimentary rocks formed by this process that are made up mostly of gravel-sized particles are called conglomerates, those composed of sand-sized particles are called sandstones, and those composed of silt and/or clay are called shales, siltstones, or mudstones. Limestones, which are sedimentary rocks composed of calcium carbonate ($CaCO_3$), usually develop in marine waters by a combination of chemical precipitation and accumulation of the hard parts of some marine organisms, such as seashells. Another group of sedimentary rocks, called evaporates, usually form as the result of evaporation of seawater under unusually high surface temperatures. However, some sedimentary rocks in this class form at great depths within the ocean, such as the salt deposits now forming at the bottom of the Red Sea. Evaporite deposits are most commonly composed of halite (NaCl) and gypsum [$CaSO_4 \cdot (2H_2O)$]. The third major rock class, **metamorphic rocks**, results from dramatic changes in igneous and sedimentary rocks affected by heat, pressure, and water that usually results in a more compact and crystalline condition. These changes usually take place at significant depths below the Earth's surface. Examples of metamorphic rocks include: 1) slate, derived from a sedimentary rock composed of silt and/or clay; 2) marble, derived from limestones; and 3) gneiss, a banded, micaceous rock typically derived from granite.]

The story of the origin of the Central California Coast is rather complicated, and there are still many unanswered questions regarding details, such as the exact age and timing of the different geological events. These events are related in a general way to the time line given in Figure 5. The following discussion is only a brief summary of this topic. For greater detail, one or all of the following books should be sufficient to satisfy your desire to learn more about this fascinating subject:

1) *Roadside Geology of Northern and Central California* by David Alt and Donald Hyndman (2000).

2) *California Geology* by Deborah Harden (Second Edition 2004).

3) *Geology of the San Francisco Bay Region* by Doris Sloan (2006).

A discussion of the geological history of the Central California Coast requires a slightly more detailed account of plate tectonics than the introduction given earlier. In order to do so, a generalized cross-section of the Earth with its three major components, crust, mantle, and core, is presented in Figure 6 and discussed below:

1) The surface material is called the Earth's **crust**. The

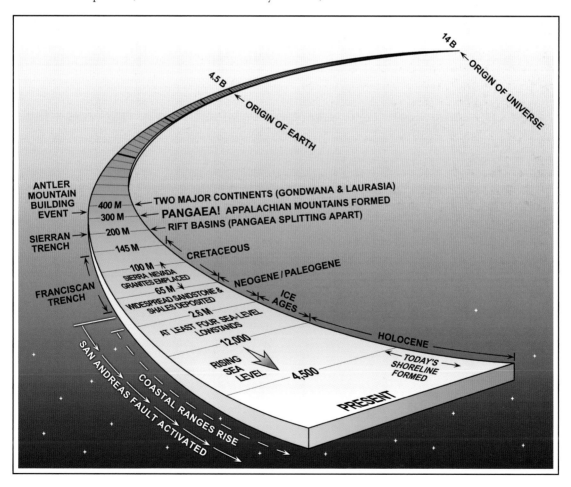

FIGURE 5. Time line for the origin of the Central California Coast. See Table 1 for nomenclature and timing of the geological periods.

crust is composed of a variety of igneous, metamorphic, and sedimentary rocks. The crust under the oceans, which averages about 6 miles thick, is composed mostly of rocks that contain an abundance of iron and magnesium bearing minerals (e.g., basalt), whereas the crust on the continents, which averages 20-30 miles thick, but can be up to 62 miles thick, is composed of rocks made up of lighter minerals (e.g., granite) (Wikipedia encyclopedia). Because the crust is relatively cold, it is made up primarily of brittle rocks (plus unconsolidated sediments in some areas). These rocks and sediments are composed predominantly of a group of minerals called **silicates**, the fundamental building block of the Earth's crust, with the atoms of silicon and oxygen combined making up about 75% of it. More information on this primary mineral group is given in the glossary. Because of their relative compactness, the rocks that make up the crust are susceptible to shearing and breaking which, of course, causes the earthquakes that occur with some degree of regularity on the California Coast.

2) The **mantle**, which contains more of the heavier elements than the crust, such as iron (Fe) and magnesium (Mg), is about 1,800 miles thick and makes up approximately 70% of the Earth's total volume. The major component of the mantle is a dark-colored, notably dense rock called **peridotite**. The most common minerals in this rock are olivine [magnesium iron silicate; $(Mg,Fe)_2SiO_4$], pyroxene (several varieties; e.g., pigeonite, $MgFeSi_2O_6$), as well as some garnet.

Another important rock type initially derived from the mantle is a dark, greenish-colored rock called **serpentinite** composed of a number of complex, usually hydrated, iron and magnesium bearing minerals. As noted by Alt and Hyndman (2000), *serpentinite forms at the crests of oceanic ridges, where seawater sinks into fractures that open as the two plates separate. The tremendous pressure of the depths forces the water down into the extremely hot mantle rocks below the ridge crest. The water reacts with hot peridotite to make serpentinite, which constitutes a large proportion of the uppermost mantle.* Because of their location at the top of the mantle, these rocks are commonly squeezed or faulted (typically in a plastic state) into the overlying rocks and sediments in the course of a mountain-building event. During one of these events that was initiated in what is now west-central California around the end of Jurassic time (150-130 million years ago), these kinds of rock masses were eventually disgorged upward to be exposed much later at the surface by erosional processes. Therefore, they now occur at the Earth's surface in the Coast Ranges, where they are easily recognized by their green color and slippery, soapy feel. Serpentinite is the state rock of

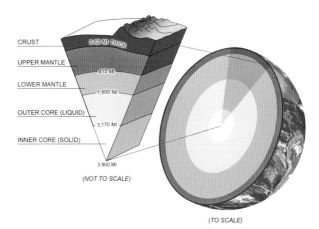

FIGURE 6. Cross-section of the Earth showing its internal structure [modified from a diagram created by Jeremy Kemp, which was based on elements of an illustration by the USGS (2005)].

California (Harden, 2004).

The uppermost part of the mantle, which is relatively rigid, is combined with the crust to form the **lithosphere**. The lithosphere is underlain by a part of the mantle known as the **asthenosphere**, which is more plastic than the overlying layer as a result of heating by radioactive decay of elements at depth. It is within this more fluid layer that the convection currents mentioned earlier transfer heat to the surface and initiate the process of the "drifting apart" of the individual tectonic plates.

3) The mantle is underlain by the Earth's **core**, which is composed mostly of iron and makes up approximately 30% of the Earth's volume. The core consists of two layers, an outer extremely hot zone, which is molten, and in inner, also very hot, inner layer that is solid because of the intense pressure at those depths within the Earth.

The lithosphere, which averages about 60 miles thick, is the material that makes up the individual tectonic plates on the Earth's surface. The plates that presently exist are illustrated in Figure 7. Innumerable individual plates have come and gone in the past, as the process of plate movement has apparently persisted far back into geological time. The general model of the evolution of an ocean basin (Figure 2) shows two oceanic plates moving away from a spreading center in the middle of the ocean called the **mid-ocean ridge**. As noted by Alt and Hyndman (2000), *these ridges mark the lines where plates pull away from each other. Every ocean has one, a long ridge with a deep rift along its crest that runs the length of the ocean.* As new lava flows spill out onto the ocean floor along fissures in the ridges (usually every few hundred years), new oceanic crust is created.

The plates on the opposite sides of the ridges move

away from each other at an average rate of about 4 inches/year according to Monroe and Wicander (1998) and 2 inches/year according to Alt and Hyndman (2000). What's an inch or two among friends? Doris Sloan (pers. comm.) observed that "the spreading rate depends on the spreading center. Moores and Twiss (1995, p. 104) divide them into slow (1-5 cm/yr), intermediate (5-9 cm/yr), and fast (9-10 cm/yr). The East Pacific Rise is fast spreading, >9 cm/yr (one inch = 2.54 cm)."

The new crust thus created, which becomes an extension of the oceanic plates, moves away from the ridges until it encounters an approaching continental plate where it dives under the oncoming barrier (a rather slow dive at 2-4 inches per year!!). As noted earlier, this area where the two plates collide is called a subduction zone, and it is commonly marked by a deep, oceanic trench that, in places, has an arc-shaped zone of volcanic islands (called an **island arc**) on its concave side. In some areas, the volcanoes are located on an adjoining continental landmass (Figure 2), as is the case for the volcanoes of the Cascade Mountains in Oregon and Washington. The volcanoes are usually

located between 50-200 miles away from the center of the trench, depending on the angle at which the source material plunges down into the trench.

As can be seen in Figure 8, which illustrates the topography and ages of the floors of the Atlantic and Pacific Oceans, the Atlantic Ocean perfectly fits the model of a central mid-ocean ridge away from which tectonic plates diverge as the main ocean body grows (see Figure 2), but the Pacific Ocean does not. In the Pacific, the ridge known as the East Pacific Rise has been overtaken and partially covered by the westward moving North American Plate. This is but one of the many factors that complicates the geology of the coast of California.

The boundaries between two lithosphere plates may be of three possible types (Figure 9):

1) **Divergent boundaries** where the two plates are moving away from each other, for example along mid-ocean ridges where the plates are pulled apart by the subducting cold slabs on either side of them, a process called "slab pull" (Doris Sloan, pers. comm.);

2) **Convergent boundaries** where the two plates collide, for example along subduction zones; and

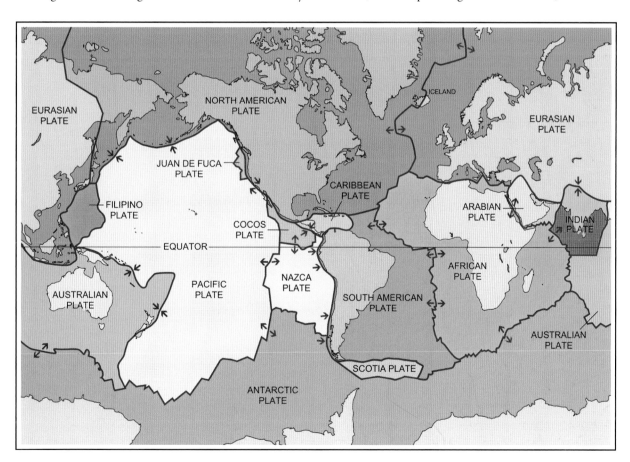

FIGURE 7. The tectonic plates that occur on the Earth and their relative movement. Note that the Pacific Plate is moving northwest relative to the North American Plate. Courtesy of Evgeni Sergeev (2009).

3) **Transform boundaries**, which are referred to as "conservative plate boundaries," where no material is created or lost. The two plates just simply slide past each other along a boundary called a **transform fault** (without colliding or pulling apart).

[NOTE: As pointed out by Alt and Hyndman (2000), *all transform boundaries connect oceanic trenches and ridges in some combination. … Transform boundaries define the edges of plates moving away from oceanic ridges and oceanic trenches.* Many of these are visible adjacent to the mid-ocean ridges illustrated in Figure 8.]

A **fault** was defined by Lahee (1952) in his classic book, *Field Geology*, as *a fracture along which there has been slipping of the contiguous masses against one another. Points formerly together have been dislocated or displaced along the fracture.* Perhaps the most common and most easily understood faults are those that show vertical movement, such as the two classes illustrated in Figure

10. In the type of fault shown in Figure 10A, the side of the fault on the left, called the hanging wall, slides down the fault surface under the influence of gravity (and/or tension) relative to the side on the right, called the foot wall. This type of motion creates a **normal** or **gravity fault**. On the other hand, where the hanging wall moves up the fault line over the foot wall under the influence of compression, the resultant motion creates a **reverse** or **thrust fault** (Figure 10B). When you take your walks along the rocky shores in Central California, you will have ample opportunity to observe both of these types of faults in the wave-cut rock cliffs.

The branch of geology devoted to the study of such features as the faults just described is called structural geology. The state of California has many structural geologists, because its geology is characterized by so many complex structural features. Examples include the mountains of the Coast Ranges, which result from

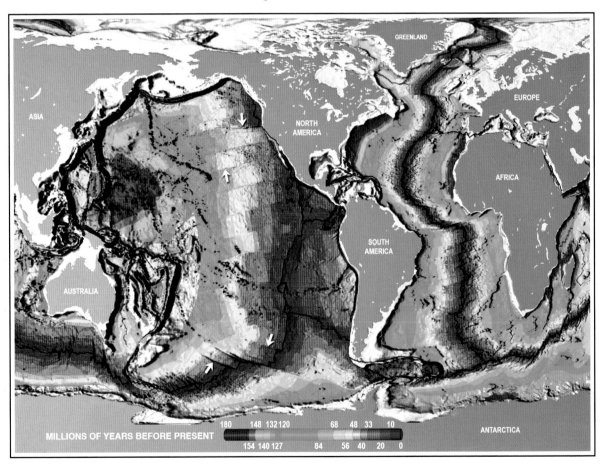

FIGURE 8. Sea floor topography, primarily of the Atlantic and Pacific Ocean Basins. The youngest lavas and rocks on the sea floors (shown in red) occur adjacent to the mid-ocean ridges away from which the ocean floors continue to spread. Note that the mid-ocean ridge is not centrally located in the Pacific Ocean; in fact, it has been overridden by the westward moving North American Plate. Arrows point to a few of the multitudes of transform faults that have occurred to accommodate the ocean floor's expansion on the spherical surface of the Earth. From Peter W. Sloss, National Geophysical Data Center (NGDC, 2009).

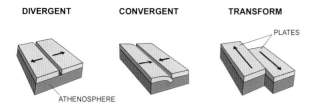

DIVERGENT   CONVERGENT   TRANSFORM

PLATES

ATHENOSPHERE

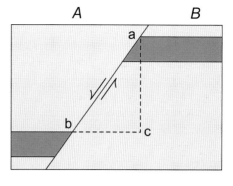

FIGURE 9. The three basic types of motion of plate boundaries: 1) Divergent – pulling away from each other, as along mid-ocean ridges; 2) Convergent – colliding with each other, as along subduction zones; and 3) Transform – where the two plates slide past each other with little resulting vertical movement. Modified after José F. Vigil, U.S. Geological Survey (1999).

extensive wrench and transform faulting; the Basin and Range province of southeastern California, generated by normal faults; and the intense combination of both normal and thrust faults within the Cascade Range and Sierra Nevada of eastern California. And, then, of course, there is the San Andreas Fault system.

The importance of transform faults was not recognized until a geologist named J. Tuzo Wilson used the concept to explain away some of the major problems facing the proponents of the theory of drifting continents. He did this in a classic paper published in 1965 titled *A New Class of Faults and Their Bearing on Continental Drift*. These newly discovered faults allowed the fresh material created along the mid-ocean ridges to conform to the spherical surface of the Earth by sliding and adjusting in such a way as to fit it. Even before Wilson, a geologist named Harry Hess, making observations of the sea floor while captaining a ship for the U.S. Navy during World War II, had concluded that somehow the floor of the Pacific Ocean was in motion and "expanding." After the war, he proposed a theory that was eventually called **seafloor spreading** (published in 1962). And by 1965, the pioneering oceanographer Bruce Heezen extended Hess's work by making detailed bathymetric maps of the floors of the world ocean. These maps showed not only the amazing lines of mid-ocean ridges with their central rift valleys, but also that the mid-ocean ridges are commonly offset into horizontal segments bounded by Wilson's transform faults (see Figure 8). Examples of these offsets occur at numerous locations along all of the mid-ocean ridges, including the Mendocino Fracture Zone offshore of northern California. Wilson's transform faults explained how these offsets are possible. He was also the first to use the term **plate tectonics** and drew the first **plate boundaries**. As a geology professor at the University of Massachusetts in the late 1960s, co-author Hayes was able to attend some of Wilson's

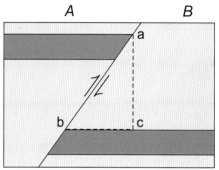

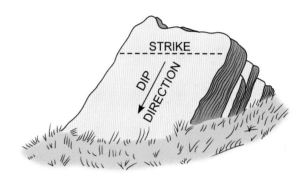

STRIKE

DIP DIRECTION

FIGURE 10. Faulted and dipping strata. (Top) A gravity, or normal fault, in which the hanging wall (A) slides down the fault plane under the influence of gravity. Thus, the orange layer is shifted lower along the fault. (Middle) A thrust, or reverse fault, in which the hanging wall (A) slides up the fault plane, over beds younger than itself, under the influence of compression (the top two diagrams are after Lahee, 1952). (Bottom) Terms used in the description of dipping rock layers, strike and dip. A strike-slip fault is one that moves laterally along the strike line without marked vertical motion.

ground-breaking lecture tours (and those by Heezen as well), and even though a mere coastal geomorphologist, it was clear to him that a new day had dawned for the geological sciences.

So what? Well, it turns out that the infamous San Andreas Fault system that has worked its way right through the middle of the Central California coastal area is a transform fault where the boundaries of the Pacific and North American Plates meet (Figure 11C). Wilson also recognized that tectonic plates don't just glide smoothly past each other, instead, over time stress builds up along the boundary between the plates to the point that accumulated potential energy is released suddenly and the rocks on the opposite sides of the fault move abruptly, commonly causing earthquakes (Wikipedia free dictionary; discussed in more detail later).

The San Andreas transform fault is a **right-lateral fault**. That is, if you are standing on one side of the fault line looking across it, the *terra firma* on the other side of the fault line has moved relatively to the right during the earthquake (as shown in Figure 9). The same thing is true if you step across the fault line and look back at the side you were standing on (i.e., that side also has moved to the right in a relative sense). A more general term is **strike-slip fault**. To understand that term you have to be familiar with the geometry of dipping rock layers. In order to measure the angle of dip of the layer, one lays a hand-held instrument called a Brunton compass on the rock surface, pointing it in the exact direction in which the rock dips and turning a pendulum-like arm within the instrument until its upper surface is level, which allows you to read the angle of dip. If you drew a line along the rock layer exactly perpendicular to the direction of dip, that would be the **strike** of the rock layer (see sketch in Figure 10-Bottom). Making such readings, one of the first things a beginning geology major learns to do, is a fundamental step in constructing the geological history of the rocks on the Earth's surface. A strike-slip fault, then, is one that moves exactly parallel to the imaginary line you "drew" on the rock layer, with little up or down motion in the process.

How Old Is It?

The most elemental question in the interpretation of the origin of rock sequences is how old are they? This information is necessary before the evolution of any significant portion of the Earth's rocks can be properly interpreted. Three methods are commonly used to answer this question: 1) applying the "law of superposition;" 2) paleontology correlations; and 3) radiometric dating.

In the 11th century, a Persian Muslim philosopher named Avicenna came up with an idea for determining the age of rock sequences that was later formulated more clearly in the 17th century by the Danish scientist Nicolaus Steno. This concept later became known as Steno's Law or The Law of Superposition, which states - *Sedimentary layers are deposited in a time sequence, with the oldest at the bottom and the youngest at the top.*

In other words, layers of sedimentary rocks are laid down in a sequence from bottom to top like you would do if baking a cake composed of three distinct layers. The first layer you put in the pan is "oldest" and the third one, which is placed on top of the second layer, is "youngest." The difference is that, while it takes only a few seconds to create the three layers in the cake, geological processes may take thousands of years to create stacked layers of rocks. However, the end product, a sequence of vertically stacked layers ranging from the oldest at the bottom to the youngest at the top, is the same. The problem becomes much more complicated if the rock layers have been tilted, or even completely overturned, as is true for many rock sequences, especially those in tectonically active areas like the California Coast.

The rock layers exposed in the sides of the Grand Canyon in Arizona are a classic example of relatively simple "layer-cake geology." Rocks at the bottom of the canyon consist of granites and schists (metamorphosed fine-grained rocks) overtopped by some tilted rock layers in places, all of which are Precambrian in age (see Table 1). These rocks were scoured off into a relatively flat surface during a major erosional event. The resultant eroded surface is called an unconformity and, because many of the rocks under the eroded surface in the Grand Canyon are tilted, that unconformity is more specifically referred to as an **angular unconformity**.

The major angular unconformity near the bottom of the Grand Canyon, which represents a significant gap in geological time, was later covered by over 5,000 feet of horizontal layers of sedimentary rocks, beginning with Cambrian sandstones more than 500 million years old at the base. Coming up the sides of the canyon, younger and younger relatively horizontal layers of sedimentary rocks are encountered until one reaches the top of the canyon, where rocks of late Triassic age (around 225 million years old) are found (Bues and Morales, 1990). These Triassic rocks, named the Chinle Formation, consist of abundant siltstones and some sandstones that were originally deposited mostly in river floodplain and lake environments. Notable examples of petrified forests and dinosaur remains are also contained within the Chinle Formation.

There are some time intervals missing from the sedimentary layers in the Grand Canyon, but this is a remarkably complete record of 300 million years of relatively steady deposition of horizontal rock layers, oldest at the bottom and youngest at the top. This technique of examining layered rocks has been used to interpret the history of the Earth's rocks since the earliest geological maps were constructed, even as far back as the 11th century!

[NOTE: Many important economic products, such

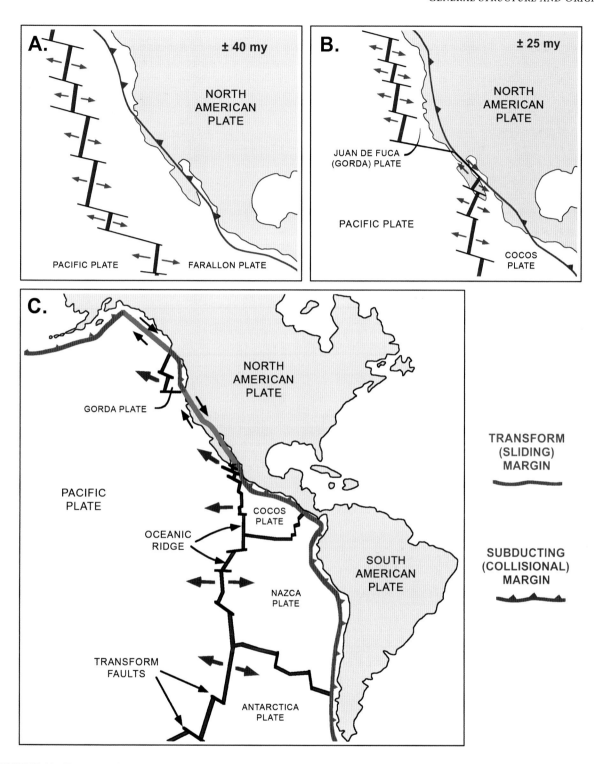

FIGURE 11. Evolution of the oceanic plates of the eastern Pacific Ocean. In diagram (A), at about 40 million years ago, the former oceanic (Pacific) ridge had not yet reached the leading edge of the North American Plate. In diagram (B), at about 25 million years ago, the oceanic ridge had met the shoreline of the Baja Peninsula and moved into what is now the Gulf of California area. The San Andreas Fault had already formed and the transform motion between the Pacific and North American Plates had begun. (C) The present configuration of the plates. Much of the Farallon Plate had disappeared under the North American Plate. The fragment of that Plate still offshore to the north of Central California is now named the Gorda Plate. The active San Andreas Fault extends inland all the way to the present Northern California along the transform margin (shown in blue). Note the subduction zones to the north and south, which are colored red. Diagrams are courtesy of Dr. Ed Clifton.

as oil, gas, and coal, have accumulated in sedimentary rock layers (or strata) similar in some respects to those exposed in the sides of the Grand Canyon. Consequently, the branch of geology known as **stratigraphy**, the study of sedimentary rock layers or strata, particularly their distribution, mode of deposition, and age, has provided professional careers for more geologists than any other branch of the geological sciences.]

Once the sequence of sedimentary rock layers had been worked out sufficiently enough to build a rudimentary time scale, it became apparent that over time thousands of organisms had come and gone through evolutionary trends and extinctions. The somewhat limited life spans of some of the most critical species were related to rock sequences around the world early in the pursuit of the science of geology. Therefore, it became possible to determine the relative ages of the sedimentary rock sequences based upon their fossil content, or their **paleontology**.

Neither the law of superposition nor paleontological correlation works for determining the ages of intrusive masses of rocks like granites (because they are not layered nor do they contain significant fossil remains), highly metamorphosed rocks, or markedly tilted and overturned strata, such as many of the rock sequences in California. Therefore, radiometric or radioisotope dating is commonly used to date igneous rocks, as well as organic remains associated with geologically young materials (e.g., by archaeologists). Some of the basic elements and concepts of this process, outlined by Gore (1999), include:

1) Naturally occurring radioactive materials break down into other radioactive materials at known rates (a process known as **radioactive decay**).

2) Radioactive parent elements decay to daughter elements. Each radioactive isotope has its own **half-life**, the time it takes for half of the radioactive parent element to decay to the daughter product.

3) Each radioactive element decays at its own nearly constant rate. Examples include Potassium-40 (radioactive parent) to Argon-40 (stable daughter) with a half-life of 1.25 billion years, and Uranium-235 (radioactive parent) to Lead-207 (stable daughter) with a half-life of 700 million years.

4) Therefore, measuring (in a rock sample) the amount of parent element and the amount of daughter element allows the geologist to estimate the length of time over which the decay has been occurring.

5) A method called Carbon-14 dating is used for preserved formerly living plant or animal remains less than 70,000 years old. All living things contain a constant ratio of Carbon-14 to Carbon-12. After death, any Carbon-14 in the tissues decays to Nitrogen-14 with a half-life of 5,730 years. The specimen is dated on the basis of the change in the Carbon-14 to Carbon-12 ratio.

Radiometric age dating, using one of the most fundamental precepts of nuclear physics, radioactive decay, is quite accurate, considering the long time spans involved in some of the measurements. The other two methods, superposition and fossil correlation, are relatively precise also, if the rocks under study are exposed well enough for detailed work to be carried out. Therefore, you can expect the age dates cited throughout this book to be fairly accurate in situations where the geological evidence is sufficient and the radiometric measurements on the samples have been accurately carried out (a safe assumption, we think, considering the amount of effort that has been put into this over the years). In some cases, these dates may be adjusted somewhat at a later time based on new data.

Evolution of the Coast

Alt and Hyndman (2000) in their book *Roadside Geology of Northern and Central California* stated that the story of northern and central California is the story of three trenches. Figure 12 shows the approximate locations of these historical oceanic trenches (with approximate ages), as well as the extension of the remnant of trench number 3 (the Cascadia subduction zone) that presently resides off the coast of northern California, Oregon, and Washington (offset oceanward along a transform fault, called the Mendocino Fault). We would add to that story the burial of the East Pacific Rise by the westward advancing North American Plate that was followed by the passage of the massive San Andreas Fault system into the area. The following listing of this chain of events, based on a number of sources [most notably Hensen and Usner (1993), Alt and Hyndman (2000), Harden (2004), Berlin (2005), and Sloan (2006)], provides a general introduction to the subject. Refer also to the time line in Figure 5.

Going back as far as the middle of the Paleozoic era, the evolution of the area of interest in this book is considerably more obscure for that time period than for later events. However, apparently the oldest of the three ancient trenches shown in Figure 12 developed about 350 million years ago (during Late Devonian and Mississippian time; Table 1). During that time, an earlier North American continental plate collided with the offshore ocean floor creating a highly complex,

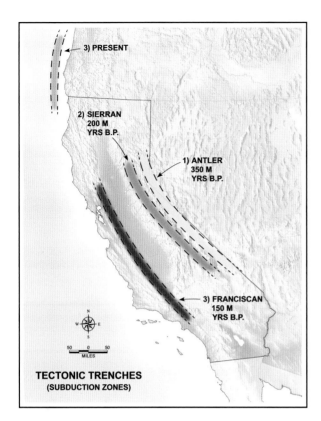

FIGURE 12. The three major trenches that have led to the gradual westerly growth of the area now known as California. Locations are only approximate. Concept suggested by Alt and Hyndman (2000). Also shown is the presently active trench off the coast of Northern California, Oregon, and Washington.

"crushed" mixture of sedimentary rocks, many of which were changed to metamorphic rocks. These "crushed" rocks had been sheared off the surface of the subducting oceanic plate. This episode is called the **Antler mountain building event**. The resulting band of what Alt and Hyndman called *abused* rocks also contains *the remains of a volcanic chain that erupted along the continental margin during the Antler event.* This mixture of diverse rocks, called the Shoo Fly terrane, now forms most of the eastern Sierra Nevada mountain belt. Give the geologists credit, they do come up with some cool names.

There is still some question about the exact timing of all this, but conventional wisdom has it that at about 400 to 300 million years ago, there were two main land masses separated by the Rheic Ocean. In the south sat **Gondwana**, a supercontinent consisting of South America, Africa, India, Australia, and Antarctica, and in the north sat **Laurasia**, made up of North America, Greenland, Europe, and part of Asia. There is still some dispute about the exact makeup of these two landmasses. One could also speculate that the "North America" that

collided with the floor of a western ocean to ultimately form the Shoo Fly terrane was part of Laurasia.

Some time around 300 million years ago, during the Carboniferous period of the Paleozoic era (see geological time scale in Table 1), these two landmasses (Gondwana and Laurasia) collided to form a single large continent called **Pangaea** (Figure 13A). In the process, the Appalachian Mountains were formed (Dietz and Holden, 1970). These mountains are said by some to have been as high as the Himalayas and by others at least as high as the modern Rocky Mountains.

[NOTE: In the winter of 1964/65, co-author Hayes led a geological expedition in Antarctica to examine a sequence of sedimentary rocks called the Beacon Sandstone of Permian age (around 300 million years old). These rocks are very similar to other Permian units in South Africa and India, containing huge petrified trees that grew in tropical forests, among other artifacts. The correlation of these very similar sandstones around the southern hemisphere was supporting evidence for the makeup of the original "southern" supercontinent Gondwana, which collided with the "northern" supercontinent Laurasia to form Pangaea, which was starting to break up into separate components during the Triassic period (Table 1). The newly created fragments of Pangaea, including the ones that originally made up Gondwana, eventually migrated to their present positions. That is, the southern parts of South America and Africa, as well as the peninsula we now call India, began to move far away from each other. By showing that the rocks of these three areas were so similar they had to at one time be joined, this study provided a small bit of supporting evidence for the theory of continental drift. Also, such studies provided some supporting evidence for the fact that the eventual wandering of the individual fragment of Pangaea now known as the separate continent of Antarctica had taken it from warmer climes to the vicinity of the present South Pole.]

Between 225-190 million years ago, during the Triassic period, the major landmass of Pangaea started to split apart along the old collision zones. This pulling apart created some **rift basins** in parts of the area now known as the eastern U.S. similar to the ones presently occurring in south-central Africa. These rift valleys filled with sediments eroded from the uplifted blocks along the rift margins. From that point onward, the continents continued to drift apart, eventually reaching their present locations (Figure 13B).

Meanwhile, on the other side of the newly created North American continent, a new trench, called the **Sierran trench** (Figure 12), appeared (about 200 million years ago). It is important to point out that after the end

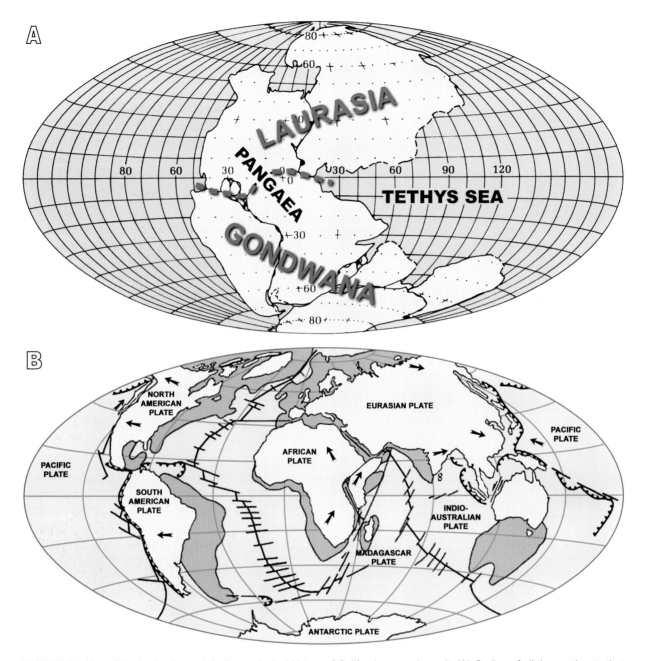

FIGURE 13. Two of the basic elements in the geological history of California as we know it. (A) Outline of all the continents that assembled to form a universal landmass as it existed at the end of the Permian period (Table 1) around 250 million yeas ago (as defined by Dietz and Holden, 1970; modified after Glen, 1975). (B) The major continents in their present position. Arrows indicate the directions the different land masses have moved since they first started "drifting apart" around 200 million years ago. The darker blue "shadows" represent previous locations of the continents as they moved along.

of the Antler event, about 350 million years ago, the western boundary of the "North American" continent was near the eastern edge of the present state of California. During the intervening period between 350 and 200 million years ago, a new coastal plain developed along that western shoreline. Later, as the continents started to separate (creating the rift basins further east), a new

collision zone developed along the western margin of the new North America. As Alt and Hyndman (2000) pointed out:

This collision first smashed the sediments that had collected along the new coast against the rocks of the Shoo Fly complex, converting them into another mass of horribly deformed and considerably recrystallized rocks, the Calaveras Complex. Included in this mix were some old volcanic islands and sediments that could only have accumulated on the deepest and most remote reaches of the ocean floor.

During Late Jurassic or Early Cretaceous time (Table 1), probably around 150-130 million years ago, another collision of the North American Plate with the Pacific sea floor took place about 60 miles west of the last one. Also, at the same time the northern end of the Sierra Nevada split off and "jumped west" about 60 miles to become the Klamath Mountains, leaving an open gap between them called the Modoc Seaway. This new trench is called the **Franciscan trench**. The crumbled and jumbled sediments and basalts that ended up in the Franciscan trench were accreted to the western boundary of the Klamath Mountains in the north and, according to Alt and Hyndman (2000), attached against *the edge of a strip of oceanic crust some 60 miles wide farther south.*

The generalized cartoon model in Figure 14, modified after Hensen and Usner (1993), Elder (2001), Berlin (2005), Sloan (2006), and California Coastal Commission (2007), is an attempt to outline the

conditions that prevailed at that time (as well as some of the structural conditions of the present). The result of these conditions was the production of three major geological units: 1) the mixture of volcanic material and sedimentary rocks that accumulated in the Franciscan trench; 2) the Coast Range Ophiolite; and 3) the Great Valley Sequence, which overlies the Coast Range Ophiolite. The elements included in this evolutionary process follow:

1) An oceanic plate, called the Farallon Plate, plunged under the western edge of the North American continent (a lighter tectonic plate).

2) As a tectonic trench located about 60 miles west of the present Sierra Nevada continued to grow, large amounts of sediment that had accumulated on the sea floor far out into the ocean toward the mid-ocean ridge were scrapped off the oceanic crust and crumpled into a thick mass along the margin of the trench. Sediment was also added to the trench from its eastern side. Some of the volcanic deposits on the sea floor were also included in this growing mass. These materials are now known as the **Franciscan Complex**.

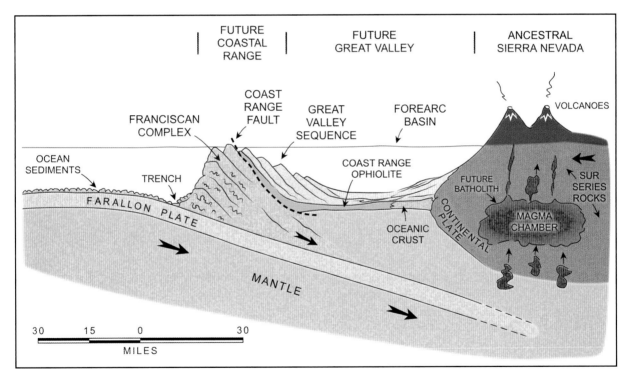

FIGURE 14. A hybrid cross-section illustrating the geology of the Central California area that incorporates elements of the Mesozoic era (the physiography) with the present situation (some aspects of the structure). The vertical scale is greatly exaggerated. This is a much simplified version of a very complex geological framework. One of the key elements relative to the coastline under discussion is the fact that the Great Valley Sequence, which was deposited within the forearc basin, overlies the Coast Range Ophiolite. Also of importance is the fact that the Coast Range Fault clearly separates the Coast Range Ophiolite (and the overlying Great Valley Sequence) from the Franciscan Complex. This diagram is significantly modified after Hensen and Usner (1993), Elder (2001), Berlin (2005), Sloan (2006), and California Coastal Commission (2007).

23

3) As the eastern end, or leading edge, of the Farallon Plate reached the proper depth, it melted as *it descended into the hot interior of the Earth. The molten rock triggered volcanic eruptions at the earth's surface and formed a series of volcanoes* (Hensen and Usner, 1993). This line of volcanoes became the ancestral Sierra Nevada. At depth, the molten material eventually formed a series of granitic rock bodies called plutons.

4) Ultimately, as the configuration depicted in Figure 14 evolved, the line of volcanoes in the ancestral Sierra Nevada was separated from the wedge of accreted material 60 miles to the west of it by an open body of water called a **forearc basin**.

5) The process of accumulation of sediments and volcanic rocks in the Franciscan trench lasted from 130 to 70 million years ago. During this same time, the plutons under the ancestral Sierra Nevada were cooling and solidifying. A little later (70-60 million years ago) these plutons were uplifted above ground and began to erode.

6) The subduction process ended about 29 million years ago *when the Farallon Plate disappeared beneath the continent and the North American and Pacific Plates met* (Hensen and Usner, 1993).

The gap between the accreted Sierran trench materials (the Calavaras Complex) and the Franciscan trench (Franciscan Complex), shown as the forearc basin in Figure 14, eventually became host to great thickness of sediment, which we know today as the Great Valley Sequence. Although most this material seems to have come from the continent, an island arc to the west may also have contributed to the sediment pile. In most of the forearc basin, these rocks accumulated on a bedrock foundation composed of a strip of oceanic crust that formerly extended westward away from the accreted Sierra Trench rocks. This oceanic crust and some associated materials evolved into what is now known as the **Coast Range Ophiolite**.

Moores and Moores (2001) stated that *in the Putah Cache bioregion, most exposed Great Valley rocks trend (strike) generally N-NW and are inclined (dip) generally east, and they are older in the west and younger in the east. The general interpretation of these rocks is that they were deposited off the edge of the continent, in approximately 3,000-5,000 feet of water* (presumably still within the forearc basin), *by pulses of sediment-rich slurries that flowed off the continental shelf in response to some disturbance (possibly individual earthquakes). These slurries are called "turbidity currents" and the deposits, such as* (many that occur within) *the Great Valley Sequence, are called turbidites* (a rock type discussed in

more detail later).

According to Martin (2009), *the Great Valley sequence consists of interbedded marine mudstone, sandstone, and conglomerate that range from Late Jurassic to Cretaceous in age (Bailey, Irwin and Jones, 1964). It crops out as thick, monotonously bedded sections of strata that generally are markedly less deformed and more coherent than sedimentary sections of the Franciscan and also have greater lateral continuity. Where most fully developed, such as along the west side of the northern Great Valley, the aggregate stratigraphic thickness of Great Valley Sequence is at least 12 kilometers (7.4 miles). The strata normally lie depositionally on Coast Range Ophiolite except where disrupted by faults, but at the north end and along the east side of the Great Valley they onlap the Nevadan and older basement terranes of the Klamath Mountains and Sierra Nevada. This enormous thickness of clastic detrital material probably represents submarine fans and turbidity deposits that formed as a result of rapid erosion of the ancestral Klamath Mountains and Sierra Nevada.*

[NOTE: The conceptual diagram given in Figure 14 should probably be called a hybrid, because it incorporates elements of the Mesozoic era (the physiography) with the present situation (the structure). The Great Valley Sequence probably accumulated as a giant wedge of sediment that later became deformed to the tilted succession shown in this Figure. Though a much more complex series of diagrams would be required to present a complete time series, we are hopeful that this approximate cross-section will adequately convey the general idea.]

Sometimes it takes a little imagination to visualize what these trenches were like as they evolved with the thick wedge of sediments and volcanic materials offshore being transported east and swallowed up into the ever-deepening trench. As described earlier, parts of this subducted material is later transformed and "belched" back up to the surface in a line of active volcanoes.

Speaking of imagination, in August 1995 the authors of this book had been camping and fishing for east-slope cutthroat trout on Big Creek in the River of No Return Wilderness in Idaho. On the climb out of the Wilderness, Michel hurt her knee, so we canceled plans to fish some other rivers in northern Idaho. Instead, we took a rest in McCall, Idaho for a few days. On our last day there, Michel suggested that we drive up a long dirt road for a view into Hells Canyon of the Snake River from the Scenic Trail Overlook. Once up there, we were near what at one time was the western edge of a tectonic trench similar in many respects to the Franciscan trench in California and of approximately the same age (possibly the same one?). After a climb from the parking area to the overlook on top of the mountain, sore knee and all, we sat for a long time in the bright sun, identifying the individual species in a flock of feeding birds that descended on the ridge top. We were all alone up there, despite the fact that it was mid-day on a Saturday. From one of the overlook benches, we could clearly see the outline of the Seven Devils Mountains to the south, a line of spectacular peaks. They looked much closer than they actually were. For us it was a beautiful time of seclusion and contemplation.

On the way back down the mountain, Michel dug out our copy of Alt and Hyndman's (1989) book the *Roadside Geology of Idaho* and started reading aloud about the Seven Devils. She read the following (p. 98): *Generation after generation of dinosaurs watched those vagrant islands rise out of the southwestern horizon, come a bit closer every year, and eventually arrive on the old west coast sometime around 100 million years ago, give or take 10 million years. The wandering islands then became much of the western fringe of central Idaho.* These welded islands form the core of the range now known as the Seven Devils, named after a legend about an Indian man who saw the seven devils there.

We often tell ourselves that the greatest joy of being geologists is that we are able to occasionally go back in time on an imaginary trip like the one described by Alt and Hyndman for the Seven Devils area.

Around 100 million years or so ago as the continents continued to drift apart (see Figure 13B, which shows the final product of this drifting), a huge arm of the world ocean, called the *Cretaceous Seaway*, covered much of what is now North America, extending north to south through the middle of the continent. All of the present U.S. south of the southern edge of the Appalachian Mountains was flooded by this large sea, which extended to the east into the ocean being created by the continents drifting apart. At that time, the western ocean (today's Pacific Ocean) extended eastward to the continental margin, presumably flooding the Modoc Seaway, as well as what is now the Great Valley area. These two areas of California were inundated with seawater on into the Neogene period, which ended about 2.6 million years ago. During that long episode of flooding within the Great Valley, an eight- to nine mile thick wedge of sediments accumulated (Alt and Hyndman, 2000).

An important point is that as the North American Plate continued to migrate to the west, it eventually caught up with the former mid-ocean ridge in the Pacific Ocean. The location of these features around 40 million years ago is shown in Figure 11A, and they are shown after they met around 25 million years ago in Figure 11B. This is a very crucial time period in the evolution of the

Central California Coast, inasmuch as the following two climactic events happened almost simultaneously (in a geological sense):

1) The East Pacific Rise met the edge of the North American Plate and "dove" under it into the Franciscan trench.

2) The San Andreas Fault zone was born.

The mountains of the Coast Ranges eventually started to rise, pretty late, around 5 to 3.5 million years ago. Consequently, during that time frame from 29 to 3.5 million years ago, the coastline of Central California was starting to take on a whole new look. As Forrest Gump would say, "for no particular reason" the Pacific Plate that had just recently adjoined the North American Plate appeared to take a slight turn to the left about this time and ever since then has been moving toward the northwest as the western limb of a long transform-fault boundary between it and the North American Plate (Figure 11C). It is important to remember that from that time onward, any geological materials located west of the San Andreas rift zone are "riding on" the Pacific Plate and any geological materials located east of it are positioned on the North American Plate. This boundary is not exactly a single, simple straight line, but rather a "swarm" of parallel faults (discussed in more detail later).

More specifically, the origin of the San Andreas system can be correlated with the time when the westernmost edge of the Farallon Plate, the plate that went under the North American Plate during the formation of the Franciscan trench (to the south of Cape Mendocino), was sucked down into the mantle near the western border of North America. The place where the Farallon Plate originated, the East Pacific Rise, also went under (Figure 11B). When this happened, about 28 million years ago according to Cole and Basu (1995), the plate boundaries changed from **convergent** during the time the Farallon and North American Plates were colliding to **transform** when the Pacific Plate joined the North American Plate. Why exactly the Pacific Plate veered to the north remains a mystery yet to be completely resolved. It may not have veered much, however, because the newly created sea floor along the northwest side of the East Pacific Rise was already moving in a slightly northwest direction away from the Rise before it went under North America, because of the north-northwest/south-southwest orientation of the Rise (Figure 8). However, it did apparently veer a little bit more to the northwest and, as a result, the Central California Coast is different from much of the rest of the convergent western shoreline of North America with the

San Andreas Fault system being largely a result of that (Figure 15).

As noted by Harden (2004), *from its southeastern end at the Mexican border, the San Andreas system follows a northwest trend for 1,350 kilometers (838 miles) to its northwestern end at the Mendocino Triple Junction.* A **triple junction** is where three tectonic plates come together, in this case the Pacific, North American, and Gorda Plates (the Gorda Plate is all that remains of the Farallon Plate; Figure 11C).

Some discrepancies exist in the descriptions of the San Andreas Fault system in the geological literature (e.g., from 10 to 30 million years suggested for its age; distance of offset variable by up to 350 miles, etc.). A relatively recent summary of the San Andreas Fault zone by Harden (2004) stated the following:

1) It began its history of right-lateral faulting 23.5 million years ago.

2) The maximum amount of offset is 195 miles. However, if you assume that the southern tip of Baja California was joined up with mainland Mexico near where the city of Puerto Vallarta, Jalisco is now located before the rifting started, that distance is 300 miles (second comment by us).

3) The long-term average slip rate is 1.34 cm/year (0.5 inches/year).

Based on matching distinctly similar geological units on opposite sides of the fault system, Sloan (2006) discovered that the different units had traveled different distances. These units are: 1) where some volcanic rocks in the Pinnacles National Monument match up with volcanic rocks on the east side of the fault in the Mojave Desert **192 miles** further south; 2) some sandstones on the peninsula south of San Francisco that are similar to sandstones on the east side of the fault in Southern California **225 miles** away; and 3) for some Cretaceous rocks now present on the Mendocino coast south of Gualala that *are composed of material eroded from rocks at Eagle Rest Peak **350 miles** southeast on the other side* of the fault. It is important to note that these distances measure how far that particular unit had moved since it was first formed and, thus, they do not measure the maximum amount of offset on the fault. The San Andreas Fault system is discussed in more detail later.

Some time around 20 million years ago or so, the Miocene marine basins in which the Monterey Formation accumulated were formed, presumably due to stresses engendered in the developing transform margin between the Pacific and North American Plates (Ed Clifton, pers. comm.). To further substantiate this point, Cardwell and Detterman (2003) reported that prior to

about 5.5 million years ago, the motion along the San Andreas Fault was slightly tensional (i.e., opposite sides were pulling away from each other) as well as parallel to each other as is expected along a transform, or strike-slip fault. As a result of the part of that motion that moved the plates perpendicular to each other (and consequently, pulled them apart somewhat), deep basins were formed and filled with sediments.

Around 17 million years ago, a huge area of flood basalts developed in the southeastern corner of Oregon. Some of these flows of molten lava advanced to the south into California. As is evident on the geological map of the state (Figure 16A), a large area of the northeastern part of the state north of Sacramento is covered with volcanic rocks of this same general age (in the area called the Modoc Plateau). These basalt flows, which ended around 15.5 million years ago, were *so overwhelming that they add an entirely new dimension to our ordinary notions of volcanic catastrophe*, according to Alt and Hyndman (2000), who postulated that this cataclysm *developed in a crater that opened where an asteroid struck the Earth and exploded.* Other explanations are available in the scientific literature for this interesting phenomenon, but their pursuit is beyond the scope of this book.

The next major event was the eventual rise of the Coast Ranges at around 5 to 3.5 million years ago. Two reasons for this rise are suggested in the literature. According to Alt and Hyndman (2000), the reason for the rise presumably was that the lighter crustal material that had been thrust down into the upper mantle when the last subduction process was in full bloom was too light to stay at that depth and gradually sought its way up to higher elevations, causing the overlying land to rise. On the other hand, Harden (2004) suggested that the mountain ranges came about as a result of the Pacific Plate shifting course slightly (and possibly also changing its rate of movement) and, as a result, *the Pacific Plate began to move obliquely past the North American Plate, rather than sliding past it. The change caused some transpression between the plates, and compression squeezed up the mountains.*

[NOTE: **Transpression** is a term that means that although the major movement is strike-slip or lateral along this major plate boundary, in some cases there is also compression (i.e., the plates move against each other in a perpendicular or oblique sense). When these two types of forces combine the result is called transpression, and the result is uplift of the adjacent land mass. Doris Sloan (pers. comm.) elaborated a bit further on this point, stating that although the dominant motion along the plate boundary is sliding, *"there is a small component of compression (about 10% according to D.L. Jones – thus*

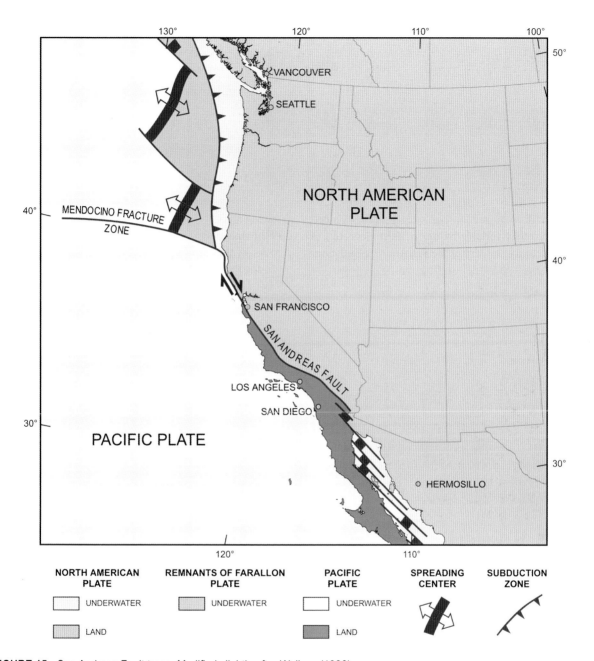

NORTH AMERICAN PLATE

PACIFIC PLATE

NORTH AMERICAN PLATE	REMNANTS OF FARALLON PLATE	PACIFIC PLATE	SPREADING CENTER	SUBDUCTION ZONE
UNDERWATER	UNDERWATER	UNDERWATER		
LAND		LAND		

FIGURE 15. San Andreas Fault trace. Modified slightly after Wallace (1990).

transpression). Our Coast Ranges are indeed quite young, but they are not a product of 'collision,' rather of bends in the San Andreas system and the small component of compression."]

Whatever the reason, the Coast Ranges did start rising, attaining heights of around 5,000 feet in places and they are still doing so. During the 1989 Loma Prieta earthquake, the Santa Cruz Mountains west of the San Andreas Fault rose approximately 3.9 feet (Harden, 2004). To complicate matters a bit, however, the whole region did not elevate all at once, only specific parts of it.

Other parts were obviously still subsiding basins in the earlier times.

Some time between 5 and 2 million years ago the massive Sierra Nevada began to rise along a major fault at its eastern margin. Before this, erosion had removed even remnants of most of the volcanoes derived from the magmas that formed at great depths some time before and during the Franciscan subduction event (Figure 14). The granitic rocks formed in those magmas that are now exposed at the surface in the Sierra Nevada, such as the famous ones at Yosemite National Park, were emplaced

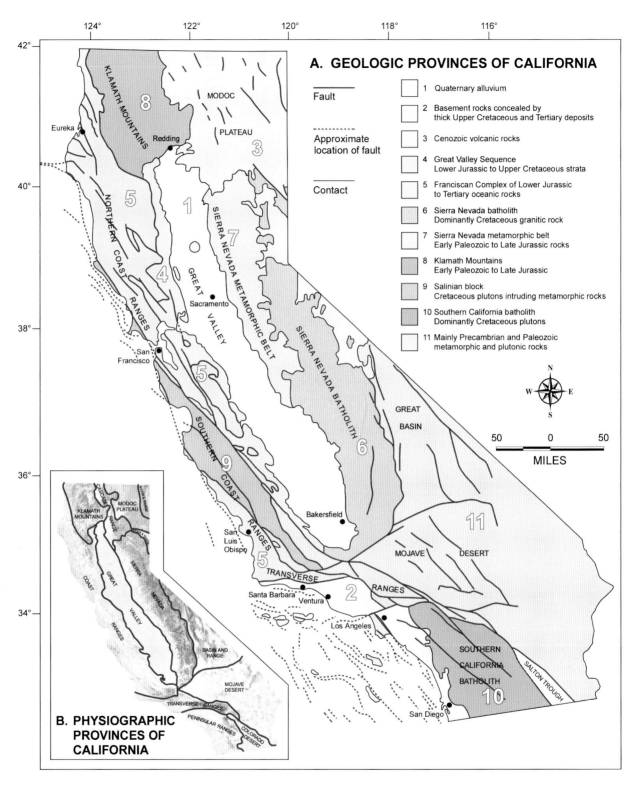

FIGURE 16. Geologic and physiographic provinces of California. (A) Geology map. Modified after Wallace (1990). (B) Geomorphic (physiographic) provinces. Courtesy of California Geological Survey.

intermittently between 225 and 80 million years ago. This large granitic mass, commonly called a **batholith**, contains more than 100 individual **plutons** (distinctive individual masses of granite formed at different times) (Alt and Hyndman, 2000).

[NOTE: In the discussion given later regarding the igneous rocks on the California Coast that were formed by slow crystallization at depth in batholiths, we will refer to them as granitic rocks. As is explained in J.D. Winter's book, *An Introduction to Igneous and Metamorphic Petrology* published in 2001, the numerous types of igneous rocks are distinguished from each other on the basis of their **texture** and **composition**. If the individual crystals of an igneous rock are readily visible to the naked eye, this rock crystallized slowly beneath the surface of the Earth, and it is called plutonic or intrusive. If, however, the individual crystals are too small to be seen by the naked eye, the rock crystallized rapidly on the Earth's surface and is called **volcanic**. In some cases, the texture of an igneous rock may display two different grain sizes that vary by a significant amount. If so, this rock is called **porphyritic**. The larger crystals, called phenocrysts, formed earlier during a period of slower cooling, and the finer crystals, which formed later, are called groundmass. Winter (2001) explained further that *because the grain size is generally determined by cooling rate, porphyritic rocks commonly result when a magma experiences two distinct phases of cooling. This is most common in, although not limited to, volcanics, in which the phenocrysts form in the slow-cooling magma chamber, and the finer groundmass forms upon eruption.* Examples of the different types of textures of igneous rocks are illustrated in Figure 17A.]

The granitic rocks in the coastal region of California are usually of one of the following four types (see also the ternary diagram in Figure 17B):

1) Granite if it is composed of 20-60% quartz and between 10 and 65% of a mixture of alkali feldspar and plagioclase;

2) Granodiorite if it is composed of 20-60% quartz and between 65 and 90% plagioclase;

3) Tonalite if it is composed of 20-60% quartz and between 90 and 100% plagioclase; and

4) Diorite or gabbro if it is composed of <5% quartz and between 90 and 100% plagioclase.

The difference in composition of these rocks is controlled by the composition of the original material the magma was created from, as well as the cooling history of the magma. Not readily apparent is the fact that different minerals crystallize at different temperatures. For example, a series of reaction,

ignoring the ferromagnesium minerals, would show the following trend as the magma cools: Calcium-rich plagioclase crystallizes first (at the highest temperature), potassium-rich plagioclase next, potassium-feldspar (e.g., orthoclase) next, muscovite mica next, and quartz last (at the lowest temperature). Consequently, the temperature of the magma when it cools to the degree where crystallization is possible determines, at least to some extent, the composition of the resulting granitic rock. However, pressure, which obviously increases with depth, and the presence of water also exert important controls on the process. This is a complex mechanism still under debate; however, the basic principles being unraveled by numerous studies of the rocks along the coast at the present time tell us a great deal more about their origin than was possible a few decades ago.

The occurrence of granitic rock exposures along the coast of Central California would be one of the most surprising geological occurrences in this area for a geologist not acquainted with the transfer of these rocks to the north by the Pacific Plate. The most noteworthy exposures of granitic rocks along the outer coast occur at Bodega Head, Point Reyes, the Devils Slide area on the Montara Headland, and Monterey Peninsula. The composition of the different granitic rocks is illustrated in Figure 17B.

As noted by Doris Sloan (pers. comm.), these granitic rocks were formed as the subducting slab started to melt and this was well inland of the trench (approximately 100 miles) as as result of the slab being carried there during the subducting process. However, the granitic rocks along the Central California Coast were formed in a location much further to the south than where they are now located. In order for granite, which is typically composed of abundant quartz (SiO_2) and orthoclase feldspar [a silicate mineral with an abundance of aluminum and potassium ($KAlSi_3O_8$)], to form in such a setting (100 miles landward of the trench), the following steps usually occur (Alt and Hyndman, 2000):

1) The process of subduction ultimately creates a chain of volcanoes that rises **parallel to an oceanic trench**. Clarifying this point further, Sloan pointed out that *the crust is dragged down into a trench landform, but the melting occurs further down the slope in the subduction zone. In a sense, the trench is a by-product of subduction and has nothing to do with the melting that forms the volcanoes.*

2) In order to form, the volcanoes depend upon the release of water by serpentinite as it is driven into the depths by the subduction process where it reverts to peridotite, like most of the remaining rocks in the mantle.

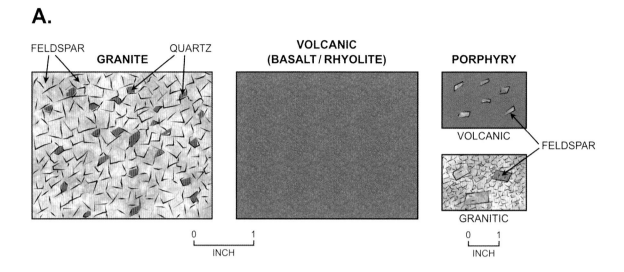

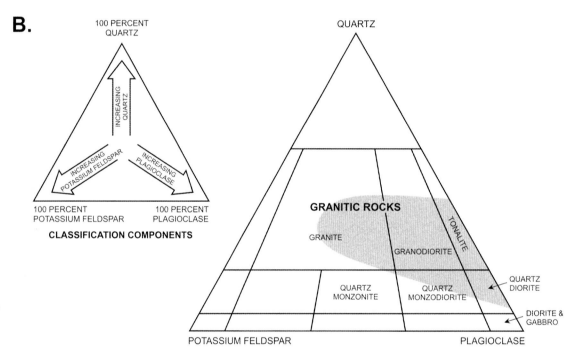

FIGURE 17. Characteristics of igneous rocks. (A) Igneous rock textures. The coarse grain size of the crystals in the granite is a result of slow cooling in the original magma. The fine texture in the two examples of volcanic rocks are the result of rapid cooling as the lavas were extruded onto the Earth's surface. The porphyries result from two different stages of cooling during the rocks' history. (B) (Left) Classification of plutonic igneous rocks (those formed at depth under conditions of slow cooling), which is based on the relative percentages of quartz, potassium feldspar, and plagioclase. (Right) Classification of the common plutonic rocks. The shaded area outlines the position on the diagram within which most of the plutonic rocks in the Sierra Nevada would plot. Modified from Harden (2004) and originally based on Huber and Wahrhaftig (1987).

Serpentinite is composed of a number of complex, usually hydrated, ferromagnesium (iron and magnesium bearing) minerals (i.e., it contains significant amounts of H_2O).

[NOTE: Alt and Hyndman suggested that the serpentinite is converted back to peridotite as a result of losing some of its water content, but such a conversion has been questioned by others. In any event, the loss of water is a key factor.]

3) It is important to note that if water is *added in any form to any kind of rock* that rock's melting temperature is lowered to the point that some of the original rock

masses become molten, forming what is known as a **magma** (a volume of molten material that can eventually cool down to the point that minerals crystallize and rocks form).

4) This process happens directly above the depth where the *sinking slab of lithosphere gets hot enough to break down its serpentinite.*

5) The water released from the serpentinite rises into the overlying mantle, resulting in it becoming partially melted and, hence, becoming basalt, which in turn rises into the overlying crustal rocks, which contain more of the lighter elements than the basalt, causing them to melt and turn into a granitic magma.

6) The magmas erupt creating the chain of volcanoes that are aligned above the zone where the sinking slab loses its water [*anywhere from 50 to 200 miles from the oceanic trench depending upon how steeply the descending oceanic crust is sinking* (Alt and Hyndman, 2000)]. The volcanic lava flows that come out of the granite magma are crystallized quickly as **rhyolite** and those from the basalt magma crystallize quickly as **basalt**.

7) After the volcanoes die and the mountaintops are eroded down, large masses of pure granite, or one of the other granitic cousins, including possibly diorite, that *crystallized from the unerupted magma in their roots,* are exposed at the Earth's surface. This process explains the presence of the large granitic batholith now laid bare in Yosemite National Park.

A major change in climate initiated the beginning of the Ice Ages (the Pleistocene epoch). Earlier workers suggested that this occurred around 1.8 million years ago, but the latest word is that that date will probably be reset to 2.6 million years ago (see Table 1). The Pleistocene epoch ended around 12,000 years ago, when the last of the major glaciations ceased. Within the Pleistocene epoch, four major glacial events occurred (there were a number of lesser ones), during which time ice covered large areas of the North American continent (as far south as the Ohio River in the eastern U.S.), as well as many of the high mountain areas in California. The approximate median peak times for these four major glacial events, expressed in years before the present, were: 650,000 for the Nebraskan glaciation; 450,000 for the Kansan; 150,000 for the Illinois; and around 20,000 for the Wisconsin. During each of these glaciations, sea level dropped to significantly lower levels than it is today; during the last glaciation period it was 350 feet lower! When sea level was lower, during what is referred to as **lowstands**, the rivers along many coastal areas, including those in California, carved deep valleys across the coastal zone and out onto the present continental

shelf. During some of the lowstands, large dune fields developed on the exposed continental shelf, with sand being blown landward into regions now above sea level.

The glaciations were separated by warming periods, called interglacials, during which sea level rose. During these warming periods, the highest sea levels were higher than it is today, with each succeeding interglacial having lower and lower **highstands**. Sea level during the last interglacial before the present one, the Sangamon (peak around 120,000 years ago) stood around 10 to 12 feet higher than it is today. Of course, it appears to be have been higher than that along the Central California Coast, about 20 feet, because of the constant rising of the Coast Ranges.

This discussion of the elevations of the different sea levels applies to those areas that are relatively tectonically stable, such as the coastline of the southeastern U.S. and the Gulf of Mexico, in order to hold the effect of tectonism constant. Along much of the California coast, on the other hand, the land has been rising almost continuously during the Ice Ages and later; therefore, the elevation of the highstands on the California coast are now higher than those on the east coast, because of this continual rise of the land. A study by Covault (2004) of uplifted, wave-cut terraces on the North-Central California Coast, located 131 miles north of San Francisco near Alder Creek (where the San Andreas Fault goes offshore), showed that the shoreline in that area has been steadily rising at the rate of 1.3 feet per thousand years during the past 410,000 years. Therefore, tectonic uplift of the Coast Ranges has had a huge imprint upon the sea-level curves along the coast of California.

The major ice-melting episode that started about 12,000 years ago marked the beginning of the **Holocene epoch**, at which time sea level was approximately 350 feet below its present position on the east coast of the U.S. If Covault's numbers are correct, points along the shore at the present mean tide level at Alder Creek are about 13-16 feet above where they were when the sea level started to rise. A considerable amount of research on this topic is still underway, even in the tectonically stable areas, and the exact dates involved are subject to change as more data are produced. Without any doubt, however, as the melting proceeded, sea level rose rapidly, as much as 2 feet/century, reaching near its present level along the southeast coast of the U.S. around 4,500 years ago (the exact date is somewhat in question, some west-coast researchers cite the dates 5-6,000 years ago). Since a near **stillstand** that occurred at that time (4,500 – 6,000 years ago), the bulk of the major Holocene landforms on the relatively stable trailing-edge coastlines of the southeastern U.S. and the Gulf of Mexico (e.g., deltas, including the modern Mississippi delta, and barrier islands) have formed. During this same time interval, a considerable amount of Holocene landforms have also developed in San Francisco Bay, including the fairly extensive Sacramento/San Joaquin delta and broad tidal flats and marshes. The smaller coastal water bodies such as Bodega Bay, Drakes Bay, Bolinas Lagoon, Elkhorn Slough, and Morro Bay have also evolved during that time frame.

The Coast Ranges

"Do you know why I live out there on the cliff? I've only told the reason to a few. I'll tell you, because you're coming to stay with me." He stood up, the better to deliver his secret. "I am the last man in the western world to see the sun. After it is gone to everyone else, I see it for a little while. I've seen it every night for twenty years. Except when the fog was in or the rain was falling, I've seen the sun set."

John Steinbeck – TO A GOD UNKNOWN

There are eleven geomorphological provinces in the state of California (illustrated in Figure 16B). Except for about 20 miles of the Transverse Ranges province along the southern border of the area under discussion (Compartment 7; Section III), the coastline of Central California is backed by the Coast Ranges province. As noted by Sloan (2006), *unlike the Sierra Nevada, which is one long mountain range, the Coast Ranges are a series of more or less parallel ranges, all running generally northwest to southeast.*

These ranges are rather new to the scene, having been raised sometime less than 5 million years ago (3-4 million according to Page, 1989), with some dispute in the geological literature on the actual cause for their formation. In addition to the two theories given earlier, another popular idea is that it was due to a pressure increase caused by a slight shift in relative plate motions as the Pacific Plate slides past the North American Plate (from Elder, 2001; citing Cox and Engebretsen, 1985).

The rocks along the shoreline of these ranges within the Central California coastal area consist of four major rock sequences – the Franciscan Complex, the Salinian Block, the Great Valley Sequence, and the western Franciscan Complex or Nacimiento Block (Figures 16A and 18). The Franciscan Complex, which can be examined at the spectacular exposures along the shoreline north of the Golden Gate, consists in its lower portions of a vertical sequence of rocks starting with seafloor pillow basalt at its base and grading upward into a beautifully laminated radiolarian chert (sometimes

called ribbon chert).

[NOTE: Chert is a fine-grained, smooth-textured rock composed of SiO_2 and formed as a sedimentary deposit.]

A **pillow basalt** typically forms when the basaltic lava flows onto the sea floor in deep water, as is common along the mid-ocean ridges. The cold water in those depths quickly cools the surface of the flow and the lava commonly takes on a pillow shape. This is a continuous process, with the hot lava breaking through the surface of the flow repeatedly to form more pillows. The size of the pillows ranges from about 4 or 5 inches up to 10 feet or so. During the period when the Franciscan trench was active, the lavas rising along the mid-ocean ridge to form new crustal material for the ever-expanding ocean floor moved persistently toward the far-away trench, forming a basal layer upon which other sediments could accumulate. Figure 19 shows an example of these pillow basalts in the Franciscan Complex. These rocks are a member of one of the oceanic terranes of the Complex (Nicasio Reservoir terrane).

[NOTE: The geological term **terrane** was defined by Sloan (2006) as *a fault-bounded body of rock, characterized by a geological history different from that of neighboring terranes.* The Franciscan Complex has been subdivided into 11 different assemblages (or terranes) in the San Francisco Bay area.]

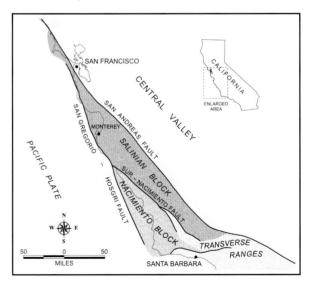

FIGURE 18. Location of the two major blocks of rock suites, Salinian and Nacimiento Blocks, and the major faults present near the coast between San Francisco and Santa Barbara. The Salinian Block contains mostly granitic rocks of Cretaceous age and older metamorphic rocks, and the Nacimiento Block contains primarily Franciscan Complex rocks of Jurassic/Cretaceous age. Modified after Henson and Usner (1993).

FIGURE 19. An outcrop of pillow basalt of the Franciscan Complex located beside the Point Reyes Petaluma Road near Lake Nicasio in Marin County.

The **chert** is composed of the siliceous skeletons of millions of microscopic-sized, unicellular marine protozoans called radiolaria. Their skeletons accumulate on the seafloor in phenomenal numbers to form a "radiolarian ooze" if the water is deep enough so that any calcareous material, such as foraminiferal skeletons, are dissolved as a result of the large percentage of carbon dioxide in the water at those depths and, hence, its high acidity (see examples of radiolarian skeletons in Figure 20B). Over the thousands to millions of years during which the Pacific sea floor moved slowly eastward on its approach toward the North American Plate, the radiolaria continued to rain down on the underlying pillow basalt. At times, fine-grained material such as volcanic ash, or possibly some land-derived muds as the shoreline became closer, accumulated as thin layers on the chert, giving the deposit a distinctly layered character. According to Sloan (2006), this deposit is about 250 feet thick in the Marin Headlands north of San Francisco. As the partly consolidated and layered "ooze" was being swept down into the trench or scraped off the deeper material and pushed against the North American Plate, it was very complexly folded (see photograph in Figure

20A).

The chert is overlain by a thick array of interbedded sandstones and shale (or mudstone). Most sandstones were deposited as **turbidites**, a sand deposit formed by a swiftly moving, bottom-flowing current called a turbidity current (Figure 21A). This rapidly flowing, dense plume contains an unusually large amount of suspended sediment. This type of flow is usually generated by a mechanism such as a landslide on a steep slope of the sea floor (there are several other possibilities). Because of its density being greater than the surrounding ocean water, this turbid mass hugs the sea floor as it flows down the slope. Once its velocity slows down, usually when the slope upon which it is flowing becomes flatter, the sediment is gradually deposited from the plume, with the heaviest (usually largest) particles settling out first. This produces what is known as a graded bed (coarse- to fine-grained sediment from bottom to top; Figure 21B), one of the principle clues used in identifying a turbidity current deposit. Figure 21C shows a photograph of the graded bed in a turbidite deposit.

This vertically consistent threesome, the origin of which is further illustrated in Figure 22, is the part of

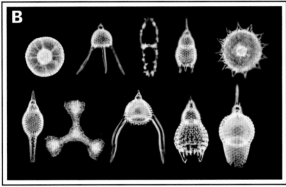

FIGURE 20. Ribbon chert. (A) The radiolarian chert of the Franciscan Complex on Conzelman Road in the Marin Uplands. This deposit of chert, which was originally about 250 feet thick (Sloan, 2006), accumulated over a period of 100 million years (from 200 to 100 million years ago), one of the longest stratigraphic sequences of (continuously deposited) chert in the world (Wahrhaftig and Murchey, 1987; quoted from Elder, 2001). According to Karl (1984), the radiolaria whose skeletons make up these chert beds lived in a portion of the sea that was located near the equator, possibly as much as 600 miles offshore. (B) Examples of the siliceous skeletons of 10 radiolaria species. Photomicrograph courtesy of the National Biological Information Infrastructure and the photographer – R. Femmer (2005).

the sequence that has not been horribly mangled in the depths of the Franciscan trench. The result of that intense alteration of the other original sediments, sedimentary rocks, and basalts that make up the Franciscan sequence is called **mélange**, which consists of a soft crushed shale or serpentinite matrix (natural material in which something is embedded) with blocks of other rocks "floating" in the matrix (Sloan, 2006). These foreign blocks may have a variety of compositions and sizes. As noted by Sloan (2006), *mélange is like a geologic chocolate pudding with*

raisins, nuts, and marshmallows (the resistant blocks) mixed into it. Many of the beautiful, soft rounded hills in Marin County, which have miscellaneous scattered blocks sticking up out of the ground, are underlain by Franciscan mélange. Mélange deposits are also common along Highway 1 in the Big Sur Country.

Alt and Hyndman (2000) pointed out further that the Franciscan Complex rocks that had been *stuffed into* the Franciscan trench (Figure 14) eventually rose to become the western part of the Coast Ranges. To explain the rocks on the eastern side of the mountains, they concluded that some of the "stuffed sediments" *had been dragged under the edge of the strip of oceanic crust* (the sea floor that formerly separated the Sierran and Franciscan trenches) *and they rose too, jacking the edge of that strip up to a steep angle to become the Great Valley Sequence along the eastern side of the Range.* Rocks in the Great Valley Sequence are rare along the shoreline in Central California, except in a few locations, where they are on the western side of the San Andreas Fault system. An example of rocks interpreted by some to be members of the Great Valley Sequence at Pebble Beach in San Mateo County and on the Big Sur coast is described in Section III.

[NOTE: To clarify somewhat the time sequence explained here, Sloan (pers. comm.) noted that *the Coast Ranges as a landform are much younger than the rocks, which are Cretaceous. The uplift that exposed the Franciscan in the Outer Coast Ranges and the Great Valley Sequence rocks on the east side of the Inner Coast Ranges occurred in the Cenozoic (Table 1). The near-vertical Great Valley Sequence rocks were pushed up as the Franciscan rocks were uplifted. Also, Great Valley Sequence rocks are found faulted in with Franciscan rocks east of the fault, for example, the ridge of conglomerate just east of the mouth of the Russian River.*]

The rocks on the Salinian Block (on the western side of the San Andreas Fault; Figure 18) contain a notable supply of Mesozoic granitic rocks. These rocks are similar in composition and age (80-110 million years old) to the granitic rocks in the Sierra Nevada. As noted earlier, these facts have led some geologists to conclude that these igneous rocks were formed as a southern extension of the Sierra Nevada batholith that was split off from it, and later transported hundreds of miles to the north aboard the Pacific Plate (Sloan, 2006). This was possible because of their position on the west side of the San Andreas Fault system. Outstanding examples of these rocks can be seen at Bodega Head, Point Reyes, Devils Slide, the Monterey Peninsula, and in several other localities. Most of these occurrences are pointed out in the discussion of the different areas in Section III.

A. TURBIDITY CURRENT

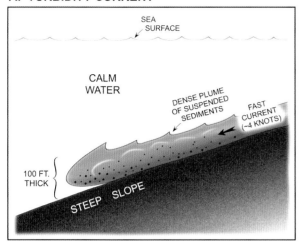

B. TURBIDITY - CURRENT DEPOSIT (TURBIDITE)

FIGURE 21. Turbidity currents and their deposits (turbidites). (A) General model. The dimensions cited here are for one measured in the Monterey Canyon by researchers at the Monterey Bay Aquarium Research Institute. These sediment plumes come in a wide range of sizes. (B) General pattern of fining upward (grading) of a turbidity current deposit. (C) Graded bed in what is interpreted to be turbidity current deposit in the Carmelo Formation at the Point Lobos State Reserve in Monterey County.

They also make up conspicuous outcrops on the Farallon Islands.

With regard to the igneous granitic rocks in the Salinian Block, other rocks of similar origin and texture to pure granite but with somewhat different composition

are also present within this sequence (e.g., diorite, granodiorite, and tonalite; defined earlier; see Figure 17B).

The western Franciscan Complex, more commonly referred to as the Nacimiento Block (Figure 18), is similar in composition to the main Franciscan Complex, but due to its position west of the San Andreas Fault system, these rocks must have originated within the Franciscan trench far to the south. They crop out along the shore in the southern one third of the Central California Coast (Compartments 6 and 7; Section III).

The rocks in these four major sequences thus far described are the older ones that eventually accumulated in either the Sierran and Franciscan trenches (Figure 12). However, long after these trenches ceased to be active, much of the area under discussion was flooded by marine waters from time to time, usually within relatively small basins. As the depths within these basins decreased, the depositional systems became more fresh-water influenced. Rocks of Paleogene/Neogene age and sediments of Quaternary age were deposited in most of these areas; therefore, these younger rocks and sediments are also present all along the coast, complicating the picture somewhat. The widespread Monterey Formation of middle Miocene age is one of the most important of these units, because of its role as a source rock and reservoir for oil.

More details on the geology of the three major rock sequences, as well as the younger ones that followed them, are presented in Section III in the discussions of the areas where they are encountered as we move from north to south in our descriptions of the coastline.

EARTHQUAKES

As noted by Harden (2004), *most of the world's earthquakes occur at plate boundaries; in fact, the zones of concentrated seismicity …*(another way of saying the rocks are faulting frequently within some limited area)… *are one of the main lines of evidence used to identify plate boundaries.* Did you notice that the contact between two plate boundaries (Pacific and North American Plates) runs right through the middle of the Central California Coast (Figure 15)?

It is certain that we do not have to remind anyone of the importance of the impact of earthquakes on the Central California Coast, but a thorough discussion of the subject is beyond the scope of this book. For further reading about earthquakes in the region, we again recommend *Roadside Geology of Northern and Central California* by David Alt and Donald Hyndman (2000), *California Geology* by Deborah Harden (2004), and

PLATE ➤ MOTION ➤

MID
OCEAN

FURTHER
EAST

TRENCH

FALLING
RADIOLARIA

ACID H_2O

TURBIDITES

TURBIDITY
CURRENTS

NEW
PILLOW
BASALTS

RADIOLARIAN
OOZE

MID-OCEAN
TRENCH

MILLIONS OF YEARS

FINAL DEPOSIT

TURBITES {

RIBBON CHERT {

PILLOW BASALT {

FIGURE 22. Origin of the first three bottom units of the rocks of the Franciscan Complex in the Marin Headlands area.

Geology of the San Francisco Bay Region by Doris Sloan (2006). Also, Michael Collier published an engaging and beautifully illustrated account of the San Andreas Fault system, and the earthquakes associated with it, in *A Land In Motion; California's San Andreas Fault* (1999).

Some of the most fundamental concepts and issues related to the earthquakes that might take place along the Central California Coast include:

1) Earthquakes occur when the strain that has accumulated in the rocks on either side of a fault is suddenly released as the rocks snap apart and slide past each other. The amount of movement along the fault is usually not more than a few feet, but an offset of up to 31 feet was measured along the San Andreas Fault during the Fort Tejon earthquake of 1857 (Collier, 1999).

2) This sudden release of energy into the surrounding Earth radiates away from the location of the fracture (place where the movement starts), called the

hypocenter (or focus) of the quake, in the form of two seismic waves. The *P*, or primary seismic wave, moves through the Earth in a fashion *similar to sound waves due to compression of the rock particles, with each particle moving backward and forward in a straight line from the origin of disturbance. P* waves are the faster moving of the two (about 4 miles per hour in the Earth's crust and 5 miles per hour in the mantle). The *S*, or secondary seismic waves, *are due to shearing or transverse motion that moves rock particles not forward and backward, but side to side or up and down the way ripples radiate out when a rock breaks the surface of a calm pond* (Collier, 1999). *S* waves, which move at about half the speed of *P* waves, are only transmitted through solids. Because they are shear waves, the *S* waves are the ones that produce the rolling motion of the ground during an earthquake and are, therefore, more damaging to structures, such as residences and office buildings, than *P* waves.

[NOTE: The **epicenter** of the quake is the point on the surface of the Earth that lies directly above the hypocenter.]

3) These seismic waves are recorded by an instrument called a seismograph, which runs continuously, recording all incoming waves. Because of their different speeds, the *P* waves from a specific earthquake will arrive ahead of the *S* waves. Although their rates of speed may vary significantly depending upon what they are moving through, the ratio of the difference between the speeds of the two waves is quite constant. *This fact enables seismologists to simply time the delay between the arrival of the P wave and the arrival of the S wave to get a quick and reasonably accurate estimate of the distance of the earthquake from the observation station. Just multiply the S-minus-P (S-P) time, in seconds, by the factor 8 km/s to get the approximate distance in kilometers* (Univ. Nevada-Reno, 2009). The smaller the interval of time between the arrival of the two waves, the shorter the distance between the epicenter and the recording seismograph. If data are recorded by several seismographs at different locations on the Earth's surface, the exact location of the epicenter of the quake can be determined by triangulation.

4) The Richter scale and the Moment-Magnitude scale are the two scales most commonly used to describe the magnitude of earthquakes. The Richter scale, invented by Charles F. Richter in 1934, is based on the calculation of the maximum magnitudes of the shear waves. Both the Richter and the Moment-Magnitude scales are logarithmic (base 10). For example, a magnitude 5 earthquake on the Richter scale would result in ten times the level of ground shaking as a magnitude 4 earthquake (and 32 times as much energy would be released) (Wikipedia encyclopedia). Quakes with Richter scale values of 3.5 to 5.4 are often felt but rarely cause damage. Those above 7.0 on the scale are classified as major earthquakes and can cause widespread damage. The Magnitude-Moment scale (expressed as M), defined as the amount of energy released on the fault, is based on several factors, including: a) how far the fault slipped (i.e., how far a fence line, railroad line, and so on was offset); b) length of the fault line that was affected; c) depth at which movement was initiated; and d) the "stickiness" of the surfaces being faulted (rock rigidity). Clearly, this measure is more difficult to calculate, but it is the one most commonly used in reporting earthquake magnitudes at this time.

A good place to get a visual image of the characteristics of the highly destructive 1906 earthquake along the San Andreas Fault is to visit the famous earthquake trail at the Point Reyes Bear Valley Visitor Center near Olema. A photo taken shortly after the earthquake by the eminent geologist G.K. Gilbert that is posted by the trail shows the linear scar in the earth that marked the location of the fault along a hillside. Rumor has it that a farmer showed Gilbert the tail of a cow sticking up through the ground that had fallen into the chasm when it opened up along the fault line during the quake. Hmmm. Pranksters? Presently, a number of blue posts have been driven into the ground near the trail to mark the fault line. The 20-foot offset of a fence caused by the fault has also been recreated.

A major fault like the San Andreas breaks up and grinds the rock fragments along the shear zone, which allows significant erosion of the associated rocks and soil resulting in a linear eroded valley. As shown in Figure 23, the valley eroded along the San Andreas Fault zone between Tomales Bay and Bolinas Lagoon is a striking example of this.

The three most damaging earthquakes to strike the Central California Coast in recent history occurred along the San Andreas Fault system. Probably the most famous one occurred in 1906, which was estimated to have an M value of 7.7. That earthquake resulted in the death of hundreds of people (exact number unknown), caused major fires in the city of San Francisco, and other calamities. The Fort Tejon earthquake of 1875, with an estimated M value of 7.8, was also very destructive. According to Collier (1999; p. 71), *together, these two earthquakes …(in 1857 and 1906)…account for fully half of all the seismic energy released along the San Andreas since 1769.* Many of us probably remember the Loma Prieta earthquake of 17 October 1989. Co-author Hayes, a "connoisseur" of the game of baseball, will never forget how, as he watched, a World Series baseball game between the Oakland A's and the San Francisco Giants was interrupted by the images of that earthquake as they were shown to the world by ABC television. The quake caused a ten day suspension of the Series, which was swept by a great A's team.

The San Andreas (1906) and Loma Prieta (1989) earthquakes are compared in Table 2.

There are a number of major faults other than the San Andreas that are also a cause for concern with respect to potential earthquake damages in San Francisco Bay and vicinity. As pointed out by Doris Sloan in her book on the geology of the San Francisco Bay area (Sloan, 2006), the San Andreas Fault system consists of two major branches: 1) The San Andreas and San Gregorio Faults which run along the coast; and 2) The so-called East Bay

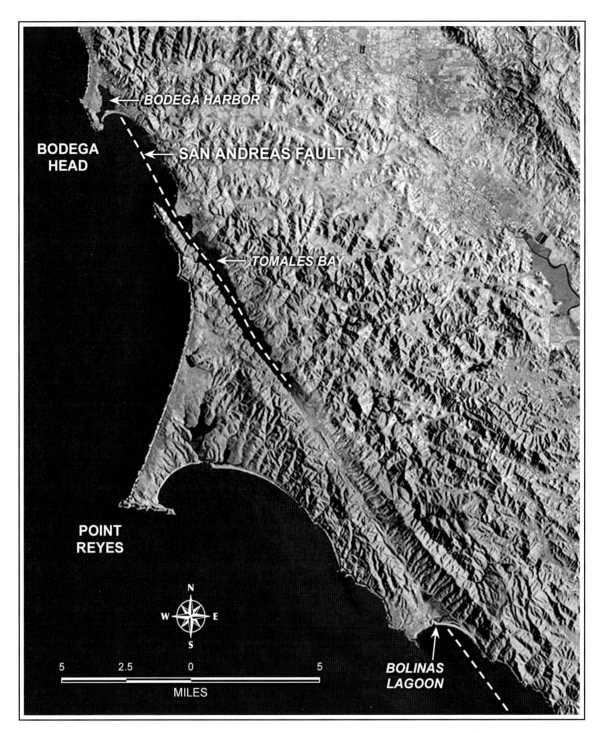

FIGURE 23. The linear valley that has been eroded along the trace of the San Andreas Fault between Bolinas Lagoon and Bodega Harbor. Image derived from National Elevation Dataset and Landsat 7 data acquired from 1999-2002. Courtesy of the California Spatial Information Library (CaSIL).

Fault system, which includes the Hayward, Calaveras, and several lesser faults. Some geologists think the next "big one" may be along the Hayward Fault, rather than the San Andreas. Figure 24 shows the locations of these Bay Area faults.

GOLD, OIL, AND WATER

No discussion of the geology of the state of California would be complete without at least some mention of the economic benefit the rocks have brought to the state,

TABLE 2. Comparison of the 1906 and 1989 Earthquakes.

	1906	1989
Date	18 April 1906 at 5:13 am	17 October 1989 at 5:04 pm
Epicenter Location (Figure 24)	South of the Golden Gate	Santa Cruz Mountains
Magnitude	7.7 M	6.9 M
Duration	45 seconds	20 seconds
Horizontal Offset (maximum)	22 feet	6 feet
Damages ($)	$400M	$6.8B
Deaths and Injuries	hundreds of deaths	68 deaths; 4,000 injuries

starting with the famous gold rush of 1849. The most notable outcrops of the bedrock that contains the gold occur along the western margin of the northern Sierra Nevada, usually closely associated with a major fault zone called the Melones Fault. Alt and Hyndman (2000) noted that somehow the juxtaposition of the fault, serpentinite (rock composed of a group of greenish, brownish, or yellowish minerals thought to be the product of metamorphism of mantle rock or oceanic crustal rock), intrusions of granite, and circulating hot water created large numbers of quartz veins that contain gold, "the fabled Mother Lode." Much of the mining of the gold was from **placer deposits** (layers of the heavier grains of gold eroded from the Mother Lode and deposited within sand bars in a stream bed) laid down by streams that crossed the gold-rich rocks during Eocene time (40-50 million years ago) and during modern times. The heyday of such mining has passed long ago, but some people still pursue the habit, especially when the price of gold rises. During a short vacation in August 1979, we were fishing for trout in the North Yuba River near Sierra City and were surprised to find the river full of people panning away for gold.

The oil-and-gas industry has been active in California for many decades. The Monterey Formation is an extensive oil-rich unit of Middle Miocene age rocks that have been productive in the Santa Maria and Santa Barbara Basins and elsewhere. This formation contains a thick interval of dark shale loaded with organic matter, phosphate rock, and diatomite [a light-colored, soft sedimentary rock formed mainly of the siliceous shells of diatoms (a microscopic unicellular algae)]. This shale is apparently the source for the oil present in the Monterey Formation.

Crude oil and gas are thought to have been formed mostly by compression and heating of ancient organic materials, such as prehistoric zooplankton and algae,

which were apparently quite abundant in the warm Monterey sea that covered much of western California around 15-20 million years ago. These shales were deposited at the same time as the flood basalt eruptions in eastern Oregon (17-15 million years ago). Alt and Hyndman (2000) concluded that *these eruptions released enough carbon dioxide into the atmosphere to cause a severe global greenhouse effect.* As Ed Clifton (pers. comm.) pointed out, the excess of CO_2 was responsible for the increase in the productivity of these ocean waters. *The extensive Miocene volcanism in the Pacific Northwest generated carbon dioxide and phosphorous – great fertilizer – which enhanced the Middle Miocene productivity.*

In the Central Valley, potential oil-bearing rocks were deposited in two sub-basins, the Sacramento Basin to the north and the San Joaquin Basin to the south. According to information provided by Victoria Petroleum, Inc., wells drilled in the mostly Cenozoic rocks (Table 1) in the San Joaquin Basin have produced 12 billion barrels of oil and 11 trillion feet of gas to date. A recent assessment by the U.S. Geological Survey estimates that another 3.5 billion barrels of oil may be added to reserves in existing fields and an additional 393 million barrels may as yet be undiscovered. All of this has made Bakersfield, California the oil-industry capitol of the state. Co-author Hayes spent some time there marketing in the 1980s and the place had the feel of a West Texas oil town, a pretty exciting place to be at that time.

The controversy over allocation of surface water in the state has been ongoing for a long time. However, another source of water, accessed by drilling wells into the geological units throughout the state, is also of great importance. This type of water, called **groundwater**, is crucial both as a source of drinking water and for agricultural purposes. There are twelve primary water-

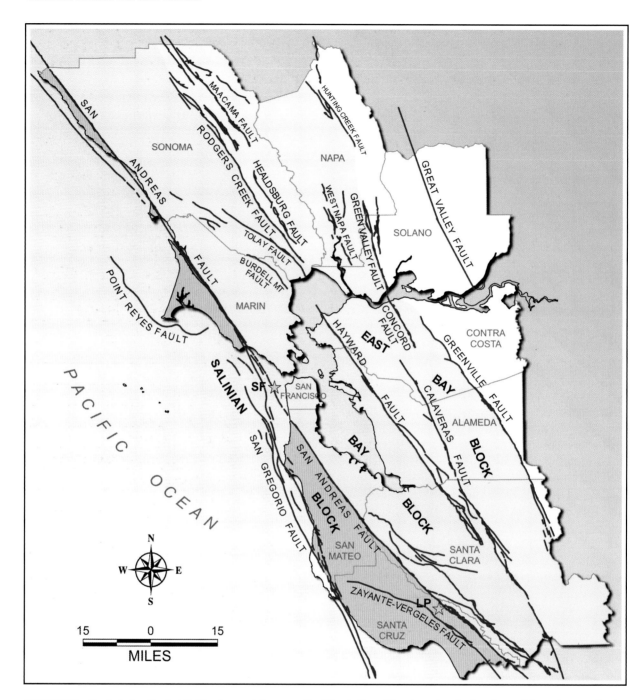

FIGURE 24. Faults in the Central California coastal area north of Santa Cruz. The stars give the locations of the epicenters for the 1906 San Francisco (SF) and the 1989 Loma Prieta (LP) earthquakes. As reported by Sloan (2006), today's faults divide the Bay Area into three large blocks: 1) the Salinian Block, which contains Cretaceous-age igneous rocks and is underlain by the Pacific Plate, west of the San Andreas Fault; 2) the Bay Block, which contains mostly Franciscan Complex rocks, between the San Andreas and the Hayward Fault; and 3) the East Bay Block, which contains abundant Great Valley Sequence rocks, east of the Hayward Fault. Both of the latter two blocks are underlain by the North American Plate. Each of these three blocks moves northwestward relative to the block to the east of it (Sloan, 2006). Basemap modified after Sloan (2006). Fault lines by U.S. Geological Survey and California Geological Survey, from the Quaternary fault and fold database for the U.S. (USGS-earthquake, 2009).

bearing geological formations (called **aquifers**) used for public water supply in the state, most of which are geologically young. They are predominantly sandy deposits that accumulated in valleys since the last ice age. These sediments, consisting of alluvial fan, river channel, and floodplain deposits, are found in all coastal basins (depressions within the Coast Ranges), such as the Salinas Valley aquifer. These types of aquifers also occur in the Central Valley area and within the Basin and Range province in the southwest. Some of the volcanic rocks in the northern region are also aquifers.

[NOTE: Most river channels run through valleys of different widths. During floods, sediment-rich (muddy) water spills out of the channel and builds up a flat plain between the channel and the sides of the valley. These flat-lying deposits occupy a zone known as the floodplain.]

The Central Valley aquifers were tapped by over 4,000 wells. The Coastal Basins aquifers, with over 1,500 wells, as well as the Basin and Range and Los Angeles/ Orange County Coastal Plain Aquifers, with over 1,000 wells each, are important sources of groundwater.

It probably goes without saying that groundwater management and monitoring are essential for the future of this resource, considering the problems already encountered with maintaining adequate surface water supplies. If treated properly, it is likely that groundwater can be a reliable and safe source of drinking water for some time into the future. However, groundwater is not a completely renewable resource. The groundwater aquifers are recharged by rainfall and snow melt, and this water is usually found at relatively shallow depths with salt water occurring deeper; therefore, if more is removed by man than is replaced by these natural processes, the wells will run dry, or more likely, start to produce salt water.

RELATIVE ABUNDANCE OF SHORELINE TYPES

The distribution of the different shoreline types (or habitats) of the Central California Coast has been mapped in detail by our group (NOAA/RPI, 2006). The shoreline types encoded on these maps, which are created using available imagery, visual inspection flying over the area at altitudes of 500 feet, oblique aerial photography, and ground inspection of critical areas, are ranked on a scale of 1-10 based upon their sensitivity to oil spills and other considerations, with those ranked 10 being the most sensitive. This type of mapping is usually done by a coastal geomorphologist (the authors of this book participated in some of the mapping of the Central California Coast and of numerous shorelines elsewhere).

The 1-10 scale we use is called the Environmental Sensitivity Index (ESI) (Table 3). This system, which was first created during a study of Lower Cook Inlet, Alaska in 1976 (Hayes, Michel, and Brown, 1977), is now used worldwide for oil-spill contingency planning purposes (Hayes, Gundlach, and Getter, 1980).

TABLE 3. Shoreline types mapped for Central California Coast (NOAA/RPI, 2006).

ESI Scale	Description
1A	Exposed rocky shores
1B	Exposed solid man-made structures
2A	Exposed wave-cut rock platforms
3A	Fine- to medium-grained sand beaches
3B	Scarps and steep slopes in sand
4	Coarse-grained sand beaches
5	Mixed sand and gravel beaches
6A	Gravel beaches
6B	Riprap
6D	Boulder rubble
7	Exposed tidal flats
8A	Sheltered rocky shores
8B	Sheltered, solid man-made structures
8C	Sheltered riprap
9A	Sheltered tidal flats
9B	Vegetated low banks
10A	Salt- and brackish-water marshes

The following is a description and brief discussion of the most common shoreline types on the Central California Coast:

1) **Exposed rocky shores (ESI = 1A):** The term exposure relates to the size of the waves breaking onshore, which in this case, is quite large because of the exposure of this shoreline to the prevailing westerly winds that blow across the open Pacific Ocean and fairly common large swell waves (waves are discussed in detail later). These shorelines are mostly vertical rock cliffs carved by the waves (Figure 25A). These rock cliffs occur along 7.6% of the total shoreline. They are most common along the Big Sur coastline (Figure 1), where they are present along 21.7% of the shore.

2) **Exposed wave-cut rock platforms (ESI = 2A):** These are also highly exposed shorelines where the waves have carved out a flat rocky platform that is usually uncovered in the lower intertidal zone during low tide (Figure 25B). These platforms occur along 19.9% of the

FIGURE 25. Rocky coasts. (A) Wave-cut rock cliffs at Point Bonita at the north side of the entrance to San Francisco Bay (arrow points to lighthouse). This photograph, taken on 1 October 2008, is courtesy of the California Coastal Records Project, Kenneth and Gabrielle Adelman. (B) Wave-cut rock platform at Montaña de Oro State Park. Arrow A points to Morro Rock and arrow B points to the Santa Lucia Mountains. Photograph taken on 5 December 2008.

total shoreline, with the Big Sur coastline once more leading the way at 33.4%. They are also common along the Central Headland between Montara Mountain and Monterey Bay, where they occur along 30.9% of the shore.

3) **Fine- to medium-grained sand beaches (ESI = 3A):** See Figure 4 for the grain-size scale. These beaches are also usually found in exposed sites on the open ocean (Figure 26A). They occur along 20.9% of the shoreline, with three areas having an abundance of sand beaches – Monterey Bay, the Central Headland between Montara Mountain and Monterey Bay, and the South Bays (Figure 1).

4) **Coarse-grained sand beaches (ESI = 4):** This type of exposed sand beach is relatively rare (4.7% of the

total shoreline), only being notably abundant along San Francisco's outer shore, where they occur along 16.4% of the coast.

5) **Mixed sand and gravel beaches (ESI = 5):** In order to be placed in this class, a sandy beach must be composed of at least 20-30% gravel. This type of exposed beach is also relatively rare (8.5% of the total shoreline), being notably abundant only along the shorelines of the San Andreas Fault Valley and on San Francisco's outer shore.

6) **Gravel beaches (ESI = 6A):** Another relatively rare shoreline type (4.6% of the total shoreline; Figure 26B). As might be expected, they are most common along the shorelines of the Big Sur Country and the Central Headland between Montara Mountain and Monterey Bay.

7) **Sheltered tidal flats (ESI = 9A):** Because these tidal flats are sheltered from wave action, they are usually composed of mud (Figure 27A). They occur along 11.2% of the total shoreline and are most common in five areas (Figure 1): 1) San Andreas Fault Valley, which contains Bodega Harbor and the head of Tomales Bay; 2) Monterey Bay, which contains Elkhorn Slough; 3) Point Reyes, which contains Drakes Estero; 4) San Francisco's outer shore, which contains Bolinas Lagoon; and 5) Morro Bay.

8) **Salt- and brackish-water marshes (ESI = 10):** Like the tidal flats, these habitats are sheltered from wave action (Figure 27B). They occur along 11.2% of the total shoreline and are present in some abundance in four areas (Figure 1): 1) San Andreas Fault Valley, which contains Bodega Harbor and the head of Tomales Bay; 2) Monterey Bay, which contains Elkhorn Slough; 3) Point Reyes, which contains Drakes Estero; and 4) San Francisco's outer shore, which contains Bolinas Lagoon.

Considering the shoreline as a whole, there are four natural combinations that clearly define the general geomorphology of this coastline: 1) the combination of exposed rock cliffs and wave-cut rock platform, which occur along 27.4% of the shore; 2) sand beaches at 25.6%; 3) combination of sheltered tidal flats and marshes at 22.4%; and 4) combination of sand and gravel beaches with pure gravel beaches at 13.2%. The presence of the sheltered tidal flats and marshes in six moderate-sized coastal water bodies in the area (Bodega Harbor, head of Tomales Bay, Drakes Estero, Bolinas Lagoon, Elkhorn Slough, and Morro Bay), which yielded a shoreline length equaling 24.1% of the total, has skewed the numbers somewhat, because the shorelines within these areas are highly complex and indented. Considering only the exposed outer shoreline, the rocky shores dominate

FIGURE 26. Beaches. (A) Wide, fine- to medium-grained sand beach at Sunset State Beach on the northeast shore of Monterey Bay. View is from the edge of a raised marine terrace and looks north. Photograph taken on 8 December 2008. (B) Gravel beach located about 0.4 miles south of the mouth of Pescadero Creek in San Mateo County. Some of the largest of these well-rounded boulders are more than two feet in diameter. There is a hint of imbrication of some of the boulders near the toe of the beachface. Photograph taken on 16 August 2008.

FIGURE 27. Mud flats and marshes. (A) Randy Imai and associates of OSPR (California Department of Fish and Game) conducting a survey across the mud flats in Morro Bay. Photograph taken on 8 May 1992. (B) Marsh near the mouth of Scott Creek about a mile north of Davenport Landing. Photograph taken on 16 August 2008.

at around 41%. About 39% of the outer shore is sand beaches, with the high numbers occurring in places like Monterey Bay and on the long straight beaches created by natural groins in the most southerly region. Gravel beaches and sand and gravel beaches, at around 20%, are still a more important component of the outer coast than might have been expected for a shoreline not along a subduction zone.

[NOTE: The term outer shore as used here excludes the shoreline of San Francisco Bay. The shorelines of the coastal water bodies along the rest of the shoreline exclusive of San Francisco Bay, such as Tomales Bay and Morro Bay, are also excluded. Another definition of outer shore could be any location open to the full Pacific swell (suggested by Ed Clifton, pers. comm.).]

3 COASTAL PROCESSES

TIDES

He consulted the tide chart in Thursday's Monterey Herald. There was a fair tide at 2:18 P.M., enough for chitons and brittlestars if the wind wasn't blowing inshore by then.

John Steinbeck – SWEET THURSDAY

Changing water levels along the shoreline, generally referred to as tides, can be caused principally by three mechanisms: 1) a process called wind set-up during which wind stress causes the water levels to increase downwind and decrease upwind; 2) a combination of wind set-up and decreasing barometric pressure that accompanies major storms, such as hurricanes (usually referred to as storm surges); and 3) the gravitational attraction of the moon and sun on major water bodies, called astronomical tides.

[NOTE: As pointed out by Ed Clifton (pers. comm.), *wind set-up on the Central California Coast is more influenced by southerly winds than by onshore winds. Sustained, strong southerlies associated with large eastward-moving storms can raise the water level in West Coast bays to the point that what should be a low tide is at the level of a normal high tide.*]

As shown in Figure 28, when the sun and moon are in line (full moon and new moon; syzygy), the gravitational attraction of the moon and sun on the water bodies of the world ocean are combined, giving rise to the maximum tides of the month (called **spring tides**). Minimum tides (called **neap tides**) occur during the first and third quarter of the moon (quadrative), when the two forces work in opposition. During some lunar cycles, the tidal range during spring tides can be almost twice that during neap tides. Because of its location so far from the Earth, the gravitational pull (tide-raising force) of the sun on the world ocean is only 0.455 that of the moon. Therefore, the moon has the greater influence on the tidal cycle, controlling the timing of high and low tides.

In the much-simplified diagram shown in Figure 28, during syzygy two bulges of the high tide occur on opposite sides of the Earth, one facing toward the moon and another one facing away from it. The bulge facing the moon is the result of the gravitational pull of the moon and sun. The bulge on the opposite side is much more difficult to explain. In actuality, the Earth and the moon revolve together around the center of mass (barycenter) of the combined earth-moon system (Figure 29A). The barycenter is displaced a distance from the center of the Earth toward the moon, and it is always located on the side of the Earth turned momentarily toward the moon, along a line connecting the individual centers of mass of the Earth and moon (Wood, 1982). To put it simply, as this system revolves around the barycenter, centrifugal force comes into play. This force is defined as the apparent force, equal and opposite to the gravitational force, drawing a rotating body away from the center of rotation, caused by the inertia of the body. Thus, the combined Earth-moon system acts as a lever as the two equal and opposing forces revolve about a fulcrum, the barycenter. Therefore, the gravitational attraction of the moon (and sun) creates the bulge facing the moon and centrifugal force creates a bulge of relatively equal size on the opposite side of the Earth (Figure 29B).

The diagram in Figure 28 shows that, as the Earth rotates about its axis, a given beach location on the Earth's surface will pass by the peak of the two bulges every 24 hours and ~53 minutes. The extra minutes are required because that is how long it takes that beach location to "catch up" with the moon, which revolves all the way around the Earth in about 28 days. Because of the intervening continents, a single bulge cannot pass all the way around the Earth. Instead, individual **amphidromic systems** of different sizes are set up within the world's ocean basins. These systems are separate, more-or-less circular, gyrating tidal systems (standing waves) with a nodal point in the middle. This is a very important consideration with respect to the tides on the coast of California.

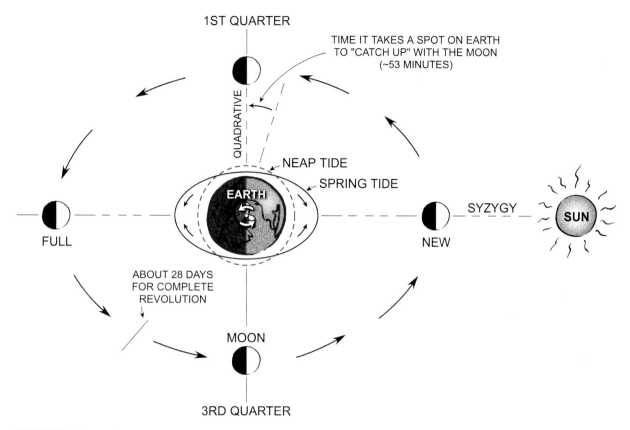

1ST QUARTER

TIME IT TAKES A SPOT ON EARTH TO "CATCH UP" WITH THE MOON (~53 MINUTES)

QUADRATIVE

NEAP TIDE

SPRING TIDE

EARTH

FULL

SYZYGY

SUN

NEW

ABOUT 28 DAYS FOR COMPLETE REVOLUTION

MOON

3RD QUARTER

FIGURE 28. Forces involved in the generation of tides.

In a personal correspondence, Ed Clifton gave the following succinct discussion of amphidromic systems that occur within the Pacific Ocean:

Amphidromic systems are large rotary tidal systems created by the behavior of tides as giant low waves that are affected by the shape of the world's oceans and the rotation of the Earth. The height of the tide depends on its location within the system. Tides at the axis of the system (nodal point) are nil, but they increase outward with distance from the axis. In the Northern Hemisphere, the tides rotate counterclockwise about the axis; in the Southern Hemisphere, the rotation is clockwise. Both diurnal (one complete rotation of the tides daily) and semi-diurnal (two complete rotations of the tides daily) can coexist in the same ocean basin. The Pacific contains two diurnal systems and eight partial or complete semi-diurnal systems.

The amphidromic systems present in the Pacific Ocean are illustrated in Figure 30.

There are many other complications to the simple story presented in Figure 28. For example, because the moon revolves in a slightly elliptical, rather than a circular, orbit around the Earth, *the distance of the moon from the Earth may vary between extreme limits*

separated by some 31,278 miles (Wood, 1982). When the moon is closest to the Earth, this *close encounter* is called the *perigee*, which is 13% closer than when it is furthest away (called the *apogee*). Thus, perigean tides may be 20% larger than normal mean tides. Furthermore, if the perigean tide corresponds with the syzygy (full and new moons), the so-called *perigean spring tides*, the tidal range can increase by 40% (Wood, 1982).

The actual final tidal curve is the result of the complex interaction of a number of **tidal constituents**, the primary one of which is the gravitational attraction of the moon. Other constituents include the gravitational attraction of the sun, the tilt of the Earth's rotational axis, the inclination of the lunar orbit, and the ellipticity of the orbits of the moon about the Earth and the Earth about the sun. Therefore, you shouldn't get too upset if the tidal predictions you read from the tide tables are off a little bit. Wind set-up is a key factor that is difficult to predict. As noted earlier by Ed Clifton (pers. comm.), the highest wind-generated tides on the California coast occur during winter storms when *strong sustained winds blow from the south, and "down-welling" occurs. An unusually high astronomical tide reinforced by a strong wind tide and the set-up associated with the abnormally large breakers is probably when most of our wave erosion*

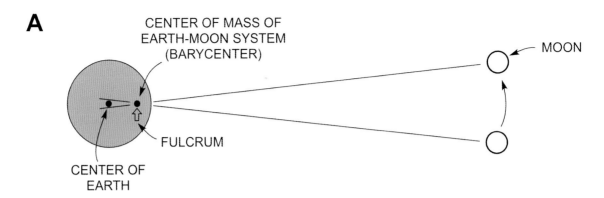

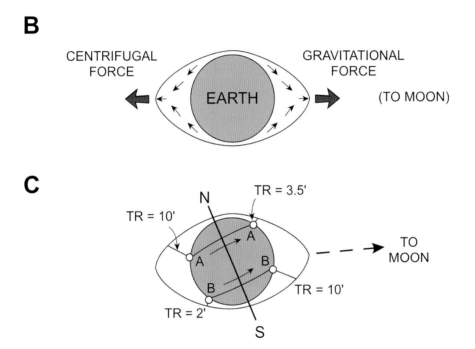

FIGURE 29. Complexities of the Earth's tides. (A) General location of the barycenter and its role as a fulcrum in generating tides. (B) Creation of tidal bulges on opposite sides of the earth. (C) One of the factors involved in creating the diurnal inequality of the tides during a complete daily tidal cycle.

occurs at Point Lobos, a location that Clifton has studied for many years.

Along most of the east coast of the U.S., the tides are termed semi-diurnal, because two complete tidal cycles occur within 24 hours and ~53 minutes. In other areas in the world, diurnal tides that have only one high and one low tide a day occur (e.g., the Texas coast during some parts of the month). The tides on the Central California Coast are called **mixed, predominantly semi-diurnal tides,** as is illustrated by the tidal curve for San Francisco in Figure 31A. Although two complete tidal cycles occur within the lunar day on the Central California Coast, and the tides are roughly semi-diurnal, there is a strong **diurnal inequality** of the different tidal levels, with four

significantly different levels occurring during one tidal day (referred to as **higher-high, lower-high, higher-low** and **lower-low** tides; Figure 31A).

A frequently cited reason for the diurnal inequality of tides is illustrated in Figure 29C. Because of the tilt of the Earth's axis, the depth of the water under the tidal bulge will be different on opposite sides of the Earth when the moon is not directly over the equator. However, on the Central California Coast, these "mixed" tides are more accurately related to the fact that this coastline is impacted by a diurnal (24 hour) amphidromic system with a nodal point located close to the middle of the Pacific Ocean (Figure 30B), and a smaller system that is semi-diurnal (12 hour) with a nodal point located a

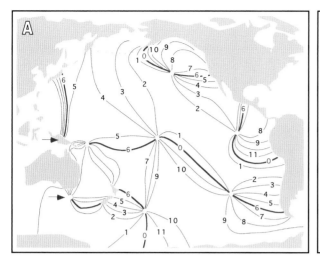
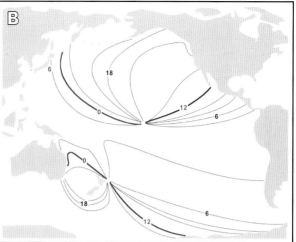

FIGURE 30. The amphidromic tidal systems of the Pacific Ocean. The tidal range is zero at the points in the middle of each system, called the amphidromic points. The tidal range increases outward from the amphidromic points. The numbers represent time in hours. The tidal phase is the same all along each line (from the amphidromic point to the end of the line). (A) The semi-diurnal systems. (B) The diurnal systems. Diagram courtesy of Dr. Ed Clifton. Diagram A is modified after Cartwright (1969) and B is modified after Dietrich (1963).

good distance straight off the coast of Central California (Figure 30A). Therefore, the resultant tidal curve is a product of the combination of the curve of the one diurnal tide with the curves produced by the two semidiurnal curves that occur during a 24 hour period (Ed Clifton, pers. comm.). Much of the time the Central California tides have a distinct diurnal inequality as was shown for 3-5 June 2008 in Figure 31A. However, Ed Clifton pointed out (pers. comm.) that *for the 28th of February 2008, only one high and one low tide occurred during that day (Figure 31B), and on September 7 that year we had the unusual situation that the day contained only a single high tide and no dead low tides* (Figure 31C). Therefore, it is clear that the combined tidal curves emanating from the two amphidromic systems well out into the Pacific Ocean create some complex tides along the California coast from time to time.

The fact that this coastline is located so far from the nodal points of the two amphidromic systems causes the **tidal range**, the vertical distance between high and low tide, to be relatively large. Keep in mind that the further a point is located away from the nodal point of the amphidromic system (the amphidromic point), the larger the tidal range. The amphidromic system is a gyrating standing wave, analogous to the wave you would create by sloshing a layer of water around the sides of a circular pan. The nodal point in the middle of the pan would maintain a constant level, with the highest vertical motion of the water layer occurring around the edge of the pan.

On the Central California Coast, the average spring tidal range increases slightly from south to north, increasing from an average of 5.2 feet at Point Arguello to 5.8 feet at Point Reyes (data from NOAA tide tables). During some spring tides, the tidal range may reach 9 feet in San Francisco Bay. This places the Central California Coastline in the mesotidal class (tidal range of 6-12 feet) during certain spring tides.

[NOTE: The explanation of why the tides are progressively later from south to north along the Central California Coast can be seen by observing the semi-diurnal amphidromic system illustrated in Figure 30A (the system that is located closest to the western shoreline of North America). The stage of the tide (high, mid-tide, low, etc.) is the same all along each of the individual lines drawn on the diagram, all the way from the amphidromic point to the end of the line. This particular system is rotating in the counterclockwise direction; therefore, the individual lines indicating similar tidal phases move progressively from south to north along the California Coast. Note that in Figure 30A, the line labeled 6 joins the shoreline at Baja California and the line labeled 0 (same as 12) joins the Alaska Peninsula. This means that when it is dead low tide on the Baja coast it is dead high tide on the Alaska Peninsula. Note that there is approximately three hours difference between the tides on the Baja coast and those off the state of Washington, with low tide in Washington being three hours (more or less) later than low tide at Baja (verified by us by checking the tide tables for the two areas). Therefore, the low tide (or high

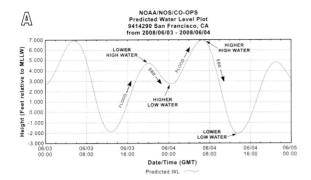

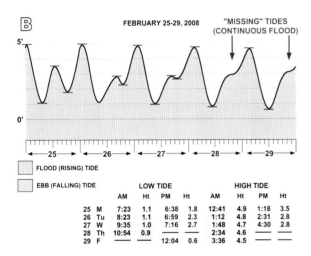

		LOW TIDE			HIGH TIDE		
	AM	Ht	PM	Ht	AM	Ht	PM	Ht
25 M	7:23	1.1	6:38	1.8	12:41	4.9	1:18	3.5
26 Tu	8:23	1.1	6:59	2.3	1:12	4.8	2:31	2.8
27 W	9:35	1.0	7:16	2.7	1:48	4.7	4:30	2.8
28 Th	10:54	0.9	—	—	2:34	4.6	—	—
29 F	—	—	12:04	0.6	3:36	4.5	—	—

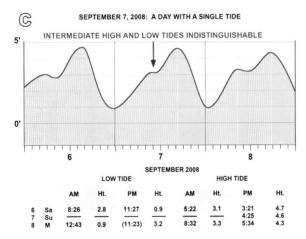

		LOW TIDE			HIGH TIDE		
	AM	Ht.	PM	Ht.	AM	Ht.	PM	Ht.
6 Sa	8:26	2.8	11:27	0.9	5:22	3.1	3:21	4.7
7 Su	—	—	—	—	—	—	4:25	4.6
8 M	12:43	0.9	(11:23)	3.2	8:32	3.3	5:34	4.3

FIGURE 31. Tides on the Central California Coast. (A) Predicted tidal curve for San Francisco on 3 and 4 June 2008, illustrating a strong diurnal inequality during that period. (B) Predicted tides for the Monterey area on 25-29 February 2008, illustrating that only one high and one low tide occurred on 28 February. (C) Predicted tides for the Monterey area on 6-8 September 2008, illustrating the unusual situation that the 7th of September contained only a single high tide and no low tides. Diagrams B and C, based on the NOAA Tide Tables, are courtesy of Dr. Ed Clifton.

tide for that matter) occurs later in Northern California than it does in Southern California.]

As a general rule, notably along trailing edge shorelines, whether the average tidal range in an area is large or small is dependent upon several factors, with the largest tides occurring where: 1) the water body has on open connection with the world ocean; 2) the continental shelf is wide; and 3) shoreline embayments occur. On the other hand, as already discussed, position within the amphidromic system is the key cause for the larger tides on the Central California Coast. Some of the amphidromic systems around the world are rather small. Ed Clifton (pers. comm.) cited the example of one in the North Sea, which has resulted in *tides on the coast of Norway that are nearly nil and those on the open coast of Great Britain that can reach 4 meters.*

With the exception of the Bay of Fundy, an area where resonance plays a role, most the larger tidal ranges the world occur in shoreline embayments, such as the Bay of Bengal (tidal range >17 feet at the upper reaches); the English Channel shoreline of France (tidal range approaching 50 feet); the head of the Gulf of California (tidal range over 30 feet); and so on. Komar (1976) gave the following explanation for this: *as the tidal front approaches a narrowing indention of the coastline, ...the enveloping shores constrict its movement and wedge the water together.*

Currents generated by the tides are very strong at the entrances of the tidal inlets that occur along the Central California Coast, most notably those that run under the Golden Gate Bridge, where tidal current velocities may reach 4 to 6 knots [one knot = one nautical mile per hour (one nautical mile = 6,076 feet)]. Figure 32 illustrates the tidal current velocities at the entrance to San Francisco Bay during ebbing and flooding tides (**ebb currents** are generated when the tide drops and **flood currents** are generated when the tide rises). Tidal currents are also strong in the major tidal channels within some of the other coastal water bodies along the coast (e.g., Drakes Estero).

WAVES

You've seen the sun flatten and take strange shapes just before it sinks in the ocean. Do you have to tell yourself every time that it's an illusion caused by atmospheric dust and light distorted by the sea, or do you simply enjoy the beauty of it? Don't you see visions?
"No," said Doc.

John Steinbeck – SWEET THURSDAY

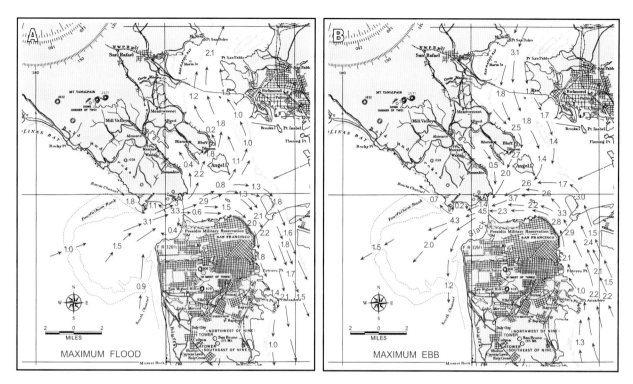

FIGURE 32. Tidal currents in and around the entrance to San Francisco Bay. Arrows show direction of flow and numbers indicate current velocity in knots. Illustrated are maximum currents during flood conditions (A) and maximum currents during ebb conditions (B). A knot is one nautical mile per hour (one nautical mile = 6,076 feet). Diagrams taken from tidal current charts published by the National Ocean Survey (1973).

General Introduction

Water waves are generated by the transfer of kinetic energy to the water surface as the wind blows over it. As shown in Figure 33A, a typical wave is defined by its **wave length** and **wave height**. The amount of time it takes two succeeding wave crests to pass a single point is called the **wave period**. The factors that determine the size of waves include:

1) The fetch of the wind (length of water surface over which the wind blows to form the wave);
2) Wind velocity;
3) Duration of the wind; and
4) Incident swell.

As indicated in Figure 33B, wind generates two different types of waves. Wave conditions in the area where the wind forms them are referred to as **"seas."** Waves that travel out of this area, perhaps ultimately breaking on distant shores, are called **"swell."** In seas, the waves can be over-steepened by the wind, causing them to break in the form of **whitecaps**. Swell typically does not break until it reaches shallow water, where it forms the familiar breakers that surfers love on the California Coast. A choppy sea surface with whitecaps indicates sea conditions; a smooth sea surface punctuated by breakers near the shoreline implies conditions of swell.

An observer standing on a beach may have difficulty distinguishing between the two types of waves, and it is very common for both types of waves to interact to form the resulting waves in the surf zone. To identify sea conditions, note the direction the wind is blowing on that day and look for the presence of whitecaps. Also, the period of swell waves is typically considerably longer than those waves generated by local onshore winds. Therefore, swell waves are usually formed by offshore storms that generate large waves. The waves generated by the storms move away from their zone of formation at a certain velocity with longer waves moving faster than shorter ones. A classic example of this escape of long waves from their zone of formation is the forerunner waves that precede the landfall of hurricanes. Commonly on the east coast, surfers flock to the shore to take advantage of these huge, long-period "forerunner" waves.

As waves approach the beach from offshore, they increase in steepness and break in different ways to be discussed later. To understand why this happens, consider a cork on the water surface as a wave passes by. The cork ascribes a circular motion as the wave passes, in the process moving a small distance in the direction

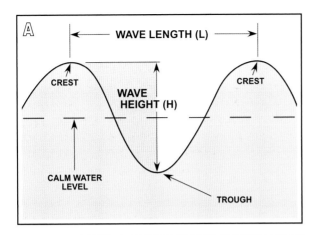

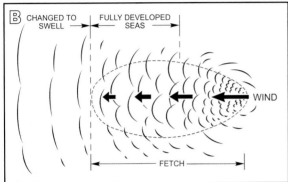

FIGURE 33. Waves. (A) Components of a typical water wave. (B) General pattern for the generation of waves by wind blowing across the water surface.

the wave is moving. As shown by the diagram in Figure 34A, the water directly beneath the wave also undergoes a circular orbit, with the size of those orbits decreasing exponentially with depth until completely ceasing to orbit at a depth of about one half the wave length.

As the wave moves onshore, the wave begins to "feel the bottom," that is, the orbital motion of the water under the wave begins to bump up against the bottom where the water becomes **shallower than one half the wave length** (Figure 34B), moving landward until they eventually break. Ed Clifton (pers. comm.) pointed out several important aspects of both breaking waves and their interaction with the sea floor:

1) *Sea waves break when they are over-steepened by the force of the wind driving the crest of the wave faster than the preceding trough is moving.*

2) *As swell advances toward the shoreline, it becomes asymmetric as it approaches the break point. Breaking swell results where the impinging bottom slows the speed of the wave in the shallow wave trough relative to the speed of the following crest.*

3) *On the swell-dominated coast of California, where*

the average wave period is around 11.5 seconds (discussed in some detail later), *the average deep-water wave length would be around 200 meters (656 feet), which means that the wave would "feel bottom" at about 100 meters (328 feet). Even in the summertime, in the absence of storms, waves with the average period of 10 seconds will "feel bottom" at around 75 meters (246 feet).*

4) With regard to the common swell waves, not storm waves, that frequent the California coast, he pointed out that *the orbital motion of the water under shoaling (10-second+ period) waves shows a shoreward asymmetry. As the waves shoal, they deform from their sinusoidal shape and the crests become sharp, separated by broad flat troughs. As a result, the time available for the onshore flow* (of the orbiting water cells, Figure 34A) *under the crest lessens and, as a result, the water moves faster than it does under the broader trough. This velocity asymmetry does several things. It drives the coarser material (on the sea floor) landward (by overcoming the threshold of transport limits) and thus tends to drive most of the sand landward (since the rate of sand transport is generally considered to be a function of a power of the velocity of the moving water), thus restoring beaches that were eroded by winter storm waves, presumably because of enhanced rip currents during storms, but there may be other factors as well, such as the presence of sea cliffs that cause the water to pile up along the shoreline.* This is an important point with regard to erosional and depositional patterns on sand beaches, which are discussed in some detail later.

The zone where the waves are breaking is referred to as the **surf zone** (Figure 34B). After breaking, the water released by the broken wave impinges on the portion of the beach known as the beachface (also referred to as the **swash zone**). The water moving up the relatively steep beachface, known as the **wave uprush**, eventually loses its momentum and that portion of the water that doesn't percolate into the beachface returns back down the slope under the influence of gravity (known as the **backwash**).

Breaking waves are usually of three distinct types – plunging, spilling, and surging. As pictured in Figure 35A, **plunging waves** take on a cylindrical shape (Hawaii Five-O type) and fall abruptly down with considerable force, usually breaking in the near vicinity of the beachface. The surfaces of **spilling waves**, on the other hand (shown in Figure 35B), start disintegrating into foaming lines fairly far offshore and continue to foam their way to shore, gradually decreasing in height as they go. **Surging waves** approach shore and wash directly up the beachface without breaking. As a generalization, for waves of any given height, the slope

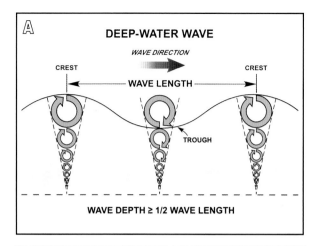

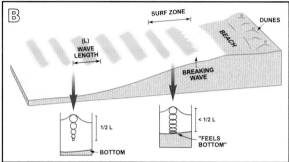

FIGURE 34. Waves approaching shore. (A) Orbital water motion at depth under a passing wave. (B) Characteristics of the surf zone under the influence of wave action. Note that in water shallower than approximately one half the wave length, the orbits of the wave-generated water motion under the waves (illustrated in A) flatten out as the wave begins to "feel the bottom." This causes the waves to slow down and become steeper and steeper until they eventually break.

of the nearshore zone determines the type of breaking wave present. Plunging waves occur more commonly where the nearshore slopes are quite steep and spilling waves where the nearshore slopes are flat, in a relative sense. Under some conditions, a single wave may have the characteristics of both plunging and spilling waves. Surging waves may occur at times, particularly during high tides, along beaches on the California coast where the nearshore zone is steep (e.g., at Monastery Beach near Monterey).

When offshore swell waves are viewed from the air, their crests are usually linear and parallel, forming what are known as wave fronts. In many situations, the wave fronts bend as they approach underwater features, this bending being referred to as refraction and diffraction. Wave **refraction** is defined by the U.S. Army Corps of Engineers (2002) as "any change in direction of a

wave resulting from the bottom contours." As noted by Woodroffe (2002), *the speed with which the wave travels in shallow water is related to water depth. Those parts of the wave which enter shallow water move forward more slowly than those in deeper water, causing the wave crests to bend toward alignment with the bottom contours.* During the process of refraction, a segment of the wave front may bend around an island or bar, as is illustrated Figures 36A and 36B. In the purest sense, **diffraction** is defined as the lateral spread of wave energy along the wave crest from a point (an example is the wave formed when a pebble is dropped into a pool of still standing water). The U.S. Army Corps of Engineers (2002) defined diffraction as *the phenomenon by which energy is transmitted laterally along a wave crest. When part of a wave is interrupted by a barrier, the effect of diffraction is manifested by propagation of waves into the sheltered region within the barrier's geometric shadow.* During incidents of wave

FIGURE 35. Two types of breaking waves. (A) A plunging wave on the Mexican coast near Acapulco. The idiot running into the surf to collect a sediment sample is co-author Hayes (summer 1961). (B) Spilling waves at Cowell Ranch Beach, California (San Mateo County) on 16 August 2008.

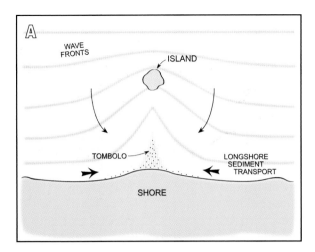

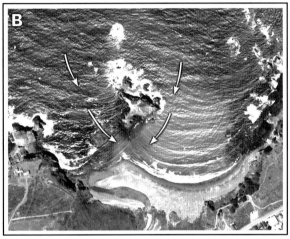

FIGURE 36. Illustrations of the creation of a tombolo. (A) Wave fronts bending around an offshore island, generating two opposing directions of sediment transport in the lee of the island. Where the two directions of transport meet, a triangular offshore projection of sand called a tombolo is formed. (B) A similar process at Gunderson Rock near Elk, California (Mendocino County). The arrows, which indicate the direction of motion of the waves, are oriented perpendicular to the wave crests, or fronts. NAIP 2005 imagery courtesy of the California Spatial Information Library (CaSIL).

diffraction near the shore, a segment of the wave front may bend into something like a navigational channel or into the lee of an offshore breakwater or reef, relative to the rest of the front. As a result, an arc, or bow, develops in the wave front as it moves into the channel or into the lee of the offshore structure.

When waves approach and break in the surf zone adjacent to the beach, they set up two very distinct, and in some cases very strong, nearshore currents – longshore currents and rip currents. As illustrated in Figure 37 (Upper Left and Lower Left), for a wave that approaches

a beach at an angle, which many do, two types of water flow are generated that have the capability of moving sediment along shore. This type of longshore sediment transport is a key consideration with regard to the stability of shorelines. An illustration of wave-generated longshore sediment transport is given in Figure 36A, which shows waves bending into a sheltered zone in the lee of an offshore island. The result of this process is the creation of a triangle-shaped sand body, called a tombolo, in the lee of the island at the point where two opposing sediment transport directions convergence. An example of this process on the California coast is shown in Figure 36B.

When waves come ashore at an angle, the wave uprush generated by the breaking wave flows at an angle to the slope of the beachface (Figure 37), which allows any sand grains moved by the shallow flow of the wave uprush to move obliquely up the beachface. On the other hand, the water in the backwash moves in a perpendicular sense directly down the slope of the beach under the influence of gravity. Therefore, with each wave sand is moved along the shoreline a certain distance, depending upon the size of the wave and its angle of approach (bigger waves and higher angles increase the distance of movement of the sand grains). Thus, the sand grain moves along the shore in a sawtooth pattern called **beach drift** (Lower Left diagram in Figure 37). This type of sediment transport is restricted to the beachface itself. At the same time, the obliquely approaching wave is piling up water in the surf zone that flows away from the direction of the approaching wave, thus generating a current that runs parallel to the shore (called the **longshore current**). On many shorelines, the longshore current is somewhat restricted to a trough that separates the beachface from a nearshore bar. These longshore currents can be quite strong, commonly exceeding a knot or two and, in concert with the beach drift, they play the major role in transporting sediment along the shore. These currents can also carry swimmers far away from where they entered the water.

Determinations of the volume of sediments that move along the shore in the longshore sediment transport system can be done in several ways, including: 1) measuring volumes of sand impounded by breakwaters or jetties; 2) determining volumes of sand dredged from harbors or other collection sites; and 3) calculations using formulas based on wave and sediment data. The longshore sediment transport rates reported for California show that a range from around 200,000 to 600,000 cubic yards of sand moves from north to south (or west to east at Santa Barbara) along the sand beaches of the outer shoreline in the state in any given

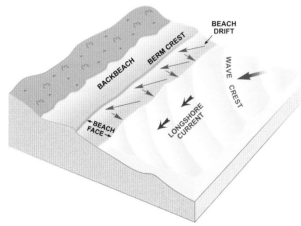

FIGURE 37. Longshore sediment transport. (Lower Left) General model for the generation of beach drift and longshore currents by waves approaching the beach at an angle. Under these conditions, the sand grains move in a sawtooth pattern along the beachface, and a strong longshore current is developed just off the beachface. (Upper Left) View looking north along St. Joe Spit in northwest Florida showing waves approaching the beach obliquely out of a southerly direction. This high-tide picture shows the swash of the waves across the beachface. Arrows indicate the direction of the longshore currents generated by the obliquely approaching waves. Note waves breaking on the offshore bar with strong longshore currents generated in the trough on the landward side of the bar. Photograph taken in April 1973. (Upper Right) Rhythmic topography along the shoreline of Matagorda Bay, Texas. Arrows indicate direction shoreline rhythms are moving through time, also under the influence of southeasterly waves. The water was relatively calm on the day this photograph was taken in August 1979. The white sediments are coarse-grained shell material (mostly oyster shells).

year. Because this number is a little difficult to grasp, we once calculated that between 150-200,000 cubic yards of sand would go a long ways toward filling a major football stadium. Two to six hundred thousand cubic yards, with an average rate of around 300,000 according to Griggs, Patsch, and Savoy (2005), does seem like a large number, but in fact, it is surpassed elsewhere. For example, the Ghana coast in West Africa has rates that exceed one million cubic yards of longshore sediment transport in one year, as a result of the large waves that obliquely approach that exposed, open-ocean coast (Nairn and Hayes, 1997).

Wang, Ebersole, and Smith (2002) determined that plunging waves create much greater sediment suspension in the breaking zone than spilling waves, as well as more sediment transport in a wider and more energetic swash zone. This helps explain the authors' general impression that beaches are more apt to accrete (build out) when the waves are spilling than when they are plunging [an observation supported by Tim Kana's Ph.D. research on the South Carolina coast, which was carried out while he was a Ph.D. student under co-author Hayes at the University of South Carolina (Kana, 1977; 1979)].

Another mode of longshore sediment transport occurs when evenly spaced bulges of sand along the beachface move down the shore like the sinusoidal loops one might throw along a rope. These bulges of sand, called **rhythmic topography** (see photograph in the

Upper Right of Figure 37), are typically spaced tens to hundreds of feet apart. They are created by waves that approach the beach at an oblique angle. The side of the sand bulge facing directly into the wave approach direction orients perpendicular to that direction, with the beachface being aligned parallel with the approaching wave crests, the most stable configuration possible. However, the distance the sediment on the exposed beachface of the bulge can extend offshore is limited by water depth in most situations. Sand is therefore either lost from the outer edge of the bulge into deeper water or is transported to its lee side. The result of this process is that the exposed face of the bulge erodes while sand accumulates on its sheltered side. Therefore, the bulge moves along the shore in what is known as the downdrift direction (direction in which the individual sand grains are moving along the shore within the longshore sediment transport system), as indicated in Figure 37 (Upper Right). These features, which occur in several places along the Central California Coast (the authors recently observed excellent examples of them at Point Reyes Beach, Marina State Beach, and Ocean Beach), move at different rates depending on the size and angle of approach of the waves. When these features were first described, some of the early authors used the term "cusp-type sand waves" in recognition of the fact that they protrude out from the beach somewhat in the same fashion as the very common, smaller perpendicular protrusions off the beach called **beach cusps** (Figure 38). These two features do not form in the same way, thus the more definitive term, rhythmic topography, is used these days.

Rip currents (Figure 39) form most commonly along shorelines where waves come straight on shore (not at an angle usually, but there are some exceptions to this general rule). Also, there is typically some degree of reflection of the waves off the beachface. These currents, which typically flow in a perpendicular direction straight off the beach, tend to be very regularly spaced. Based on extensive research, most rip currents owe their formation to the occurrence of edge waves – standing waves with their crests normal to the shoreline, according to Bowen and Inman (1969) and Komar (1976). Therefore, the direction of motion of the water within these waves is parallel to the shore and their crests are perpendicular to the shore. Furthermore, they are "trapped" against the beach, decreasing in amplitude exponentially in the offshore direction. If, indeed, they are standing waves, it would follow that the rip currents would be positioned at equally spaced distances where the waves have their maximum amplitude, not at the nodal points. Oh well, it would take a Ph.D. in physical oceanography and at least

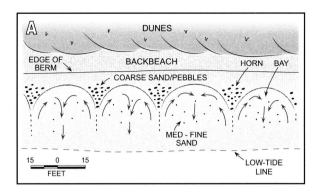

FIGURE 38. Beach cusps. (A) General model. (B) Beach cusps at Point Reyes Beach on 18 August 2008.

a B.S. in physics to completely understand edge waves. Rip currents may also be associated with some types of nearshore bars and off certain man-made structures (e.g., groins, which are discussed later), but the most common ones appear to be related to wave reflection and edge waves. At any rate, one of the authors of this book, Hayes, swears he has seen edge waves with his own eyes off a steep beach in Ghana, but Michel is a non-believer in edge waves, because she says "I have never seen one." [NOTE: Ed Clifton (pers. comm.) mentioned that *some of the most prominent rip currents I have seen have been on beaches where the wave approach was oblique. He particularly noted those he observed while working*

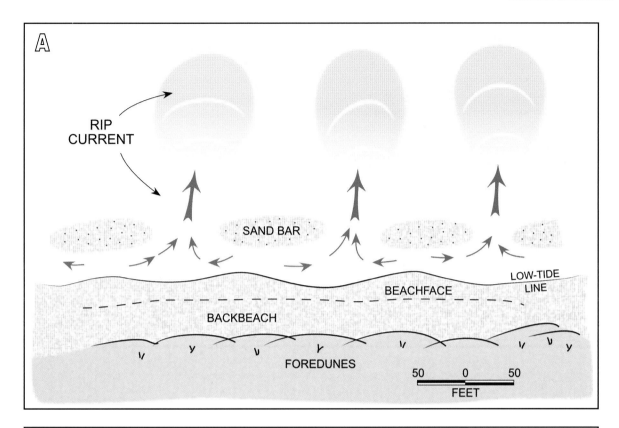

FIGURE 39. Characteristics of rip currents. (A) General diagram. (B) Rip currents along the shore of Cape Cod, Massachusetts in April 1972. Note the onshore welding of sand bars in the space between the rip currents.

on a project on the Oregon Coast. However, most of the rip currents we have seen have been produced by waves coming straight onshore. Another topic we are not covering in detail in this book is the occurrence of sand bars in the nearshore zone off the beach, as well as those that weld to the beach. In some cases, such bar systems that have been formed by oblique waves will deflect strong currents offshore at an angle to the beach, which could also be considered to be rip currents (i.e., they have similar dynamic qualities, except for their orientation). There is a myriad of nearshore bar types, some of which project perpendicular to the shore.]

Beach cusps are relatively small-scale features (a few tens of feet maximum) that develop on the scalloped seaward margin of an mound of sand deposited on the intertidal beach called a berm (Figure 38). They consist of a regularly spaced sequence of indented bays and protruding horns. They are thought to be primarily depositional features (i.e., on the seaward flank of a prograding beach) that form as a result of the interaction of shore-normal standing waves (edge waves) with shore-parallel, reflective waves, the same general process that produces rip currents (but at a smaller scale). This mode of origin was originally proposed by Guza and Inman (1975). There are other theories that challenge the edge-wave solution, believe it or not.

Both rip currents and beach cusps are very common on the Central California Coast, and even a casual observer touring along the coastal roads will see some. If you want to study them at home, check out the imagery of several of the long, straight sand beaches in Central California (on Google Earth) and you will see plenty.

Rip currents can be very dangerous. If by chance you ever get caught in one while swimming off the beach, make sure you swim parallel to the beach so as to escape the current, which has the potential of carrying you far offshore. According to a web site of BeachCalifornia. com (2009) in a discussion of Huntington Beach, *the prevalence and severity of rip currents is greatest in California...*(no doubt true in comparison with the beaches on the east coast).... *According to NOAA, there were over 40,000 rescues in California in 2003 with over 80% due to rip currents. The drownings were almost 8 times higher at unguarded beaches (California State Parks reported 71% of all drownings, because state parks and beaches don't always have lifeguard service).*

Waves on the Central California Coast

All of the aspects of waves and related beach morphology discussed in the preceding section can be seen in abundance on the Central California Coast.

Wind patterns along a coast and offshore storms determine the average size of the waves, which have a major input in shaping the coast. Coastal storms during *El Niños* generate by far the largest waves that occur on the beaches in Central California. The specific effects of such storms, both with respect to the waves and shoreline erosion, are presented in the next section.

The daily sea conditions on most coastlines are related primarily to global wind patterns. Inman and Jenkins (2005a) made the following points regarding the interaction of global wind patterns with the world's oceans:

1) A tremendous amount of kinetic energy is expended through momentum exchange between the atmosphere and the ocean (at the sea surface).

2) The prevailing wind systems cause all large bodies of water to have windward (prevailing wind blowing onshore) and leeward (prevailing wind blowing offshore) coasts.

3) The Pacific coasts of the Americas are, in general, windward coasts and those on the Atlantic are leeward coasts.

4) As is illustrated in Figure 40, the general wind flow in the Northern Pacific Ocean region is clockwise around a semi-permanent, mid-latitude area of high pressure called the North Pacific high. Consequently, when these winds reach the coast of Central California, they are blowing out of the northwest. As also shown in Figure 40, in the Southern Hemisphere, *circulation patterns are essentially counterclockwise flowing mirror images of those in the Northern Hemisphere* (Inman and Jenkins, 2005a). Consequently, when these winds reach the coasts of Chile and Peru, they are blowing out of the southwest.

The northwesterly wind pattern thus experienced along the Central California Coast is clearly responsible for the generation of deep-water waves that come dominantly out of the northwest toward the shore (Figure 41).

An up-to-date treatment of this subject was discussed by Beyene and Wilson (2006). They pointed out that *typically, California wave parameters are measured offshore in deep water using sensors mounted near the bottom or attached to oil drilling platforms.* One of these, a National Data Buoy Center buoy (46028) stationed offshore of Cape San Martin (in Compartment 6, Figure 1) provided nearly 19 years of wave data (1983-2002; National Data Buoy Center, 2009).

Before reviewing this information, a new term, **significant wave height**, should be defined. This

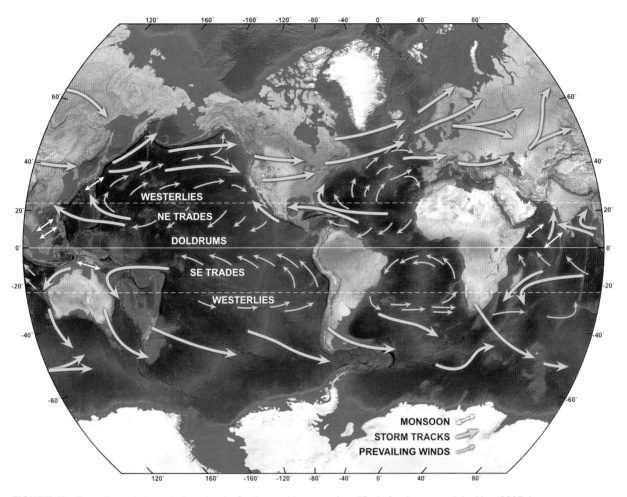

FIGURE 40. Prevailing winds and storm tracks for the world oceans (modified after Inman and Jenkins, 2005a).

measure portrays the average height of the waves that comprise the highest 33% of waves in a given sampling period. As it turns out, an experienced person making visual observations will most frequently report heights equivalent to the average of the highest one-third of all waves observed. Furthermore, it is the larger waves that are of the most concern to sailors, coastal engineers, and the like. Wave roses for two of the NDBC buoys for the periods 1993-2000 in Figure 41B show the expected dominant northwest waves.

Based on these 19 years worth of data presented by Beyene and Wilson, the average significant wave height for that part of the Central California Coast is 7.6 feet, with an average standard deviation of 3.0 feet. The dominant wave period is 11.49 seconds. Beyene and Wilson (2006) also noted that wave heights greater than 23 feet are common in that area. The following general summary of these wave-direction statistics for the Central and Northern California coasts was given in that same paper:

From November through March, the dominant wave direction is west with the northwest sector receiving significant wave energy. The November-March period is characterized by very energetic waves with significant wave height averages approaching 3 meters (9.8 feet) and dominant wave periods averaging above 10 seconds. The Normidcal (Central California) summer season is from April to October and the dominant wave direction is from the northwest, west being the second most energetic direction. Surprisingly, the Normidcal area receives some significant "southern swell" in the summer from the southwest directions, presumable from storms south of the equator – the Southern Hemisphere winter. The long-term average annual significant wave height in the Normidcal summer season is approximately 2.5 meters (8.2 feet) and the average wave period is over 10 seconds.

These time-series data also show that 46-52.5 feet significant wave heights are possible every few years in the area north of Point Conception. Guess we do not have to point out that these are pretty big waves!!

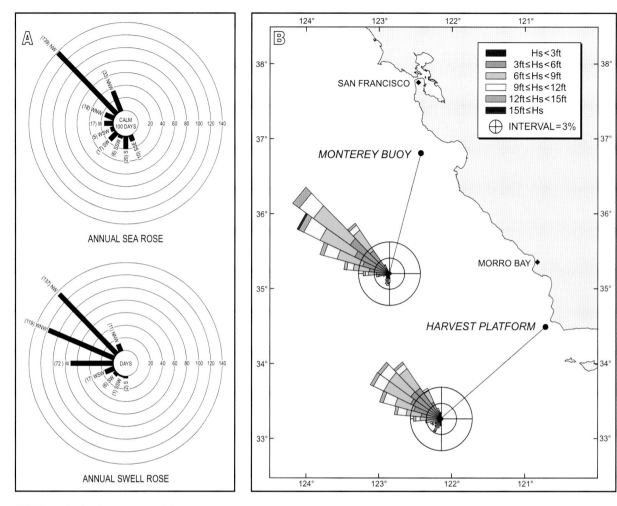

FIGURE 41. California waves. (A) Annual sea and swell roses developed for the California coast by National Marine Consultants (1960) (from Domurat, Pirie, and Sustar, 1979). These graphs show directions only. (B) Wave roses of significant heights (highest one third of all the waves measured for heights of <1 meter to ≥5 meters) off Monterey (1995-2000) and off Point Conception (1993-1995) (from Bottin and Thompson, 2002).

To add another wrinkle to observations of this type, Allen and Komar (2000) found that average wave heights (not significant wave heights) in the north Pacific increased 20% between 1975 and 1999 (from about 5.2 feet to 6.2 feet). Similar results exist for the north Atlantic area. The reason for this increase is still a matter of debate, with ocean-wide oscillations and increased water temperatures in the last quarter of the century being possibilities. But, as Inman and Jenkins (2005a) pointed out, it is unclear what portions of these increased wave heights are also associated with global warming (Graham and Diaz, 2001).

MAJOR COASTAL STORMS

In the North Pacific Ocean, the most intense wave action results from **cyclogenesis** (development

or intensification of a low-pressure center, or cyclone) over the poleward-flowing western boundary current, the Kuroshio Current, which is matched by its twin, the Gulf Stream Current, along the western boundary of the Atlantic Ocean. As noted by Inman and Jenkins (2005a), *these warm water jets carry equatorial heat to higher latitudes and produce strong temperature gradients that spin-up intense storms.* In the wintertime in the North Pacific, cold and dry Siberian winds blow out of the west across the Kuroshio Current. And, as further noted by Inman and Jenkins, *these winds produce a series of cyclonic cold fronts that collectively have long fetches and generate the high waves for the Pacific coast of North America.* When they are far out in the ocean, these storms produce swell along the California coast, and once they approach the shore, they generate sea waves that strike the coast along their tracks.

Referring to the time intervals between the occurrences of coastal storms (the typical seasonal patterns), Griggs, Patsch, and Savoy (2005) pointed out the effect of the strength and position of the North Pacific high. During the summer months, the high-pressure area offshore, which has migrated as far north as San Francisco (38 degrees north), either breaks down the large storms moving east from the western Pacific, or deflects them northward so that they completely miss the California coast. This lack of storms also highly diminishes the likelihood of rainfall in the summer time. On the other hand, during the winter, as the high weakens and moves southward as far as the Hawaiian Islands (20 degrees north), *the California coast is no longer protected, and large storms forming in the western Pacific regularly reach the coast along with rain and large waves.*

This kind of storm generation takes place, at least to some extent, about three out of every four winters (Griggs, 1998), which clearly accounts for the larger average wave heights that occur during winter months (average significant wave heights of 9.8 feet between November and March and 8.2 feet from April to October off the Central California Coast). However, the largest storms by far occur during the *El Niños*, which take place when the typical weather patterns described above and illustrated in Figure 40 are abruptly and severely modified on a global scale. Conditions during *La Niñas* are quite the opposite from those during *El Niños*.

These two globally coupled, ocean-atmosphere phenomena are complex and the subject of intense research at the present time for obvious reasons. A brief outline of the characteristics of the two events is given below based principally on three sources: 1) Inman and Jenkins (2005b); 2) Griggs, Patsch, and Savoy (2005); and 3) the Wikipedia free encyclopedia.

First, the *El Niño-Southern Oscillation (ENSO),* the technical name of the *El Niño:*

1) The name is from the Spanish for "the little boy," in reference to the Christ child, because its appearance is usually noticed by the arrival of unusually warm water off the coast of Peru around Christmas time. Warm water also occurs off California during these events.

2) They are signaled by anomalies in the pressure fields between the tropical eastern Pacific Ocean and Malaysia. These anomalies yield an atmospheric signature called the Southern Oscillation Index (SOI) (which reflects the monthly or seasonal fluctuations in the air pressure between Tahiti and Darwin, Australia). If the index is negative (i.e., lower pressure at Tahiti and higher pressure at Darwin), then an *El Niño* or warm ENSO event occurs.

3) Another associated condition that develops during *El Niño* events is a generally high-pressure anomaly over the northern half of North America and a huge low-pressure anomaly over the eastern North Pacific Ocean that hugs fairly close to the North American shoreline (mostly in the winter time). This forces storm-track enhancement along the southern edge of the low that aims the storm tracks straight at California!

4) Three factors, the relatively low pressure over the Pacific Ocean, the thermal expansion of warm water, and a weakening of the **trade winds** (during normal conditions, these strong winds generate a very strong westerly flow of the surface waters to the west in the equatorial region) cause an increase in water level on the California coast during *El Niños* (as much as 8-12 inches). Such an increase in water level leads *to significant coastal inundation and sea cliff erosion during the intense storms that are common during El Niño events* (Inman and Jenkins, 2005b).

5) During these events, unusually heavy rains typically accompany the increased number of winter storms.

6) Major *El Niño* events occurred in the years 1790-93, 1828, 1876-78, 1891, 1925-26, 1982-83, and 1997-98 (Davis, 2001). The most recent occurrence of *El Niño* started in September 2006 and lasted until early 2007. Lesser ones have occurred in the last few decades, including in 1986-87, 1991-92, 1993, 1994, 2002-2003, and 2004-2005.

On the other hand, some of the characteristics of the *La Niñas* include:

1) *La Niñas* generally show effects opposite to those of the *El Niños*. For example, a decrease in water levels along the California coast is possible. Rainfall is greatly diminished and the trade winds are usually strengthened.

2) Another associated condition that develops during *La Niña* events is a generally high-pressure anomaly over the eastern North Pacific that also hugs fairly close to the North American shoreline. This forces storm tracks to come ashore further north than those that occur during *El Niño* events.

3) There was a strong *La Niña* episode during 1988-1989, and some lesser ones in 1995 and 1999-2000. Another moderate one started developing in mid-2007.

Inman and Jenkins (2005b) noted that *hypotheses explaining the causes of unusual ENSO events are numerous and range from the extrusion of hot lava on the sea floor to variations in intensity of the sun's radiations. Most climate modelers favor a complex set of changes in the interaction of the atmosphere and ocean that cause fluctuations in trade winds, monsoon intensity, and sea surface temperature* (e.g., Pierce, Barnett, and Latif, 2000). No kidding?! Well, anyway, enough is now known about the two systems that it is possible for weather forecasters to predict their arrival. However, to complicate things a bit more, increases in global temperatures that we are now experiencing may make the ENSO events more intense and less predictable (Federov and Philander, 2000).

According to Storlazzi and Griggs (1998) and Griggs (1998), twenty-two *El Niño* events occurred on the California coast during the years between 1910-1995. To that total we can add the big one in 1997-1998. During that 85 year period, Griggs noted that *75 percent of the storms that produced major damage along the central coast (of California) occurred during El Niño winters, adding that it is important to realize that these global climatic events are not uncommon acts of God, and that they didn't start in 1983.*

A considerable amount of data has been accumulated regarding the last two major *El Niño* events, which occurred in 1982-83 and 1997-98. The following information describes the 1982-83 event [from a variety of sources including Griggs, Patsch, and Savoy (2005); Dingler and Reiss (2002); and Griggs and Brown (1998)]:

1) During January, February, and March of 1983, eight major storms struck the coast of California.

2) Because of the southerly displacement of the North Pacific high, these storms approached the shore from the west and southwest, impacting shorelines formerly protected from northwesterly approaching waves by rocky headlands to the north and also creating a wave refraction pattern the beaches were not adapted to.

3) Heavy rains that weakened the soil, higher than normal water levels because of the increased water temperatures, etc., and the storms crossing the coast during higher than normal astronomical tides (the two largest storms hit during high spring tides) all combined to expose the beaches and cliffs to extreme erosion.

4) These storms produced offshore significant wave heights between 16 and 22 feet.

5) More than 3,000 oceanfront homes and 900 businesses were damaged, at a cost of around $100 million.

The following information from the same sources describes the 1997-98 event:

1) During this interval, numerous major storms struck the coast of California bringing some very heavy rains with mudslides and shore erosion in places.

2) These storms also approached the shore from the west and southwest.

3) However, the largest waves from the two biggest storms hit during neap tides, which significantly decreased the erosion of the backshore areas.

4) Also, in the late 80s and early 90s, seawalls and revetments had been built along many of the areas that were eroded heavily during the 1982-83 *El Niño* event, which also contributed to decreasing this event's costs in comparison to the one in 1982-83.

Therefore, the 1997-98 *El Niño* did not do nearly as much structural damage along the coast as the 1982-83 event. Another important consideration is that the heavy rains that accompany these storms cause the rivers to flood and, as a result, a new pulse of sand is added to the beaches. The issue of how sand supply relates to the erosion and deposition of sand beaches is discussed in some detail later.

TSUNAMIS

On 26 December 2004 an earthquake off the coast of Sumatra, Indonesia produced the largest trans-oceanic tsunami in over 40 years, killing more people than any tsunami in recorded history (225,000). We probably all remember the videos of people running from the advancing waves, hanging onto tree trunks, and so on. That tsunami was restricted mostly to the Indian Ocean, but the Central California Coast is not completely exempt from the effect of tsunamis.

A tsunami, a Japanese word translated as *harbor wave*, is a large ocean wave usually generated by a submarine earthquake, although they may also be generated by volcanic eruptions, landslides (both those initiated on shore and submarine), and even by a comet or asteroid impact (not by meteorites; too small). Historically, these waves have been called tidal waves, but astronomical tides are not involved in their formation. *Seismic seawaves* is another term that scientists commonly use.

These are typically very long waves (up to several hundred miles) with relatively small wave heights (commonly <3 feet), and long periods (over an hour in some cases) that have sufficient energy to travel across entire oceans, as was noted during the Sumatran earthquake in 2004 when the waves crossed the total width of the Indian Ocean. They move very fast in the open water at speeds measured in hundreds of miles per hour; however, as they approach shore, they usually slow down to 40-50 miles per hour. Like their smaller wind-generated cousins, when they enter shallow water, they abruptly peak and break, reaching heights of well over 100 feet in some cases.

Inasmuch as a tsunami has a much smaller amplitude (wave height) offshore, and a very long wave length, they generally pass unnoticed at sea, forming only a slight swell in the normal sea surface. Obviously, a tsunami can occur at any stage of the tide, but even at low tide, it will still inundate coastal areas if the incoming wave's surge is high enough.

As noted in the Wikipedia encyclopedia, tsunamis are most commonly generated by earthquakes where converging plate boundaries generate faults in which at least one side of the fault line undergoes abrupt vertical movements that displace the overlying water. It is very unlikely that major ones can form at divergent plate boundaries or along transform faults because those types of features do not typically have significant vertical movement. Therefore, faults generated by these types of movements do not usually displace the water column vertically. Subduction zone related earthquakes generate the majority of all tsunamis (i.e., where major plates converge as along major oceanic trenches around the Pacific Ocean "ring of fire").

On 1 April 1946 a Magnitude 7.8 (Richter Scale) earthquake occurred near the Aleutian Islands, Alaska. It generated a tsunami that inundated Hilo on the island of Hawaii with a +40 foot high surge. The area where the earthquake occurred is where the Pacific Ocean floor is subducting (or being pushed downwards) under Alaska (Wikipedia encyclopedia).

Such a subduction zone is located off the coast of northern California, Oregon, and Washington (Figures 11, 12, and 15). USGS geologist B.F. Atwater has spent much of his career studying the likelihood of large earthquakes and tsunamis taking place in that region. A description of his work is given in the following source (Brian Atwater, 2009):

In 2005, he published a book with others, "The Orphan Tsunami of 1700," that summarizes the evidence for a moment magnitude 9 earthquake in the Northwest on 26 January 1700, known as the 1700 Cascadia Earthquake. The earthquake produced a tsunami so large that contemporary reports in Japan noted it, allowing Atwater's team to assign a precise date and approximate magnitude to the earthquake. Its occurrence and size are confirmed by evidence of a dramatic drop in the elevation of Northwest coastal land, recorded by buried marsh and forest soils that underlie tidal sediment, the deposition of a layer of tsunami sand on the subsided landscape, the death or injury of affected trees, and descriptions of the earthquake and tsunami in regional Amerindian legends.

The authors of this book have been conducting research along the earthquake/tsunami prone shoreline of Alaska for many years, thus we have observed the effects of these phenomena up close in many localities along that shoreline. As noted by Dr. George Pararas-Carayannis (2009a) in his detailed account the "Good Friday" Alaska earthquake (on 27 March 1964), that earthquake generated catastrophic tsunami waves that devastated many towns in the Prince William Sound area of Alaska, along the Gulf of Alaska, along the West Coast of Canada and the U.S., and in the Hawaiian Islands. In Alaska, the tsunami run-up measurements varied from 20 feet at Kodiak Island, 30 feet at Valdez, 80 feet at Blackstone Bay, and 90 feet at the native village of Chenega in Prince William Sound. This 1964 earthquake caused 119 deaths in Alaska alone, with 106 of these due to tsunamis that were generated by tectonic uplift of the sea floor, and by localized upland-generated and submarine landslides. According to Dr. Pararas-Carayannis, there were two different types of "tsunami generation mechanisms" associated with this earthquake, one along the continental shelf bordering the Gulf of Alaska, and the other in the Prince William Sound region:

1) Tectonic movements of the sea floor caused the tsunamis along the open shoreline of the Gulf of Alaska; and

2) Tectonic movements that uplifted some of the shoreline over 30 feet and the resulting landslides were responsible for most of the local tsunami waves which caused the destruction and deaths within the Prince William Sound region. These tsunamis destroyed local towns and fishing villages, killing 82 people. The maximum wave height recorded in the Sound was over 200 feet.

In a write-up on historical tsunamis in California, Pararas-Carayannis (2009b) stated that, in this century, there have been numerous tsunamis generated from distant earthquakes that have reached California, referring to the ones in 1946, 1952, 1960, and 1964. He also noted that the tsunami waves generated by the Good Friday earthquake in Alaska in 1964, which was just discussed, affected the entire California coastline. These tsunami waves *were particularly high from Crescent City to Monterey with heights on the open coast ranging from 7–21 feet.*

The impacts of this tsunami were particularly noteworthy at Crescent City, where the following occurred (Hartness, 2001):

1) *There was eight million dollars worth of damage and eleven people died, yet no damage was reported to either the immediate north or south of the city.*

2) *The topography of the sea floor outside of Crescent City itself is such that it concentrated the wave energy onto*

a narrow stretch of shoreline. This caused the wave height to grow very tall and thus inundate more land (than in the areas to the north and south).

3) *The original quake happened at 3:36 AM Universal Time (UT, Greenwich Mean Time). At 5:36 AM UT a warning was issued to California that a tsunami was possible. By 6:44 AM UT the sheriff of Crescent City had been notified that it was definitely on its way. Evacuation was begun at 7:08 AM UT. All of the coastal front areas were evacuated. One hour later the waves arrived. The first wave crested four meters (13 feet) above low tide. The second wave was smaller. The third and fourth waves damaged low lying areas around the south beach, with the third running inland 500 meters (546 yards).*

4) *Thirty city blocks were flooded by these waves. Single story wood framed houses were either badly damaged or destroyed. After the first two waves had passed some of the townspeople believed the threat was over, and returned to their homes and businesses.*

5) *One such group of seven returned to a tavern to begin cleanup. While working, they decided to have a drink. They were trapped in the building by the third wave and five of them drowned.*

6) *Other people returned to the shore to watch the ocean. They were surprised by a fifth wave, which crested six meters (20 feet) higher than the low tide level mainly because the tide was at its high level by this time.* More people lost their lives in this wave, but the actual number seems to be in doubt.

At Santa Cruz Harbor, the tsunami wave (this same Good Friday, 1964 tsunami) *reached as high as 11 feet, sinking a hydraulic dredge and a 38 foot cabin cruiser and causing minor damage to the floating docks.*

Pararas-Carayannis (2009b) made the following points regarding the tsunami threat in California:

1) Tsunamis from distant earthquakes pose the greatest threat for California, particularly those generated in the Aleutian Islands and mainland Alaska (an intense subduction zone). In general, the Northern California coast is more susceptible than the Southern California coast.

2) Areas such as Half Moon Bay and the northern portion of Monterey Bay are potentially vulnerable to tsunami waves from distant sources setting up resonant oscillation, which may amplify the height of tsunamis along one side of the bay. The April Fool's Day tsunami in 1946 from the Aleutian Islands produced no noticeable waves at Monterey, located on the south side of Monterey Bay, but these waves exceeded 10 feet at Santa Cruz on the north side.

3) Generally speaking, the threat for local tsunamis in California is relatively low *because of the low recurrence frequencies from these disasters.* The frequency of large tsunamis has been estimated to be once every 100 years.

4) *Only earthquakes in the Transverse Ranges, specifically the seaward extensions in the Santa Barbara channel and area offshore from Point Arguello, can generate local tsunamis of any significance.* The faults in that area have substantial vertical displacement, unlike the ones associated with the San Andreas system that have mostly horizontal displacement (last statement by us). The subduction zone off Northern California would be another possible source for a tsunami.

5) In conclusion, Ed Clifton (pers. comm.) pointed out the *potential for tsunamis caused by large-scale mass failures on the continental slope or on the walls of submarine canyons like Monterey Canyon.*

Two other web sites with useful information on tsunamis include My Highlife (2004) and American Geological Institute (2002).

4 MAJOR LANDFORMS OF THE COAST

ROCKY COASTS

Introduction

Rocky coasts are some of the most breathtakingly beautiful shorelines in the world, with those on the Central California Coast taking a backseat to few others, if any (see examples in Figures 25A and the cover of this book). The combination of exposed rock cliffs and wave-cut rock platforms occur along 27.4% of the shoreline of Central California. However, as mentioned in the discussion of young mountain range coasts, rocky coasts occur along more than 50% of many areas on the leading-edge shorelines of the continental plates of North and South America. We think one reason for the lower number along the Central California Coast is the long-term effect of the San Andreas Fault system's location along a transform fault at the boundary between the Pacific and North American Plates, where the motion along the fault is mostly horizontal rather than vertical (with an occasional exception to this general rule). Ed Clifton (pers. comm.) pointed out that another factor may be *that in this tectonically active setting we have a lot of young sedimentary rocks cropping out along the coast. Most of these are fronted by sand beaches and are classified as such* (on our maps, which leads to the relatively low number of 27.4%). Higher numbers are to be expected along subduction zones, where the North and South American Plates converge with the Pacific and Nazca Plates (see Figure 7), with high, steep mountainous shorelines being common.

Glacial rebound coasts, such as those on the eastern shoreline of Canada, shorelines of volcanic islands, and numerous other miscellaneous localities (e.g., the famous white cliffs of Dover in England) also have significant percentages of rocky coasts. According to Davis and Fitzgerald (2004), rocky coasts make up 75% of the world's shoreline, a number that may be a little high, in our estimation. Whatever the exact number, no doubt this is a common shoreline type, a fact that people living on trailing-edge shorelines, like those in the southeastern U.S., might be surprised to learn.

First, we will start with some excuses as to why coastal geomorphologists do not know more than they do about these types of shorelines:

1) The changes that take place on some of these rocky coasts may be almost imperceptible over a human lifespan (Davis and FitzGerald, 2004), unlike sand beaches, where significant changes may take place within a few minutes. However, noticeable changes may occur rapidly during a major *El Niño*, especially if the rock cliff is composed of easily erodable rocks or is subjected to a landslide.

2) Therefore, the most significant changes usually occur during relatively infrequent coastal storms.

3) Difficulty in making wave measurements and diving observations, because of the rough terrain and the necessity of making these measurements during storms.

4) Difficulty of access to the high and precipitous cliffs.

Those are the reasons there is not a proliferation of published papers on rocky coasts, especially in comparison with sand beaches, wetlands, and several other coastal environments. However, if, after reading this section, you still hunger for more knowledge about rocky coasts, we refer you to books by A.S. Trenhaile, *The Geomorphology of Rock Coasts* (1987) and T. Sunamura, *Geomorphology of Rocky Coasts* (1992), the only two books we know about that are devoted entirely to this topic.

Rock Cliffs

Nearly vertical rock cliffs mark the landward boundary of the shoreline along most rocky coasts. With a few exceptions, the cliffs are created by wave erosion and they erode at rates of less than an average of 3 feet per year globally, according to Stephenson and Kirk (2005). Usually, erosion rates differ spatially along the shore, with the erosion typically taking place in brief spurts during storms.

Assuming a relatively constant wave size and storm frequency along a given stretch of shoreline, as is the case for the Central California Coast, the rate of retreat is dependent for the most part on the **type of rocks** that make up the cliff. Griggs and Tenhaile (1994), in a succinct chapter on coastal cliffs and platforms in a book titled *Coastal Evolution*, made the following comment about cliff erosion rates on the Central California Coast:

Within a distance of 60 kilometers .. (37.2 miles) .. along the shoreline of Central California, a variety of materials including unconsolidated Pleistocene dune sands, Pliocene sandstones, Miocene mudstones, and Mesozoic granites are exposed in the coastal bluffs. Erosion rates in these materials range from 2.5 meters/ year .. (8.3 feet/year) .. in the dune sands to imperceptible rates in the granites.

The principal physical process involved in creating rock cliffs along shorelines is wave-induced erosion, which acts in the following ways:

1) A process called **wave hammer**, the high shock pressures generated against the cliff by the breaking waves.

2) At high tide during storms, the breaking waves may trap air between the cliff face and the oncoming water. *As the wave collapses, extremely high pressures are instantaneously produced and the air is compressed on the rock surface.... This process is particularly important when air pockets are compressed into the crevices of rocks, leading to the enlargement of the crack and ultimately to a shattering of the rock* (Davis and FitzGerald, 2004).
[NOTE: Ed Clifton (pers. comm.) pointed out that *the compression of air in the rock pockets/caves is responsible for an interesting phenomenon on rocky coasts: the deep (and sometimes literally ground-shaking) booms that sometimes occur. I have seen docents at Point Lobos* (State Park) *think they experienced an earthquake* (after experiencing one of these booms). *It is an awesome experience and specific to the rocky shores.*]

3) **Quarrying**, a process whereby pieces of rock, which may vary in size from tiny fragments to large blocks, are removed from the cliff face by the breaking wave. Griggs and Trenhaile (1994) observed that *the presence of large, angular debris and fresh rock scars has convinced many workers that wave quarrying is usually the dominant erosive mechanism in vigorous wave environments.* Ed Clifton also pointed out that *during a major set of storm waves in early January 2008, a 6-ton sandstone block was plucked (quarried) from the top of a 15 foot cliff and left upside down on top of the cliff.*

4) If the water in the breaking waves contains a lot of suspended sediments, ranging in size from sand to pebbles, or perhaps even up to small cobbles, the result is intense abrasion of the rocks in the cliff, especially if they are soft sedimentary rocks.

5) Because the waves tend to break further offshore during low tide, the zone of greatest wave erosion is most likely located between neap and spring high tides.

Perhaps just as important as wave action with regard to cliff formation and retreat is an associated process called **mass wasting**, which is defined as a mechanism whereby weathered materials are moved from their original site of formation to lower lying areas under the influence of gravity. Examples of mass wasting relevant to this discussion include: a) when pieces of rock of miscellaneous sizes tumble down the rock scarp (called rock fall); and b) deep seated landslides, which *require suitable geological conditions* to occur (Griggs and Trenhaile, 1994; discussed in more detail later). The results of such mass wasting are debris piles of fragments of rocks up to boulders in size (and soil in some cases) at the base of the cliffs. This accumulation of material, sometimes referred to as **talus**, is typically moved away either down the beach by normal longshore sediment transport or offshore during big storms.

A noteworthy example of talus accumulations at the base of eroding rock scarps occurs at Waddell Bluff south of Point Año Nuevo. The high, near vertical bluffs there are composed of Santa Cruz Mudstone of Miocene age (Table 1). Before 1940, the talus material that accumulated along the base of the bluffs was dispersed by the waves and wave-generated currents and moved alongshore (and offshore). After construction of Highway 1 along the shore in the 1940s, the new roadway was situated between the bluffs and the beach, preventing the talus material from being naturally dispersed by waves. These days, when talus accumulates at the base of the bluff on the east side of the road, workers from the highway department periodically remove this material and dump it on the beach. Annually, they remove about 30,000 cubic yards from the foot of the Waddell Bluffs and dispose of it along 3,000 feet on the adjacent beach (between 15 October and 31 December, presumably in the expectation that processes generated by winter storms will remove the material; information from a kiosk created by CALTRANS located near the Bluff).

Another example of rock accumulation at the base of rock cliffs occurs along the Plio-Pleistocene rocks south of Fort Funston, which have been subjected to some giant landslides. Houses along the cliff-tops there commonly must be abandoned (Ed Clifton, pers. comm.).

Once such material is removed from the base of

the cliffs by wave action, the waves can impact the cliffs directly during later storms. Obviously, for cliffs where such talus or a wide beach, either sand or gravel, typically is present, the retreat rates will be slower than that experienced by cliffs composed of similar material but without such protection. Furthermore, mass wasting will be accelerated where the waves cut an eroded notch into the base of the cliffs.

With regard to the timing of episodes of rock-cliff erosion, Griggs and Trenhaile (1994) offered these words of caution – *little erosion or retreat may occur for years and then, under the right combination of cliff weakening or weathering (perhaps due to excess groundwater or pore pressure) and wave attack and tidal elevation, major failure and significant erosion may take place virtually overnight.*

The tallest of the rock cliffs on rocky coasts that we have observed usually occur where steep mountains border the coast and where the waves are relatively large, for example along the coasts of Peru, Chile, Baja California, northern California, and parts of the Alaskan coast. The size and proximity of the mountains behind the shore also affect the height of the cliffs along the Central California Coast. Examples include the tall cliffs where the leading edge of Montara Mountain abuts the shore at Devils Slide, and along the Big Sur coast in several places.

As the rock cliffs retreat, a dazzling array of erosional remnants are sometimes left behind on the rock platforms offshore of the cliffs, namely sea stacks and sea arches. **Sea stacks** are isolated pinnacles of rock that are typically taller than they are wide with tops at a somewhat lower elevation than the top of the adjacent rock cliff (Davis and FitzGerald, 2004). The stack is usually composed of rocks that are significantly more resistant to wave erosion than their neighboring rocks in the formerly existing cliff. Figure 42 shows three examples of sea stacks. It is important to point out that sea stacks are one of the favorite nesting and roosting sites for birds, most notably cormorants and brown pelicans (Figure 42B), as well as haulout spots for California sea lions and harbor seals in some areas.

The presence of **joints** in the rocks (plane surfaces along which the rocks have split apart, usually during one or more of the ever present tectonic movements along this coast), especially vertical joints oriented perpendicular to the shoreline, favor the formation of sea stacks. Joints also speed up the erosion process along the face of the cliff. During one of our field surveys on 15 August 2008, we stopped at a pullover a little north of Bonny Doon Road (one half mile south of Davenport), where there is a spectacular little embayment located

FIGURE 42. Sea stacks. (A) Sea stack in a small bay one half mile south of Davenport in Santa Cruz County (arrows point to sea caves). This bay is located in the center of Figure 43A. (B) Brown pelicans and cormorants on two sea stacks off Westcliff Drive in Santa Cruz. Photographs in both A and B taken on 16 August 2008. (C) Sea stacks composed of the Porphyritic Granodiorite of Monterey in Garrapata State Park, about a mile south of Carmel Highlands in Monterey County. Arrows point to sea caves in the granodiorite, a rock type defined in Figure 17B.

within an indention in some highly jointed sedimentary rocks. The large sea stack pictured in Figure 42A is located just offshore of the embayment. As illustrated in the image and sketch map in Figure 43, the positions of all of the features in the embayment, including a little pocket beach, are controlled by the presence of joints in the sedimentary rocks (Santa Cruz Mudstone). A couple of sea caves are also present in these rock cliffs.

Sea arches and **sea caves** are most commonly formed in flat-lying sedimentary rocks, or layered volcanic deposits, because the separate rock layers commonly have different degrees of resistance to wave erosion. In a situation where the top layer is more resistant than some underlying layers, parts of the underlying layers may be removed by wave action to form a sea cave. Sea caves are common throughout the whole Central California Coast (see example in Figure 44). In the situation where a narrow peninsula composed of layered rocks with a more resistant top layer projects offshore, wave refraction around the headland may focus the wave erosion such that two caves eroded into the opposite flanks of the peninsula meet, forming a **sea arch**. As the waves continue to erode the peninsula and enlarge the sea arch, eventually the top of the arch collapses, creating a sea stack. This has happened at several localities along the Central California Coast, probably most notably at the Natural Bridges State Beach in Santa Cruz (see comparison photographs in Figure 45). Point Lobos State Park, an easily accessed shoreline located just south of the Monterey Peninsula, also has some spectacular sea caves and sea arches in granodiorite, formed where the waves attacked a sheared zone in the rock (Ed Clifton, pers. comm.). In Section III, a few of the most notable sea arches and sea caves are illustrated and discussed as they are encountered during our treatment of the 7 major Compartments of the coast.

Rock Platforms

As the eroding rock cliff retreats, a flat rock platform is left behind, some of which have sea stacks, as has already been discussed (Figure 25B). In the geological literature, several pleas have been entered to call these features **shore platforms**, because when non-conforming, unrefined field people like ourselves call them **wave-cut rock platforms**, we are implying that we know how they originated. Not to suggest that we were the first ones to do this in California. In the "old days," it was a common practice (Bradley and Griggs, 1976).

We have been doing this (calling these things wave-cut platforms) during mapping projects along many shorelines since 1976, and the only place we felt

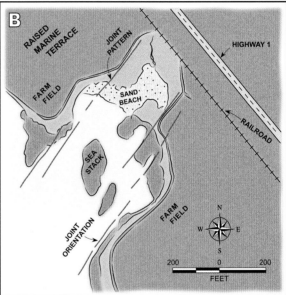

FIGURE 43. Small embayment one half mile south Davenport in Santa Cruz County. (A) Note the straight sides of the embayment, which parallel joint patterns in the bedrock. Figure 42A is a view looking southwest of the area, taken from near the railroad. The upland area is a very flat, raised marine terrace. 2005 aerial photograph courtesy of the California Spatial Information Library. (B) General plan view sketch of embayment showing the orientation of the major joints in the bedrock.

uncomfortable about calling them "wave-cut" was for the very wide platforms in some of the relatively sheltered portions of Lower Cook Inlet, Alaska, where it is probable that the waves are augmented considerably in the erosion process by the abrasion of the rocks by sediment-bearing ice layers (during the winter time). A

FIGURE 44. Sea cave located 0.4 miles north of Martin's Beach in San Mateo County. This cave is eroded into the wave-cut scarp on the margin of a spectacular raised marine terrace. There is a clear contact between two different types of geological materials about a third of the way down from the top of the scarp (arrow). The upper horizontal layers are Quaternary marine terrace deposits, but the lower unit, into which the cave has been carved, consists entirely of the sandstones, mudstones, and siltstones of a sedimentary rock unit thought (by us) to be the Pliocene Purisima Formation, which is thought (by others) to have been deposited in a shallow marine basin around 2-6 million years ago (Alt and Hyndman, 2000). This photograph, taken on 1 October 2008, is courtesy of the California Coastal Records Project, Kenneth and Gabrielle Adelman.

good case can be made that rock platforms are formed, at least in part, by other processes in the coldest polar regions and in the tropics. However, a discussion of those features in this book seems unwarranted.

Besides the obvious fact that the waves have to be pretty big to erode rock platforms, it is also necessary that **sea level remains fairly constant** for a significant amount of time. How significant? That depends on how fast the cliff is retreating and what processes may account for the downcutting of the platform (also mostly related to wave action, in our opinion). In any event, we do know for sure that sea level along the Central California Coast has been relatively constant for the past 5,000 years or so, except for the impact of the continuing uplift of the Coast Ranges, which is examined elsewhere in the discussion of raised marine terraces.

The widest platform we have surveyed in Central and Northern California is 400 feet wide at Point St. George

in Northern California. There are also wide platforms at Duxbury Reef near Bolinas (maximum width of about 1,100 yards with an average width of 300 yards) and at the Fitzgerald Marine Reserve, north of Half Moon Bay (maximum width of about 300 yards with an average width of 150 yards). Both of these rock platforms, which are referred to locally as "reefs," are illustrated in Figure 46. Amazingly, according to Griggs, Patsch, and Savoy (2005), wave-cut platforms with average widths of a half mile to a mile are present off the cliffed coasts in the Santa Barbara area, but none that wide occur in Central California.

We don't have a good summary of detailed width measurements along the Central California Coast, but widths of 160 feet are probably about average along cliffs that are not composed of granite or some extremely hard metamorphic rocks. If that is a workable average, we can probably conclude that the average retreat rate of the cliffs

FIGURE 45. A natural bridge in the Natural Bridges State Beach in Santa Cruz. A beautiful, fairly long arch was present in the late fall of 1978 (photograph A). However, by 16 August 2008, the date when photograph B was taken, the arch had fallen down. The arrow in B points to an outer arch, partially hidden in a shadow, which remains in place. This is just one of several sea arches and sea stacks that have suffered a similar fate along this shoreline. More details on the occurrence of this phenomenon on the coastline in and around Santa Cruz are given in Griggs and Ross's book, *Santa Cruz Coast* (2006).

in this area is around 0.4 inches per year, which seems a little slow [almost an order of magnitude less than the global rate given by Stephenson and Kirk (2005)]. But, as already noted, cliffs retreat in spurts related to storms, so maybe that number is not too far off. An interesting observation on cliff retreat rates was made by Griggs, Patsch, and Savoy (2005) in their discussion of the cliff at Duxbury Point (near Bolinas). Referencing data gathered between 1913 and 1950, the cliff retreated between 0.4 inches and 7 inches per year during that time interval in the area where the rocks (Santa Cruz Mudstone) in the cliffs are less disrupted and support more vegetation. However, to the south, where the mudstone is more fractured and less competent, the rate increased to over 36 inches per year. These observations once again point

out the importance of the composition of the cliff in controlling the rate of its retreat.

Rock platforms have a much more diverse population of biota and plants than sand beaches. There are some great places for viewing and studying the life in the tide pools on the rock platforms (at low tide). Some of the most noteworthy areas to do this are Duxbury Reef near Bolinas (Figure 46A), the Fitzgerald Marine Reserve (Figure 46B), the platform between Bean Hollow Beach and Pebble Beach, and most anywhere else that the rock platform is wide enough to contain tide pools and access is possible (and permitted). More details on this fascinating pastime are given in Section II.

Raised Marine Terraces

One of the more intriguing aspects of the morphology of the California coast is the presence of elevated marine terraces located landward of presently eroding rock cliffs (see examples in Figure 47). These flat or slightly seaward dipping terraces usually are at least a few hundred yards wide, and they are flanked on their landward sides by a somewhat degraded, former wave-cut rock cliff. Several of these raised terraces, which resemble giant stairsteps up the side of the mountains, occur in a number of areas, some of which are discussed in Section III. Griggs and Patsch (2004) observed as many as 13 raised terraces on the Palos Verdes Peninsula in Los Angeles County. However, on the Central California Coast, raised terraces are usually considerably fewer in number than that.

A process for the formation of elevated terraces like the ones on the Central California Coast was suggested by Davis and FitzGerald (2004) and Griggs, Patsch, and Savoy (2005), based on studies in New Guinea by Chappell (1983). According to this hypothesis, the following steps are required for the formation of a series of raised platforms on shorelines subject to episodes of uplift like this one (see general model in Figure 48):

1) A **period of stable sea level** occurs during which the waves erode the rock cliffs, the cliffs retreat, and a wide rock platform is created. During the relatively stable highstand of the past 5,000 years, rock platforms up to a few hundred yards wide have been carved in some places along the Central California Coast, but most of them are considerably narrower than that (average probably around 160 feet).

2) **Sea level drops** to a lowstand for thousands years, as has happened at least four times during the major glaciations that have occurred within the past 650,000 years.

3) In order for a systematic arrangement of raised

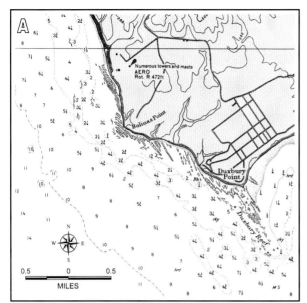
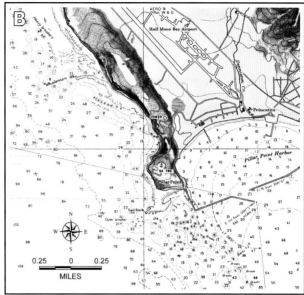

FIGURE 46. Two of the widest and most visited (to view and study the tide pools) intertidal rock platforms on the Central California Coast. (A) Duxbury Reef near Bolinas in Marin County, which has a maximum width of about 1,100 yards and an average width of 300 yards. Depths are in fathoms (one fathom = 6 feet). (B) The shallow waters around Pillar Point, at the north end of Half Moon Bay in San Mateo County, and along the shoreline to the north, where the Fitzgerald Marine Reserve is located. Depths are in feet. The very famous and much visited rock platform at the Reserve has a maximum width of about 300 yards with an average width of 150 yards. The area shaded blue has water depths <18 feet on both maps.

marine terraces to develop as is illustrated in the general model in Figure 48, it would be necessary for the adjacent land to continue to rise at a fairly consistent rate, during both lowstands and highstands. However, calculation of the actual rate of uplift is difficult to do and results differ among different researchers. To name only three of these: a) Sloan (2006) gave rates of **1 to 3 inches per century** for the mountains in the Bay Area; b) Page, Thompson, and Coleman (1998) estimated **3.9 inches/century** for the central and southern parts of the Coast Ranges; and c) Covault (2004), studying uplifted wave-cut terraces on the North-Central California Coast located 130 miles north of San Francisco near Alder Creek, concluded that area is presently rising 1.3 feet per thousand years (**1.56 inches/century**).

4) In fact, it is virtually impossible to estimate general rates of uplift that apply for the entire length of the Coast Ranges. The following complications of this phenomenon were pointed out by Ed Clifton (pers. comm.): a) The coast north of Santa Cruz and south of Monterey has been rising during the Pleistocene, creating the stair-step terraces, but the Monterey Bay plain has been subsiding during this time; b) *At Point Lobos, I put the current rate of uplift at about 7 inches/1,000 years, but it increases to the south and decreases to the north.* The lowest terrace at Monterey lies substantially lower than at Point Lobos; and c) At Fort Funston, the Merced Formation accumulated

in a subsiding basin until a few 100,000 years ago, then it warped up to the south, creating the current sea cliffs. Therefore, it appears that the general model given in Figure 48 holds for places like the coast north of Santa Cruz, but not everywhere along the coast.

5) Getting back to the general model, when sea level comes back during the next highstand, the sea level at that time is considerably lower relative to the adjacent land mass than when the first rock terrace was created, because the landmass has continued to rise. By this time, the original wave-cut rock platform has been raised as much as a hundred feet or maybe considerably more, and it is located landward of this later highstand shoreline. If we use Covault's number of 1.3 feet/thousand years, obviously if sea level were to be down 100,000 years before it comes back up, the elevated terrace would be 130 feet above the new shoreline. The youngest of the raised terraces in three locations along the Central California Coast are presently positioned at the following elevations: a) 180 feet at Bolinas; b) 150 feet a little north of Waddell Bluff; and 140 feet a few miles northwest of Santa Cruz.

6) This **cycle** can be **repeated** over and over as long as sea level continues to go up and down in a somewhat regular fashion, as it did during the Pleistocene epoch and, of course, the land continues to rise at some relatively consistent rate.

FIGURE 47. Raised marine terraces. (A) Raised marine terrace located just north of Sand Dollar Beach in Monterey County. It is over 700 yards wide at this point. Arrow points to some remnant sea stacks still preserved on top of the terrace. The bedrock in the wave-cut rock cliffs are highly faulted rocks of the Franciscan Complex that contain some serpentinites. Photograph taken on 23 September 2008. (B) Raised terrace located about 4 miles southeast of the Diablo Canyon Nuclear Power Plant in San Luis Obispo County. It is over 600 yards wide at this point. Note the wave-cut rock platform at the bottom of the image. Most of the materials that compose the wave-cut scarps are most likely Quaternary marine terrace deposits, and the underlying greenish colored rocks are Paleogene/Neogene volcanics, according to the geological map of the state. Photograph taken on 28 October 2005. Both photographs are courtesy of the California Coastal Records Project, Kenneth and Gabrielle Adelman.

This general model is illustrated in Figure 48. The exact correlation of sea-level curves with the present terrace levels along the Central California Coast is a research project that is beyond the scope of this book, but it is indeed an interesting one. However, the general process suggested by this hypothesis seems to adequately account for the continued elevation of many of the terraces.

However, the hypothesis illustrated in Figure 48 does not take into account the possibility of an abrupt, sizeable elevation of the land during a major earthquake. In the course of our work on the south-central coast of Alaska, we observed a wave-cut rock terrace that was raised over 30 feet on Montague Island during the Good Friday earthquake of 1964. However, Montague Island is located along a subduction zone, whereas the Central California Coast is not normally subjected to such strong vertical movements during earthquakes at the present time. On the other hand, the Santa Cruz Mountains west of the San Andreas Fault did rise approximately 3.9 feet during the 1989 Loma Prieta earthquake. Therefore, such abrupt uplift during earthquakes may play some role in the creation of the raised terraces on this coastline, but probably not a very substantial one, at least compared to the global fluctuations in sea level.

SAND BEACHES

A River of Sand

A beach is an accumulation of unconsolidated sediment transported and molded into characteristic forms by **wave-generated** water motion. The landward limit of the beach is the highest level reached by average storm waves, exclusive of catastrophic storm surges, and the seaward limit is the lowest level of the tide. In Central California, the beach's landward limit typically is either a line of wind-blown sand dunes, or at the base of a rock scarp. Along the approximately 842 miles of the outer shoreline under discussion (linear distance including all the nooks and crannies), the landward side of the beach abuts against a man-made structure along 36 miles of it (vertical concrete seawalls and riprap).

Beaches will form on any shoreline where clastic particles somewhat resistant to wave abrasion are available and there is a site for sediment accumulation protected from extreme daily erosive wave action, such as wave reflection off rocky cliffs. On the Central California Coast, those clastic particles range from sand-sized fragments of rocks with a wide range of composition up to boulders. The discussion in this section will focus on the sand beaches.

Most sand beaches in the area under consideration are a virtual "river of sand" with the sand being in constant motion up and down the beachface and alongshore (see example in Figure 26A). We call the zone within which the sand moves back and forth the **zone of dynamic change** (ZODC), which is illustrated in Figure 49 for areas where the beach is backed by a line of wind-blown sand called

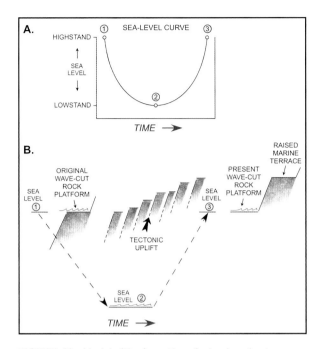

FIGURE 48. Model of the formation of raised marine terraces. (A) Typical sea level curve showing the change from a highstand to a lowstand and back to another highstand through time (after Davis and FitzGerald, 2004). (B) The different steps required for the formation of a raised marine terrace on the Central California Coast. During the first highstand (1), a wide wave-cut rock platform is carved by the waves. The land is slowly uplifting, but the downcutting action of the waves compensates for it. During the interval between the two highstands, while the sea level is lower, eventually retreating all the way to the lowstand level (2), the land continues to rise, either steadily or sporadically. By the time the second highstand occurs (3), the land has uplifted a considerable vertical distance (100 feet is common on the Central California Coast). Finally, a second wave-cut rock platform develops 100 feet or so below the former one that has by then been converted to a raised marine terrace.

foredunes. The foredunes act as a storage area for sand between storms. Thus, the landward limit of the zone of dynamic change is the point at which the foredunes are eroded during "normal" storms (not extreme *El Niño* storms). Accordingly, during these "normal" storms, the sand formerly deposited on the intertidal beach, as well as up to several feet of the foredunes, is usually transported into the subtidal area which, in some areas, contains an offshore bar (see Figure 49; the red zone represents the portion of the beach/dune sand that potentially could be eroded during a typical, moderate storm). The seaward limit of the zone is what engineers call the closure point, the most seaward distance for sand to move in an offshore direction under relatively

normal adverse conditions (again, not during extreme *El Niño* storms!!). As indicated in Figure 49, all of the sand grains in motion do not stay within the zone of dynamic change at a specific geographic site on the beach for an infinite amount of time, but most likely will eventually "escape" from the zone by one of three primary types of "leakage:"

1) **Alongshore** out of the zone, under the influence of longshore sediment transport by wave-generated currents;

2) **Offshore** beyond the point of closure during major storms, such as during extreme *El Niño* events; and

3) **Onshore** by wind action, beyond recovery by offshore winds or by erosion during storms.

Another way sand grains may "escape" from the "river of sand" for a long period of time is for the shoreline to build out seaward to the extent that more lines of foredunes develop until they are located far enough landward to be out of the reach of normal storm waves. However, shorelines of this type are very rare on the Central California Coast.

The Beach Profile

When coastal geologists or engineers study beaches to determine such issues as their potential for erosion, they usually measure a topographic profile across the

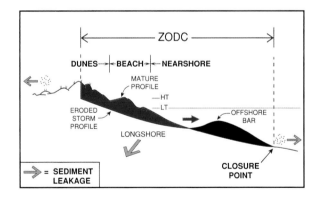

FIGURE 49. Zone of dynamic change (ZODC). The area in red depicts sand that may be eroded from the beach area and deposited offshore during storms. Much of this sand is deposited on the offshore bar (area in black). In between storms, much of the eroded sand is returned to the beach/dune area. The orange arrows indicate the three mechanisms by which the sand might be lost: 1) carried offshore beyond the closure point during large storms, such as hurricanes or extreme *El Niños*; 2) blown landward out of the foredune area during periods of high wind activity; and 3) transported alongshore by beach drift and longshore currents as illustrated in Figure 37.

beach surface using a variety of surveying techniques. Some studies use a remote sensing system called **LIDAR** (Light Detection and Ranging) to produce topographic images of the beach and nearshore zone. This system is more expensive than simple horizontal-leveling surveys carried out by two or three junior surveyors; therefore, they are usually spaced at least six months apart.

Beach surveys are usually repeated over a period of months or years in order to determine the cycle of erosion and/or deposition on that particular beach. Such studies have been carried out along many of the coastlines of the world, including the coast of Central California. For example, U.S. Geological Survey scientists, under sponsorship by the State of California Department of Parks and Recreation, surveyed nine beaches along the shoreline of Monterey Bay 34 times between 1983 and 1998. Results of that survey are discussed later.

For purposes of discussion, it is necessary that the different morphological components of the intertidal beach be defined. In the early days of research on the California coast, the profile shown in Figure 50A was proposed as typical for that area. Under that scheme, the beach was divided into two segments: 1) Backshore – the landward sloping surface of a broad depositional feature called a **berm**; and 2) Foreshore – the seaward face of the berm also alluded to as the **beachface** in that scheme. The subtidal area was divided into the surf zone, breaker zone, and offshore. This type of profile most commonly forms on a windward shore subject to ocean swell and a relatively large average wave height, such as occurs on the California coast.

A key aspect of any beach profile is the slope of the beachface. Beach research as early as during World War II focused on this issue, due to its relevance to beach landings by amphibious watercraft. Those early researchers found that the slope of the beachface in any given area will attain an equilibrium profile (i.e., a nearly constant slope of the beachface) under what they referred to as "steady" conditions, that is, relatively unchanging, day-to-day wave characteristics. During storms, sand beaches are commonly eroded down to a flat surface that essentially parallels the water table in the beach with sand typically being transported off the beach to an offshore bar (as is suggested in Figure 49). When sediment returns to the beach after the storm, the depositional profile attained by the beachface is steeper than the beach was immediately following the storm. Komar (1976) explained why:

Due to water percolation into the beachface and frictional drag on the swash, the return backwash tends to be weaker than the forward uprush. This moves sediment onshore until a slope is built up in which gravity supports

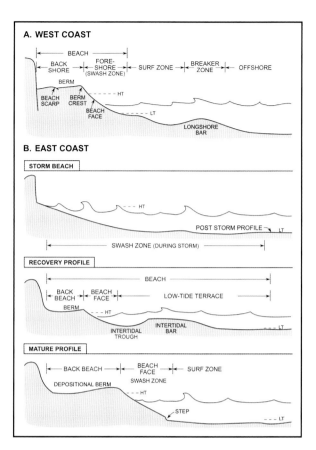

FIGURE 50. Representative beach profiles from the west (A) and east (B) coasts of the U.S.

the backwash and offshore sand transport.

Several field studies and wave tank experiments have shown that coarser-grained beaches, especially those composed of gravel, have steeper slopes than those composed of finer-grained sediments (Wiegel, 1964). Coarser-grained beaches allow greater percolation of the water brought to the beachface in the wave uprush than finer-grained beaches, thus reducing the strength of the return of the backwash. Consequently, a steeper beachface is required to maintain an equilibrium profile on coarse-grained beaches (gravity-supported transport). On sandy beaches of comparable grain sizes, those beaches consistently exposed to large waves tend to be flatter than sheltered beaches exposed to smaller waves, because larger waves lose relatively less of their total volume to percolation into the beach sediment than smaller waves, therefore they produce stronger backwash, which tends to flatten the profile.

The Beach Cycle or "Normal Seasonal Changes"

The concept of the cyclical change of beaches from a flat, erosional profile in winter to a wide, depositional berm in summer had been well ingrained in both the popular and scientific literature for many years following research carried out in the 1940s. The concept originated from detailed studies on the coast of Southern California (e.g., Shepard, 1950b; Bascom, 1954). The idea was that, generally speaking, erosional storm waves are more common in the winter, and flatter, depositional swell waves are more common in the summer on the California coast – hence the terms summer and winter beaches. Beach profiles measured at Scripps pier in La Jolla in the 1940s commonly showed wide beaches in the late summer and fall and narrow beaches in the winter and spring.

However, this is an idea that has not withstood the test of time for other parts of the world. In 1969, co-author Hayes and his colleague Cy Galvin observed a striking contrast between beach cycles on the U.S. east coast and those presented in the literature for the west coast. Eroded winter profiles and strongly depositional summer profiles are not a dominant theme on northern Atlantic beaches, particularly in the New England area. The beach cycle in that area is controlled by the passage of individual storms. This also appears to reflect, at least to some degree, less severe wave climate and relatively smaller seasonal changes in wave climate on the northern Atlantic coast. Mean wave heights and periods are 19 inches and 6.9 seconds along the northern Atlantic coast, considerably smaller values than those for the west coast (discussed earlier). These dissimilarities in wave climate result to some extent from differences in their locations on western and eastern edges of oceans with predominantly westerly winds, with the winds blowing predominantly onshore on the California coast and offshore on the east coast. However, Ed Clifton (pers. comm.) commented that *local winds are less a factor than the fact that the largest storms (typhoons) develop on the western side of the ocean, and that every large weather system in the western Pacific (and the Gulf of Alaska as well) generates long period swell which arrives on the U.S. west coast.* As alluded to earlier, the largest waves that occur along the Central California Coast are generated in one of two ways: 1) when major *El Niño* storms cross the coast; or 2) when some of the largest swell from especially huge offshore storms reach the coast.

In the 2002 Coastal Engineering Manual, published by the U.S. Army Corps of Engineers, the following statement appears: *the seasonal onshore-offshore exchange of beach and nearshore sediments … is now known to apply best to swell-dominated coasts, such as the U.S. west coast, where wave climate changes seasonally.* Therefore, the general consensus today is that the terms "winter beach" and "summer beach" should not be used with reference to erosional and depositional cycles on beaches, although many published discussions of the beaches in California still use these terms, or else refer to "normal seasonal changes," which presumably means more-or-less the same thing.

Based on six years of research between 1964 and 1969, co-author Hayes and his students at the University of Massachusetts found that, on the northern New England coast, northeasterly storms play the dominant role in the generation of cycles of erosion and deposition of the sand beaches in that area (Hayes and Boothroyd, 1969). Observations at 40 beach profile stations surveyed every two weeks over that six year period revealed the following stages of low-tide morphology relative to storm occurrences (these profiles are illustrated in Figure 50B):

1) **Early post storm** (up to 3 or 4 days after the storm) – Beach is flat to concave upward and the beach surface is generally smooth and composed of medium- to fine-grained sand. Severest storms leave erosional scarps along the dunes.

2) **Early accretional, or constructional** (usually 2 days to 6 weeks after storm) – Small berms at the high-tide line and intertidal bars and troughs are quick to form.

3) **Late accretional or maturity** (6 weeks or more after storm) – Landward-migrating intertidal bars weld onto the backbeach to form broad, convex berms.

Paul Komar, working on the Oregon coast, suggested the terms **storm profile** (flat beach) and **swell profile** (well-developed berm).

Two Studies of California Beach Profiles

Many scientific studies have been carried out on the geomorphology of the California beaches, far too many to be discussed individually in this book. However, two such studies are discussed briefly in order to give the reader an insight into this general approach.

In 1992 and 1993, the authors of this book directed field surveys of 81 beach stations on the central and northern coastline of California as part of a joint effort between NOAA's Hazardous Materials Response and Assessment Division and the Office of Oil Spill Prevention and Response (OSPR) of the California Department of Fish and Game. This activity provided ground-truth data for the aerial surveys for the ESI mapping program discussed earlier. Also, the OSPR staff gained more familiarization with the beach habitats of the state and practiced beach profiling and other beach description techniques.

The initial survey of the 81 stations was carried out on the central coast between 6-12 May 1992, and on the northern coast between 20-25 October 1992. A follow-up survey of thirty of the stations, fifteen on the central and fifteen on the northern coasts, was completed between 3-6 March 1993 by several teams of researchers.

For this study, we prepared a somewhat more complex "typical" California sand beach profile than the standard one in Figure 50A. As shown by the cross-sectional view in Figure 51A and plan-view sketch in Figure 51B, this model beach is made up of four primary components – foredune ridge, high berm, active berm, and low-tide terrace, with the active berm and low-tide terrace having some subdivisions. Definitions of these features follow:

1) **Berm:** A wedge-shaped sediment mass built up along the shoreline by wave action. It typically has a relatively steep seaward face and a gently-sloping landward surface (2-3 degrees). A sharp peak (berm crest) usually separates the two oppositely slopping planar surfaces on the top of the berm. There can be two berms present during all tidal phases except unusually high spring tides: a **high berm**, the most landward, oldest berm, and an **active berm**, the most seaward and most recently activated berm.

2) **Beachface:** The zone of wave uprush and backwash during mid- to high-tide. It usually slopes seaward at an angle of 5-10 degrees and is commonly the seaward face of the berm. However, the berm may be washed away during a storm. In that situation, the beachface would be a flattened-out, eroded surface subject to the uprush and backwash of the waves.

3) **Beach Cusps:** The scalloped seaward margin of a berm consisting of a regularly spaced sequence of indented bays and protruding horns. They are thought to be primarily depositional features (i.e., on the seaward flank of a prograding depositional berm) that form as a result of the interaction of shore-normal standing waves (edge waves) with shore-parallel, reflective waves, the same general process that produces rip currents (but at a smaller scale) (Guza and Inman, 1975). See an example and the general model in Figure 38.

4) **Low-Tide Terrace:** A relatively flat surface that is located seaward of the toe of the beachface and slopes offshore at a small angle (2-3 degrees). The landward margin is usually near mean sea level and the seaward margin near mean low water. The sand that composes the terrace is commonly water saturated throughout, essentially representing the intersection of the water table with the beach surface. During periods of recovery of the beach following an episode of erosion, **intertidal bars** migrate across the low-tide terrace in a landward direction. An **intertidal trough** is commonly located on the landward side of the elevated intertidal bar.

The results of the follow-up surveys in March 1993 indicated that fifteen of the thirty stations profiled showed little to no change between the surveys. Twelve stations showed erosion and three showed deposition at the time of the second survey. Beaches composed of gravel (including granules) and those with prominent intertidal bedrock platforms showed the least change, whereas those composed of sand showed the greatest change (measured in tens of yards of horizontal change and tens of inches of vertical change in places).

Along the Central California Coast, six of the fifteen beaches resurveyed in March 1993 were erosional relative to the May 1992 survey. For example, one of the more erosional beaches, the relatively coarse-grained sand beach at Point Sur (Figure 52), was 70 feet narrower in March than it had been the previous May.

An example of a beach on the Central California Coast that showed essentially no differences between the surveys (similar to seven of the others) was at Montaña de Oro State Park near Morro Bay (Figure 53). The profile of this moderately steep, coarse-grained sand and granule beach (hence the steep slope of the beachface) was essentially unchanged over time.

Thirteen cyclonic storm centers crossed the California coast between May 1992 and March 1993. Seven storm centers crossed the coast during January and February 1993, which no doubt accounts for the erosion observed on 40% of the beaches during the repeat surveys. As a gross generalization, the beaches that face in a more southerly direction showed a slightly greater tendency to be erosional in comparison to those facing north.

Of the fifteen resurveyed beaches in northern California, six of them were considerably eroded in comparison with the original survey, the most extreme case being on the North Spit at Humboldt Bay, where a depositional sand berm 131 feet wide was completely missing during the second survey. Seven of the profiles did not appear to have changed much between the survey periods. All of these beaches showing little change were composed of coarse-grained sand or gravel or a mixture of the two, and four of them were associated with intertidal rock platforms. Only two of the beaches had prograded significantly.

Two of the beaches in northern California that illustrate the contrast between a beach that built out versus one that eroded during the winter months (or, more properly, during the days or few weeks before the

A.

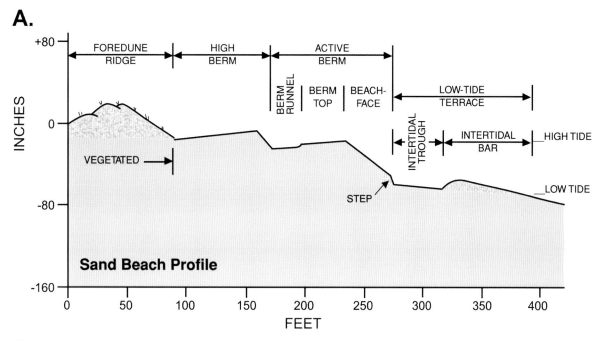

B.

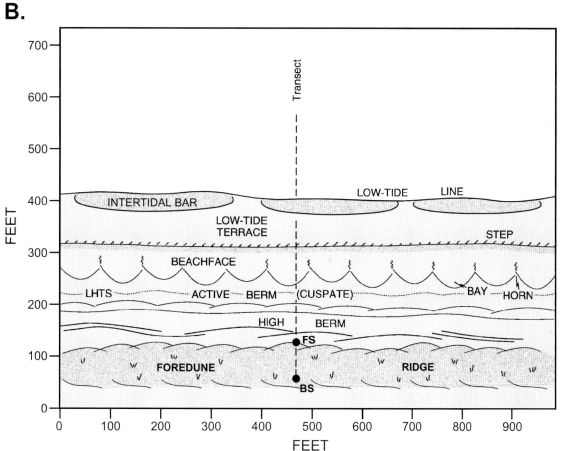

FIGURE 51. The beach morphology of the Central California Coast as envisioned by Montello, Hayes, and Michel (1993). (A) Nomenclature for the sand beaches in California presented in a topographic profile. (B) Typical beach configuration in plan view. FS and BS represent permanent front and back stakes for surveying topographic profiles such as the ones shown in Figure 52B. LHTS = last high-tide swash line.

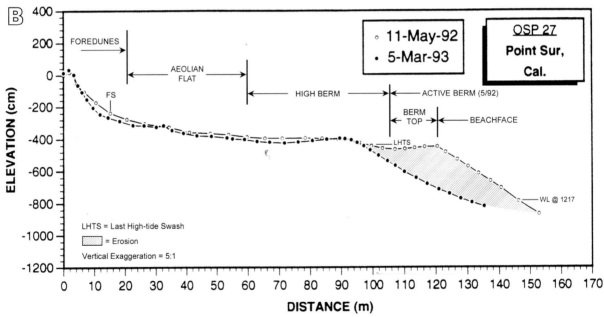

FIGURE 52. The wide, medium- to coarse-grained sand beach north of Point Sur (Monterey County). (A) View looking southwest with the rock headland at Point Sur visible in the upper left. Photograph taken in the early 1990s. (B) Topographic beach profiles measured on 11 May 1992 and 5 March 1993, which show that an active berm, such as the one on the beach in May 1992 (20 meters or 66 feet wide), was not present during the follow-up survey in March 1993.

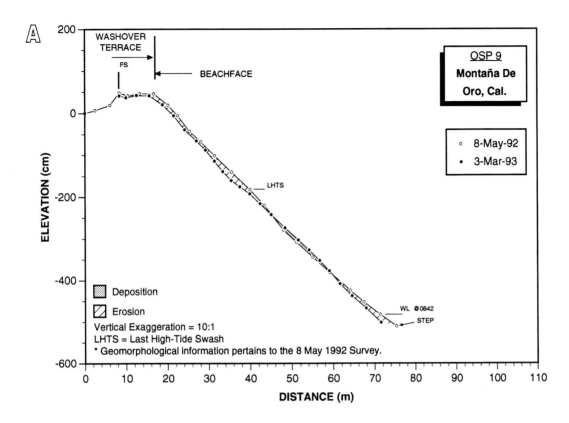

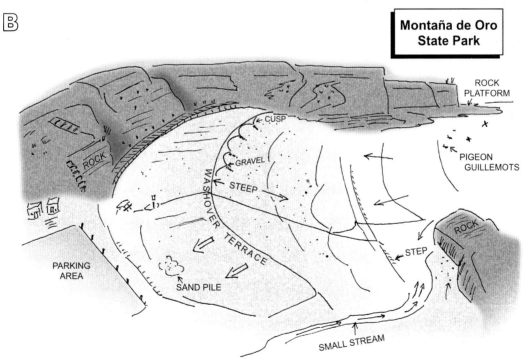

FIGURE 53. Coarse-grained sand and granule beach at Spooner's Beach in Montaña de Oro State Park (San Luis Obispo County). (A) Topographic beach profiles surveyed on 8 May 1992 and 3 March 1993, which showed very little change for the two surveys. (B) Field sketch drawn on 8 May 1992 that illustrates some key features, such as a moderately steep beachface, a washover terrace at the top of the beach, and a step at the base of the beachface.

profiles were measured) are illustrated in Figure 54. Station OSP 71 at Elk showed increased sand deposition at the time of the second survey relative to the first one (Figure 54A). During the 24 October 1992 survey, a 230-foot-wide zone of multiple berms composed of medium grained sand was present. On 5 March 1993, a new 72-foot-wide berm had formed seaward of the zone of multiple berms that, presumably, had been deposited during the "winter" season. Offshore of this beach, a large sea stack is present around which the waves refract creating this depositional beach. The beach at station OSP 49, located near the mouth of Wilson Creek, shows the opposite trend (Figure 54B). In this case, a depositional intertidal beach berm/washover terrace also about 72 feet wide (horizontal scales are different in the two plots in Figure 54) surveyed on 20 October 1992 had been completely removed by the time of the 2 March 1993 survey. This station had been somewhat unwisely located across a washover terrace (wide, landward-dipping surface on top of the berm that is frequently washed over during the highest spring tides) that had closed off the mouth of Wilson Creek, which was essentially dry at that time. By the time of the 2 March 1993 survey, the stream had flooded as a result of winter rains and the original berm/washover terrace was washed away leaving the beach with an eroded flat profile.

The fact that 60% of the beaches surveyed in both Central and Northern California at the end of the winter season were not erosional indicates that the process is more complicated than a simple seasonal erosion/deposition trend. Commenting on these results, Ed Clifton (pers. comm.) observed that *your data show that the classic summer/winter beach profiles are an idealized concept and that the real world is far more complex,* indicating that the results were not surprising to him, based on years of observation along the California and Oregon beaches.

The second study we will discuss under this heading is a fifteen year project to measure the profiles of nine of the sand beaches on the eastern shoreline of Monterey Bay, summarized by Dingler and Reiss (2002). This study was sponsored by the U.S. Geological Survey, with supporting funding from the State of California Department of Parks and Recreation. All but one of the profiles were surveyed 34 different times during the fifteen years, an average of about two times per year.

At the start of the project in Monterey Bay in early 1983, the nine beaches had been severely eroded during the strong *El Niño* of 1982-83, which was discussed in some detail earlier. Figure 55, which plots beach width versus the dates of the surveys, gives the results for the

profiles at Sunset State Beach (see photograph in Figure 26A). This plot is fairly typical of the other beaches along the Bay shoreline, although there is some very interesting variation among the nine profiles. These differences, as well as other informative observations in this paper on beach widths and slopes, beach cusps, rhythmic topography, longshore sediment transport, and cliff retreat, have added significantly to the understanding of the beach processes in Monterey Bay. The curve in Figure 55 aptly demonstrates the fact that as soon as the major storms of 1982-83 ended, the beaches began to rebuild to significant widths, an average of about 160 feet, and they stayed in the vicinity of those widths until the next major *El Niño* season of 1997-98. The last survey reported in that paper was completed in the spring of 1998. In summary, this study appears to indicate that seasonal fluctuations in the beach profiles along the beaches on the eastern shore of Monterey Bay, though sometimes impressive, are relatively subdued in comparison to the impact of extreme *El Niño* storm events.

Smaller-scale Physical Sedimentary Features

Because of the relatively large tides in Central California, when you visit the sand beaches at low tide you may be able to walk across an intertidal zone that contains an abundance of smaller-scale physical sedimentary features that will no doubt arouse your curiosity. Although geology has long been considered to be primarily a field science, the greatest breakthrough in the understanding of these sedimentary features was the result of laboratory flume studies carried out by hydraulic engineers (e.g., Simons and Richardson, 1962). These workers demonstrated that the sediment bed of a laboratory flume (box-like trough with a sand bottom and a mechanism to generate water flow across the sand bed) could be made to pass through the following sequence of features (called **bedforms**) by simply increasing the velocity of the water flowing across the sand bed or changing the depth of the water:

> *flat bed → small ripples → large ripples (megaripples) → washed-out megaripples → flat bed → antidunes*

While doing this work, they came up with an empirical relationship that dictates which of these different bedforms are active at any given time. It is called the **Froude number**, which equals:

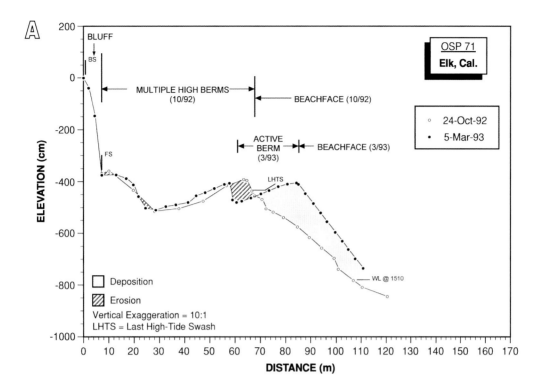

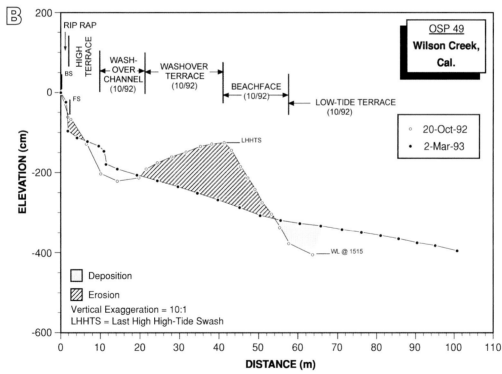

FIGURE 54. Topographic beach profiles of two Northern California beaches. (A) Comparison of 24 October 1992 and 5 March 1993 surveys of a medium-grained sand beach near Elk. The beach showed significant deposition at the time of the March 1993 survey. (B) Comparison of 20 October 1992 and 2 March 1993 surveys of a sand and gravel beach near the mouth of Wilson Creek. In March 1993, the beach showed extensive erosion, with the complete removal of a steep-faced depositional berm that was there in October 1992, as a result of the high runoff of Wilson Creek during the interval between the two surveys.

$$\text{Froude number} = \frac{V}{\sqrt{gD}}$$

V = velocity
 g = gravitational constant
 D = water depth

Therefore, as the water velocity increases, the Froude number also increases, and as depth increases, the Froude number decreases. Incidentally, in flowing water, sand grains on the underlying sediment surface start to move down current at velocities of about 10 inches/second.

The first part of the sequence up to the washed-out megaripples occurs when the value of the Froude number is <1.0. This part of the sequence is called the **lower flow regime**. By increasing the Froude number to values >1.0, by either increasing the velocity of flow or decreasing the water depth (see formula), the sand bed goes into the **upper flow regime**, where either flat beds or antidunes are formed.

The lower flow regime bedforms are asymmetrical, with the sand grains moving in discreet steps by rolling or bouncing up the gently sloping upcurrent side of the bedform and sliding down the steeply dipping surface on the downcurrent side (called the slip face). Thus, these forms move along in the direction the current is flowing, sometimes several feet during a single tidal cycle. On a beach, it is rare to see any bedforms larger than **ripples** (spacing between crests less than 2 feet) and they usually occur in the intertidal troughs on the landward side of the intertidal bars, where the water depths are greater than elsewhere on the beach (Figure 56A). **Megaripples** (spacing between crests 2-18 feet) are seen more commonly in tidal channels, where the water is deeper and the currents are stronger (Figures 56B and 56C). The mode of sediment transport over these types of bedforms is illustrated in Figure 57. Larger bedforms called **sand waves** (spacing between crests greater than 18 feet) do occur on this coastline, but they are mostly found in deep tidal channels or tidal inlets.

As you walk across the beach and tidal sand flats at the coast, you will notice some dramatic changes in the nature of the lower flow regime bedforms (at the scale of ripples), even though the changes in the topography are quite subtle. As the diagram in Figure 58 shows, there is a progression of the three-dimensional shapes from straight-crested to undulatory ripples at conditions of relatively low flow strength. As flow strength increases, the ripples take on a cuspate shape. This shape is very

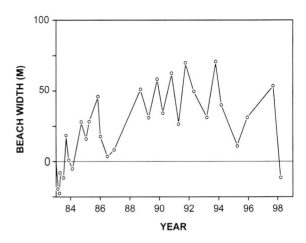

FIGURE 55. Plot of beach width versus the dates of the surveys reported by Dingler and Reiss (2002) at Sunset State Beach on the northeastern shoreline of Monterey Bay. Figure 26A is a photograph of this beach. This plot illustrates the fact that as soon as the major storms of the *El Niño* season of 1982-83 ended, this beach, as well as eight others in the area (not illustrated here), began to rebuild to significant widths, an average of about 50 meters (164 feet), and they stayed in the vicinity of those widths (with some significant fluctuations) until the next major *El Niño* season of 1997-98.

common in the intertidal troughs on the beach. As the flow approaches the upper flow regime (Froude number >1.0), the ripples convert to a planed-off rhomboid shape just before the bed goes flat. The photographs in Figure 59 illustrate these bedform types. You will undoubtedly see the pattern illustrated in Figures 58 and 59 repeated time and time again during your walks over the intertidal bars during low tide. The pattern will almost always be present where the high-tide waters were flowing down very subtle slopes on the sand bars.

The most common bedform type under upper flow regime conditions is the **flat bed**, over which the flow of the sand grains is more-or-less continuous, streaming in long streaks across the sediment surface. This is the most common type of sand surface occurring on sand beaches, where water depths are typically small and wave-generated water flow is generally rapid. Thus the Froude number is typically >1.0 (see formula again). Under conditions where the velocity is more rapid than normal, for example, when the water flows down a subtle slope (e.g., over the crest of an intertidal bar or down a more steeply sloping high beachface), a feature called an **antidune** is formed. Under this condition, the sand surface develops a sinusoidal shape and a relatively thin layer of water flows in phase across the top of these features, which are linear and parallel with a typical

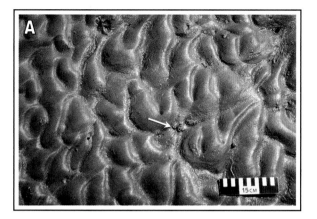

FIGURE 56. Ripples and megaripples. (A) Ripples. Arrow points to exposed ghost-shrimp burrow. Scale is ~ 6 inches. (B) Megaripples viewed from a bridge across a tidal inlet. Arrow points to the megaripple shown in C. The ebb current that created these megaripples was flowing from the lower left to the upper right. (C) Ground view of the megaripple shown in photograph B. Note the high angle of dip of the beds created by the migration of the bedform by the mechanism illustrated in Figure 57. The machete is about 2.5 feet long. The ebb current that created this bedform was flowing from left to right.

spacing of about 15-20 inches. Actually, a more accurate definition of the shapes of antidunes is that of a trochoid (Hand, Wessel, and Hayes, 1969), a word created by Gilles de Roberval for the curve described by a fixed point as a circle rolls along a straight line (Wikipedia encyclopedia).

As shown in Figure 60A, the water flowing down the slope on the downcurrent side of the sinusoidal (trochoidal) antidune picks up sand grains from that relatively steep slope only to drop them out again as the water flows up the slope on the upcurrent side of the next antidune in line. A slice through a preserved antidune, shown in Figure 60B, illustrates the deposited sand on the steep upcurrent side of the antidune. At first glance, as you watch antidunes in motion, you experience an optical illusion that the sand grains are moving upstream against the current. Actually, you are viewing the individual sinusoidal forms of the antidunes that are moving against the current, because of erosion on the downcurrent side and deposition on the upcurrent side. Although they move in a series of stops and starts, the sand grains always move in the direction the water is flowing.

If you look carefully, you will almost always see antidunes forming on the Central California beaches in one of three places:

1) Where water draining from intertidal trough outlets cut through and across the intertidal bar located just seaward of the trough.

2) Anywhere along the upper or lower part of the beach where the backwash flow gains velocity as it moves down the beach slope.

3) Where the outlets of small freshwater streams flow through shallow channels cut through the upper portion of the beach (best seen during low spring tides). An example of this type of flow is shown by the photograph in Figure 60C.

It has been our experience that students walking in the swash zone get pretty excited when they feel the antidunes move by past their feet. Sand moving uphill!! Not exactly, but close.

The shallow, rapid flow of water over antidunes, or flat beds for that matter, tends to sort the sediment by composition, moving out the larger and lighter quartz and feldspar grains and leaving behind more dense and smaller grains. These more dense minerals, called **heavy minerals**, are darker in color than the lighter minerals, hence they may leave linear traces that mark the position of the antidunes in which they were deposited. Another place where you will see relatively thick deposits of heavy minerals is at the base of an eroding dune scarp

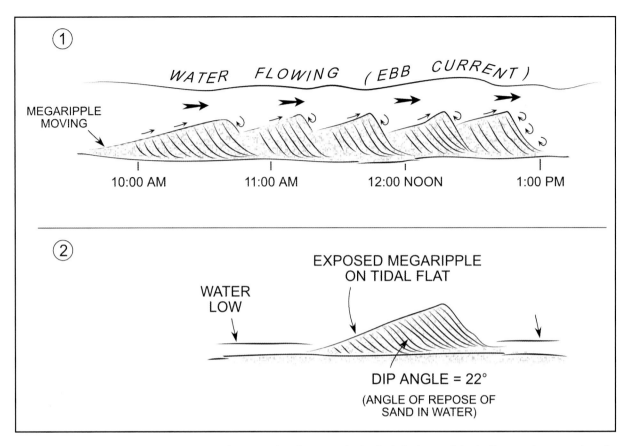

FIGURE 57. Mechanism for the movement of megaripples. Compare the implied slip faces of the bedform as it moves along (in the sketch) with the ones exposed in the photograph in Figure 56C.

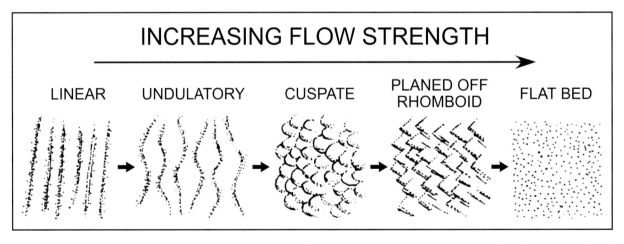

FIGURE 58. Changes in linearity of crests and 3-dimensional shape of ripples under conditions of increasing flow strength. Based on data of Allen (1968), Boothroyd (1969), and numerous field observations by our group on the intertidal bars and sand flats around the world. In all cases, the water current that created the ripples was flowing from left to right.

FIGURE 59. Photographs of the ripple types illustrated in Figure 58. (A) Linear ripples. Scale is 1 foot. The ebb current that created these ripples was flowing from left to right. (B) Ripples with undulatory crests, formed by an ebb current flowing from left to right. Scale is 15 centimeters (~ 6 inches). (C) Cuspate ripples, formed by an ebb current flowing from the upper right to the lower left (courtesy of Jon C. Boothroyd). (D) Small rhomboid ripples. The current that formed these ripples was flowing from left to right. Scale is 1 foot. This ripple form is the last to develop before the bed passes to the flat-bed stage in the transitional zone between lower and upper flow regime.

at the back of the beach, where the wind has carried the lighter grains away and left the heavier ones behind. Occasionally, the high part of the beachface is covered with a layer of them, usually after a storm. You may also see such heavy mineral deposits in the dune fields landward of the beach. A variety of minerals occur in these black layers on the Central California beaches and dunes, such as hornblende, magnetite, and hematite. In some places, such as at Pfeiffer Beach in the Big Sur Country, the layers are pure **garnet**, a member of a complex suite of silicate minerals that are commonly reddish in color.

[NOTE: In the usual mix of heavy and light minerals on a beach, the heavy minerals will be smaller in diameter than the lighter minerals, because of their hydraulic equivalence (e.g., if dropped into a cylinder of water, they would all reach the bottom of the cylinder at nearly the same time). That is, as they fall through the water in a breaking wave, the smaller, heavier minerals fall

at the same rate as the larger lighter minerals, because the buoyancy of the relatively dense water the grains are falling through creates more resistance on the larger grains than the smaller grains. Once the sand grains are laying on a flat beach, the larger, lighter mineral grains, such as quartz and feldspar, will be transported away more readily by the shallow sheet flow of the water in the wave swash than the smaller heavy mineral grains, such as magnetite and garnet, because they protrude further up into the water than the smaller heavy minerals (a process called pivotability, which is discussed in more detail elsewhere). Over time, this action may leave behind a surface layer of the heavy minerals, because they do not stick so far up into the water column as the large lighter grains, which are carried away. The process of the formation of heavy minerals at the toe of a dune scarp or on the surface of the wind-blown sand dunes is different because, in this case, the influence of gravity takes over again. In a personal communication, Ed

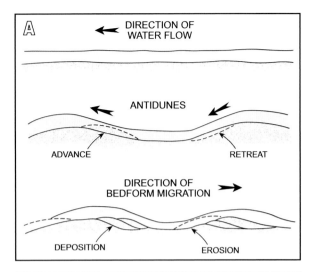

Clifton suggested a different process for the formation of heavy mineral layers on the "back edges of beaches" based on his observations on Oregon beaches. He commented that *sand transport during wave uprush is mostly in the form of suspension, whereas during the backwash, it moves in a flowing bed along the bottom where the grains are colliding with each other in something akin to* **grain flow**. *In this flow, "collisional sorting" causes the largest particles to move to the top and front of the flow. The small size of the heavy minerals promotes their transport in suspension up the beach. But during the modified grain flow under backwash, the small size and their density work together to inhibit transport. So the heavy grains are carried in suspension with the large light grains to the back of the beach and the larger, lighter grains are selectively carried back down the beach, leaving the heavies behind.* In some ways, we are both saying the same thing, except we did not distinguish between suspended sediment transport and bottom creep in the uprush and backwash. So what is more important, grain flow or pivotability of the individual sand grains? Do both processes work, maybe at the same time? We have contacted several other sedimentologists about this phenomenon, and we are still waiting for the final answer to completely explain this very common occurrence.]

[NOTE: **Grain flow** is defined as the movement of sediment under gravity where the sediment is supported by direct grain-to-grain contact (encyclopedia.com). This differs from turbidity flow, where the sediment moves under the influence of gravity in a turbulent flow of water.]

GRAVEL BEACHES

Introduction

In this book, the term gravel is defined according to the Wentworth (1922) scale, which includes four classes – *granules* (median diameter = 2-4 millimeters), *pebbles* (median diameter = 4-64 millimeters), *cobbles* (median diameter = 64-256 millimeters), and *boulders* (median

FIGURE 60. Antidunes. (A) Formation and growth. As the shallow sheet of water flows across the linear form of the antidune, water flowing down the downcurrent side of the antidune tends to erode sand from that slope. As it flows up the slope of the next antidune down current, the water slows down and sand is deposited on that slope, causing the feature to migrate up current. This bedform is called an antidune, because it moves in the opposite direction to the current in which it forms. Lower flow regime ripples, on the other hand, move in the same direction as the current (compare this figure with the diagram in Figure 57). However, the sand grains themselves move down current in both situations. (B) Trench through antidunes on the beach on Seabrook Island, South Carolina (photograph taken in June 1975). Water was flowing from right to left when the antidunes were forming. Compare the internal sand layers in the preserved antidune with the sketch in diagram A. Scale is one foot. (C) Antidunes (smaller features near people by channel) and standing waves in the outlet channel of the San Lorenzo River in Santa Cruz (at low tide). Photograph taken on 8 November 1992.

diameter >256 millimeters) (see also Figure 4). The general term gravel clast is used in this discussion when referring to gravel particles irrespective of their size class. Gravel beaches composed of all these size ranges are present on the Central California Coast. As noted earlier, these types of beaches, in combination with sand and gravel beaches, occur along about 20% of the length of the shoreline of the outer coast in Central California (see example in Figure 26B).

Although gravel beaches are widely distributed around the shorelines of the world, they have not been the subject of nearly as much research by coastal geologists, sedimentologists, and coastal engineers as sand beaches, presumably because of the intense usage of sand beaches by recreationalists as well as their propensity to erode. The useful insights established by the classic gravel beach papers of Bluck (1967) and Orford (1975) have been embellished in recent years by numerous other studies, so maybe the gap is closing. The authors of this book have mapped and surveyed gravel beaches in numerous localities, including much of the Pacific coast of Alaska (e.g., Hayes and Michel, 1982; NOAA/RPI, 1993-2001), New England (e.g., Hayes, 1969; NOAA/RPI, 1999), and the 930 miles of the shoreline of Oman (Hayes and Baird, 1993). However, it was not until the problem of the lingering subsurface oil at the *Exxon Valdez* (1989) oil spill site evolved that it became clear to us that any coastal geologist involved in oil-spill response should have a thorough understanding of the characteristics of gravel beaches (Hayes and Michel, 1999; Michel and Hayes, 1999; Hayes, 1999).

Gravel beaches are the most difficult of all the beach types to clean and restore after an oil spill (Hayes, Michel, and Fichaut, 1990). Specific characteristics that make them so include: 1) high porosity and permeability that allow deep penetration of oil from the surface (>3 feet in some places); 2) potential for deep and rapid burial by clean sediments; 3) complex patterns of sediment reworking during storms (in contrast to sand beaches); and 4) slow rates of natural replenishment (should the contaminated sediment have to be removed during cleanup). Cleanup techniques presently employed on gravel beaches are usually disruptive to the environment, expensive, and logistically complex.
[NOTE: **Porosity** is defined as the ratio of the open pore spaces between the individual sand grains or gravel clasts to the space taken up by the sand and gravel clasts themselves (when they are stacked together). **Permeability** is a measure of the ease with which a fluid flows through the pore spaces between the sand grains or gravel clasts. During an oil spill, oil will penetrate more deeply into a gravel deposit with great permeability than

it will into a mud or very fine-grained sand deposit, which typically have much lower permeability than the gravel.]

Occurrence of Gravel Beaches

In order for waves to form a gravel beach, clearly the shoreline has to be either exposed to a significant fetch and/or subject to relatively large swell waves. However, the size of the waves is not necessarily the deciding factor. The presence of a source for the gravel is even more essential. Hayes (1967a) and Davies (1973) recognized that the common occurrence of gravel in continental shelf and shore zone sediment budgets in the higher latitudes (>40 degrees north and south) is primarily the result of **glaciation** at those latitudes. Furthermore, the shoreline classification data from the Environmental Sensitivity Index (ESI) maps produced by our team in many parts of the world verify the fact that gravel beaches are very common along coastlines that have been subject to glaciation, either under present conditions or during the ice ages of the Pleistocene epoch. Glaciers have the competence to carry huge volumes of coarse material to the coastline, where it is redistributed along the shore by wave action (Hayes et al., 1976a).

As noted earlier in the discussion of leading-edge coasts, gravel beaches are also common along rocky, mountainous leading-edge coasts, such as those on the west coast of Baja California, and the coasts of Peru and Chile. On the other hand, the shorelines on the trailing edge of the North American plate (e.g., North and South Carolina; NOAA/RPI, 1996a; 1996b) average only 2.4% gravel and mixed sand and gravel beaches (primarily shell beaches in this case).

Highly Exposed Coarse-grained Gravel Beaches

A general model of highly exposed (to wave action) coarse-grained gravel beaches based on our field experience on the open-ocean coasts of Baja California, south-central Alaska, and northern New England, as well as California, is presented in Figure 61. By coarse-grained we mean beaches made up primarily of cobbles and boulders, with a minimum of pebbles and granules (Figure 4). The most landward portion of these beaches commonly consists of an elevated, linear pyramidal-shaped mound referred to as the **storm berm**, which is composed typically of coarse pebbles and cobbles. This berm is only activated during major storms and, unless the storm is unusually severe (e.g., during a hurricane or major *El Niño* storm), the berm is built up even higher during the storm event and not eroded away as is

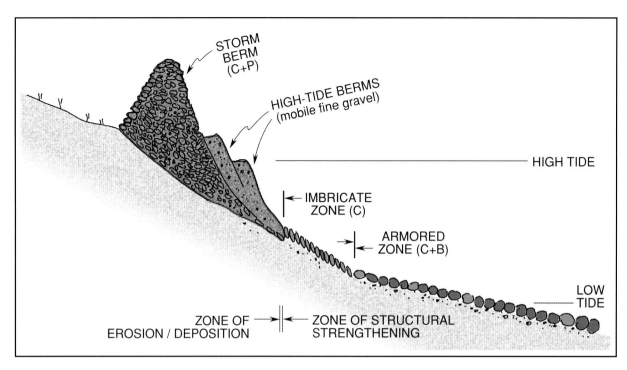

FIGURE 61. Example of gravel beach morphology. Typical topographic beach profile and sediments of an open ocean (exposed) coarse-grained gravel beach. The landward half of the profile is primarily depositional, with the storm berm being relatively stable during storms and the minor, high-tide berms eroding during storms and reforming after they pass. The seaward half of the profile is an area where structural strengthening occurs by either clast imbrication or by the formation of a surface armor of coarse gravel. B = boulders; C = cobbles; P = pebbles.

common for depositional berms on most sand beaches during storms (Shepard, 1950a; Hayes and Boothroyd, 1969).

The issue of the landward transport of the gravel clasts to form the storm berm has been discussed at some length in the literature. According to Lorang (1991; 2000), the height of the crest of the storm berm is controlled by the incident breaker height and period, combined with the gravel size. The larger the waves and the coarser the gravel, the higher the berm. Furthermore, it is generally agreed that, on a gravel beachface, the influence of infiltration of water into the sediments is more extensive during the backwash than during the uprush, considerably reducing the strength of the backwash in comparison (Packwood, 1983). This weaker **flow competency** (a term that accounts for the ability of the moving water to transport sediment) of the backwash relative to the uprush brings about a net landward transport of the gravel clasts up the beachface, stranding particles too large to be transported back down slope by the weaker backwash. Also, the loss of the volume of the backwash through percolation aids in the generation of the characteristic steep slopes of gravel beachfaces [average of around 15 degrees in Prince William Sound, Alaska (Michel and Hayes, 1991)]. In

discussing the slope of the beachfaces on sand beaches, Komar (1976) explained that, because of the discrepancy between the competence of the uprush and backwash to transport sand, the uprush moves sand onshore until a slope is built up in which gravity supports the backwash and offshore sand transport. That is, a steeper slope thus created compensates for the decreased water volume in the backwash (relative to the uprush). A similar explanation must apply to gravel beaches, which means that at some point in its evolution, the gravel beach attains an equilibrium slope considerably steeper than those on even coarse-grained sand beaches. The greater permeability of the gravel allows a greater loss of water into the beachface than would occur on a sand beach subject to the same size of waves.

Numerous papers have pointed out that disc-shaped clasts are common on gravel beaches, especially on the storm berm (e.g., Bluck, 1967; Orford, 1975). The gravel beaches we have studied show a very distinct segregation of gravel shapes along the beachfaces of storm berms – discs on the top, spheres and blades further down, and spheres and rollers at the base. The different gravel shapes – which you will no doubt see the entire spectrum of if you spend much time on the gravel beaches of Central California – are illustrated in Figure 62. This pattern develops because once the uprush of the wave has moved the pebbles or cobbles up the beach, disc-shaped ones remain preferentially up high on the storm berm for two reasons: 1) discs stay in suspension longer than the other more compact shapes because of their greater maximum projection areas (Bagnold, 1940; Folk, 1955; see also Figure 63A) and, therefore, they can be carried further landward during the uprush than more spherical (or compact) clasts of equal volume which settle quicker; and 2) discs have a lower pivotability than the other shapes, so they are less likely to roll back down slope. The terms maximum projection area and pivotability are explained graphically in Figure 63. You may wonder why we are elaborating so much on shapes of pebbles, cobbles, and boulders. Well, if you spend some time wondering up and down the beachfaces of gravel beaches during your excursions to the Central California Coast, you will be amazed how well the waves have sorted the shapes of the pebbles and cobbles during storms.

Komar and Li (1988) discussed the selective transport of sediment particles by water (called **entrainment**) of varying sizes within a sediment accumulation. A key factor determining whether a particle would be moved is its relative projection above the bed, called its *pivotability* (Figure 63B). The fact that coarser-grained sand particles composed of light minerals, such as quartz and feldspar, are readily transported over beds of placer deposits composed of finer-grained heavy minerals is well studied (Komar and Wang, 1984; Slingerland, 1977), and was discussed in some detail earlier. Sand beaches with a variety of sand sizes, such as is common for California beaches, may experience a similar size sorting, with the medium- to coarse-grained sand being preferentially transported offshore during storms, leaving behind a flat, fine-grained post-storm beach composed of finer-grained quartz as well as an increase in the abundance of heavy minerals (Hayes and Boothroyd, 1969). As Komar and Li (1988) so eloquently put it, *"first movement occurs when the fluid forces of drag and lift overcome the grain's immersed weight, pivoting it over the underlying particles so that it is rotated out of resting position."* Therefore, on a gravel bed of a storm berm composed of a mixture of Zingg's (1935) basic shapes (Figure 62) of relatively equal volumes – discs, blades, rollers, and spheres – the discs will be moved last of all the shapes by a backwash-generated current because of their lower pivotability. Therefore, once you reach the top of that storm berm on your favorite gravel beach, there should be a lot of discs up there. However, in many places on the Central California Coast, the base of an eroded cliff at the backside of the beach may only contain an ill-sorted array of coarse cobbles and boulders, rather than a well-

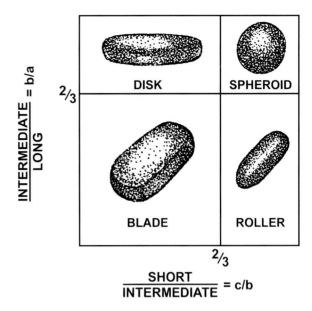

FIGURE 62. Gravel shapes (from Zingg, 1935).

sorted storm berm, because many of these clasts are deposited as piles of talus that are only rarely reworked by the waves. Also, the reflection of waves off the rock cliffs during storms may disrupt the size sorting.

Rollers or rods, with two short axes and one long axis (Zingg, 1935; Figure 62), are particularly susceptible to the downward push of the returning waters of the backwash and the pull of gravity because of their high pivotability (as well as relatively large maximum projection areas; the largest outline of the cobble or pebble, which is the portion of the clast most subject to push by running water; Figure 63A). Therefore, they tend to congregate at the base of the storm berm, assuming such a variety of clast shapes are present on the beach. Spheres will move easily down the slope as well. In some cases, of course, the shape originally inherited from the source material may not contribute such a diverse combination of shapes.

As shown by the general model in Figure 61, one or more normal high-tide berms composed of mobile fine gravel occur on the lower half of the storm berm. The highest of these high-tide berms will have been deposited during a spring tide; lower high-tide berms are deposited as the tide falls from spring to neap conditions or vice versa. During storms, high-tide berms can be eroded and redeposited over a time frame of minutes to hours. However, in some instances, the coarser gravel in these berms will be added to the storm berm proper. During the post-storm recovery period, the finer gravel returned from down the beach or from offshore is redeposited as normal high-tide berms.

Some exposed coarse-grained gravel beaches have

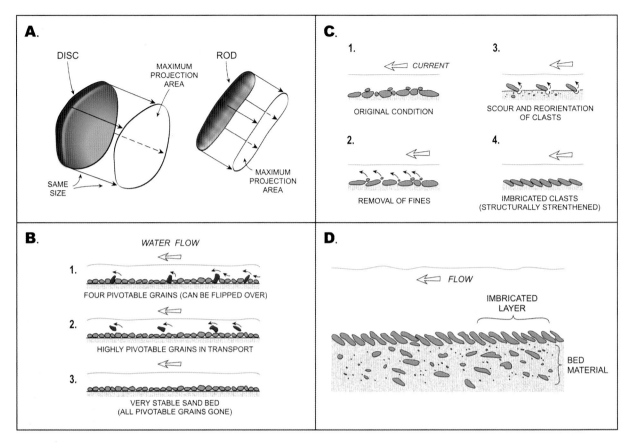

FIGURE 63. Characteristics of gravel clasts. (A) Maximum projection area. This is an important trait of a clast in that, among other things, it determines the rate of settling of the clast through the water column (e.g., the disc in this diagram would settle slower than the rod, even though they have similar maximum lengths). (B) Pivotability. If one of the clasts sticks up into the current higher than its neighbors, it may be pivoted out of its position and moved along with the current. (C and D) Imbrication. Clasts, such as disc-shaped cobbles, tend to tilt into the current (dip seaward on gravel beaches because the uprush is stronger than the backwash), their most stable configuration, because their maximum projection areas face against the force of the current.

a zone of **imbricated clasts** located at the toe of the beachface (Figures 63C and 63D). On exposed coarse-grained gravel beaches, the imbricated clasts are most commonly in the cobble size range and dip seaward at angles of around 20-25 degrees. Imbrication is best developed by platy or disc-shaped clasts, because imbrication is a process whereby the maximum projection areas of the clasts face (dip toward) the strongest force acting upon them (Figure 63A and 63D), in this case the uprush of the breaking wave, because the force of the backwash is diminished as a result of loss of some of the water volume available for the backwash by percolation into the gravel. Once imbrication occurs, the gravel on that portion of the beach becomes very stable, to be disrupted only during the strongest of storms. The role of pivotability is greatly diminished once imbrication is achieved. Currents as much as 5 to 6 times greater than that required to entrain spheres of

equal volume may be needed to move the imbricated clasts (Komar and Li, 1988). Figure 64 is a photograph of this type of imbrication on a gravel beach on the Central California Coast.

The next zone seaward of the imbricate zone is a more gently sloping, low-tide terrace, or bedrock platform in some areas, covered on the surface by a well-sorted blanket of the coarsest clasts on the beach (usually coarse cobbles and boulders). The process that forms this surface of coarser sediments, termed *armor* by fluvial hydrologists, is well studied for gravel bars in rivers (e.g., White and Day, 1982), and is commonly observed on the surfaces of braid bars on fan deltas (Hayes, 2005). Wilcock and DeTemple (2005) noted that hydrological modeling indicates that stream armor typically persists in a flood-stage environment; therefore, the armor is extremely stable.

The armoring process has not been studied in as

FIGURE 64. View looking south of a gravel beach located 7.9 miles north of Ragged Point Inn in the Big Sur Country on 6 December 2008. Many of the boulders at the toe of the beachface in the lower right are imbricated (they dip seaward), their most stable configuration (refer to Figures 61 and 63D for illustrations of imbrication). These clasts dip offshore, turning their maximum projection areas (Figure 63A) into the oncoming wave uprush, which generates the strongest current they are subjected to, because of the loss of velocity in the backwash due to percolation of the water into the beachface gravel.

much detail on beaches as it has on river bars. Because of its importance in preventing the removal of lingering subsurface oil from the *Exxon Valdez* (1989) oil spill, its presence was documented and emphasized heavily in the follow up studies by the authors of this book (Hayes and Michel, 1999; Michel and Hayes, 1991).

During our earliest studies in Prince William Sound, Alaska, we observed that armoring appears to develop best on gravel beaches that have a wide range of clast sizes, illustrating the process with the two diagrams in Figure 65. Because of the variation in wave conditions, beaches typically have constantly changing velocities of the swash-generated currents, depending upon the frequency of storm waves. Therefore, the threshold and maximum transport conditions for the particles of different sizes within the gravel can be achieved at different times repeatedly throughout the mid- to high-

tide periods, with the largest particles not susceptible to transport being left behind. Also, in the process of sediment transport, smaller particles are sometimes shielded by larger particles as illustrated in Figure 65A. These factors combine to allow intermediate-sized gravel, or fine gravel not shielded by the larger particles, to be removed from the surface of that part of the beach at velocities too low to transport the largest clasts. The overpassing of the intermediate-sized particles is enhanced as the space between the largest clasts become smaller, allowing no downward escape between the larger clasts. Therefore, an armor of the coarsest gravel available develops on top of the finer particles (Figure 65B). In Prince William Sound, the finer particles that underlie the coarse surface layer are on the order of 4 to 6 times smaller than the clasts in the armor. Armoring does not occur in the high-tide berms, where the gravel

Major Landforms of the Coast

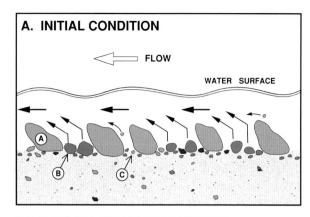

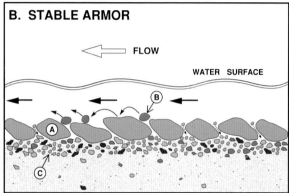

FIGURE 65. Schematic model of the processes involved in the development of an armored surface of coarse clasts on a gravel beach. (A) During storms, the largest clasts (size A) are too large to be removed by the prevailing swash-generated currents, the intermediate-sized ones (size B) are readily transportable, and some of the smallest ones (size C) are sheltered by the larger clasts and are not removed. (B) At a later stage after the surface is completely armored. With the coarser clasts so close together, the intermediate-sized grains in transit are rejected and pass over the armor. In some cases, finer particles in transit may filter between the armored clasts to become part of the substrate.

is repeatedly turned over and moved around by breaking waves at high tide. It only occurs on the flatter portions of the profile from the base of the high-tide berms (lower edge of the beachface) to the low-tide line.

Once armoring is achieved on gravel bars in rivers, a process known as *structural strengthening* occurs, such that a stronger current (at least one fourth greater) is required to transport the armored clasts (White and Day, 1982). The same type of structural strengthening no doubt occurs on beaches. In the general model given in Figure 61, the combined imbricated and armored portions of the beach profile are labeled as the *zone of structural strengthening*.

In locations where sand is available in abundance to the longshore sediment transport system, a sand zone may form seaward of the storm berm, such that during periods of maximum sand deposition, only the upper portion of the storm berm itself will be exposed. This is a fairly common beach type on the Pacific coast of Baja California and the Central California Coast (e.g., Sand Dollar Beach, California; Montello, Hayes, and Michel, 1993). Sand Dollar Beach, located near Cape San Martin in Compartment 6 (Figure 1), is discussed in considerable detail in Section III. As noted by Ed Clifton (pers. comm.), there is also a possibility that some gravel beaches may be a post-storm phenomenon. He observed some beaches north of San Diego that *were composed of sand most of the time, but developed into steep gravel beaches in the aftermath of a storm, not a storm berm as such, but the whole beachface was composed of gravel.*

LITTORAL CELLS

The term **littoral cell** is one that comes up frequently in both written and verbal communications about the sand beaches in California. It is a concept first suggested by Inman and Chamberlain (1960) as a way of defining a sand budget for the sand beaches of the Southern California Coast. As Inman (2005) noted, the boundaries of the cells *delineate the geographical area within which the budget of sediment is balanced.* The budget is "balanced" if the amount of sand leaving the cell is the same as the amount that comes into it. Obviously, if more sand is lost from the beaches than comes into the cell, beach erosion problems will most likely ensue.

Littoral cells have notably different characteristics among the diverse shorelines of the world (e.g., trailing-edge coasts, polar regions, etc.), with those on the leading edges of continental plates, such as the ones in California, having the following defining characteristics (Griggs, 1986; Inman, 2005; Griggs and Patsch, 2006):

1) The so-called updrift, or up coast, side of the cell, the direction from which the sand comes, is typically composed of a **rock headland** around which very little sand passes to the beaches located downdrift, or down coast, from it. In California, the rock headland that blocks sand moving along the shore from entering a littoral cell is usually located at the northern side of the cell.

2) The sand that reaches the beaches located within one particular littoral cell may be derived from streams (dominantly in most cases), eroding rock cliffs, and eroding bluffs composed of less resistant material than

rocks (and of course by artificial beach nourishment, but that is a relatively new entry into the equation).

3) Once the sand is on the beach, it moves alongshore in response to prevailing wave conditions by the process of **longshore sediment transport** illustrated in Figure 37. The prevailing direction of sand transport appears to be north to south in most cells on the California coast, except for the Santa Barbara cell, where it is predominantly west to east.

4) The sand within a cell can be removed from it primarily by one or more of three natural processes: a) funneled down submarine canyons; b) blown landward into dune fields; and c) carried so far out onto the continental shelf during *El Niño* storms that it cannot be returned by normal onshore sediment transport. Some sand might also be lost into major bay/estuarine systems, but their numbers are too small on the Central California Coast for them to be significant sediment sinks (engineering term for the catchment area into which the sediment is lost).

Figure 66 shows a general model for the most common type of littoral cell on the California Coast.

According to Inman (2005), *"the littoral cell now plays an important role in the U.S. National Environmental Protection Act (1974) and the California Environmental Quality Act (1974), and it has become a necessary component of environmental impact studies. In the realm of public policy and jurisdictions, the littoral cell concept has led to joint-power legislation that enables municipalities within a littoral cell to act as a unit (Inman and Masters, 1994)."*

These legislative acts no doubt have occurred because an understanding of littoral cells has substantial practical applications. For example, Griggs (1986) pointed out the following important consequences related to the location of marinas and harbors within a littoral cell:

1) Those built either between cells or at the updrift end of a cell (more northerly end in most cases) *have been relatively maintenance free,* because of the small amounts of sand brought into those areas by the longshore sediment transport system.

2) Those built in the middle reaches or the most down coast end of the cells *have had expensive dredging problems,* because of so much sand being transported into the harbors and marinas. The classic case of this problem is the Santa Barbara Harbor, where the annual dredging rate averaged around 300,000 cubic yards of sand a year between 1963 and 2001 (Griggs and Patsch, 2002).

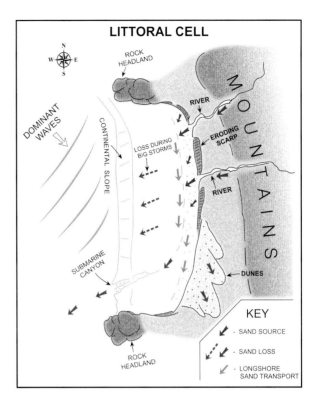

FIGURE 66. General model of a littoral cell, a concept first suggested by Inman and Chamberlain (1960) as a way of defining a sand budget for the sand beaches of Southern California. Sand is delivered to the cell by rivers and streams or by the erosion of shoreline bluffs. It is lost down submarine canyons, into coastal sand dunes, and possibly, in some cases, onto the outer reaches of the continental shelf during storms.

An in-depth treatment of the subject of littoral cells is given by Patsch and Griggs (2006). In that publication, they gave budget numbers for ten of California's littoral cells. As an example, for the Santa Cruz cell, the total sand supply for a year averages 223,000 cubic yards. Of that, 85% is brought in by rivers and 15% is derived from bluff erosion. However, anthropogenic impacts to the sand supply have recently reduced the annual volume by 14,000 cu yards, 6,000 because of dams on rivers and 8,000 because the base of some of the bluffs have been armored with seawalls and/or riprap. This is a relatively small number, but the cell in southern Monterey Bay has a deficit of 33% of the amount of its original annual supply (a whopping 237,000 cubic yards a year deficit), because dams have been built on the rivers that originally brought in the sand.

There are a number of additional littoral cells on the Central California Coast, some of which are better defined than others.

EASILY RECOGNIZABLE, REPETITIVE SHORELINE FEATURES

Crenulate Bays

The California coastline contains a number of morphological features known as **crenulate bays**, which are illustrated in Figure 67A. With a few exceptions, most of these crenulate bays are oriented north-south on the California Coast. They typically have a fishhook shape, with the shank of the hook pointing in a southwesterly direction in response to the dominant northwesterly approach of the larger waves and swell along this coast. Three of the most prominent ones on the Central California Coast are Bodega Bay, Drakes Bay, and Half Moon Bay. There are several others that are discussed in Section III. The sandy beaches in the crenulate bays are most commonly located between two segments of rocky shoreline. The shape of the bay is the result of refraction/diffraction of the dominant northwest waves around a rock headland area located north of the bay (in California). This general pattern of wave-front bending

is given in Figure 67A, and the calculated refraction/diffraction pattern of swell approaching Drakes Bay, which has a rare east-west orientation, from the west is shown in Figure 67B. Note that the wave-front crests of the refracted/diffracted, nearshore swell waves in Drakes Bay are almost parallel to the shoreline. The waves arriving at the shoreline refract and diffract into the **shadow zone** in the lee of the up coast bedrock headland, creating longshore currents that remove sediment from that zone and transport it toward the opposite end of the bay (except for some reversals, minor in some bays and major in others, in which the finer-grained sand moves back toward the updrift headland), eventually producing a stable equilibrium beach with a crenulate, or fishhook, shape.

Two of the large crenulate bays in California, Drakes Bay and Half Moon Bay, were the subject of some of the earliest studies on crenulate bays. For example, Krumbein (1944) found an increase in beachface slope and grain size from the shadow zone to the **tangential end** (far end of the bay away from the up coast rock headland) of Half Moon Bay. He also concluded that some of the

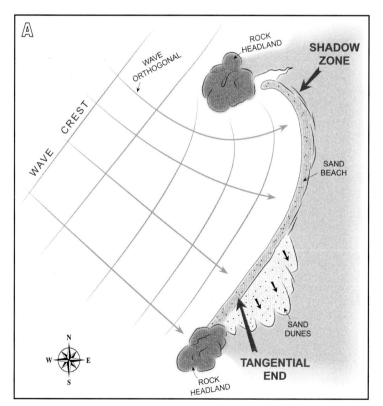
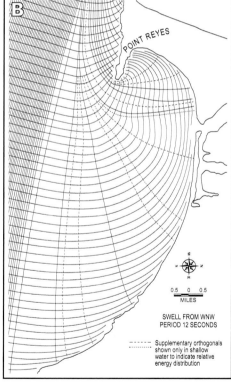

FIGURE 67. The crenulate bay. (A) General model. On the California Coast, the shape and orientation of the bay is the result of the bending (refraction/diffraction) of the dominant northwest waves and swell around a rocky headland located to the north of the bay in many areas. (B) Wave refraction/diffraction pattern for Drakes Bay in Marin County, which has an east-west orientation, with the dominant waves approaching the bay from the west (after Cherry, 1966). Note that north is to the right in diagram B.

finer-grained sand was transported into the shadow zone. Bascom (1951), who also studied Half Moon Bay, concluded that the grain size of the sand, beachface slope, and "wave energy" increased from the shadow zone to the tangential end. Trask (1959) found a net longshore transport toward the tangential end of Drakes Bay, with some minor reversals toward the updrift rock headland, somewhat similar to the results of Krumbein's study of Half Moon Bay. On the other hand, Cherry (1965), who also studied Drakes Bay, concluded that little net longshore movement of sand occurs in the bay and that the beaches are "generally in equilibrium."

Despite the somewhat minor discrepancies among the aforementioned California studies, Finkelstein (1977; another co-author Hayes student), who studied the crenulate bays on Kodiak Island, Alaska, was able to make the following generalizations about the most important aspects of crenulate bays (based to some extent on the pioneering work of Silvester, 1974):

1) The shape of the bay is controlled by the predominant wave approach direction, with erosion in the protected shadow zone in the lee of a bedrock headland at the entrance to the bay (in the early stages of its developmental history). Meanwhile, deposition takes place at the long straight shank of the fishhook, the tangential end of the bay.

2) These bays may eventually develop a stable morphology that results in the wave crests breaking more-or-less simultaneously around the bay with no net erosion/deposition or longshore sand transport occurring [as Cherry (1965) apparently found in his study of Drakes Bay]. Of course, this "equilibrium" may be upset by a major storm and it could take years for the bay to return to its original, more stable condition, which may account for some of the minor differences in the California studies cited earlier. A major disturbance of the general, long-term trend would occur if a large *El Niño* storm, or series of *El Niño* storms, were to generate waves that approach the bay from the south or southwest, negating the dominant wave-refraction/diffraction pattern of the northwest approaching waves responsible for the original formation of the bay and, therefore, completely upsetting the "equilibrium" pattern that took so long to evolve (this latter comment added by us).

3) As a generalization, the shadow zone has smaller waves, smaller grain size, gentle beachface slopes, and a generally eroding shoreline relative to the rest of the bay. In some bays, the finer sand is transported back toward the head of the bay forming long, northwest projecting spits (e.g., at Bodega Bay and San Diego Harbor). Whether or not these particular spits obtained their impetus (and major spurts of growth) during *El Niño* storms would appear to be a legitimate question. Nonetheless, the beach at the center of the bay is characterized by bigger waves, larger grain size of the sediments (than the sand in the shadow zone), moderate beachface slopes, and some fluctuations between erosion and deposition. The tangential end, which is relatively straight and stable, has somewhat smaller waves than the central area (in some cases, depending on the size of the bay and the dominant refraction/diffraction pattern), the beachface slopes are steeper, and the shoreline is depositional overall.

Crenulate bays are present all around the world on coastlines with patchy bedrock distribution, and they are one of the most definitive clues of long-term sediment transport directions that exist. When we did the overview study of the leading-edge coasts of North and South America, we saw literally hundreds of them while studying the Google Earth images along those shorelines. One striking contrast we observed was the difference in the orientation of the crenulate bays in the northern hemisphere, particularly those under the influence of the dominant northwesterly winds, from those in the southern hemisphere, which are mostly under the influence of dominant southwesterly winds. The shanks of the northern hemisphere bays point toward the southwest, and the shanks of the southern hemisphere bays point toward the northwest, a very striking pattern, one of the most arresting trends we saw during that survey.

Tidal Inlets

On a shoreline dominated by barrier islands, like those in North Carolina, South Carolina, Georgia, Florida, and Texas, **tidal inlets** are usually defined as major tidal channels that intersect barrier islands, or separate two barrier islands, usually to depths of tens of feet (Hayes, 2009). The tidal inlets are the focus of the most dynamic changes that occur on any component of these barrier-island systems. Because tidal inlets connect the open ocean with more sheltered backbarrier settings, including some potential harbor sites, engineers have long been engaged in efforts to control them, commonly by constricting their migration with a set of two parallel jetties.

When a storm surge floods a barrier island or barrier spit, water flowing through breaks in the dunes is capable of scouring channels across the island or spit. This is one mechanism by which many new tidal inlets are formed in barrier islands and barrier spits (Hayes,

1965, 1967b; Shepard and Wanless, 1971); however, most of these new channels are closed off by wave-transported and wind-blown sand within a few weeks after the storm. Needless to say, some survive if they are deep enough and other conditions, such as altered tidal current patterns, promote the longevity of the inlet.

The ultimate morphology, or shape, of a tidal inlet through a barrier spit is the product of the constant interaction, or contest if you will, between tidal and wave-generated forces. Longshore currents and beach drift generated by wave action move sediment along shore in such volumes that would normally fill the inlet in the absence of tidal currents. Sediment carried into the inlet throat (see general model in Figure 68A) by the wave-generated currents is constantly swept away by tidal currents that scour the deep inlet throat. Sediment transported from the scoured-out throat of the inlet by flood currents may be deposited as a lobe of sediment on the landward side of the inlet called the **flood-tidal delta**, and that carried seaward by the ebb currents may form a seaward lobe called the **ebb-tidal delta**.

The general model of tidal inlets given in Figure 68A, which shows relatively equal volumes of sediment in the two tidal deltas, is rare in nature. More typically, in microtidal, wave-dominated areas like the Texas Coast, the flood-tidal delta is much larger than the ebb-tidal delta, because of the influence of wave action in both moving sediment to the inside of the inlet and in erosion of the outer lobe (among other factors). On mesotidal, more tidally influenced coasts like those in South Carolina and Georgia, ebb-tidal deltas tend to be much larger than flood-tidal deltas. In fact, flood-tidal deltas are almost non-existent in those two states. This strange phenomenon (i.e., missing flood-tidal deltas) has been studied in some detail by FitzGerald, Nummedal, and Kana (1976), and FitzGerald (1977). In the simplest terms, because of the complex nature of the backbarrier regions of the barrier islands in South Carolina and Georgia (extensive marshes and tidal flats cut by numerous tidal channels), water "running uphill" into these areas during the flood portion of the tidal cycle takes longer to fill the backbarrier region with water at high tide than it takes that region to empty when the tide falls. This means that the flood phase is considerably longer than the ebb phase, which results in the ebb currents being stronger than the flood currents. Any fresh water that a river or rivers may bring into the estuary/bay system landward of the tidal inlets may also add some impetus to the ebb flow. However, even with essentially the same amount of water going into the inlet as comes out during a single tidal cycle, much stronger currents are generated during the shorter time of the

falling tide on the ebb. These strong ebb currents are able to sweep sand brought into the inlet by the wave-generated and flood-tidal currents out of the inlet and deposit it on the ebb-tidal delta. The large ebb-tidal deltas on the South Carolina coast contain huge volumes of unconsolidated sand (77% of the total volume on the coastline; Hayes, Sexton, and Sipple, 1994; Sexton and Hayes, 1996; Hayes and Michel, 2008).

A general model of ebb-tidal deltas is given in Figure 69. The morphological components of ebb-tidal deltas include a main ebb channel (with slightly stronger ebb currents than flood currents) flanked by linear sand bars on both sides and a terminal lobe at the seaward end. The main ebb channel is bordered by a platform of sand (swash platform) dominated by swash bars (intertidal bars built up by wave action). The swash platform is separated from adjacent barrier beaches by marginal flood channels (which are dominated by flood currents).

As fate would have it, the largest ebb-tidal delta in the world, as far as we know, occurs on the Central California Coast. This feature, parts of which are known as Four Fathom Bank and Four Fathom Shoal, is situated just outside the entrance to San Francisco Bay. As shown by the diagram in Figure 70, the general configuration of the ebb-tidal delta off the Golden Gate Bridge is strikingly similar to the general model given in Figure 69. This is not a big surprise, because the ebb-tidal model is quite consistent around the world, especially in areas with moderate-sized tides (mesotidal; tidal range between 6 and 12 feet), such as San Francisco Bay, which is mesotidal during spring tides. Also, there is no flood-tidal delta present inside the Bay, probably for the same reason as the one just provided for their absence in South Carolina and Georgia. The main ebb channel, the outer entrance to San Francisco Bay in this case, is floored by large waves of sand created by the strong currents that flow in and out of the Bay (Figures 32 and 71).

In this discussion, we will broaden the definition of tidal inlets a bit to include any channel-like opening that connects the sea with the mainland. Of the 170 tidal inlets (thus defined) that are present on the California Coast, the most common inlets are the ones where a relatively small stream mouth and its associated valley, meets the sea.

These 170 tidal inlets are focal points for designing strategies to protect the vital resources of the state's coastal water bodies from oil spills (e.g., salt marshes, key bird feeding areas, etc.), because it is through such conduits that oil spilled on open ocean waters could reach these resources. Consequently, the Marine Spill Response Corporation (MSRC) and Office of Oil Spill

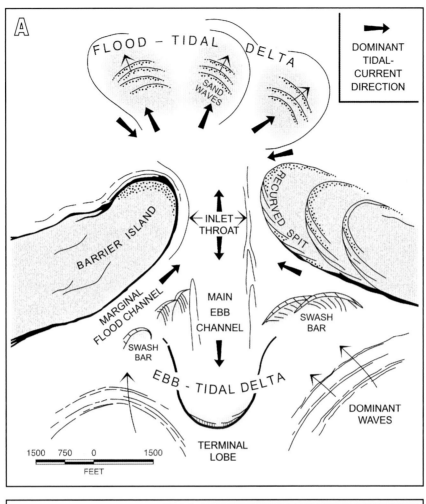

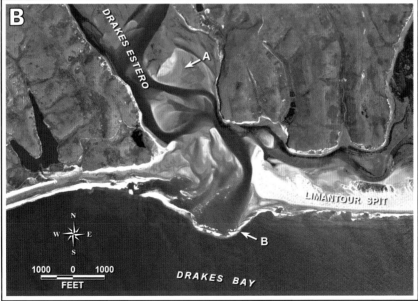

FIGURE 68. Morphology of tidal inlets. (A) General model. (B) Drakes Estero Inlet, located in Marin County, which conforms fairly closely to the general model. Arrow A points to the flood-tidal delta and arrow B points to the ebb-tidal delta. NAIP 2005 imagery courtesy of the California Spatial Information Library (CaSIL).

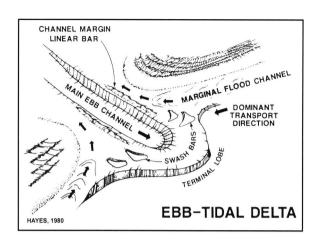

FIGURE 69. Typical morphology of ebb-tidal deltas in mesotidal settings (after Hayes, 1980). Arrows indicate dominant direction of tidal currents. This general model was derived for the ebb-tidal deltas of the New England area, but it applies equally well to tidal inlets around the world.

Prevention and Response (OSPR) of the California Department of Fish and Game hired our company, RPI, to design protection strategies (oil boom configurations, sand dikes, etc.) for each of the inlets. This project was completed in 1993, with Volumes I and II of the final reports (Hayes and Montello, 1993a and 1993b) covering all of the inlets on the Central California Coast.

According to our newly proposed definition, there are seven types of tidal inlets in California. There are a number of inlets in barrier-spit systems that fit the morphological pattern based on the barrier island shorelines on the east coast that is illustrated in Figure 68A. The entrance to Drakes Estero in Marin County, which has well developed flood- and ebb-tidal deltas, is a fairly close fit (see Figure 68B). The Tijuana Estuary Inlet in San Diego County and Eel River Inlet in Humboldt County also match the general model.

However, California's coast is dominated by bedrock shores, not barrier islands, primarily because of its location on the tectonically active, western shoreline of the North American Plate. This fact accounts for a relatively wide variety of other types of tidal inlets in California, that include:

1) **Entrances to flooded major alluvial valleys.** During periods of increased ice formation during the Pleistocene epoch, sea level was lowered and stream valleys were carved to near that level. When sea level started to rise, as it did at the end of the last glaciation about 12,000 years ago, the valleys were flooded and estuaries were produced. Where the valleys were narrow, because of underlying rocks resistant to erosion, narrow

entrances to the estuaries were formed. The best example of this type of inlet in California is the entrance to San Francisco Bay at the Golden Gate (Figures 32, 70, and 71). And, for the record, this is the most difficult tidal inlet in all of California to deal with regarding potential protection strategies against oil incursion during a major oil spill, because of its strong currents, large size, and exposure, at times, to the large waves of the open ocean. Co-author Michel spent many days in the Bay Area as part of the NOAA team providing scientific support to the U.S. Coast Guard during the *Cosco Busan* oil spill in 2007. This 53,000 gallon spill of heavy fuel oil resulted in oiling of 64 miles of rocky shoreline, 46 miles of sand beaches, and 10 miles of salt marshes, from Point Reyes to Half Moon Bay, including Central San Francisco Bay.

2) **Stream-mouth/bar systems.** By far the majority of tidal inlets in California occur at the mouths of somewhat small, relatively steep-gradient streams, with 54 inlets of this type occurring on the Central Coast. The stream mouth usually has a sand or sand and gravel spit built across it. The spit, which is typically a few tens of yards wide, is frequently overwashed during high spring tides and storms, and it may be removed or highly modified during the larger floods on the stream. Minor wetlands commonly occur along the stream channel landward of the spit. The frequently visited inlet at Rodeo Cove in Marin County is a good example of this type of inlet (Figure 72). Some of the larger stream-mouth/bar systems may stay open to the ocean for a significant period of time during some rainy winter seasons (e.g., the mouth of the Salinas River in Monterey County). Anadromous fish, such as steelhead trout, depend on such winter breeching for access to upstream spawning grounds.

[NOTE: The term **spit** is defined as a linear projection of sediment across an embayment, stream outlet, and so on that builds in the direction the longshore transport system is moving the sediment. Some spits grow in spurts, with the welding beach berm on the downdrift end curving most of the way around the extension. After a period of little growth, the next spit that follows projects beyond the end of the previous one, creating a series of the recurving spit ends. This type of spit is called a recurved spit, but most of the spits in the stream-mouth/bar systems in California are straight rather than recurved (see example in Figure 72).]

3) **"Half inlets."** Because of the omnipresent bedrock on the Central California Coast, some of the tidal inlets abut eroding bedrock scarps. In situations where the inlet is located in an asymmetric valley, and there is room opposite the eroded scarp for some of the morphological components of the inlet (such as at least part of an ebb-

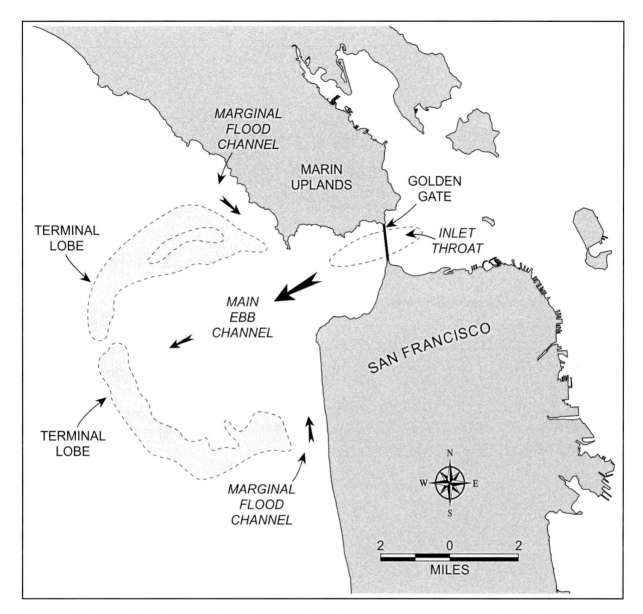

FIGURE 70. The ebb-tidal delta outside of the Golden Gate Bridge. Compare it with the general model of ebb-tidal deltas presented in Figures 68A and 69.

tidal delta) to form, it may take on the configuration of the "half inlet," with the scarp bisecting the typical inlet morphology. An example of this type of inlet in Central California is Tomales Bay Inlet in Marin County (Figure 73). This inlet has a moderately well developed "half" of a flood-tidal delta with a long sand beach on the northeast side (right arm of the inlet), but the left (southwest) side of the inlet is the rocky Tomales Point extension of the Inverness Ridge, so half of the inlet, according to the general model in Figure 68A, is missing.

4) **Fault zone inlets.** Some of the coastal water bodies and their associated tidal inlets are located in valleys along fault zones. Examples in Central California include the already discussed Bolinas Lagoon Inlet (Marin County; Figure 23) and Bodega Harbor Inlet (Sonoma County). Both of these inlets are located between rock headlands and long sandy spits in the "shadow zones" of crenulate bays. The spits are elongated in a westerly direction under the influence of up coast (northwest to west) longshore sediment transport created by waves refracted/diffracted into the crenulate bays, or possibly during *El Niño* storms (as noted earlier). The entrance to Tomales Bay is also a classic fault zone inlet, as well as a good example of a "half inlet."

5) **Crenulate bay inlets.** Because of the refraction and diffraction of the dominant northwest approaching

97

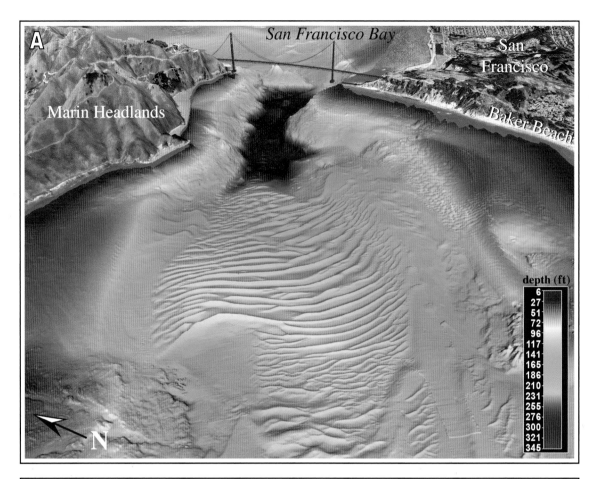

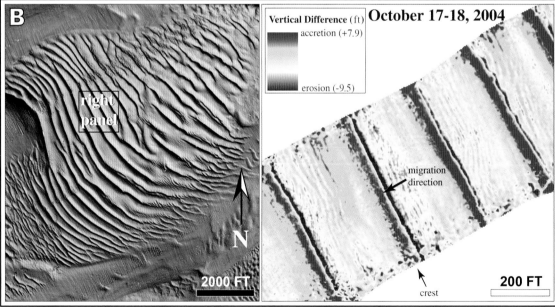

FIGURE 71. Sand waves in the main ebb channel of the Golden Gate ebb-tidal delta. (A) View along of the main ebb channel (see Figure 70) looking toward the Golden Gate. This color-coded bathymetric map shows large ebb-oriented sand waves (i.e., formed by the ebb current) that cover the bottom of the channel. The inlet throat is shown in dark blue. (B) Detailed views of the big sand waves (wave length over 200 feet) on the bottom of the channel. The blue zones are the accreting slip faces on the sand waves as they advance toward the southwest. Images courtesy of Barnard et al. (2006).

FIGURE 72. The Marin Headlands area. The northward oriented, coarse-grained spit (Rodeo Beach), composed of a mixture of granules, pebbles, and sand, that closes off Rodeo Lagoon is only breached during storms. NAIP 2005 imagery courtesy of the California Spatial Information Library (CaSIL).

waves on the California coast, the lee areas south of the rocky headland at the head of a crenulate bay is an obvious place to locate harbors, creating in some cases a man-made tidal inlet (e.g., Pillar Point Harbor at the head of Half Moon Bay in San Mateo County; Figure 46B). Silver Strand Beach and its accompanying tidal inlet in the lee of Point Loma at San Diego Harbor in San Diego County is another example (in addition to the ones at Bolinas Lagoon and Bodega Harbor) of a sand spit growing toward the northwest (in the lee of the rock headland) and closing off a coastal water body that develops its own natural tidal inlet.

6) **Jettied inlets.** A few of the originally natural inlets on the Central California Coast have been stabilized with jetties, which are defined as two parallel, linear structures extending into a body of water (in this case

FIGURE 73. "Half Inlet" at the entrance to Tomales Bay (Marin County). Arrow points to the well-developed flood-tidal delta. NAIP 2005 imagery courtesy of the California Spatial Information Library (CaSIL).

the Pacific Ocean) that are designed to provide access to a harbor or marina. They are usually composed of large pieces of rock of boulder size called riprap. Two prominent examples on this coastline are the ones that stabilize the inlet at Moss Landing and the entrance to Morro Bay.

As we were describing the geomorphological characteristics of the 54 tidal inlets on the Central California Coast classified as *stream-mouth/bar systems,* we were surprised to learn that 27 of them have spit forms built across the stream mouths that had been formed by longshore sediment transport to the north. That is, at these inlets a spit form projects across the stream mouth with the stream's outlet being pushed up against the northerly side of the valley. Prograding spits like these are one of the geomorphological clues often used to determine dominant longshore sediment transport directions on a coastline.

The stream outlet at Rodeo Cove in Marin County illustrates this northerly transport trend (Figure 72). Two other excellent examples occur at the entrances to Estero de San Antonio and Estero Americano in Marin County (discussed and illustrated in Section III). The spits at only six of the remaining inlets showed a definitive southerly orientation. Twenty one of the inlets showed no clear-cut trend in either direction, because of their small size, more central location of the stream outlet across the spit, and so on. The problem with this observation is that several lines of evidence indicate that

the prevailing sand transport direction on this coastline is from north to south, including:

1) Prevailing wind and wave approach directions are from the northwest (Figure 41).

2) The standard thinking, based on some pretty good data, is that the sand transport direction in the littoral cells on the California coast is mainly from north to south (Figure 66).

3) The orientation of the sand dunes in several coastal dune fields clearly shows that they are created by northwest winds.

4) The orientation and shape of the crenulate bays (Figure 67).

Yet, despite all this, half of the spits across the entrances of this type of tidal inlet on the Central California Coast indicate sediment transport to the north, with only eleven percent of them showing a southerly trend.

Some important clues that might help explain the northerly migration of the spits follow:

1) All but three of the northerly oriented spits are composed of coarse-grained sediment (very coarse sand, granules, and pebbles with even some cobbles at a couple of sites).

2) The three northerly oriented sand spits are located in large crenulate bays, where large-scale wave refraction/diffraction plays a role in spit orientation (this is not counting the two associated with Bodega Harbor and Bolinas Lagoon).

3) All of the southerly oriented spits are composed of sand.

4) At some of the inlets, offshore rocks are causing local inversions of wave-approach directions which, no doubt, contribute at least in some degree to the "reversal" in sediment transport direction (back toward the north). This is clearly the case at the Arroyo de la Cruz Inlet in San Luis Obispo County, which is illustrated in Figure 74.

Here is our working hypothesis. The northerly oriented spits, which are universally coarse-grained (except those in the large crenulate bays), are activated only by the large and widely-spaced (in time) *El Niño* storms which, in many cases, generate waves that approach the coast from the west or southwest, causing a pulse of northerly transport of the gravel-dominated sediment resulting in the activated spits forcing the stream outlet to the north. Also, the heavy rains that normally accompany such storms may have generated a flood on the stream that eroded away at least part of the former spit. As the flood receded, the waves approaching

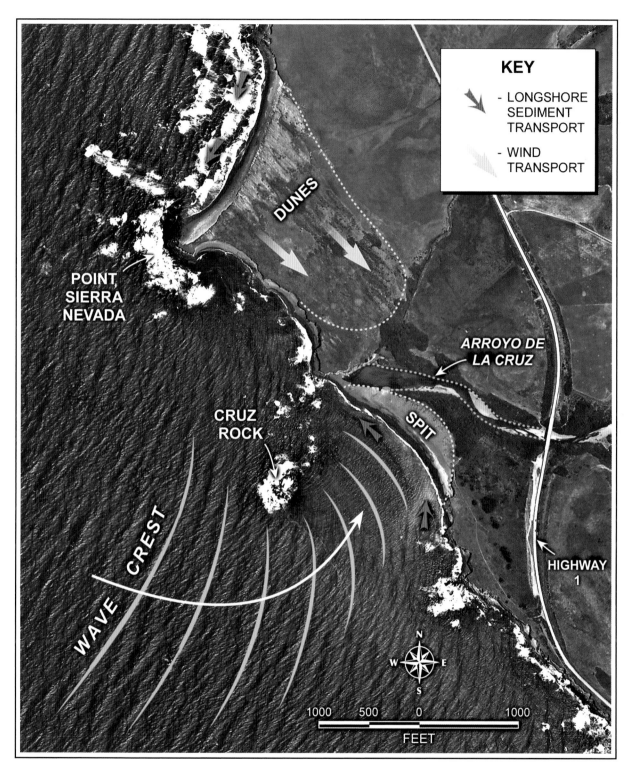

FIGURE 74. The dynamic coastal processes in the vicinity of the Arroyo de la Cruz Inlet (San Luis Obispo County). The key elements include: 1) The natural groin effect of the bedrock of Point Sierra Nevada, which has created a sand beach facing directly into the dominant northwesterly waves; 2) A dune field that extends to the southeast away from the beach created by the headland; 3) Wave refraction/diffraction around the Cruz Rock topographic high that causes a reversal of sediment transport direction (toward the north); and 4) A northward oriented spit composed of coarse-grained sediments, including some pebbles, that closes off the mouth of Arroyo de la Cruz, a moderate-sized stream that drains the southern margin of the Santa Lucia Mountains. NAIP 2005 imagery courtesy of the California Spatial Information Library (CaSIL).

101

the inlet from the south and west could have restored the spit in short order. As discussed elsewhere, the coarser-grained beach profiles we measured during our beach surveys on the California coast in 1992-93 showed very little seasonal changes in response to the "normal" winter storms (see Figure 53A). The prevalent waves from the northwest are most effective in transporting sand rather than gravel. Therefore, we conclude that the six southerly oriented sand spits are created by the less energetic northwesterly waves. Any different ideas you may have would be welcome at this time.

The tidal inlets on the Central California Coast are pointed out and discussed in some detail where appropriate in the treatment of the different coastal regions in Section III.

Rocky Headlands

Some of the most impressive features on this coastline are the imposing rock headlands that play a prominent role in: a) segregating the littoral cells; b) impacting wave-refraction/diffraction patterns; c) providing roosting and breeding areas for millions of seabirds; and d) furnishing haulout areas for marine mammals, not to mention providing outstanding viewing areas for all of the above. The primary reason why these headlands form is the fact that they are underlain by rock formations that are more resistant to erosion than those in neighboring areas. For example, Bodega Head, Point Reyes, and the headland at Devils Slide are underlain by a substantial amount of granitic rocks, which are much more resistant to erosion than most sedimentary and metamorphic rocks because of the abundance of tightly "interwoven," crystallized associations of relatively hard minerals, namely quartz and feldspar, that they contain, as well as their typically more massive character.

These rocky headlands show a wide range of sizes and shapes. All of the major ones are described and briefly discussed, with accompanying photographs presented for some of them, in Section III.

Natural Groin/Sand Beach Systems

A man-made **groin** is a shore protection structure built perpendicular to the shoreline intended to trap sand being carried along the coast in the longshore sediment transport system by the processes illustrated in Figure 37, its purpose being to retard shoreline erosion. A natural, rocky protrusion out into the longshore sediment transport system can serve the same function, that is, trap sand and form a relatively permanent sandy beach. We call such rock barriers **natural groins**. There

are a number of these natural groins along the Central California Coast, some forming very short beaches and some that create beaches that extend several miles. Three of the larger ones include: a) Point Reyes, which traps the sand now present on Point Reyes Beach; b) Point Sur, which traps a wide coarse-sand beach on its northern side (see Figure 52A); and c) Franklin Point in Año Nuevo State Preserve. In some situations, as at the southern end of San Luis Obispo Bay, with its natural groin at Seal Rocks, and the next bay to the south, with its natural groin system at Purisimo Point, the natural groins are located at the southwest ends of long beaches oriented perpendicular to the direction of the dominant northwest winds. These two natural groin/sand beach systems are described in more detail in Section III.

Coastal Dune Fields

Wind-blown sand dunes are most commonly present on the Central California Coast where there is a length of shoreline oriented perpendicular to the dominant onshore wind direction (the northwest). Also required is a large supply of sand moving along the shoreline and a topographic area landward of the beach that is relatively low. Where the beach has this orientation, the rate of longshore sand transport is slowed down compared to beaches with a more north/south orientation, thus excess sand may accumulate on these beaches. In this case, some of the surplus sand is available to be transported landward if the backshore topography is low enough, creating wide dune fields in some areas.

Most of the sand transported by the wind into the dune fields is moved by a process called **saltation**, whereby the sand grains bounce off the sand bed and "fly" for some distance at heights usually not more than a couple of feet (illustrated in Figure 75). Some of the grains roll slowly along the ground (a process called

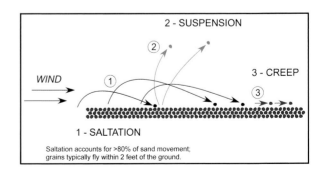

FIGURE 75. The mechanisms by which sand grains are transported by the wind – saltation, suspension, and creep. From Guadalupe Restoration Project – Dunes Center Manual (2008).

creep).

The four morphologic types of dunes that occur in these dune fields – transverse, blowout (usually lobate), barchan, and parabolic – are sketched in Figure 76 (see also detailed definitions in the glossary). Photographs of some of these dune types are presented in Section III. Where the dunes are vegetated, blow-out dunes, or possibly more commonly parabolic dunes, are the most usual type. In fact, parabolic dunes are probably the most common dune type on this coastline. The other three types are present depending on variables such as sand supply, consistency of the wind, viability of the vegetation, and so on. And, finally, these dune fields are probably the second most important mechanism by which beach sand is lost from the littoral cell, after loss to submarine canyons (Figure 66).

A classic book by W.S. Cooper titled *Coastal Dunes of California* was published by the Geological Society of America in 1967. In it, he presented maps of all the significant dune fields on the whole California Coast. The dune fields he mapped on the Central California Coast, all of which are discussed in some detail in Section III, include:

1) Compartments 1, 2, and 3 (Figure 1) – Bodega Head, Dillon Beach, Point Reyes, San Francisco, and Laguna Salada.

2) Compartments 4, 5, and 6 (Figure 1) – Año Nuevo, Monterey Bay, Monterey Peninsula, Point Sur, Arroyo de la Cruz (Figure 74), and Point Piedras Blancas.

3) Compartment 7 (Figure 1) – Morro Bay, Santa Maria River, Purisima Point, Santa Ynez River, and Point Conception.

The largest of the dune fields are present in San Luis Obispo and Santa Barbara Counties (Compartment 7; Figure 1), where rock headlands, such as Mussel Rock/ Point Sal, Purisima Point, and Point Pedernales, have served as natural groins with the resulting creation of

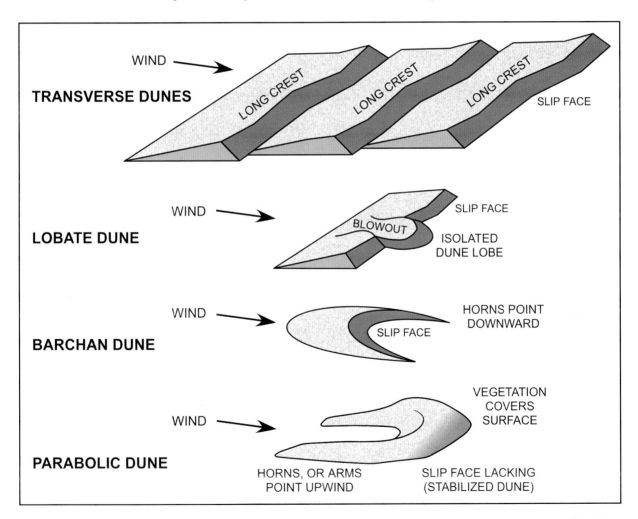

FIGURE 76. Four sand dune types typically present in the coastal dunes of the Central California Coast. From Guadalupe Restoration Project – Dunes Center Manual (2008).

updrift sand beaches fed historically by a number of rivers, such as the Santa Maria and Santa Ynez Rivers. These beaches, which are long and straight, face directly into the northwest winds. The dune fields in this area are illustrated in Figure 77, which is based on a combination of the work by Cooper (1967), Orme (2005), and Orme and Tchakerian (1989). The map in Figure 77, which is considerably simplified from the original, shows three phases of dune formation:

1) The oldest dunes shown on the map, although there are some older ones still in evidence, are called Late Pleistocene dunes. These are extensive dune fields that formed during the ice ages, with the detailed unraveling of their history being beyond the scope of this book. The field of Late Pleistocene dunes on the south flank of the Santa Maria Valley extends for over 12 miles in a downwind direction (toward the southeast).

2) The next oldest dunes, termed here as **older modern (or older Holocene) dunes**, were formed after sea level started to rise following the latest glaciation. Cooper referred to these dunes as Episode I dunes and Orme called them older Flandrian dunes. During this time interval, sea level rose from as much as 350 feet below its present level, starting to rise sometime between 16,000 and 12,000 years ago, and reaching its present level about 5,000 years ago. These older modern, or older Holocene, dunes are probably composed of at least some sand brought to the shoreline when it was located as far offshore as the edge of the modern continental shelf. Some of that sand advanced landward in dune fields spread across the exposed former continental shelf as sea level rose, with some of the sand reaching the level of the present shoreline. According to Cooper (1967), these dunes are at present mostly stabilized (by the dune shrub community). He also pointed out that there are some cases where these older modern dunes are perched on top of a presently eroding wave-cut scarp and, "there being no receptive shore" (scarp too high and steep), the younger modern dunes cannot form there. An example of this occurs between Mussel Rock and Point Sal in Santa Barbara County.
[NOTE: The term Flandrian refers to the segment of geological time that has occurred since the end of the last glaciation (also called the present interglacial or Holocene time). Flandrian time has continued to the present (as defined by some geologists).]

3) The youngest of the dunes are here termed **younger modern (or younger Holocene) dunes**. Cooper referred to these dunes as Episode II dunes and Orme called them younger Flandrian dunes. Once sea level stabilized, sand was added to those original dunes as rivers continued to feed sand into the longshore sediment transport system and the northwest winds blew some of this new sand onshore. The more modern dunes are active, and bearing little vegetation, they extend away from the beach, invading the masses of older Holocene dunes if they are present.

Figure 77 is presented here to acquaint you with the magnitude of the dune fields along the southern shore of the Central Coast. Some of these dune fields are spectacular natural features, definitely worth a visit.
[NOTE: Some of the information on sand dunes presented in this section and in Section III is based on the Guadalupe Restoration Project – Dunes Center Manual (2008) carried out in the Guadalupe Oil Field at the Nipomo Dunes. Although we did not participate in the production of that particular report, co-author Hayes was part of a NOAA-sponsored field team that carried out research in the summer of 1994 on both the dunes and the beach sand in the Guadalupe Oil Field while investigating a large spill of diluent from pipelines through the dunes. That field study resulted in a follow-up memorandum by the co-authors – Hayes and Michel (1994). The diluent was a mixture of kerosene and diesel, and its purpose was to make the viscous crude oil thinner and easier to pump from the wells and through pipelines. An estimated 12 million gallons leaked over decades. There was also some crude oil mixed into the spilled liquid.]

River Deltas

An irregular bulge of the shoreline into a standing body of water at the mouth of a sediment-laden stream is normally referred to as a **delta**. Many deltas have an inverted pyramid shape, or perhaps more appropriately, an inverted version of the upper class Greek symbol delta. The distinction between deltas and estuaries is not readily apparent in some situations, as is disclosed in more detail in the discussion of estuaries.

In their classic summary paper on deltas, Coleman and Wright (1975) discussed over 50 parameters that have an impact on river delta morphology. Factors such as characteristics of the drainage basin, river slope, and coastal climate were acknowledged. Most present-day workers, however, simplify matters by focusing on three basic controls – sediment supply, wave energy, and tidal current energy – in their attempts to classify deltas. However, that point will not be pursued any further here, because there is only one major river delta on the Central California Coast, the Sacramento/San Joaquin Delta east of San Francisco Bay, an area not covered in

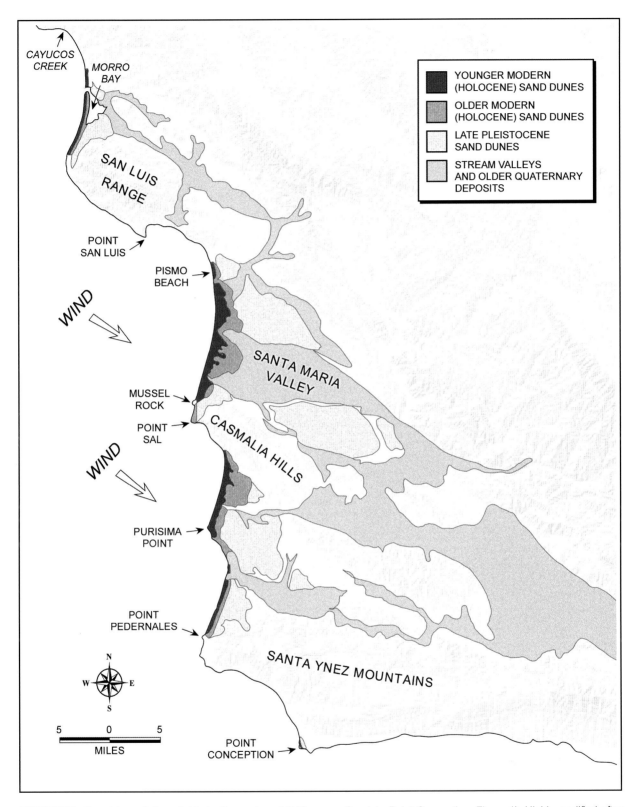

FIGURE 77. Coastal sand dune fields in Compartment 7 (Cayucos Creek to Point Conception; Figure 1). Highly modified after Orme (2005).

this book. Some geologists would refer to this delta as a **bayhead delta**, a type of delta that is fairly common in the rock record (apparently). Otherwise, there are only three small river deltas of any consequence on the coast of Central California, two of which occur in Tomales Bay. Walker Creek, which enters Tomales Bay on the north/northeast side, builds a small delta with some fairly extensive salt marshes and tidal flats a small ways out into the Bay. Olema Creek has deposited a modest-sized delta, which also has noteworthy salt marshes and tidal flats, at the very southeast end of the Bay. Chorro Creek has built a small delta (diameter of about one mile) into Morro Bay, which also has fairly extensive salt- and brackish-water marshes on its surface, as well as some tidal flats. All of these deltas, which occur in bays sheltered from large waves, are accessed easily by driving and they are described in more detail in the discussions in Section III.

5 HABITATS OF THE COASTAL WATER BODIES

INTRODUCTION

The Central California Coast as we have defined it (Figure 1) contains seven coastal water bodies separated from the open ocean by tidal inlets – Bodega Harbor, Tomales Bay, Drakes Estero, Bolinas Lagoon, San Francisco Bay, Elkhorn Slough, and Morro Bay. These water bodies contain virtually all of the sheltered tidal flats (mostly mud flats) and most of the salt and brackish water marshes on the Central California Coast, because of their protected position relative to open ocean waves.

These water bodies have a variety of origins, most of which do not match the numerous estuarine systems on the east coast, because of their location on the tectonically active western side of the North American Plate. Accordingly, this section is only a brief outline of the components of these coastal water bodies, using the classic estuarine system as a model. The detailed characteristics of several of these water bodies are treated in some detail in Section III, with the exception of San Francisco Bay, the discussion of which is not a part of this book.

DEFINITION OF AN ESTUARY

According to the original definition, which was basically a description of Chesapeake Bay (Pritchard, 1967), an **estuary** has three defining qualities: 1) a **flooded river valley** that was formed during the lowstand of sea level that culminated about 20,000 years ago; 2) a water body with a **substantial freshwater influx**; and 3) a water body **subject to tidal fluctuations**. The second and third criteria are self explanatory, but the first requires some understanding of sea-level fluctuations during the Pleistocene epoch. As suggested earlier, when the sea level was low, streams flowing across the mainland to the sea carved deep valleys that extended out onto the continental shelf in some areas. In many cases, this type of erosion was confined to the same valleys during

each of the four major drops in sea level. When sea level started to rise at the beginning of the Holocene epoch about 12,000 years ago, those valleys were flooded as the shoreline advanced across the then-exposed continental shelf. When sea level essentially stopped this sudden rise around 5,000 years ago, the valleys were flooded with salt water for some distance inland. Where the valleys were eventually filled with sediment and a bulge of sediment protruded out into the ocean away from the general shoreline trend, the original "estuaries" were converted to what we now call "deltas." A second type of delta, that some call a bay-head delta, may form further up the flooded valley far short of the open ocean. As noted in the previous chapter, the delta at the head of San Francisco Bay is of this type. Bay-head deltas formed in situations where the volume of sediment supplied by the river was not sufficient to completely fill the valley out to the open ocean shoreline within the time allowed.

Few of the coastal water bodies in California fit Pritchard's definition. San Francisco Bay is by far the largest *bona fide* estuary in the state, following that definition. Of the other coastal water bodies on the Central California Coast, Drakes Estero, Estero Americano, Estero de San Antonio, and to some extent, Elkhorn Slough conform to the classic definition. Bodega Harbor and Bolinas Lagoon are flooded fault troughs closed off by sand spits. Morro Bay is within a major shoreline embayment, with the water body also closed off by a sand spit. It is not a flooded river valley, but it does have some freshwater inflow and tides, so it is commonly referred to as an estuary. Tomales Bay, a flooded fault trough, likewise does not conform to the usual definition perfectly because it is not a flooded river valley. However, it is tidal and has some freshwater inflow. Ed Clifton (pers. comm.) pointed out the fact that many of the coastal water bodies of Central California *have significant input of fresh water only in the winter,* which adds to Pritchard's definition a certain element of seasonality.

ESTUARINE CHARACTERISTICS

In order to compare the coastal water bodies we do have on this shoreline with the classic estuary, its general characteristics are reviewed. A key factor in defining the physical structure of an estuary is the way in which the salt and fresh water mix. As shown in Figure 78 (Upper Diagram), a highly stratified estuary with a well-defined **salt wedge** occurs where a significant freshwater stream enters a flooded valley with a small tidal range, with the more dense salt water hugging the bottom of the channel during a rising tide. Mixing of the salt and fresh water is enhanced by an increase in tidal energy; therefore, with a larger tidal range, the two water masses become partially mixed (as indicated by the arrows in the Middle Diagram in Figure 78). With an even larger tidal range, the water within the estuary may become vertically homogeneous from top to bottom, completely fresh top to bottom at the head of the estuary and pure salt water from top to bottom at the entrance (Lower Diagram in Figure 78).

Figure 79A shows a plan view model of an estuary of the type found along the mid-Atlantic states, and Figure 79B shows a cross-section illustrating the circulation within these typically partially mixed estuaries. We consider the most landward boundary of an estuary to be the place where the tide "stops going up and down," the so-called "head of tides." In the San Francisco Bay area, that point can be a hundred miles or more from the nearest salt water, as it is in the Sacramento River (Ed Clifton, pers. comm.). In the upper portions of the estuary, the primary hydrodynamic process is the river flow, whereas at the entrance, waves and tides dominate. As far as sediment sources are concerned, the sediments in the middle and upper portion of the estuary are provided by the incoming river, but in some estuaries, a considerable amount of sediment comes into the entrance of the estuary from offshore under the influence of flood-tidal and wave-generated currents.

Considering the different components of the system as illustrated by the plan view map in Figure 79A, the upper estuary is a zone of relatively diminished tidal flow, and the river flow is still strong in a seaward direction. In the middle of the estuary, tidal flow is increased, the flood tide reverses the river currents when the tide rises, and a complex two-way, two-layer circulation occurs, creating a zone of mixing of the two water masses. In that zone of mixing, demonstrated in the cross-section in Figure 79B, an important process known as **flocculation** of the clay particles in suspension occurs. Flocculation takes place because the abundant clay particles in suspension brought in by the river flow slow down their seaward movement at the point of mixing as a result

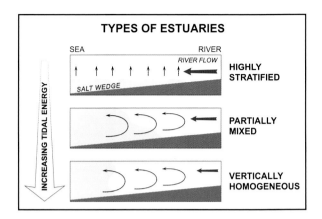

FIGURE 78. Effect of tides on the mixing of salt and fresh water in estuaries (modified after Biggs, 1978).

of the reversing of tidal flow directions twice a day. This allows an increase in the number of clay particles in the water column to the extent that they begin to collide with one another. These colliding particles tend to adhere together under the influence of the increased concentration of cations, such as Na+, Ca++ and Mg++, derived from the seawater. As a result, the clay particles, which may have diameters as little as one micron, cluster into groups (**flocs**) that may have diameters measured in hundreds of microns. As a result, many of the flocs sink to the bottom at slack tide (Figure 79B). Once on the bottom, some of the flocs are resuspended by ebb- and flood-tidal currents. The net result is the formation of a zone in the middle reaches of the estuary that contains abundant fine-grained sediment in suspension known as the **zone of the turbidity maximum**. This zone of turbidity can migrate up and down the estuary over time, moving downstream during periods of flood on the river and upstream during low-flow periods. A side effect of this process is the creation of a "mud zone" on the tidal flats and in the tidal channels in the middle of the estuary (Figure 79A).

At the estuary entrance, conditions take on an entirely different aspect in that:

1) Tidal currents are more dominant than river currents with common differentiation between ebb-dominated and flood-dominated channels.

2) The waters are more mixed, with diminishing estuarine stratification.

3) Wave activity becomes a more important dynamic process.

4) Moving of sediments along the bottom by wave- and tide-generated currents causes the tidal flats and channel bottoms to become much more sandy than those in the central portions of the estuary in the "mud zone."

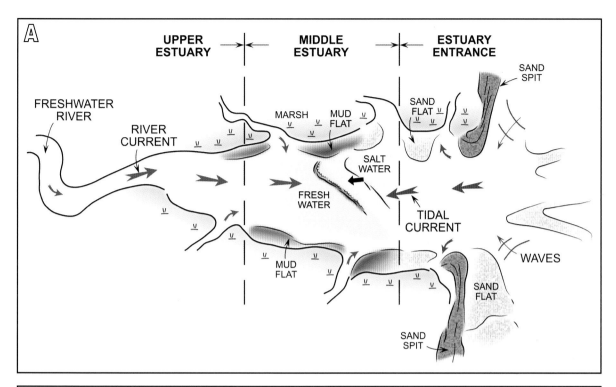

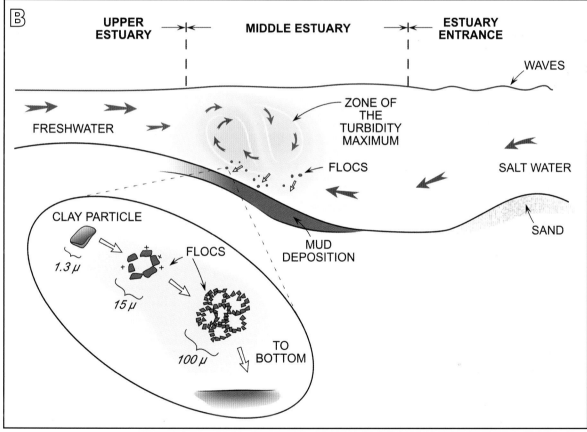

FIGURE 79. Estuaries. (A) Plan view of a typical estuary. (B) Cross-section of a typical estuary showing predominant processes and circulation patterns. Especially noteworthy is the formation of *a zone of the turbidity maximum* where the clay particles (<0.004 millimeters in size; Figure 4) brought in by the river undergo a process called *flocculation*.

MARSHES

The coastal water bodies of the Central California Coast are host to fairly extensive salt, brackish, and fresh water marsh systems, which are well documented as being important food sources for the coastal and nearshore ecosystems of the region (see example in Figure 27B). Marshes are flooded with water on a regular basis.

Marshes in estuarine systems occur in areas that were originally tidal flats where sediment accumulated in the early stages of evolution of the system. As the flat was built up to or slightly above the appropriate sea level (variable between mean sea level and neap high tide), marsh grasses began to take root. Once the grasses grew on the flat, the sedimentation process was accelerated because of the baffling effect of the plants on the tidal currents. These present marshes are, in effect, intertidal flats vegetated with **halophytes** (plants that are adapted to grow in salty soil; Basan and Frey, 1977). Marshes can expand very rapidly, up to a yard or so a year, if the slope is flat and sediments are abundant.

For purposes of description, we have divided the typical estuary in the mid-Atlantic states into upper, middle, and lower (entrance) zones (Figure 79). As noted above, each of these zones is characterized by distinct hydrodynamic and sediment characteristics. These lateral salinity changes up and down the estuary, most notably the decrease in salinity toward the head of the estuary, have a striking impact on the plant communities. **Fresh water marshes** are most common in the upper estuary, where there are tides but very low salinities; **brackish marshes** occur most commonly in the middle estuary, where salinities generally average less than 15 parts of "salts" per thousand parts of water (ppt); and **salt marshes** occur in the lower estuary, where salinities range from 15-36 ppt. The marsh communities of the coastal water bodies present on the Central California Coast, excepting San Francisco Bay, are described in some detail in Section III.

TIDAL FLATS

Definition

Tidal flats are nearly flat, unvegetated intertidal surfaces that are relatively sheltered from direct wave attack. In most places, the sediment grain size decreases landward, away from main tidal channels, because the sediments are carried to their final resting place by tidal currents that decrease in velocity away from the channels as the water flows up onto the flats. The depositional and erosional rates on tidal flats are much slower than on

beaches on the open-ocean front, where the sediment is moved readily by wave-generated currents.

Exposed Sandy Tidal Flats

In the entrances to some of the larger estuaries in the mid-Atlantic area, broad intertidal flats up to several hundred yards long and wide occur (Figure 79A). These flats are usually composed of sand, which indicates that tidal currents and waves are strong enough to mobilize the sediments from time to time, removing the muddy sediments. Even so, sedimentation rates are considerably slower than on the more exposed beaches. On the lower portions of some of these flats, waves may have formed low ridges that migrate very slowly across the flat and ripples and even larger bedforms may be present in the vicinity of tidal channels that cut across the flat. Despite this kind of sediment motion lower down, most of the sand and muddy sand on the upper half of these flats is relatively stable; therefore, it usually contains a huge population of infaunal organisms (those that burrow in the sediment).

Sandy tidal flats are present in some parts of San Francisco Bay, but mud flats are by far the most common type of tidal flat in the Bay. Sandy tidal flats are also fairly abundant in Bolinas Lagoon.

Sheltered Mud Flats

For the most part, intertidal mud flats occur in the most sheltered portions of coastal water bodies, as well as on the most landward portions of wide tidal flats that are sandy on their lower reaches (see example in Figure 27A). The problem of why large quantities of mud accumulate on these more sheltered tidal flats is an interesting one. Several Dutch workers have commented on this issue. Van Straaten (1950) and Postma (1967) concluded that the mud is deposited during the last stages of the flood flow, during slack water, or at ebb just before water is drawn away. The finest particles in the mud are composed of clay minerals, which have a thin, sheet-like shape (like a thin coin). This shape slows down the process of settling of the particles from suspension, which allows them to be transported landward by the waning flood currents beyond the point where more spherical particles with a similar mass would settle out. Furthermore, a stronger current is required to pick the mud back up than the one that deposited it. Once the flat particles have settled on the bottom, their thin edges provide little resistance to the outflowing water because of their lower pivotability, as was explained for disc-shaped clasts on gravel beaches. Therefore, they are not easily picked back up and set into suspension. This process of permanent deposition is aided further by the fact that floccules of organic matter and suspended fine-grained sediment (see Figure 79B), as well as slime-secreting diatoms, create other particles (in addition to individual silt and clay particles) that readily settle out in the quiet water. Once this particular mud is deposited, it tends to remain in place as well because: 1) diatoms move up through the mud and deposit slime; 2) the mud dries out somewhat during low tide, and 3) burrowing organisms tend to stabilize the mud. In some cases, mud is trapped by plants on the flat.

Another factor of prime importance to mud deposition is the super abundance of **suspension-feeding** organisms, such as oysters and clams. During the filtering process, these organisms compress the finely divided clay particles and bind them together in their intestines as fecal pellets, which are excreted onto the flats. Another part of the suspended matter is coagulated in their gills and pushed back into the water as pseudo-feces. These feces and pseudo-feces are easily deposited, even in comparatively turbulent water.

Dominant Biogenic Features

The tracks, trails, and types of burrow structures of the animals that live on tidal flats and in their sediments have fascinated geologists for decades, because such evidence is useful for interpreting the depositional environments in ancient sediments (commonly used by geologists interpreting rock layers in their search for economic deposits such as oil and gas). Knowledge on this topic has been greatly advanced by studies on the Georgia coast, specifically by researchers at the University of Georgia (e.g., Frey and Howard, 1969; Howard and Dorjes, 1972). Generally speaking, the animals that live in the sediment on tidal flats (called infauna), such as clams and burrowing worms of various kinds, live in either vertical tubes or U-shaped burrows, which maintain less variable temperatures and salinities than exist on the surface of the flats. You may see some of these infauna and their burrows as you walk the intertidal regions of the coastal water bodies of the Central California Coast (WITH SHOVEL IN HAND!!).

TIDAL CHANNELS

Before delving too far into this subject, a brief review of channel systems in general is in order. On a global scale, there are four basic styles of channel morphology that occur in natural river and tidal systems, which are illustrated in Figure 80. **Straight channels**

are rare in general, so be suspicious if you see one in the coastal region of Central California, because it has most likely been dug for navigation or other purposes, such as mosquito control. As a rule, **braided channels** occur under conditions of: 1) relatively steep channel slope; 2) high bedload content (sedimentary material on channel bottom; e.g., gravel and coarse sand, that moves along the bottom by rolling and bouncing), and 3) flashy discharge (high flow during flooding periods and low flow during the rest of the time). **Meandering channels** tend to have: 1) flatter slopes than braided channels; 2) more sediments suspended in the water column than moving along the bottom; and 3) a more steady discharge. Meandering channels are by far the most common channel type in the coastal water bodies of Central California. **Anastomosing channels** typically occur on extremely flat slopes in waters with a very high ratio of suspended to bedload sediment.

Ed Clifton (pers. comm.) pointed out that *sinuous dendritic channel systems* (branching like a tree) *occupy muddy tidal flats and can be used to identify them from aerial photographs.* He also alluded to some not clearly understood *large, low-amplitude sediment waves that occur on many sandy tidal flats,* the origin of which possibly has to do with *a combination of wave effects and runoff patterns.* Clearly, the channel systems on the tidal flats in this area are a topic in need of further study, which is true on a global scale, as far as that goes.

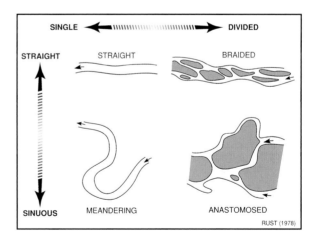

FIGURE 80. Channel types, classified as to whether they are straight or sinuous and whether they are single channels or a complex divided channel (after Rust, 1978).

6 CONTINENTAL SHELF AND SUBMARINE CANYONS

Though not a place you are apt to visit during your naturalist explorations on the Central California Coast, unless you own or hire an ocean-going vessel, the **continental shelf** represents the seaward end member of the coastal zone continuum. The continental shelf is defined as a wide and gently sloping submerged surface that connects the shoreline with an abrupt increase in slope, or **shelf break**, where the shelf edge joins a steeply descending planar surface called the **continental slope** (Figure 81).

Another of the major contrasts between the leading edge and trailing-edge coastlines of North America is the width of the continental shelf. Off the coast of the southeastern U.S., the shelf is quite wide, extending an average of about 60 miles off the beach, with the shelf break located at depths of around 600 feet The wide shelf off the east coast is underlain by relatively horizontal sedimentary rock layers that range in age back to the late Triassic or early Jurassic, the time when the present ocean began to take shape as a result of the opening of the mid-Atlantic rift zone.

On the other hand, if we use that same depth as the edge of the continental shelf (600 feet) as is used for the east coast, the width of the shelf off the shoreline of the Central California Coast, which is illustrated in Figure 81, is highly variable but generally quite narrow compared to the shelf on the east coast. The shelf is widest in the north, with a width of about 27 miles off Duxbury Point. However, this part of the shelf contains the Farallon Islands. South of San Francisco off Pillar Point, at the head of Half Moon Bay, the shelf is around 22 miles wide. It narrows to around 14 miles off Año Nuevo. A short distance north of Monterey in Monterey Bay it is about 9 miles wide. The shelf is noticeably narrower to the south, especially south of Monterey in Big Sur Country where the high mountains abut the shore (e.g., the shelf is only 5 miles wide off Castle Rock a few miles north of Point Sur). The shelf widens a bit at the south end of the Central Coast north of Point Arguello (compartment 7, Figure 1). And, finally, the 600 foot depth contour is

less than 2 miles offshore at the landward edge of the Monterey Submarine Canyon at Moss Landing.

Of course, these numbers are for our artificially chosen depth of 600 feet (the 100 fathom depth). Others use different numbers, for example a write-up on the Gulf of Farallones Marine Sanctuary (2008) makes the following comment.

Offshore of Tomales Bay and Bolinas Lagoon, the sanctuary is characterized by a gently seaward-sloping continental shelf area that is approximately 55 kilometers (34 miles) wide. The western edge of the shelf is currently in 90 meters of water depth (295 feet) and coincides with the shoreline of the last ice age.

The shelf off the east coast of the U.S. is quite wide because the trailing eastern boundary of the North American Plate along the ever expanding ocean's edge has been allowed to accumulate sediments in the continental shelf area for the past 200 million years or so. The reason the California shelf is much narrower is a little more complicated than it would first seem. Of course, this coastal area is a location where subduction of the sea floor along the leading edge of the North American Plate has been taking place for a long time. However, this process ceased around 30 million years ago. Even so, this continental margin has continued to be very tectonically active, and it is "young" relative to the Atlantic margin (e.g., the formation of the Coast Ranges), having changed greatly in the last 5 million years, thereby not allowing enough time for a wide continental shelf to develop (possible cause suggested by Ed Clifton).

A detailed account of the geology of the California continental shelf seems unwarranted for this coastal-oriented book. However, we can say that, for the most part, the rocks on the shelf are athwart the Pacific Plate; therefore, many of them originated in the far south. North of Santa Cruz, the granitic rocks carried north as part of the Salinian Block crop out along the shore at Bodega Head, Point Reyes, and Devils Slide (where Montara Mountain meets the beach), as well as on the somewhat isolated, and resistant, remnant that eventually

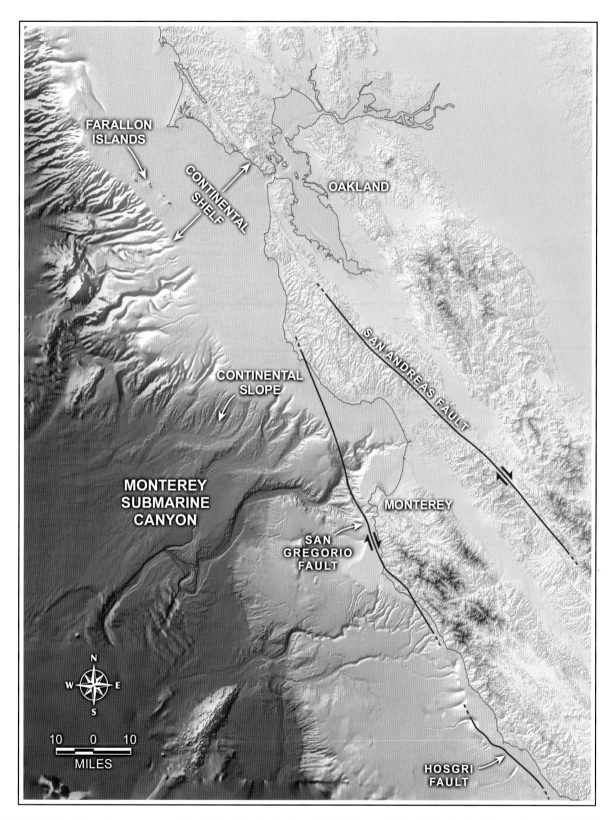

FIGURE 81. Submarine canyons off the Central California Coast. Note the variability in the width of the continental shelf. The approximate location (and relative motion) of the parts of the San Andreas and San Gregorio Faults located near the Monterey Submarine Canyon are also shown (courtesy of Ed Clifton). Map image courtesy of NOAA National Marine Sanctuaries (MBNMS-Maps, 2009).

became the Farallon Islands (part of a segment of the Salinian Complex called the Cordell-Farallon Islands Block). These granitic rocks are apparently mostly of Late Cretaceous age and, as noted earlier, many geologists think these rocks share their origins with the core of the Sierra Nevada. Generally speaking, however, most of the oil-and-gas producing rocks on the shelf are Paleogene/Neogene in age. For example, the offshore Ventura-Santa Barbara Channel field produces mainly from Pliocene, Miocene, Oligocene, and Eocene sandstones and fractured Monterey chert of Miocene age (ages of these Paleocene/Neogene periods are given in Table 1).

Of more relevance, no doubt, as far as the coastal zone is concerned, is the role of numerous **submarine canyons** located along the edge of the continental shelf (see Figure 81). Submarine canyons are steep, V-shaped valleys eroded into the continental slope. They typically head at the shelf break and end near the bottom of the continental slope. There has been much speculation in the scientific literature regarding their origin. Many of them are located off major rivers, such as the Mississippi and the Ganges/Brahmaputra, where the largest submarine canyon on Earth, the "Swatch of No Ground," feeds the largest submarine fan in the world, the Bengal Deep-Sea Fan. One idea is that the rivers played a role in carving the canyons during the lowstands of the Pleistocene epoch, when the river mouths were near the edge of the shelf. According to this reasonable explanation, the sediment dumped at the river mouth was then carried deeper by **turbidity currents** (that are illustrated in Figure 21). The fact that some canyons are not located off major rivers (at the present time) implies that other processes may be involved, such as seafloor slumping and other forms of mass wasting. In any event, there is considerable evidence that at the present time turbidity currents help to keep the canyons eroding and that many submarine deposits (fan-like lobes in some cases) were deposited by such density currents. The coastal scientists who originally conceived the idea of littoral cells for the California coast (Inman and Chamberlain, 1960; Bowen and Inman, 1966) cited turbidity currents and other mechanisms, such as sand slumps or debris flows, carrying sand into submarine canyons as the principle way sand is lost from these cells.

On the Central California Coast, the most famous submarine canyons occur within the Monterey Bay National Marine Sanctuary, with the Monterey Canyon being the largest, closest to shore, and most heavily studied of them all (see Figure 81). Other notable canyons in the Sanctuary include: Pioneer Canyon (offshore of Half Moon Bay); Ascension, Año Nuevo, and Cabrillo Canyons (offshore of Davenport); Carmel and Soquel Canyons (which feed into Monterey Canyon); and Sur, Partington, Mill Creek, and Lucia Canyons (offshore of Big Sur).

Several sources note that the Monterey Submarine Canyon is comparable in depth, gradient, and length to the Grand Canyon in Arizona. Ed Clifton (pers. comm.) made the following comments regarding this comparison:

The Monterey Canyon compares to the Grand Canyon of the Colorado in width, depth (relief), and steepness of slope, but I think the Grand Canyon at 277 miles is substantially longer than Monterey Canyon which, strictly speaking, is the part incising the continental slope (about 80 miles long). If you add the deep sea channel portion across the Monterey deep sea fan, the total length is about 190 miles. What is different ... is that the gradient of the submarine canyon floor is about 20 times steeper than the average fall of the Colorado River, so that sediment gravity flows like turbidity currents (see Figure 21) are much more important in submarine canyons (as the driving mechanism for the continued erosion of the canyon, in contrast to surface water river flow in the Grand Canyon).

Recent measurements of the turbidity currents in the Monterey Canyon, which verify the role of such currents in maintaining the Canyon, were reported by the Monterey Bay National Marine Sanctuary (2008). Researchers at the Monterey Bay Aquarium Research Institute measured a large turbidity current that carried an estimated over two and a half million cubic yards of gravelly sediment approximately 1,814 feet down the Monterey Submarine Canyon. These researchers thought that high wave conditions set off the current. If true, such sediment gravity flows do not require exceptional triggering events such as earthquakes. This source states further that scientists with the U.S. Geological Survey *documented the first ever in-situ measurements of velocity profiles for turbidity currents in Monterey Submarine Canyon. The maximum along-canyon velocity of a turbidity current was 190 centimeters per second, which is equivalent to 4.25 miles per hour* (around 4 knots, a very strong current that tidal currents in tidal inlets occasionally, but rarely, exceed). *The currents were generally confined to within 50 meters (164 feet) of the seafloor, and as they moved down the canyon, their heads moved closer to the canyon floor. All of the turbidity current events recorded during this study occurred in winter and may have been triggered by large storm waves and high stream flows from the nearby Salinas River. This* important finding adds substantial supporting evidence for the earlier ideas put forward on the role of turbidity currents (illustrated in Figure 21) in the maintenance of submarine canyons, as well as in the generation of many of the rock formations present in the cliffs along the Central California Coast. One of the most notable examples of such turbidity-current deposits is the "upper" sequence in the Franciscan rocks that crop out along the shore south of the Golden Gate (Figure 22). It is not clear whether the currents that created these sandstones were associated with submarine canyons or not; however, the notion that they were deposited by turbidity currents appears to be beyond serious dispute.

7 SHORELINE EROSION AND LANDSLIDES

INTRODUCTION

Shoreline erosion is a common phenomenon on many of the coastal areas of the globe. In fact, almost no coastal area of the world that has been developed by man is free of erosion problems. Various publications have given numbers on coastal erosion rates; for example, Dean and Walton (1975) reported an average erosion rate of between 1-3 feet per year for the U.S. shoreline. Another study by the U.S. Army Corps of Engineers (National Shoreline Survey) in 1971 concluded that nearly 43% of the U.S. shoreline is undergoing "significant" erosion. Maybe for the Central California Coast, a distinction should be made between shoreline erosion and **shoreline retreat**. For example, after a major *El Niño storm, a sand or gravel beach eroded during the storm may come back, but an eroded part of a wave-cut rock cliff will not* (Ed Clifton, pers. comm.).

In a paper published in 1985, co-author Hayes made the following somewhat pompous assertion:

What becomes evident in this discourse is the conclusion that man himself has caused most of his own beach erosion problems. That being the case, perhaps he can do something to solve them. Lessons learned from the mistakes of the past should be applied in the planning phases of every new development in the coastal zone. The costs involved in preventing erosion by wise planning are orders of magnitude less than those required to solve an erosion problem once the structures are in place.

Now, in reviewing what has been happening in the 24 years since that pronouncement, one would have to conclude that either: 1) not more than a dozen policy makers read it, or 2) if they did, they had a good laugh, ignored it, and went on about their business.

Getting back to the numbers, a later estimate by Bird (1996) in a book on beach management concluded that 70% of the world's beaches are now eroding. Therefore, it should come as a surprise to no one that at least some of the beaches in Central California are eroding. In fact, in

a recent report sponsored by the U.S. Geological Survey, Hapke et al. (2006) noted that 66 percent of California's "sandy shorelines" are indeed retreating, with significant associated coastal land loss in some areas.

A book entitled *Living With the Changing California Coast* by Griggs, Patsch, and Savoy (2005) made the following observation on the rates of shoreline erosion and retreat (not just the erosion of sand beaches) in California:

1) *Variations in the physical forces that produce cliff erosion or retreat (wave exposure, rainfall and runoff, earthquakes, etc.), as well as in rock strength or the resistance of seacliff materials to failure, have resulted in wide-ranging erosion rates along California's 1,100 miles of coastline.*

2) *The hardest materials stand out as points or promontories, such as Point Conception, Point Sal, and others, and where the materials exposed along the shoreline are weak or where the topography is very low, we see embayments such as Monterey Bay.*

3) *At some sites cliff erosion has been negligible for the 75-100 years of reliable records, simply because the rocks are so hard and resistant to wave attack. At other localities the average rate of retreat may be as high as 5 to 10 feet per year. At Point Año Nuevo, in southern San Mateo County, the erosion rate of the low bluffs has averaged about 9 feet per year for the last 300 years, one of the highest natural rates along the state's entire coastline.*

4) *Coastal erosion also tends to be an episodic process, with much of the long-term failure or retreat taking place with a few severe storm events every 5 or 10 years.* [NOTE: These events were reviewed earlier in the discussions on major storms and the beach cycle.]

5) *As an example, one section of seacliff in the city of Santa Cruz eroded about 25 feet in the interval from 1931 to 1982. During the January 1983 storms, waves removed about 46 feet of cliff top. This single storm increased the "long-term" average erosion rate from about 6 inches per year to over 16 inches per year! Such observations suggest that average erosion rates be used with great caution and*

that every effort be made to research the historical changes from maps and aerial photographs as far back in time as possible.

MAJOR CAUSES OF SHORELINE EROSION

Introduction

The first part of this discussion of the causes of shoreline erosion is limited to **depositional coasts** so that some of the basic principles can be presented in a more general way. The causes for erosion on tectonically active shorelines like the Central California Coast are a bit more complex; thus, they are presented after this first, more generic review.

With that given somewhat limited focus, we can say that for any given depositional coastal area not unduly influenced by man, it is the interaction of **sediment supply** and **water-level changes** that controls beach erosion. In this case, we are discussing sand beach erosion primarily, because most depositional coasts have an abundance of sand beaches, which in many areas have been developed extensively for housing and recreational purposes.

Along some coastlines, fluctuations in sediment supply prevail, and on others, rapid changes in water level are most important. The most serious erosion problems occur either where man interferes with sediment supply or where some phenomenon causes an unusually rapid rise in water level (e.g., abrupt lake level rise in Lake Michigan or around sinking abandoned delta lobes on the Mississippi Delta).

Deficits in Sand Supply

To understand which natural sources are most important in providing sand to the shoreline at any given site, a **sediment budget** would have to be calculated in which the sediment gains (credits) and losses (debits) are determined and equated to the net gain or loss. This kind of sediment budget has been attempted for parts of the California coast (e.g., Bowen and Inman, 1966; discussed later).

There are a number of natural processes that could produce a deficit of sand at a particular beach, including:

1) **Switching of the mouth of the main river** in a delta system. This is a common process on river-dominated deltas such as the Mississippi Delta.
[NOTE: A river-dominated delta is one in which the sediment supply provided by the river overwhelms the effects of waves and tides such that a bird-foot-like lobe projects well out into the open water.]

2) **A variety of processes at tidal inlets.** For example, as a tidal inlet migrates, the shoreline on the side of the inlet in the direction the inlet is migrating is eroded away and part of the sand on the beach on that side of the inlet is carried offshore where it is stranded on the ebb-tidal delta for a period of time, which may be years in duration.

3) **A number of natural processes that remove sand** from the immediate vicinity of the beach zone (Figure 49), such as: a) transport of sand offshore during storms beyond the depth from which the sand is normally returned to the beach during calm periods; b) wind transport of the sand landward out of the normal zone of the beach cycle; and c) transported alongshore out of the immediate vicinity of the beach in question. The processes for sand removal that operate in the littoral cells on the California Coast were discussed previously (loss to submarine canyons, loss to coastal sand dune systems, and carried far offshore onto the continental shelf during storms; Figure 66).

There are also a number of man-induced changes that could produce a deficit of sand at a particular beach, including:

1) **Dams on rivers.** Sediment contribution of any river discharging on the coast is of importance to the sediment budget of that coast. If a dam is built on the river to create a reservoir for storage of water, the current velocities in the river are reduced to such an extent that practically all sandy sediment carried by the river settles in the reservoir. The water discharged from the dam contains very little sand; therefore, the sand supply formerly delivered to the coast by that river is cut off. This is an overwhelming cause of beach erosion on shores with narrow or absent coastal plains, such as Japan and California (discussed in more detail later).

2) **Diversion of rivers.** Some engineering projects call for the diversion of rivers away from their original channels. In addition to the construction of a dam across the Santee River, South Carolina in 1942, approximately 90% of the river flow below the dam was diverted into the Cooper River, which flows into Charleston Harbor (about 40 miles down the coast). This assured that even most of the fine-grained sediment carried by the river was diverted from the river's mouth. Needless to say, the dam and diversion of the river's flow brought about erosion of the shoreline of the Santee/Pee Dee Delta, the largest river delta on the east coast.

3) **Sand mining.** This is not a practice that is normally accepted on the beaches in the U.S., although the coastal

sand dunes and beaches along the shore in southern Monterey Bay were mined extensively between 1906-1990. This practice was stopped because of fears it was causing beach erosion. A study by Thornton et al. (2006) to determine if this sand mining actually had increased the beach erosion rate along that stretch of shoreline was inconclusive, partly because of complications related to the 1997-98 *El Niño* storms. However, sand mining is a common practice in many other parts of the world. The extraction of sand from sand bars in the nearshore zone, such as ebb-tidal deltas, for beach nourishment projects is another practice that could be called sand mining. If such sand is left in its natural configuration, some of it might eventually end up on the shore and its removal could expose the beach to erosion in places. Even mining of sand from shoals located further offshore can possibly influence beach erosion. Dredging of these shoals may possibly have two side effects: a) alteration of depth contours, which changes wave-refraction patterns and may cause a focusing of wave energy at different locations on the shore not previously so heavily impacted; and b) elimination of a natural breakwater, which acts to reduce wave energy arriving at the shoreline.

4) **Tidal inlet stabilization.** As a general rule, the ultimate geomorphic form of tidal inlets is the result of an adjustment to the dynamic action of both tidal currents and waves. Also, the volume and pathway of littoral drift (volume of sand moving along the shore) contributes to the configuration of a tidal inlet. From the studies of O'Brien (1931), it is well known that the stability of an inlet is governed by the relative strength/volume of the littoral drift and the tidal prism volume (amount of water that flows in and out of the inlet during a single tidal cycle). Sand moving along the shore bypasses the

inlet by sand-bar migration under the influence of waves or by tidal current transport of individual sand grains. On coasts with a large littoral drift, there is a tendency for the inlet to migrate downdrift (direction along the shore the sand in the littoral drift is moving) due to the infilling of the mouth of the inlet on the updrift side (direction along the shore the sand in the littoral drift is coming from). Almost any man-made modification to a tidal inlet, therefore, may have significant influence on the erosional/depositional patterns on adjacent beaches. For example, artificial deepening of an inlet to develop a harbor could change the bar-bypassing nature of the inlet, depriving the downdrift coast of its supply of sand and thus leading to erosion.

5) **Construction of shore perpendicular structures.**

Jetties are two long parallel structures built at a river mouth, a tidal inlet, or even an artificially dug harbor, in order to stabilize the channel, prevent its shoaling by littoral drift, and protect its entrance from waves (Figure 82A). Jetties are also designed to direct or confine the flow to help the channel's self-scouring capacity. Usually, jetties extend through the entire nearshore zone to beyond the breaker zone in order to prevent sand deposition in the main channel. Therefore, the jetties act as barriers to the longshore transport of sand, causing a significant volume of the sand to accumulate on the updrift side of the jetties. At the same time, on the downdrift side, the sand transport processes continue to operate and cause the sand to move away from the jetties, resulting in the erosion of the shoreline on that side. For example, at the jetties for the small craft harbor in Santa Cruz, the sand beach on the updrift (west) side of the jetties is 200 feet wider than the sand beach on the downdrift (east) side (see Figure 82A). In some instances, the updrift jetty may

FIGURE 82. Engineered shorelines. (A) Jetties at the entrance to Santa Cruz Harbor. (B) Groin on the shoreline of Santa Cruz. View looks north. Dominant longshore sand transport is from west to east (arrow). Note wide sand accumulation on the west side of the groin. Both photographs, which were taken on 1 October 2008, are courtesy of the California Coastal Records Project, Kenneth and Gabrielle Adelman.

have a weir system whereby sand is trapped in a confined area so it can be either hauled offshore or passed across the jetties to the downdrift side of the inlet.

Groins are another structure built perpendicular to the beach, in this case for the purpose of widening the beach by trapping a portion of the sand in the littoral drift. They are relatively narrow in width and may vary in length from about 30 to several hundred feet (see example in Figure 82B). Though their function differs from jetties, they are also barriers to longshore sand transport. If they trap a significant portion of the sand in transport, naturally the beach will erode on the downdrift side. Groins may be used in a series (groin fields) to protect a large area, but the zone of erosion will shift down the coast through time. Once the sand builds out to near the end of the groin, some or all of the sand in transit will pass on downdrift. Poorly designed groins may deflect sand past their ends into deeper water, resulting in its loss to the longshore transport system. Kana, White, and McKee (2004) discussed the utility and management of groins on the South Carolina coast. They made eight planning guidelines and recommendations, including: 1) only use groins on the open coast where the erosion rate exceeds about six feet per year; and 2) construct the groins of a groin field in the downdrift to updrift direction, and nourish the "cells" between the groins "to capacity."

6) **Construction of seawalls and revetments.** Seawalls are vertical, hard structures very commonly built to protect man-made structures. Thousands of examples could be cited to demonstrate the effectiveness of such structures in saving property, at least in the short term. However, erosion of the sand on the beach itself is usually accelerated in front of these features because of wave reflection from the hard, vertical faces of the seawalls. According to Silvester (1977), when waves are obliquely reflected from such seawalls, energy is applied *"doubly"* to the sediment bed and *"hence expedites the transmission of material down coast."* The waves reflect off the seawall at 90 degrees to their original approach direction. Once reflected, a wave shortly comes in contact with the next wave approaching from offshore. At the point where the waves meet, the two waves "combine forces" to create a strong current that runs parallel to the shore, creating the "scour zone" illustrated in Figure 83A. Over the past few decades, there has been a hue and cry by a number of concerned scientists against building seawalls along open ocean beaches, to the extent that some states now ban them (e.g., North Carolina). Revetments usually serve the same purpose as sea walls, but they are typically made of materials such as boulder-sized chunks of rock, called riprap. These features do not reflect waves as severely as sea walls, but

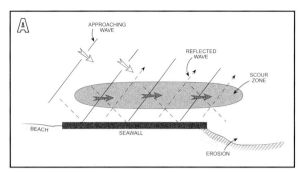

FIGURE 83. Seawalls and beach erosion. (A) Illustration of erosion in front of and on the downdrift side of a seawall as a result of waves reflecting from the seawall (highly modified after Silvester, 1977). The approaching wave crests meet the reflecting wave crests at approximately right angles, generating a flow parallel with the shore in the downdrift direction. This current scours a channel a few yards seaward of the wall. The beach downdrift of the seawall will also erode due to loss of sand into the scour zone. (B) Waves reflecting off a seawall at Hampton Beach, New Hampshire. Solid arrow indicates approaching wave and dashed arrow indicates reflected wave moving away from the seawall at a 90-degree angle to the approaching wave. Photograph by A.D. Hartwell taken circa 1969.

beaches seldom build out in front of them, for a variety of reasons, including wave reflection.

Sea-level Rise

He said, "I'm surprised they don't lock you up – a reasonable man. It's one of the symptoms of our time to find danger in men like you who don't worry and rush about. Particularly dangerous are men who don't think the world's coming to an end."

John Steinbeck - SWEET THURSDAY

When coastal geomorphologists like ourselves use the term sea-level rise, we normally insert the word **relative** as a modifier, because there are two possible factors that create major changes in sea level: 1) changes in the volume of water in the world ocean; and 2) the moving up and down of the Earth's surface. Changes in the volume of water in the world ocean are mostly related to major changes in the world's climate. For example, during the peak of the last ice age, around 20,000 years ago, the surface of the world ocean was about 350 feet below its present level, because so much of the Earth's hydrosphere was taken up in the huge ice sheets that covered large land areas of some of the continents.

A number of reasons can cause a land mass to rise or fall, including (to name but a few):

1) Tectonic activity such as mountain building. In areas of Alaska and Baja California, we have observed former beach lines elevated hundreds of feet up the sides of the mountains along the shoreline. The origin of the raised terraces along the Central California Coast with respect to sea-level changes and tectonic uplift was discussed earlier.

2) Elevation of the land as a result of the melting of the large ice sheets present during the Pleistocene epoch, because the Earth's surface was depressed under the heavy load of the ice. On Herman Melville Island in northern Canada, we observed numerous lines of beach deposits elevated many tens of feet back of the shoreline as a result of a process called **glacial rebound** as the land rose slowly in response to the removal of the weight of the ice.

3) Sinking of land where large rivers have dumped huge sediment loads on the continental shelf. This kind of sinking, aided by compaction of the sediments, has caused relative sea level to rise up to 4 feet per century in areas of Louisiana, where former lobes of the Mississippi River Delta were abandoned as the river shifted positions. Once the delta lobe was abandoned, the river no longer supplied sediment to continue the delta-building process, and as a consequence, the abandoned lobe began to sink into the abundant underlying mud deposits.

Available data suggest the following historical rates of sea-level rise along the tectonically stable shorelines, such as those that occur on many trailing-edge coastlines (partly from a summary by Pilkey and Pilkey-Jarvis, 2006):

1) Between 12,000 and 6,000 years ago – Overall rate of **3 feet per century**, with possible "blips on the curve" of as much as 10 feet per century. Also, there were some fairly well-documented stillstands during this rise, at which time river deltas, barrier island systems, and so on

were deposited, the remnants of which are still present on the continental shelf in places.

2) Between 6,000 and 4,500 years ago – probably around **1.5 feet per century**.

3) Between 4,500 and the present – This is the time interval we refer to as the present **"stillstand"** of sea level. It is during this period of time that most of the modern deltas and barrier islands on the trailing edge coastline of the North American continent have been formed. Others have suggested 5,000 to 6,000 years ago as the start of the stillstand on the California coast. Numerous sea-level curves show that sea level has bumped up and down several times during this "stillstand" period, sometimes more than 3 feet.

Having been working on the shorelines around the world for the past 35 years, it is clear to us from simple observation that sea level is rising at the present time. How much? Probably around one foot per century on the tectonically stable east coast.

The theory that a rising sea level is the major cause of beach erosion worldwide prevails in both scientific and popular literature. A careful analysis of beach erosion in a number of settings casts considerable doubt on this assumption, particularly with regard to problems of concern in the near future. In fact, we think that, in most cases, man's impact on sediment supplies and unwise construction practices cause most beach erosion problems. However, the opinions of the experts differ markedly on this topic.

Our friend and coastal engineering *guru*, Cyril J. Galvin, challenged a report published in 1981 by a group of coastal geologists, who cited rising sea level as a primary cause of beach erosion (Concerned Coastal Geologists, 1981). In his challenge, Galvin stated, *"sea-level change has negligible effect on shore erosion, compared to fluctuations in longshore sand transport rate."* On a site-specific basis, his assertion is demonstrably true. In South Carolina, for example, beaches that are eroding tens of yards per year are located within a few hundred yards of beaches accreting (building seaward) tens of yards per year. Such differences are usually related to sediment bypassing at major tidal inlets during which a sand deficit is created on one side of the inlet. Obviously, both the eroding and depositing beaches in these instances were subject to the same conditions of sea-level change. On the other hand, changes in water level clearly impact shoreline erosion in the Great Lakes, which show 11 year cycles of major water-level changes (Hands, 1977).

Water-level changes not related to the aforementioned global patterns also affect beach erosion trends. On the South Carolina coast, with its considerable tidal range (near 10 feet during some spring tides), the most severe erosion on the beach occurs at high tide. During spring tides, higher levels of the beach are exposed to wave action than during neap tides, so erosion is at a maximum during spring tides under similar wave regimes. Consequently, coastal storms usually cause maximum rates of beach erosion when they cross the coast during high spring tides. As noted earlier, the *El Niño* storms of 1982-83 caused more erosion on the California coast than the 1997-98 *El Niño* storms partly because those in 1982-82 occurred during spring tides (Griggs, Patsch, and Savoy, 2005; Dingler and Reiss, 2002; Griggs and Brown, 1998).

Water-level changes also show seasonal variations. On the east coast of the U.S., sea level is generally lowest in the spring and highest in the fall. This fact is well known by east coast developers, who have built too close to the beach, as the high tides in the fall commonly cause them problems. According to Komar (1976), these annual sea-level changes can be attributed primarily to seasonal variations in climate and ocean water properties. However, much is yet to be learned about this process.

Why the issue of **global warming** and the potential for it to cause sea level to continue to rise, possibly at more accelerated rates in the future, is such a controversial political issue is a complete mystery to us, because the evidence is indisputable that: a) global warming is occurring, based on evidence such as melting glaciers (which we have witnessed first hand in Alaska) and actual temperature measurements; and b) the concentration of CO_2 in the atmosphere is steadily increasing, based on studies such as at the Mauna Loa Observatory in Hawaii over the past 40 years. The relationship of these two issues is treated in detail in Al Gore's book (2006), *An Inconvenient Truth*.

Pilkey and Pilkey-Jarvis (2006) also discussed this topic in an informative way. They summarized the issue succinctly in the following statement:

The evidence for the human connection to global warming is the correspondence of the massive production of excess carbon dioxide produced by burning fossil fuel in the last few decades and the simultaneous atmospheric increase in carbon dioxide. On a purely physical basis, the additional atmospheric CO_2 requires that some greenhouse warming must occur, but how much remains a question.

So what does the future look like with respect to sea level on a worldwide basis? Pilkey and Pilkey-Jarvis (2006) had the following assessment that relates to that issue as well:

The current most widely accepted prediction of sea-level rise is that its rate will be two to four times the present rate by the year 2100, and at that time the sea level will be two to three feet above its present state. At the same time, atmospheric temperatures will rise four to five degrees Fahrenheit. But these are numbers with a lot of leeway.

The problem, as with a lot of models used to predict change in nature, is the assumptions, or data inputs, used to create the model. In the earlier models, the data inputs related to rising sea level were the melting of glaciers, which is well documented as noted above, and the physical expansion of water as it warms. They did not take into account large-scale melting of the big ice fields in Greenland and Antarctica, which is pretty scary when you consider that *"in 2002, Antarctica's 1,255 square mile Larsen B ice shelf broke off and disappeared in just 35 days. And recent NASA data shows that Greenland is losing 53 cubic miles of ice each year – twice the rate it was losing in 1996."* (from a 29 January 2007 article in The State Newspaper). That type of major melting was addressed as one of the possible doomsday scenarios in Al Gore's book.

All of that being said, there is no doubt that an abrupt rise in sea level of two or three feet would affect an unimaginable amount of beach homes, hotels, golf courses, and other structures that are now built close to the beach in many areas. While writing this section of the book on 17 September 2008, we received by e-mail 300 oblique aerial photographs (courtesy of coastal geologist Dr. Richard Watson) of the coastline impacted by hurricane Ike on the Texas Coast (notably Galveston Island and the Bolivar Peninsula). The storm surge on the Bolivar Peninsula during hurricane Ike was 12-15 feet and it is "estimated" that 80% of the houses there were "destroyed," but that is a pretty high percentage. The ocean sides of these islands were swept clean of hundreds of beachfront houses in many places.

As pointed out by Griggs, Patsch, and Savoy (2005), speaking with respect to sea-level rise on the California coast, *"the question is not whether sea level is rising, but how high will it rise, when will it reach this level, and how this rise will affect individual coastal communities and development."*

Question is, what can be done to prevent or slow the future rise in sea level? The answer to that question is way beyond the scope of this book.

METHODS COMMONLY USED TO PREVENT SHORELINE EROSION

Engineers have attempted to curtail shoreline erosion for centuries. Their success ratio has been variable, depending upon the vagaries of the sediment supply, changes in water levels, and storm-wave conditions. For purposes of discussion, we divide the techniques used to prevent such erosion into two classes, "hard" engineering solutions and "soft" engineering solutions.

In the past, engineers have usually dealt with shore erosion by building resistant, permanent features that reflect or dissipate incoming waves. Some of these "hard" solutions include:

1) **Seawalls, revetments, and bulkheads.** In places where fixed property, such as highways and large hotels, are threatened by erosion, these three types of features are commonly built. As noted earlier, seawalls, massive concrete structures designed to hold the line against storm-wave erosion (illustrated in Figure 83), are a double-edged sword, so to speak, because, while they tend to succeed in keeping the road or hotel in place, at least until a major storm hits, the beach sand itself is usually sacrificed, even under normal wave conditions. A bulkhead is made of pilings, composed of a wide range of materials, driven into the ground. They are usually built in areas of moderate waves. Revetments are constructed by armoring the slope or face of a dune or bluff with one or more layers of rock, concrete, or asphalt. The armor stones of revetments tend to dissipate waves and inhibit reflection better than vertical walls, thus sand removal by wave reflection is not as drastic as it is for seawalls. These types of structures are severely scrutinized these days by public officials in some states.

2) **Groins.** These features (Figure 82B), which are discussed in some detail in the section on causes of beach erosion, are commonly made of rubble stone, but they may consist of wood, sand bags, gabions (rocks or gravel in wire mesh), or other materials. They work best where waves approach the coast obliquely. Whereas they work well if installed properly in the right place, groins have not proved to be an effective solution to beach erosion in many localities, especially where the dominant wave crests approach parallel to the beach.

3) **Offshore breakwaters.** These features, which are usually composed of riprap or heavy concrete blocks of miscellaneous shapes, are built offshore, detached from the beach in segments several tens of feet long oriented parallel to the beach. They reduce wave energy on their landward sides, which causes sand moving along shore in the longshore sediment transport system to accumulate in their lee, commonly forming tombolos (see example of tombolos in Figure 36). Such an obstacle to the movement of the sand along shore will commonly cause erosion on their downdrift side. Thus, many engineers recommend placing nourished sand in the shelter of these structures so that sand can continue

moving along the shore. Offshore breakwaters have been used successfully to curtail erosion in a number of areas, such as on the shoreline of Israel.

There are many workers in the area of coastal erosion, particularly coastal geologists, who prefer to use solutions to shoreline erosion that do not involve hard structures. Two of the lines of reasoning used to support this position are that hard structures, such as seawalls, accelerate sand loss, and once in place, hard structures are difficult to remove, making it virtually impossible to correct a mistake. Some examples of these types of "soft" solutions include:

1) **Setback lines.** The "softest" of the soft solutions is the construction of a setback line behind the shoreline seaward of which building is prohibited. They work best and are easiest to implement in areas that have not been developed as yet. The criteria used to establish such lines include historical analysis of shoreline trends, preservation of the line of foredunes, defining areas of flooding and storm wave uprush, and so forth. Where such setback lines have been judiciously applied, they usually work. Our group designed setback lines for a development on Kiawah Island, South Carolina in 1974, and no serious beach erosion has occurred on the island since that time except in one place where the setback was not adhered to (Hayes and Michel, 2008).

2) **Sand bypassing.** As noted earlier, jetties constructed to stabilize navigational channels usually block the flow of sand along the shore, with severe erosion commonly developing downdrift of the jetties. This is one of the more common major causes of beach erosion problems around the world. This type of erosion problem may be at least partly solved by installing mechanical sand by-passing systems, such as land-based dredging plants, pumping systems, and so on, which move the sand from the updrift to the downdrift sides of the jetties. Such systems have been established in many areas. The sand by-passing system at the Santa Barbara Harbor, California is an example.

3) **Relocating a tidal inlet.** Migrating tidal inlets erode the shore as they move down the coast. Relocating the inlet back up the coast in the direction from which it came would relieve the down shore area of erosion, at least until the inlet migrates back. To our knowledge, one of the major examples, and maybe the only one up to that time, of this was carried out by our group in March 1983 at Captain Sams Inlet on the South Carolina Coast, when the inlet was moved to a new dredged channel approximately 4,000 feet up the coast (to the northeast). This stopped the erosion of the inlet into a development to the south, as well as provided a huge volume of sand to

an eroding beach to the south when the abandoned ebb-tidal delta of the original inlet was driven ashore by wave action. At an erosional area along a golf course that had been fortified by sand bags and riprap, the sand beach built out 1,000 feet as a result of this project (Hayes and Michel, 2008). Compared to establishing hard structures or jetties, this was a relatively inexpensive process. The downside is that it has to be repeated about every 14 years or so at this particular site, but it would take many repeats to reach the costs of other types of protection.

4) **Beach nourishment.** Beach nourishment with sand is the most commonly used of the "soft" solutions. Numerous projects of this type have been carried out around the world, some of which involve moving millions of cubic yards of sand (e.g., at Miami Beach, Florida; California projects are discussed later). Sources for such sand include dredging sand deposits on the continental shelf, dredging sand bodies associated with tidal inlets, moving sand by various mechanisms from adjacent beaches that have an abundance of sand, and hauled or pumped in from land-based sources. The major objection to beach nourishment schemes is that they are not permanent. Watching millions of dollars worth of sand wash away within a few years does not appeal to either public servants or even casual observers. Therefore, such projects require a careful analysis of monetary costs and benefits balanced against the aesthetic, economic, and recreational value of maintaining a sand beach in place.

LANDSLIDES

The retreat of a shoreline is augmented considerably when a landslide moves a mass of debris from the coastal cliff into the intertidal zone. Landslides are an example of the process called mass wasting, which is defined as *a mechanism whereby weathered materials are moved from their original site of formation to lower lying areas under the influence of gravity.* Harden (2004) illustrated the general characteristics of landslides with a diagram we have modified slightly in Figure 84A, and made the following cogent points about these features:

1) In addition to feeding the longshore sediment transport system, landslides have dammed rivers, destroyed hundreds of homes, and killed several people *in the last 30 years.*

2) A slow movement of material, called creep, occurs *on virtually all hill slopes.*

3) When mass movement occurs abruptly, *a visible landslide scar is left in the area where debris was detached, which are typically bowl-shaped, with a steep slope at the head of the slide area* (Figures 84A and 84B).

4) The lower end of the slide debris is a *lumpy and disrupted mass.*

5) Debris flows, which are created *when rock and soil become completely saturated, sometimes becoming so mobile that they can obtain speeds of hundreds of kilometers per hour.* Needless to say, these flows are exceedingly dangerous and houses should never be built on slopes prone to such slippage.

Because of its location on the tectonically active western side of the North American Plate, several conditions enhance the probability of landslides along the Central California Coast, such as abundant steep slopes, uplifted naturally weak materials, including unconsolidated sediments or relatively young sandstones and shales, and the common occurrence of extremely sheared and faulted rocks. Other notable favorable conditions for the generation of landslides include the frequent earthquakes, wet winter storms, and big waves, especially during the *El Niño* storms, that under cut the rock cliffs back of the beach. The Franciscan rocks that crop out all along the shores of the Coast Ranges (Figure 16A), with their serpentinites, faulted shales and sandstones, and abundant mélange, are especially susceptible to landslides.

Perhaps the most famous of the coastal cliff landslides in California occurred in the Portuguese Bend area in the Palos Verdes Hills. Major slides in 1956 and 1976 cost many millions of dollars in damages and over 150 houses were either badly damaged or completely destroyed. According to Griggs, Patsch, and Savoy (2005), *because marine erosion around the base of the cliffs of the Palos Verdes Peninsula continues, and because material delivered to the shoreline by the landslides is removed about as quickly as it arrives, the cliffs will continue to be oversteepened and thus unstable. The litany of failure-prone cliffs and slopes should be warning enough against building too close to the cliff edge.*

CALIFORNIA'S SHORELINE EROSION PROBLEM – AN ASSESSMENT

The coastal erosion problem in California is the subject of a recent book, *Living With the Changing California Coast* by Griggs, Patsch, and Savoy (2005) which, in addition to providing thorough discussions on the processes involved in such erosion, gives detailed maps of the hazards levels and erosion rates for the entire coastline of the state. Anyone possessing coastal property or with any kind of interest in the shoreline erosion issues throughout the state should own a copy of this book. A very important observation the authors made in the book is that the *"conflict between coastline development and the hazards associated with it"* has

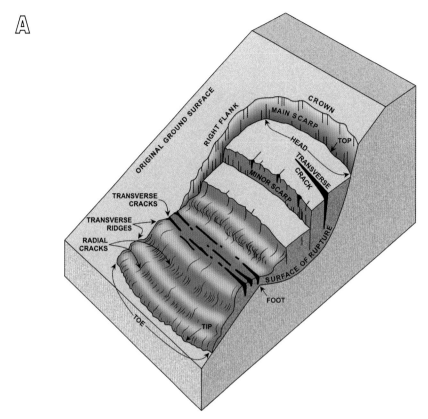

FIGURE 84. Landslides. (A) The features of a typical landslide (modified after Harden, 2004; original source: Highway Research Board, 1958). (B) Landslides near Point Sal (Santa Barbara County). These slides closely mimic the model given in A. It is 270 yards between arrows A and B. This oblique aerial view, taken in January 1989, is courtesy of the California Coastal Records Project, Kenneth and Gabrielle Adelman.

become more evident in recent years for five reasons:

1) An increased migration to coastal communities and the desirability of owning oceanfront property;

2) The progressive erosion of oceanfront yards and vacant property causing structures and utilities to be undercut and or threatened;

3) The human-induced acceleration of seacliff erosion due to cliff-top construction with its associated roof, patio, driveway, and street runoff and its landscape watering;

4) An era of more frequent and severe *El Niño* events beginning in 1978, bringing heavy rainfall, elevated sea level, and larger waves; and

5) Coastal engineering projects such as groins, jetties, and breakwaters have directly or indirectly accelerated erosion rates in adjacent areas, principally by trapping sand and starving down-coast beaches.

Another sobering bit of information was provided by a U.S. Geological Survey report titled "Coastal Beach Erosion on Increase in California" released on 18 September 2006, in which the following was revealed:

The net shoreline change in the short-term (25-40 years) indicates that 66 percent of California's beaches are eroding. Central California's beaches, which covers the area from Point Reyes to just north of Santa Barbara, shows the highest percentage of erosion.

As discussed in considerable detail in the earlier section on major coastal storms, it is during the *El Niños* that the most dramatic shoreline erosion occurs along the Central California Coast. More discussion of that effect is not warranted here, but specific erosion events attributed to such storms are described where appropriate in the discussion of the different coastal areas given in Section III.

Studies on the California coast led the way in determining sediment budgets for sand beaches (Inman and Chamberlain, 1960; Bowen and Inman, 1966; Inman and Masters, 1991). Applying the concept of the littoral cell, which is illustrated in Figure 66, the budget is calculated by determining how much sand is brought into the cell by rivers and cliff and bluff erosion as opposed to how much is lost to wind-blown dunes, submarine canyons, and so on. Such determinations allow engineers to calculate the potential annual losses, an important factor in determining where artificial beach nourishment or shore protection structures might be desired.

As also noted earlier, the stabilization of coastal cliffs with riprap and seawalls has reduced the volume of sand added to the beach from that source, as well as created some erosion problems downdrift. According

to Stamsky (2005), approximately 10% of California's coastline is currently armored. But, the percentage of the shoreline on the Central California Coast that is armored is considerably less than that at 4.3% (NOAA/RPI, 2006).

The chief cause of loss of sand formerly provided to the littoral cells is the construction of dams on the streams, which restrict the volume of sand reaching the shore in two ways: a) trapping the sediment behind the dams; and b) reducing volumes of the peak floods that formerly were the most important mechanism for getting the sediment to the shore. Griggs, Patsch, and Savoy (2005) made the following observations on the history and impacts of these dams:

1) Since the process began, there have been 539 dams built within the coastal watersheds in California that drain directly into the Pacific Ocean.

2) These dams have reduced the average annual sediment supply by more than 25% to the 20 major littoral cells in California.

3) On the Southern California Coast, where sand beaches are in great demand, sediment supply has been reduced by over 50% to half of the littoral cells.

4) The northern California sand supply is considerably less altered.

Another practice that has deceased the amount of sand the streams carry to the shore is mining the sand from the channels and flood plains of the streams before it gets to the shore, a common routine almost everywhere.

Griggs, Patsch, and Savoy (2005) also pointed out that, in Southern California, *artificial beach nourishment associated with a number of large coastal construction projects essentially kept pace with sediment losses from dam construction over much of the twentieth century* (e.g., sediment dredged from harbors, coastal dunes, etc. and placed on the beach). However, today such construction projects are over and the two big *El Niño* storms since the early 1980s have caused some serious erosion problems. Suggestions have been made for removal of some of the dams and, of course, artificial beach nourishment projects are another possibility. In 2001, a major nourishment project in San Diego County placed two million cubic yards of sand on 12 beaches at a cost of $17.5 million. According to Griggs, Patsch, and Savoy (2005), *beach surveys indicated that within a year most of the sand had moved alongshore or offshore.* The problem is, such nourishment projects can obviously be quite costly, so where does the money come from? A longshore sediment transport rate of 300,000 cubic yards of sand per year on many of the beaches in the state exacerbates the problem. As far as we know, no major beach nourishment projects near the scale of the one in San Diego County have been carried out on the Central California Coast.

CONCLUDING REMARKS

Getting back to the issue of the numerous methods that have been tried to stop shoreline erosion, the material presented up to this point is a far from complete list. But as you may have guessed by now, this issue goes beyond a simple engineering or geological problem, having blossomed into a socio-economic, political, and even cultural phenomenon. The fact that a large percentage of the world's population lives near the coast amplifies interest in the subject.

Over simplifying the story quite a bit, one could say that the scientists and engineers who are supposed to know something about shore erosion fall into two major camps. On the one hand, we have environmentally oriented marine scientists, particularly coastal geologists, who are more inclined to let nature take its course and not crowd the ever narrowing beaches with beach cottages, hotels, etc. They think most sand beach erosion is caused by rising sea level or bad engineering practices and are more-or-less convinced that the problem will get worse with an increase in global warming. One of the leading spokespersons of this faction is another one of our friends, Dr. Orrin H. Pilkey, Emeritus Professor at Duke University, who has been known to use catch phrases such as *"New-Jersey-fication of the coast"* and *"let the lighthouses fall into the sea."* He has also authored and edited many books on living with the coast. He is not even much of a fan of some "soft" solutions, such as beach nourishment, heavily criticizing some of the models engineers use to predict the life of the nourished sand on the shore. Needless to say, Dr. Pilkey is not idolized by many coastal engineers. However, some of his arguments have been heeded by policy makers in different states of the U.S. (e.g., North Carolina's ban of seawalls on the open beach). If sea level does rise abruptly, many of Pilkey's arguments will be awfully difficult to refute.

On the other hand, we have the coastal engineers, who are trying to perform a very difficult job. Our friend, Bill Baird, a world-renowned coastal engineer, has founded a company based in Canada (Baird and Associates) with the following motto – "making the world a better place to be!" His company does that by designing ports in developing countries to fuel their economy, preventing coastal villages in Africa from falling into the sea, and so on.

Where do we stand on this issue? To put it simply,

our hearts are with the notion that the shorelines should be left free of so much development, but our heads love the science of dealing with the problem. Of course, there are numerous classic cases of misunderstanding and mismanagement of shore erosion issues in coastal developments. We have owned and still own coastal property in the state of South Carolina, but never on the open beach. We were once offered a lot on one of the barrier islands in South Carolina as "payment in kind," but didn't take it because of concerns about erosion during hurricanes. However, when faced with a tricky scientific problem to solve, such as moving a tidal inlet or keeping an African village from falling into the sea, our heads win and we enthusiastically tackle the problem. In fact, we feel pretty good about some of the projects we have done around the world.

Artificial beach nourishment is one of those coastal engineering practices burdened with controversy. In our naturalist's guide to the South Carolina Coast (Hayes and Michel, 2008), we presented a previously published debate in a business magazine on projects of this type, specifically those carried out at Myrtle Beach, South Carolina. The two main debaters were Dr. Orrin Pilkey, already introduced, and Dr. Tim Kana, who did his dissertation at the University of South Carolina under co-author Hayes' supervision.

In 1976, Hayes and Kana co-authored a book, *Terrigenous Clastic Depositional Environments*, which presented some of the earlier data on coastal processes and sedimentation. Also, before he formed his present company, Tim was our partner at RPI for several years. Since that time, he has recommended and supervised numerous beach nourishment projects. As professional geologists and members of the same professional societies, as well as through University contacts, we have had many interactions with Orrin Pilkey over the years.

Back to the debate, two excerpts from it that relate specifically to the recent beach nourishment projects at Myrtle Beach follow (this "debate" took place in the fall of 2007):

Pro (Kana) – *Their lots didn't wash away, their seawalls didn't fail, and now they look out over a band of dunes to a white, sandy beach that accommodates thousands of visitors each day.*

Con (Pilkey) - *Beach nourishment encourages high-density development immediately adjacent to the beach. In a time of rising sea level in a location where big storms are a certainty, beachfront construction is at least irresponsible and at worst can only be considered a form of societal madness.*

The very morning we are writing this, 24 September

2008, co-author Hayes had observed on CNN an owner of a beachfront house on southern Galveston Island, who had just seen what hurricane Ike had done to his former abode, say to the reporter "I'll need some help to replace this." Presumably he meant financial help. Societal madness indeed.

Kana and Pilkey, who both have an abiding love and respect for the natural environment, obviously approach the subject of beach nourishment in a different way. Kana has a practical approach, which we certainly sympathize with, and he strives to provide practical solutions to problems presented by his clients whenever possible. When RPI followed up on a suggestion our group made to relocate Captain Sams Inlet, South Carolina in 1983 to provide a major, never-before-tried method of beach nourishment, Tim Kana was the on-the-ground supervisor of the project. He also continued to monitor the project in later years and has published several papers on the results (e.g., Kana and McKee, 2003).

Obviously, Pilkey is different, preferring a "let it be" philosophy. He has an agenda, which we also sympathize with, and he sticks to it, much to the chagrin of many coastal engineers who question the authenticity of some of his data and suppositions (as noted earlier). The clearest exposition of his philosophy we can find in his writings appears in the final chapter of his artfully illustrated and informative book, *A Celebration of Barrier Islands,* published in 2003. In this book, he suggested that barrier islands lead a Gaia-like existence. As you probably know, the Gaia hypothesis is an ecological theory that proposes that the planet Earth functions as a single living organism. Beliefs somewhat similar to this go back at least to the great Spinoza in the mid 1600s, whose philosophy was cited, but not necessarily believed, by Einstein. Actually, this concept goes back thousands of years (under different names), but you probably get the point.

To quote Pilkey's summary of this matter exactly, he said:

> *Putting barrier islands in a Gaia context does provide a basis for thinking about how to live with them in a way that will keep islands existing and evolving, a way to preserve them for future generations. Perhaps more important, the distinction between life and death, as it applies to a barrier island, should provide guidance to distinguish good and bad development practices.*

Strangely enough, in co-author Hayes' book entitled *Black Tides,* published by The University of Texas Press in 1999, a sentiment similar to Pilkey's was expressed. In that book, which deals with our experiences responding to most of the major oil spills that have occurred since 1974, reference was made in the last chapter to a short article entitled "Ensoulment of Nature" by Gregory A. Cajete. That paper was published in the book Native Heritage in 1995. Cajete stated in the article that the importance the American Indians put on connecting with their place of origin (i.e., the environment) is not just a romantic notion out of step with the times, but rather the quintessential ecological mandate of our times. He stated further that the native peoples experienced nature as a part of themselves and themselves as a part of nature. Under this concept, ensoulment means that the human participates with the Earth as if it were a living being. Hayes concluded this subject with the following comment:

> *If we all adopted this frame of mind, I'm sure it would be one more step in the honorable direction toward making the Earth a better place to be, which is something all of us tree-huggers and right-thinking engineers want to do, correct?*

That final chapter in *Black Tides* concluded with a quote from the Bureau of American Ethnology Collection (published in Brown, 1970), which follows:

> *The old men*
> *say*
> *the Earth only*
> *endures.*
> *You spoke*
> *truly.*
> *You are right.*
> *The Earth only endures.*

Of course, the animals and plants that lived on the Earth have not fared so well at times. An example of those hard times being when the Earth was struck by a large asteroid or comet, such as the one proposed to have caused the Cretaceous/Tertiary extinction event about 65 million years ago. But, sure enough, the Earth itself has endured, so far.

And there are times when the people and the hills and the Earth, all, everything except the stars, are one, and the love of them all is strong like a sadness.

John Steinbeck – TO A GOD UNKNOWN

SECTION II
Coastal Ecology Overview

But when the tide goes out the little water world becomes quiet and lovely. The sea is very clear and the bottom becomes fantastic with hurrying, fighting, feeding, breeding animals.

John Steinbeck – CANNERY ROW

8 INTRODUCTION

The material presented in this section is based primarily on a detailed synthesis of the coastal environments and biological resources of the Central California Coast conducted by our group (RPI) and published as a coastal resource atlas in 2006 (NOAA/RPI, 2006). The project was funded by the NOAA Monterey Bay National Marine Sanctuary, California Department of Fish and Game Office of Spill Prevention and Response, and the Monterey Bay Sanctuary Foundation. In addition, support also came from the NOAA Office of Response and Restoration in Seattle, Washington.

The part of the project related to the types of shoreline present along the coast was discussed in some detail in Section I (the shoreline types are listed in Table 3). Section III includes maps showing the distribution of those shoreline classes in each of the seven major compartments of the coast (Figure 1). Maps showing the distribution of the key biological resources in several of the more ecologically diverse localities are also included in Section III; Figure 85 (Point Reyes) is an example of one these maps. Specific details on sites for the best potential birding, whale watching, tidal-pool observation, and so on are discussed in Section III.

The information presented in the atlas and the associated database were developed to provide summary information on sensitive natural and human-use resources for the purpose of oil and chemical spill planning and response; however, these materials are also very useful for other environmental and natural resource planning purposes, as well as providing key basic information for this naturalist's guide.

The biological information presented in the atlas was collected, compiled, and reviewed with the assistance of biologists and resource managers from the following agencies:

- California Department of Fish and Game
- National Oceanic and Atmospheric Administration (NOAA), Monterey Bay National Marine Sanctuary

- NOAA National Marine Fisheries Service
- Moss Landing Marine Laboratories
- U.S. Geological Survey
- NOAA Gulf of the Farallones National Marine Sanctuary
- National Park Service, Point Reyes National Seashore, and Golden Gate National Recreation Area
- University of California Santa Cruz (UCSC), Long Marine Laboratory
- Point Reyes Bird Observatory
- U.S. Fish and Wildlife Service
- UCSC, Santa Cruz Predatory Bird Research Group
- Vandenberg Air Force Base
- H.T. Harvey and Associates
- Pacific Eco Logic
- Ventana Wildlife Society
- Pepperdine University
- Partnership for Interdisciplinary Studies of Coastal Oceans (PISCO)
- University of California Santa Barbara
- NOAA National Centers for Coastal Ocean Science Biogeography Team
- Elkhorn Slough National Estuarine Research Reserve

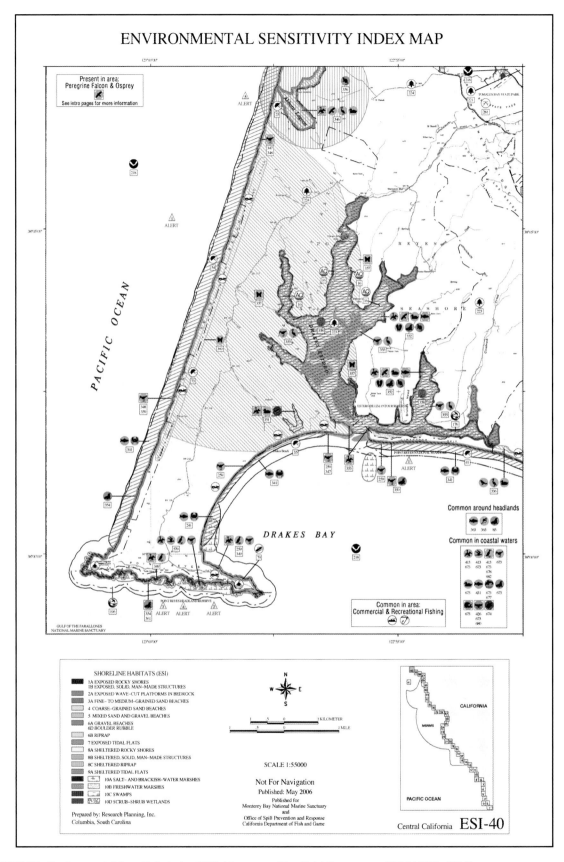

FIGURE 85. Environmental Sensitivity Index (ESI) Atlas map of the Point Reyes area (NOAA/RPI, 2006).

9 WHY THE CENTRAL CALIFORNIA COASTAL WATERS ARE SO BIOLOGICALLY RICH

The Central California coastal waters support a rich biological community, starting with the smallest of plankton and ending with the largest of whales. This rich community is a result of the combination of a steep offshore continental slope, ocean currents that come very close to shore, seasonal wind patterns, and the shape of the coastline. The California Current flows along the shore from north to south, bringing cool water from the north. Starting in March, winds blow consistently from the northwest. Because of the Coriolis effect, the water currents moving south along the shore veer to the right (offshore), and deep, cold, and nutrient-rich waters "upwell" to the surface. The nutrients in the water, plus exposure to sunlight, spurs blooms of phytoplankton (tiny plants that are the basis of the food chain in the ocean) that support zooplankton (small animals that are the next level in the food chain, such as krill and the larvae of many fish and shellfish). They, in turn, are eaten by other plankton, fish, and filter-feeding animals such as shellfish, jellyfish, and whales. Fish and shellfish are eaten by other fish, birds, sea otters, seals, sea lions, dolphins, and whales. The cold water at the surface is also the reason for the many foggy mornings along the coast in the summer. This upwelling season lasts until about September, when the winds die down some and the water surface becomes relatively warm.

From November to February, the winds blow mostly from the southwest. Again, because of the Coriolis effect, the winds create a current that flows towards the shore and to the north, inland of the California Current. The water in this current is quickly depleted of nutrients, so the biological productivity is shut down until March, when the process starts again. The diversity of fish, birds, and marine mammals along the Central California Coast is a direct result of the seasonal upwelling of nutrient-rich waters. However, this pattern is disrupted during what is known as the *El Niño*. During *El Niño* periods, unusually warm, nutrient-poor water from the south flows close to shore. These *El Niño* periods can be catastrophic to marine communities, causing phytoplankton production to drop, fisheries to decline, seabirds to starve, and marine mammals to stop breeding. Some of the effects of an *El Niño* can last for decades. Thus, there is much concern about how climate change could alter the seasonal upwelling pattern that is the key to the productivity of the California coastal waters. Research indicates that upwelling will be stronger, but delayed (Snyder et al., 2003). More frequent and stronger *El Niño* periods could be very disruptive as well.

10 KEY BIOLOGICAL RESOURCES OF CENTRAL CALIFORNIA

FEATURES ON THE ECOLOGICAL MAPS PRESENTED IN SECTION III

Animal and plant species are represented on the maps shown in Section III as polygons, points, or lines (see Figure 85, the map of Point Reyes as an example). Species are organized into groups and subgroups based on their behavior, morphology, taxonomic classification, and spill vulnerability and sensitivity. The icons below are used to represent this grouping on the maps:

MARINE MAMMAL

- Dolphin/Porpoise
- Pinniped
- Sea Otter
- Whale

BIRD

- Diving Bird
- Gull/Tern
- Passerine Bird
- Raptor
- Seabird
- Shorebird
- Wading Bird
- Waterfowl

TERRESTRIAL MAMMAL

- Small Mammal

REPTILE

- Amphibian/ Other Reptile
- Turtle

FISH

- Fish

INVERTEBRATE

- Bivalve
- Cephalopod
- Crab/ Other Invertebrates
- Gastropod
- Insect

HABITAT

- Eelgrass

Polygons, points, and lines are color-coded based on the species groups, as shown below:

- Marine Mammal
- Bird
- Terrestrial Mammal
- Fish
- Invertebrate
- Reptile
- Kelp
- Eelgrass
- Multi-element Group

For species that are found throughout general geographical areas or habitat types on certain maps, displaying the polygons for these species would cover large areas or would obscure the shoreline and biological features, making the maps very difficult to read. In these cases, a small box is shown on the map which states that the species are "Common in ..." (e.g., "Common around headlands" or "Common in coastal waters"). Figure 85 shows all of these mapping conventions.

In the following sections, we provide an overview of the major groups of animals and habitats that are present along the Central California Coast. Site-specific details for some areas are provided in Section III.

MARINE MAMMALS

Introduction

Marine mammals that can be seen along the Central California Coast include whales, dolphins, porpoises, pinnipeds (seals and sea lions), and sea otters.

Whales Along the Central California Coast

Sighting whales from the coast or on a whale-watching boat trip is one of the greatest thrills you can experience. You most often first see the spout, the moist air being exhaled from the blowhole. In fact, expert whale watchers can identify the whale by the size and shape of the spout. You can even hear the sounds of spouting, if the whales are close. Listed below are the fifteen species of whales that are present off the Central California Coast, attesting to the richness of these coastal waters. However, you will most likely see humpback, gray, blue, and minke whales. Also, the chance of seeing each species varies by season. So, if seeing a whale is an important objective of your visit to the coast, you will need to plan your visit during the peak seasons. A whale-watching boat trip is the best way to really make sure you see whales. There are boat charters operating out of Bodega Bay, San Francisco, Half Moon Bay, Santa Cruz, and Monterey, among others. Figure 86 shows the gray whale migration routes, from their summer feeding area in the Bering Sea to their winter breeding and calving area from San Diego to Baja, Mexico.

Baird's beaked whale *(Berardius bairdii)*: They are present May-October in deep offshore waters, but have been rarely sighted.

Blue whale *(Balaenoptera musculus)*: Federally listed as endangered. They are present June-November, feeding mostly on krill (shrimp-like planktonic crustaceans).

Cuvier's beaked whale *(Ziphius cavirostris)*: They are present year-round in deep offshore waters, but have been rarely sighted.

Dwarf sperm whale *(Kogia simus)*: They are present year-round in deep offshore waters, but have been rarely sighted.

Fin whale *(Balaenoptera physalus)*: Federally listed as endangered. They are present year-round, but are rarely seen because they occur in deep offshore waters.

Gray whale *(Eschrichtius robustus)*: They are present November-May. They migrate close to shore and are often seen. The entire population migrates from northern feeding areas December-February with a January peak; they migrate from southern breeding and calving areas February-May with a March peak (Figure 86). On their way north late in the season (March-May), the females and calves tend to hug the shore to avoid killer whales, which are known predators of the calves.

Humpback whale *(Megaptera novaeangliae)*: Federally listed as endangered. They are present March-November,

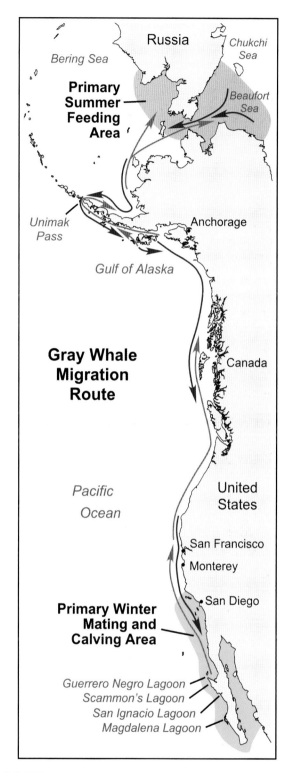

FIGURE 86. Map showing the migration routes of gray whales from their summer feeding areas in the Bering Sea to their winter mating and calving areas in southern California and Mexico. They migrate north from February to May, with a peak in March, and south from December to February, with a peak in January. Map is courtesy of MTYcounty.com (2002).

133

but most commonly seen in summer and fall. They are among the most commonly sighted whales in Central California.

Killer whale (*Orcinus orca*): They are present year-round and are most frequently sighted in Monterey Bay from January-May and from September-November.

Mesoplodont beaked whales (*Mesoplodon spp.*): They are present year-round in deep offshore waters, but have been rarely sighted.

Minke whale (*Balaenoptera acutorostrata*): They are present year-round, calving March-May, occurring in moderate concentrations from Bodega Head to Pescadero Point, and low concentrations south to Point Conception.

Northern right whale (*Eubalaena glacialis*): Federally listed as endangered and considered to be the most endangered whale species. They are present March-May, although there have been only seven sightings off Central California.

Pygmy sperm whale (*Kogia breviceps*): They are present year-round in deep offshore waters, but have been rarely sighted.

Sei whale (*Balaenoptera borealis*): Federally listed as endangered. They are present year-round, calving September-March. They occur in deep offshore waters and have been rarely sighted off Central California.

Short-finned pilot whale (*Globicephala macrorhynchus*): They are present year-round but have been rarely sighted.

Sperm whale (*Physeter macrocephalus*): Federally listed as endangered. They are present year-round, with calving June-October. They are found far offshore in very deep water.

Some of the leading places to observe whales on the Central California Coast are at Bodega Head, Point Reyes, Montara State Beach, Pillar Point, Julia Pfeiffer Burns State Park, and San Simeon State Park. A kiosk at Bodega Head advises the observers to *watch for spouting as gray whales migrate south from late November to January. Just beyond the surf zone you might see their frolicking, lazing about, or lifting their tails, flukes, or even their heads high out of the water, completing a more than 10,000 mile journey–the longest migration of any mammal. The gray whales are born black but, as they age, barnacles and whitish scars give them a gray appearance. When feeding, they use their down-turned snouts and bodies to plow along the bottom. From the turbid water they filter out crustaceans, molluscs, and bristle worms through rows of comb-like plates called baleen.*

If you go on a whale-watching trip, you will also see large numbers of dolphins and porpoises, including bottlenose dolphin, Dall's porpoise, harbor porpoise, long-beaked common dolphin, northern right-whale dolphin, Pacific white-sided dolphin, Risso's dolphin, and short-beaked common dolphin. You will also see pinnipeds (sea lions and seals), sea otters, sea turtles, and sharks.

Dolphins and Porpoises

There are seven species of dolphins and porpoises in Central California: Bottlenose dolphin, long-beaked common dolphin, northern right-whale dolphin, Pacific white-sided dolphin, Risso's dolphin, short-beaked common dolphin, Dall's porpoise, and harbor porpoise. Short-beaked common dolphins are the most abundant cetacean off California; the less abundant long-beaked common dolphin was recognized as a different species only in 1994. The northern right-whale dolphin occurs off California primarily during the colder water months and shifts northward into Oregon and Washington as water temperatures increase in late spring and summer.

Risso's dolphins and Pacific white-sided dolphin are commonly seen in slope and offshore waters of California, Oregon, and Washington. In Central California NOAA has identified three distinct stocks of harbor porpoise: San Francisco-Russian River stock (~8,500 animals), Monterey Bay stock (~1,600 animals), and Morro Bay stock (~1,600 animals). Their justifications for this were: 1) fishery mortality of harbor porpoise is limited to Central California (mostly from the halibut set gillnet fishery); 2) movement of individual animals appears to be restricted within California; and consequently 3) fishery mortality could cause the local depletion of harbor porpoise if Central California is not managed separately.

In 1997, NOAA implemented a Pacific Offshore Cetacean Take Reduction Plan, which included skipper education workshops and required the use of pingers and minimum six-fathom extenders. As a result, cetacean entanglement rates in the drift gillnet fishery dropped considerably. Scientists are not quite sure why pingers reduce dolphin and porpoise by-catch. Possible mechanisms are: 1) the pingers may operate as acoustic alarms that alert animals to the presence of fishing gear so they can avoid it; 2) the sound may repel the animals away from the gear; and 3) the pingers may disperse the prey that the marine mammals feed on (NMFS, 1997).

Due to its exclusive use of coastal habitats, the California bottlenose dolphin population is susceptible to fishery-related mortality in coastal set net fisheries.

In 2002, a ban on set gill and trammel nets inshore of 60 fathoms from Point Reyes to Point Arguello became effective, greatly reducing the potential mortality of coastal bottlenose dolphins in the California set gillnet fishery. So, through active management and coordinate with the fishery, the fishery by-catch of marine mammals has been greatly reduced.

Pinniped Haulouts and Rookeries

Pinnipeds are a group of carnivorous marine mammals with fin-like limbs. The species found along the Central California Coast include the following:

California sea lion *(Zalophus californianus)*: The most abundant marine mammal in California. Present year-round. Often seen in nearshore waters and hauled out on land in colonies, which can include thousands of animals (mostly males). Large colonies are located at the Farallon Islands, Bodega Rock, Point Reyes, Point Lobos, Año Nuevo, Monterey Peninsula, Point Piedras Blancas, and Point Sal, among others. They breed mostly on offshore islands from the Channel Islands to Mexico; however, pups have been born on Año Nuevo Island and the Farallon Islands. These are often the trained circus "seals."

Guadalupe fur seal *(Arctocephalus townsendi)*: State and federally listed as threatened and rarely seen in California. The only current breeding area is on Guadalupe Island, 180 miles west of Baja California.

Pacific Harbor seal *(Phoca vitulina)*: Commonly seen in the surf zone "watching" you on the shore. They prefer nearshore waters and are often seen hauled out on sandy beaches and tidal flats.

Northern elephant seal *(Mirounga angustirostris)*: Seen during pupping (December-March) and molting (April-July) season in a few areas, such as Point Reyes, Farallon Islands, Point Piedras Blancas area, and Año Nuevo (up to 10,000 individuals!). The rest of the time, they are far offshore.

Northern fur seal *(Callorhinus ursinus)*: Not common in Central California, though common in Southern California and throughout the North Pacific. They come ashore during pupping season (May-August) on the Farallon Islands and have been reported in low numbers at Point Sal.

Steller sea lion *(Eumetopias jubatus)*: Federally listed as threatened. They come ashore during pupping season (May-August) on Año Nuevo Island and the Farallon Islands.

Photographs of pinnipeds are given in Figure 87 and pinniped haulouts and rookeries along the outer coast of Central California are mapped in Figure 88. In Section III, the numbers of pinnipeds at haulouts in each area are provided. These are based on several years of survey data from 1998-2004. The ranges in numbers of animals represent the maximum single day count from the year with the lowest count and the maximum single day count from the year with the highest count.

FIGURE 87. Common pinnipeds of Central California: (A) Stellar sea lion; (B) northern elephant seal; and (C) California sea lions (photo credits: National Oceanic and Atmospheric Administration).

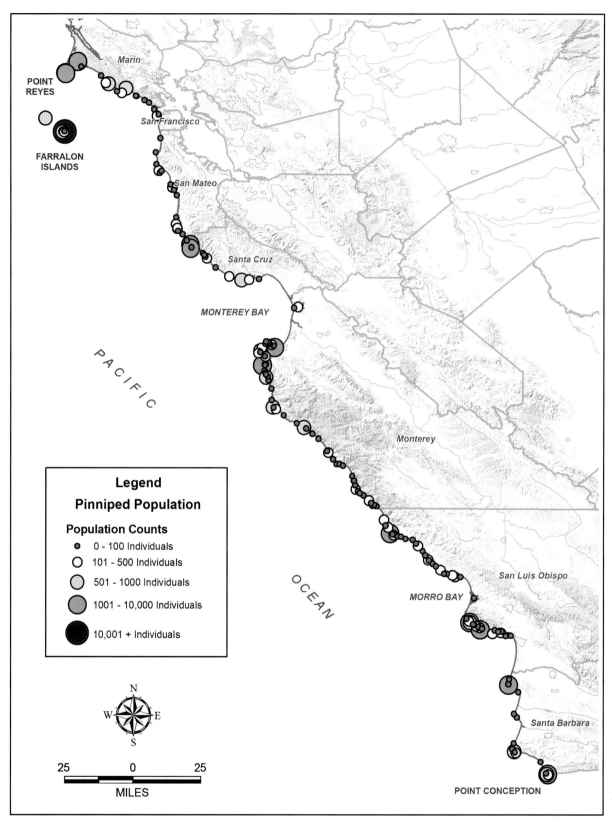

FIGURE 88. Map of the location and size of haulouts on the Central California Coast for the following pinnipeds: California sea lion, Steller sea lion, Guadalupe fur seal, northern fur seal, northern elephant seal, and harbor seal. Many of these marine mammals have made dramatic recoveries after enactment of protection plans.

Sea Otters

The southern sea otter (*Enhydra lutris nereis,* federally threatened) is the smallest marine mammal in North America. Unlike other marine mammals that have blubber to keep them warm in water, sea otters rely on their thick fur to survive. Therefore, they are at great risk from oil spills, which can foul their fur and quickly result in hypothermia and death. Sea otters eat 25% of their body weight each day because of the energy needed to keep warm in cold water. They almost went extinct in California due to hunting pressure, and the California Sea Otter State Game Refuge, extending from the Carmel River to Cambria, was established in 1941. Now, sea otters extend along 300 miles of the California coast, from Tunitas Creek in the north to Coal Oil Point in the south. The 2006-2008 three year average spring population was 2,826 animals, quite the success story for an animal that many considered extinct as recently as the 1930s.

MARINE AND COASTAL BIRDS

Introduction

Central California has a large and diverse group of marine and coastal birds along its coast. It is an important wintering and migratory area, and many species nest on the offshore islands and mainland from February through November. You will see birds everywhere at any time of year! Birds mapped in the coastal resource atlases are grouped into assemblages as listed in Table 4. The maps highlight species that are federally and state listed and coastal nesting, roosting, and rafting locations. Representatives of all of these groups are shown on the maps illustrated in Section III. Information on colony size and species present is discussed for each area. It is important to note that the last comprehensive surveys of all the colonies were conducted in the late 1980s/early 1990s. There have been several upgrades in the 2000s, and the data presented include these latest data.

Threatened and Endangered Bird Species

Ten coastal and marine birds in the coastal Central California region are state or federally listed as threatened or endangered:

Marbled murrelet *(Brachyramphus marmoratus)*: Present year-round in nearshore waters in low numbers south of Monterey and in moderate numbers north.

Xantus' murrelet *(Synthliboramphus hypoleucus)*: Show up in Central California July-October after breeding in Southern California.

TABLE 4. Bird assemblages on the Central California Coast.

ASSEMBLAGE	SPECIES EXAMPLES
Seabirds	Auklets, murres, murrelets, puffins, storm-petrels, albatrosses, shearwaters, and guillemots
Diving birds	Pelicans, cormorants, grebes, and loons
Raptors/Vultures	Condor, osprey, falcons, hawks, and golden eagle
Shorebirds	Plovers, oystercatchers, phalaropes, yellowlegs, sandpipers, willet, tattlers, killdeer, stilt, avocet, curlews, whimbrel, godwits, turnstones, surfbird, knot, sanderling, dunlin, dowitcher, and snipe
Wading birds	Rails, bitterns, herons, and egrets
Waterfowl	Brant, dabbling ducks, diving ducks, and geese
Dabbling ducks	Mallard, gadwall, wigeons, teals, shoveler, and pintail
Diving ducks	Canvasback, ring-necked duck, scaup, scoters, bufflehead, goldeneye, mergansers, and ruddy duck
Gulls	California gull, western gull, herring gull, glaucous-winged gull, and Sabine's gull
Terns	Caspian tern, elegant tern, common tern, Forster's tern, and California least tern

Brown pelican *(Pelecanus occidentalis)*: Present year-round, proposed for delisting in 2008.

California least tern *(Sterna antillarum browni)*: Nesting May-August on beaches along Vandenberg Air Force Base, Oceano Dunes, and Guadalupe-Mussel Rock and dispersing to Avila Beach and Monterey.

California condor *(Gymnogyps californianus)*: Present along the shoreline and inland at reintroduction sites in San Luis Obispo county. Also commonly seen along the shore south of Point Sur in Monterey County.

American peregrine falcon *(Falco peregrinus anatum)*: Uncommon along the coast north of Point Arguello.

Western snowy plover *(Charadrius alexandrinus nivosus)*: Nesting March-September on sand beaches throughout the region, but in decline due to habitat loss.

California black rail *(Laterallus jamaicensis coturniculus)*: Known outer coast locations include Morro Bay, Waddell Creek at Big Basin Redwoods State Park, Bolinas Lagoon, Olema marsh at the head of Tomales Bay, Abbotts Lagoon at Point Reyes National Seashore, and the marshes in the southeast region of Bodega Harbor.

California clapper rail *(Rallus longirostris obsoletus)*: Mostly in San Francisco Bay but also present in Walker Creek marshes near the entrance of Tomales Bay.

Bank swallow *(Riparia riparia)*: Nesting March-August at Fort Funston.

REPTILES AND AMPHIBIANS

Leatherback sea turtles *(Dermochelys coriacea,* federally endangered) occur in coastal waters throughout Central California from May-November. They nest in Indonesia, and the adults swim thousands of miles to the California coast to feed on the abundant jellyfish, on which they almost exclusively feed. The best chance to see leatherbacks is during a birding or whale-watching boat trip. However, you would still have to be lucky to see one. Researchers conducted aerial surveys for leatherback sea turtles along nearly 20,000 miles of transects off Central and Northern California during the period 1990-2003; they only saw 100 animals the whole time (Benson et al., 2007)! Some years, they see only two animals.

The coastal resources maps show a few sensitive terrestrial and fresh water/brackish water species, including California red-legged frog *(Rana aurora draytonii,* federally threatened), San Francisco garter snake *(Thamnophis sirtalis tetrataenia,* state and federally endangered), western pond turtle *(Clemmys marmorata)*, Santa Cruz long-toed salamander *(Ambystoma macrodactylum croceum,* state and federally endangered), and California legless lizard *(Anniella pulchra)*.

FISH AND INVERTEBRATES

Marine and Estuarine Fishes

Fish concentration areas and some general distributions of coastal (e.g., California halibut, *Paralichthys californicus*), kelp-bed associated (e.g., rockfish), sandy habitat associated (e.g., California grunion, *Leuresthes tenuis*), and rocky habitat associated species (e.g., seaperch), particularly those that spawn nearshore during part of the year (e.g., surfperch), are shown on the coastal resource maps and cited in the discussion of the different compartments in Section III. Species associated with important estuarine systems (e.g., Elkhorn Slough, Morro Bay) are also highlighted on the maps and discussed in Section III.

Anadromous Fishes

Fish that live in salt water but migrate to fresh water habitats to spawn are called anadromous, such as salmon, steelhead (a unique form of rainbow trout), and Pacific lamprey. The Central California Coast coho salmon *(Oncorhynchus kisutch,* federally and state endangered) migrate to spawn in streams and rivers from Salmon Creek north of Bodega to Monterey Bay. The Central California Coast (federally threatened), South-Central California Coast (federally threatened), and Southern California (federally endangered) steelhead *(Oncorhynchus mykiss)* occur in many of the coastal streams in Central California. Adult steelhead and coho congregate at river mouths from late October to mid-June as they wait for the spits that close off the stream mouths to breach as a result of the winter rains. Thus, they are called "winter-run." Migration from creeks to the ocean occurs around June 15. During the summer and fall, juveniles are rearing in lagoons and adults and juveniles may be nearshore. Steelhead used to be abundant, but were listed as threatened or endangered because of habitat destruction and modification from dams, water diversions, urban development, livestock grazing, gravel mining, logging, and agriculture. Catch and release fishing for steelhead is allowed in most streams during the winter steelhead season. If you plan to fish for salmon or steelhead in California, you need to consult the California Department of Fish and Game regulations; there are many restrictions even in season, including day of the week, section of stream, gear, license, and possession of report cards.

Invertebrates and Tide Pool Communities

Invertebrates are animals that do not have an internal skeleton made of bone; examples include anemones, sea stars, jellyfish, clams, snails, crabs, shrimp, and worms. Invertebrates are an important component of the very diverse tide pool communities along the coast. As discussed in Section I, 41% of the outer coast shoreline of central California is composed of rocky shores, reflecting its position on the tectonically active western shoreline of North America. Abundant rocky shores, in combination with nutrient-rich coastal waters from upwelling and a tidal range of 6-7 feet, gives the California coast a rich intertidal community of plants and animals. The rocky shore habitats are quite varied, ranging from wave-cut rock platforms with numerous depressions that hold water during low tide (called **tide pools**), to exposed cliffs that are pounded by continuous surf, to protected nooks and crannies that teem with organisms that cannot survive in the harsh surf.

To see the lush communities on rocky shores, it is important to time your visit during low tides. You will discover that the distribution of animals and plants varies by tidal elevation and degree of exposure to waves (Figure 89). The low-tide zone is dominated by a thick cover of marine algae or seaweeds, but also includes surf grass, which is a flowering plant. In this low-tide zone, there will be numerous sea stars, urchins, anemones, chitons, crabs, snails, etc., animals and plants that cannot endure long exposures to the sun and air. In the mid-tide zone, there is a change in the seaweed species to those that can tolerate longer exposures to sun and air. The same is true for the animals, which now include different species of chitons and limpets, numerous anemones, hermit crabs, dense clusters of snails and mussels, barnacles, and fewer sea stars. At the high-tide zone, only animals and plants that have developed qualities that tolerate exposure to the sun and air for an entire day are present. This zone has few algae, and the animals are dominated by barnacles, snails, mussels, isopods, marine flies, and shore crabs.

There are many guides to intertidal plants and animals; they come laminated so you can use them while exploring the shore. Some of the leading locations for exploring tide pools are: Duxbury Reef at Bolinas, Fitzgerald Marine Reserve at Moss Beach, and many areas along the Monterey Peninsula. The California Coastal Commission provides these rules and etiquette for tide pooling:

• Watch where you step. Step only on bare rock or sand.

• Don't touch any living organisms. Most tide pool animals are protected by a coating of slime. Touching them with dry hands can damage these animals.

• Don't poke or prod tide pool animals with a stick. Don't attempt to pry animals off of rocks.

• Leave everything as you found it (or cleaner by picking up any garbage you come across). Collecting organisms will kill them and is illegal in most tide pools.

INTERTIDAL HABITATS: BIOLOGICAL COMMUNITIES

The shoreline types along the Central California Coast are described in Table 3. In this section, we summarize the dominant biological communities living in or on the most common shoreline types.

Rocky Shores

The intertidal biota of rocky shores along the U.S. west coast are among the most studied and most diverse. There can be hundreds of species present! They occupy niches based on ability to survive direct wave action, tidal elevation, light, predation, and associations with other species that provide habitat, such as algae, mussel beds, and kelp. As succinctly described on the Monterey Bay National Marine Sanctuary web site, intertidal rocky communities have very distinct distributions by tidal elevation (MBNMS, 2009).

The **splash zone** is almost always exposed to air, and has relatively few species. The periwinkle, *Littorina keenae,* is used in some cases as an indicator of this zone, and microscopic algae are common in winter months when large waves produce consistent spray on the upper portions of the rocky shore. **The high intertidal zone** is exposed to air for long periods twice a day. The barnacle, *Balanus glandula,* and red algae, *Endocladia muricata* and *Mastocarpus papillatus,* are used as indicators of this zone, but these species are also found in other areas of the rocky shore. **The mid-intertidal zone** is exposed to air briefly once or twice a day and has many common organisms. At wave-exposed sites, the mussel, *Mytilus californianus,* can dominate the available attachment substratum. **The low intertidal zone** is exposed only during the lowest tides, and the presence of the seagrass, *Phyllospadix* (commonly known as surf grass), is a good indicator of the mean lower low water tide level (0.0 meters). This zone is also where sponges and tunicates are most common.

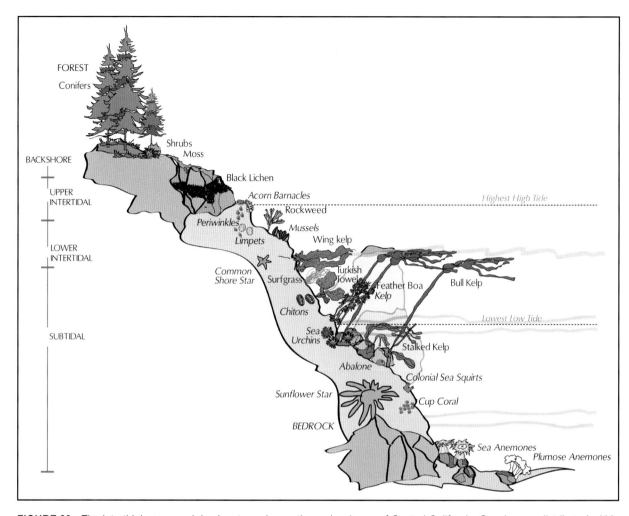

FIGURE 89. The intertidal zones and dominant species on the rocky shores of Central California. Species are distributed within the intertidal zone based on their ability to withstand exposure to air and sun during low tide. Modified illustration from the "Coastal Shore Stewardship: A Guide for Planners, Builders, and Developers" (SCBC, 2004), courtesy of the Stewardship Centre of British Columbia.

One of the most popular activities when visiting rocky shores is exploring the tide pools. There are many good field guides to help you identify the most common species. Just be careful not to trample them.

Sand Beaches

The outer coast sand beaches are inhabited by a variety of invertebrates and insects adapted to living under high wave energy conditions. At the high-tide line, the beach wrack provides habitat and food for kelp flies, wrack flies, beetles, amphipods, and isopods. Some of these remain buried in the sand during the day to escape the heat, desiccation, and predation, coming out at night to feed on the wrack. Pick up any pile of wrack and you will see a flurry of activity as the animals scurry for protection. In the swash zone, the relatively

few species, such as clams, amphipods, isopods, mole crabs, and tube-building worms, have evolved as rapid and/or deep burrowers, to prevent buffeting by waves and being eaten by birds. The dominant species is the mole crab, *Emerita analoga;* it feeds by extending a feathery antenna into the water to filter out food as a wave recedes, then scrapes the food items into its mouth. They are highly active, moving up and down the beach with the tide and along the beach with the currents. Mole crabs can be very abundant, but highly patchy. They are an important food source for coastal birds, fish, and sea otters. At low tide, where the sand stays wet, microscopic diatoms and zooplankton live between the sand grains and are important food for clams and shrimp. Along the lower tidal levels, razor, bean, and pismo clams are the dominant burrowers, along with the spiny mole crab and amphipods.

Tidal Flats

Intertidal sand and mud flats are highly productive habitats, because of the large amount of particulate organic material derived from adjacent wetland plants, benthic microalgae, and phytoplankton (called primary productivity), as well as a great diversity of animals that live in and on the sediments (called secondary productivity). The fine-grained sediments are water saturated and not subject to a lot of reworking, thus they provide a stable and rich habitat for worms, clams, amphipods, snails, and shore crabs. Some clams, such as gaper and Washington clams, filter food from the overlying water through siphons; others, such as *Macoma* clams, are deposit feeders, extracting food particles from sediment pumped through their siphons and rejecting most of the sediment as psuedofeces. Mud flat sediments are usually black below the surface (indicating the lack of oxygen, or anoxic conditions), because oxygen cannot penetrate far into the fine-grained sediments. Bacteria use up all the oxygen, degrading the abundant organic matter, so below the depth of oxygen penetration, a different type of bacteria (called anaerobic) uses other sources of oxygen, including sulfate ions. This process creates hydrogen sulfide, which gives the mud a rotten egg odor and a black color from precipitation of iron sulfides. So, animals who live deeper in the sediments have burrows through which they can circulate oxygenated water and food.

With so much food, tidal flats are important feeding areas for birds and fish. Along the California coast, estuaries with tidal flats are few and far apart, making each one an important stopover for migrating shorebirds. When the flats are covered with water, fish and crabs enter to look for food. Estuarine tidal flats are particularly important nursery habitat for juvenile fish, including threatened and endangered species such as salmon, trout, and tidewater goby. However, proliferation of invasive, non-native species such as green crab, slipper snail, and Japanese and Atlantic oyster drill have disrupted the ecological balances of tidal flat communities, out competing and preying on native species and affecting community dynamics and food webs. There are major efforts in all California estuaries to control invasive species, with limited success.

Salt Marshes

Like all salt marshes, the types of vegetation in salt marshes of Central California are controlled by tidal elevation. Pickleweed (*Salicornia virginica*) is the dominant native plant in the low marsh, at tidal elevations that are regularly flooded. The low marsh has less variation in the salinity of the marsh soils because of this regular flooding. At higher elevations, saltgrass (*Distichlis spicata*) is the main species in the mid-marsh plain, although this community is more diverse and can include heliptrope, gumplant, alkali heath, and others because of the higher variations in soil salinity caused by less-frequent flooding and seasonal differences (higher in summer and lower in winter). The high marsh is seldom flooded and transitions to upland vegetation. Salt marsh vegetation has adapted to grow in saline soils, and each species has developed different strategies to regulate salt. Some species exclude salt uptake in the roots. Pickleweed is a succulent that stores water and salt in new growth. Saltgrass controls the salt content by secreting salt through salt glands; if you look closely you should be able to see small salt crystals on the leaves. Coastal salt marshes in Central California have undergone degradation from landfill, channelization, decreased freshwater inflow, increased sedimentation from the watershed, poor water quality, and many other stessors. There are many efforts under way to restore coastal salt marshes, with varying success. You will note kiosks in many areas that describe these efforts.

SUBTIDAL HABITATS: KELP FORESTS AND EELGRASS BEDS

The **kelp forests** along the California coast are very productive offshore habitats, mostly because of the cold, nutrient-rich waters brought to the nearshore by upwelling. Kelp is a marine algae that consists of three components: a holdfast that attaches to the rocky seafloor; a stipe that is similar to a plant stalk; and fronds that attach to the stipe. Kelp forests require wave energy and currents to bring in nutrients, clear water to allow photosynthesis, and cool water temperatures. There are two dominant canopy-forming species: giant kelp (*Macrocystis pyrifera*) and bull kelp (*Nereocystis leutkeana*). Other kelp species occur inshore of these forests and in shallower water – feather boa kelp (*Egregia menziesii*), intertidal giant kelp (*Macrocystis integrifolia*), and *Cystoseira osmundacea,* a seaweed which is a form of sargassum. Kelp forests provide a rich and complex habitat that supports many fish and invertebrates. Birds and marine mammals (particularly the sea otter) use the kelp canopy for forage and refuge. The kelp canopy is most often present from March to November. Storm waves often rip the kelp from its holdfast during winter months. The shoreline adjacent to kelp forests can be covered with wrack that provides a habitat for kelp flies, maggots, and small crustaceans on which many birds feed. There is a limited amount of kelp harvest for abalone aquaculture feed.

Eelgrass (*Zostera marina*) beds along the outer coast occur in Tomales Bay, Drakes Estero, Elkhorn Slough, and Morro Bay. Eelgrass beds provide protection, food, and nursery habitat for many important fish and shellfish. Tomales Bay has 965 acres of eelgrass beds mostly in the upper bay near the mouth where water quality is best. These beds provide important areas for herring spawning. Drakes Estero has around 642 acres of eelgrass beds, covering 36% of the estuary. In Morro Bay, there are 279 acres of eelgrass beds that provide significant nursery habitat for English sole and California halibut. Large eelgrass beds are found in only a few estuaries in California, and many species are entirely dependent on them for a part of their life cycle.

11 NATIONAL MARINE SANCTUARIES

Nearly 70% of the coastal and offshore waters along the Central California Coast are designated as National Marine Sanctuaries managed by the National Oceanic and Atmospheric Administration (NOAA). There are three of them that extend contiguously from Bodega Bay to Cambria, as described on the NOAA websites:

1) The **Gulf of the Farallones National Marine Sanctuary** encompasses 1,255 square miles of water off the California coastline west of San Francisco and includes offshore marine regions of the Gulf of the Farallones and the nearshore waters of Bodega Bay, Tomales Bay, Estero de San Antonio, Estero Americano, and Bolinas Lagoon. The sanctuary is a nursery for harbor seals, elephant seals, harbor porpoises, Pacific white-sided dolphins, rockfish, and seabirds including the tufted puffin. Twenty percent of California's harbor seals breed in the Gulf of the Farallones Sanctuary. It has the largest breeding concentration of seabirds in the contiguous U.S.; more than 400,000 seabirds breed in the Sanctuary. The Visitor Center is located at Crissy Field in San Francisco. More information is available at GFNMS (2009).

2) The **Cordell Bank National Marine Sanctuary** protects an area of 529 square miles off the California coast. The centerpiece of the sanctuary is an offshore granitic bank 4.5 miles wide by 9.5 miles long. This rocky submerged knoll emerges from the soft sediments of the continental shelf, with the upper pinnacles reaching to within 115 feet of the ocean's surface. The continental shelf depth at the base of the Bank is roughly 400 feet. More information is available at CBNMS (2009).

3) The **Monterey Bay National Marine Sanctuary** stretches 276 miles from Marin to Cambria and covers 5,322 square miles of ocean, extending an average distance of 30 miles from shore. At its deepest point, the Monterey Bay National Marine Sanctuary reaches down 10,663 feet (more than two miles). Its natural resources include the largest kelp forest in the U.S., one of North America's largest underwater canyons, and the closest-to-shore deep ocean environment in the continental U.S. In 2009, the Davidson Sea Mount, which includes extensive deepwater coral and sponge fields, was added to the Sanctuary. It is home to one of the most diverse marine ecosystems in the world, including 33 species of marine mammals, 94 species of seabirds, 345 species of fishes, and numerous invertebrates and plants. The Visitor Center is located in Santa Cruz. More information is available at MBNMS (2009).

SECTION III
Major Compartments

12 INTRODUCTION

In this section, we discuss and identify opportunities for naturalists to learn about and enjoy the diverse ecosystems that exist within the **seven major geomorphological Compartments** of the shoreline of the Central California Coast. These Compartments occur within the boundaries of eight counties – Sonoma, Marin, San Francisco, San Mateo, Santa Cruz, Monterey, San Luis Obispo, and Santa Barbara (Figure 1). All of the major Compartments are subdivided into Subcompartments (called Zones) in order to focus more sharply on individual features and locations that illustrate unusually compelling aspects of the major Compartment itself. The discussion of each Zone includes details on the most interesting and unique biological components that are present, as well as specific information on its beach morphology and sediments and beach erosion issues, where appropriate. Interesting rock outcrops that might be present are also described and related to the regional geological framework. A list and discussion of the key places to visit conclude the coverage of each of the Zones. The seven primary Compartments are labeled on Figure 1 and listed below:

- The San Andreas Fault Valley
- Point Reyes
- San Francisco's Outer Shore
- Central Headland
- Monterey Bay and Vicinity
- Big Sur
- The South Bays

13 COMPARTMENT 1 – THE SAN ANDREAS FAULT VALLEY

INTRODUCTION

The valley eroded along the San Andreas Fault zone is the centerpiece for this Compartment (Figure 23). As a result of this topographic depression, two bay systems are present, Bodega Harbor to the northwest and Tomales Bay to the southeast (Figure 90). Other prominent features include a significant granodiorite-based headland, Bodega Head, a crenulate bay shoreline with a westward migrating spit in the shadow zone at the northwest end of Bodega Bay, a half inlet with a dune field on its northeast flank at the entrance to Tomales Bay (Figure 73), and a long, linear complex shoreline around Tomales Bay that contains numerous pocket beaches and small stream outlets. Ecological highlights include one of the most noteworthy whale-watching sites in the state at Bodega Head, and excellent birding along tidal flats in Bodega Harbor and the fringes of the two deltas in Tomales Bay (at the mouths of Walker Creek and Olema Creek; Figure 90).

For purposes of discussion, we have further subdivided this Compartment into four Zones (as mapped in Figure 90):

A) **Bodega Harbor and Vicinity** is located within the valley eroded along the San Andreas Fault zone. In addition to the major rocky headland at Bodega Head, it includes the bay and harbor complex (Bodega Harbor) in the lee of a major west-projecting sand spit (Doran Beach) in the shadow zone of the crenulate bay that forms the northern shoreline of Bodega Bay.

B) **Central Uplands** is completely devoid of any man-made structures, and contains a phenomenal number of sea stacks offshore of eroding rock cliffs. Two modest-sized creeks flow into narrow flooded valleys creating two classic estuaries – Estero Americano and Estero de San Antonio.

C) **Dillon Beach and Dunes**, which contains the picturesque coastal village of Dillon Beach, is primarily a sandy beach attached to the eroding rocky shore along the southeast side of Bodega Bay. The sandy beach, the

right limb of the half inlet at the entrance to Tomales Bay (Figure 73), is backed by an extensive field of active sand dunes.

D) **Tomales Bay** is a twelve mile long, narrow water body situated within the valley eroded along the San Andreas Fault system. Two small deltas are host to fairly extensive tidal flats and salt marshes. The beaches are mostly short pocket beaches composed of a variety of sediment types.

BODEGA HARBOR AND VICINITY

Introduction

As shown in the image in Figure 91 and the map in Figure 90 (Zone A), the dominant geomorphological features of this area are: a) the two mile long bedrock headland to the west (Bodega Head); b) the shallow bay/harbor located exactly over the valley eroded along the San Andreas Fault (Bodega Harbor); c) a westward extending sand spit that nearly closes off the Harbor (Doran Beach); and d) the northernmost end of an open water bay about eight miles long (Bodega Bay) that flanks a broad crenulate bay shoreline (Figure 90). This is a very rich biological area, with world-class whale watching and birding opportunities.

Geology of Bodega Head

The oblique aerial photograph in Figure 92A gives a broad overview of Bodega Head, and a ground view looking northwest from the top of the wave-cut rock scarp is shown in Figure 92B. This is one of the more interesting geological localities along the whole Central California Coast, because of its excellent exposure of the granitic rocks of the Salinian Block so far to the north (Figures 16A and 18). Between the headland and the hills of the mainland lies the valley that has been eroded along the San Andreas Fault system, meaning that the rocks in these cliffs have been moving north aboard the Pacific

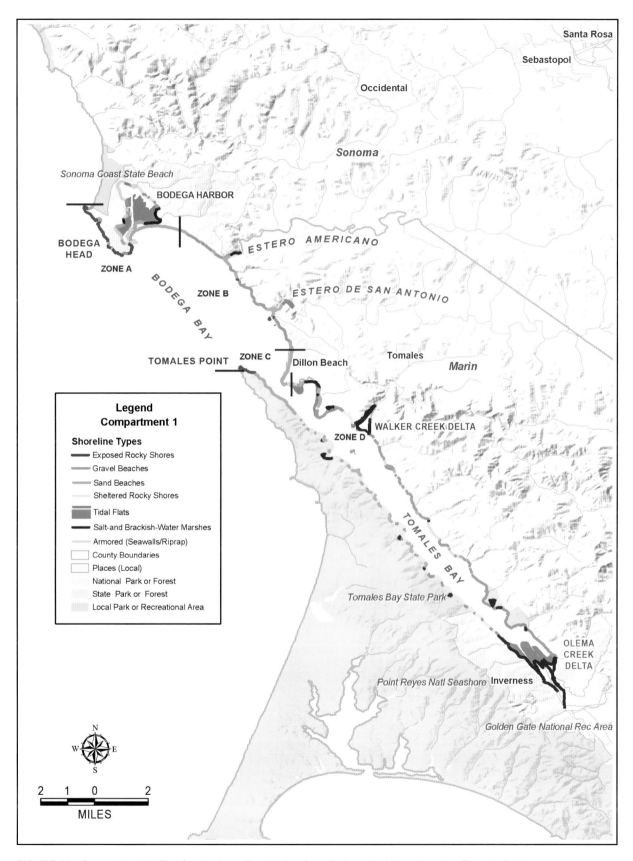

FIGURE 90. Compartment 1, The San Andreas Fault Valley, from Bodega Head (north end) to Tomales Point.

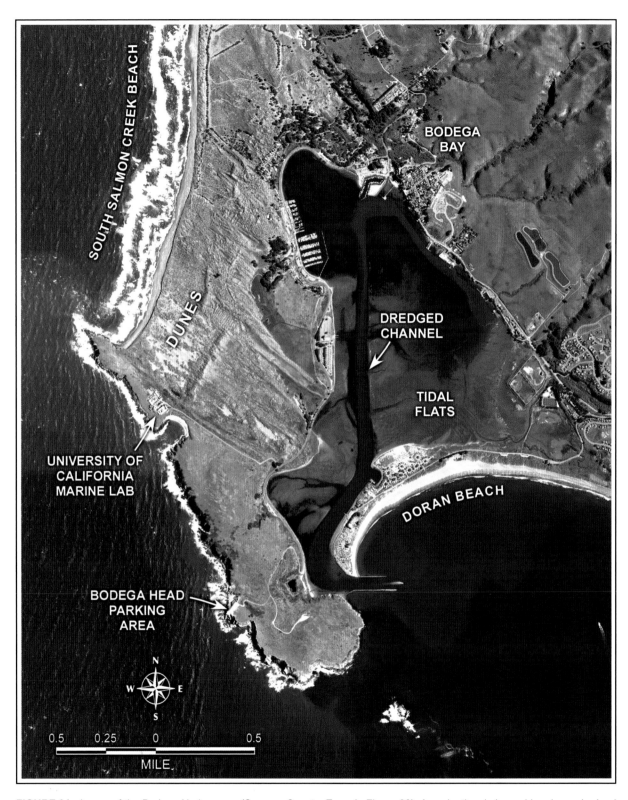

FIGURE 91. Image of the Bodega Harbor area (Sonoma County; Zone A, Figure 90). A navigational channel has been dredged through extensive tidal flats. Note the wide dune field between the Harbor and the outer South Salmon Creek Beach. NAIP 2005 imagery courtesy of the California Spatial Information Library (CaSIL).

FIGURE 92. Bodega Head (Sonoma County; Zone A, Figure 90). (A) Oblique aerial view taken on 5 October 2005, courtesy of the California Coastal Records Project, Kenneth and Gabrielle Adelman. (B) View looking north from near the parking area shown in A. Photograph taken on 17 August 2008.

Plate for the past 25-30 million years or so. During the 1906 earthquake, Bodega Head and the adjoining harbor moved 15 feet to the north relative to the mainland on the other side of the fault.

The igneous rocks in these cliffs are thought to have been originally formed about 100 million years ago during the Cretaceous period as part of a batholith complex very similar to the one that now forms the core of the Sierra Nevada. Similar rocks are located on the mainland of Mexico far to the south, a disputed distance that could be as much as several hundred miles. A photograph of these grayish granitic rocks is given in Figure 93. Actually, in this case, these rocks have been called variously **diorite** and **granodiorite**, rocks similar in texture to granite that are made up largely of white to light gray plagioclase feldspar and hornblende, with less than 10% quartz (SiO_2), in the case of diorite. Granodiorites may have up to a little over 50% quartz (Figure 17B).

Wright (1996) gave the following description of the cliff face near water level just to the north of the beach shown in Figure 92. *Standing in the small sandy slot near the water on the north side of the beach, you can see a smooth wall of granodiorite. Light colored minerals are glassy grey crystals of quartz, and white, opaque crystals of plagioclase feldspar. Dark flecks of biotite and hornblende cause the pepper effect. The pock-marked surface here results from plucking out weaker dark minerals by crashing waves.* Remnants of rocks possibly as much as a billion years old are included as black metamorphic fragments in the granodiorite.

In some places, relatively narrow bands of pink granite, called **pegmatite dikes,** have pierced through the granodiorite, clear evidence that the granite is younger in age (though still in the 100 million year range) than the granodiorite. In order for granite, which is typically composed mostly of quartz and orthoclase feldspar (a silicate mineral with an abundance of aluminum and potassium) to do this means that the pure granite, formed at great depths in the Earth, as was the granodiorite, was the last of the two to crystallize.

In other areas, especially in the higher portions of the cliffs, horizontal layers of much younger sedimentary rocks are present. More details on the geology of Bodega Head are available in a self-guided, illustrated field trip available on the internet by Wright (1996).

The Bay and Harbor

There may be some confusion about the names being used in this write-up, because there is some overlap and inconsistency in the usage of the names

FIGURE 93. A small sea arch in the Salinian Block granodiorite located in the wave-cut rock cliff just to the south of the parking area shown in Figure 92A. Photograph taken on 17 August 2008.

in general. Anyway, there is a town called Bodega Bay with a population of 1,423 at the time of the 2000 census that is located on the banks of the enclosed water body landward of the Doran Beach spit that we refer to as Bodega Harbor. The inlet that provides an entrance into Bodega Harbor is stabilized with a set of two parallel jetties composed of riprap. Bodega Bay proper, if the commonly used maps and charts are reliable, is the open water area between the spit and the entrance to Tomales Bay (Figure 90). The northwestern part of Bodega Bay is the site of a crenulate bay referred to in several places in Section I. At low tide, the Harbor area is mostly tidal flats. A major navigation channel has been dug through the tidal flats to the uppermost Harbor site, and a Coast Guard station is located in a small basin dug into the landward side of the spit (Doran Beach; Figure 91).

Beaches and Dunes

The arcuate Doran Beach, located at the northwest end of the crenulate bay on the shoreline of Bodega Bay, is a popular recreational beach composed of medium- to fine-grained sand (Figure 91). Note the large dune field, now mostly vegetated, on the other side of the Harbor behind South Salmon Creek Beach. The sand moving toward the south along that beach, some of which was no doubt delivered to the coast by Salmon Creek, has accumulated against the rocky natural groin at the north end of Bodega Head. This sand accumulation, along with a relatively low hinterland, have allowed the dune field to extend further to the southeast (over a mile) in the area adjacent to Bodega Head than elsewhere (Figure 91).

Beach Erosion

Savoy et al. (2005) mapped the outer coastline in this area as having a *Caution: moderate risk* for beach erosion. However, no man-made structures are close enough to the beach to be in peril at this time.

General Ecology

With exposed rocky shores, sand beaches, and sheltered mudflats all in close proximity, the Bodega Bay area has a wide diversity of marine animals. The offshore islands and isolated rocky shores provide nesting sites for Brandt's cormorant, pigeon guillemot, rhinoceros auklet, western gull, and black oystercatcher, and haulouts for California sea lions, Steller sea lions, and harbor seals. Loons, grebes, scoters, murres, murrelets, auklets, cormorants, phalaropes, and shearwaters can be seen in the nearshore waters. The tidal flats support many waterfowl, wading birds, and shorebirds. You should see peregrine falcons, bald eagles and, if you are lucky, California black rails in the marshes in the southeast corner of the Harbor. Western snowy plovers nest on South Salmon Creek Beach and Doran Beach. With so many birds, it is one of the most popular birding locales in the region. Bodega Head is the northern extent of the Gulf of the Farallones National Marine Sanctuary. There are boat charters for bird and whale watching and places to rent kayaks to explore the Harbor and nearshore waters.

Places to Visit

And sometimes, starting to turn over a big rock in the Great Tide Pool – a rock under which he knew there would be a community of frantic animals – he would drop the rock back in place and stand, hands on hips, looking off to sea, where the round clouds piled up white with pink and black edges. And he would be thinking, What am I thinking? What do I want? Where do I want to go? There would be

wonder in him, and a little impatience, as though he stood outside and looked in on himself through a glass shell. And he would be conscious of a tone within himself, or several tones, as though he heard music distantly.

John Steinbeck – SWEET THURSDAY

During your visit to this Zone, you should probably go first in the direction of the overlook at Bodega Head, which is a little tricky to find. You have to drive all the way around Bodega Harbor to get there. While driving north along Highway 1, you will eventually pass into the town of Bodega Bay, the setting for the 1963 Alfred Hitchcock movie, *The Birds*. Once through the main part of the town, you will encounter a sign located near the landward end of the Harbor (near the primary port area), that gives directions to Bodega Head. Take a left turn at that sign and continue along the shoreline on Bayflat Road. The road curves around the Harbor and eventually turns toward the southeast. Along the way, you will pass two zones of heavily loaded boat berths, the most seaward of which is shielded by a long offshore breakwater. These boats appear to be an assorted mixture of professional fishing boats and pleasure craft. By the way, this is the largest and busiest harbor area between San Francisco and Fort Bragg. Another half mile along you will drive across the end of the huge dune field discussed earlier (see Figure 91), which is now vegetated, At one time, the dune field you are driving across had active dunes that transported sand all the way into the waters of the Harbor.

At the edge of the vegetated dunes a small clump of trees mark the intersection with a road that leads up the hill to the University of California's Bodega Marine Laboratory, which is no doubt a wonderful place to visit. Only problem is, you can only do that between 2 and 4 pm on Friday (as a tourist). In any event, co-author Michel has visited there several times on business, and she says that they have a nice aquarium and lab facilities. During visiting hours, you can walk to a little slot beach, which is about 250 yards long and 200 yards deep in a strange rectangular embayment in the wave-cut rock cliffs (Figure 91), where you can observe intertidal communities.

Back on Bayflat Road, continue southeast around the edge of the Harbor. If you go there at low tide, you will see very wide tidal flats exposed which contain numerous feeding birds. On 17 August 2008, we saw a large flock of white pelicans on the flats, as well as numerous shorebirds. Just to drive around this Harbor area during any low tide would be pretty good birding. In fact, this is probably one of the best birding spots for tidal flats in the Central California area, except for San Francisco Bay, because it is so easy to get to and the flats are so wide. Except for the dredged channels, this Harbor area is almost all tidal flats during low spring tides.

Continuing for a few more hundred yards you will eventually reach the point where the main entrance channel to the Harbor comes close to the road. A little ways beyond that point, turn right up the hill and drive to the end of the road where you will reach the main overlook at Bodega Head, one of the most prominent landmarks along the Central California Coast. It is a beautiful spot with great views of the rocky cliffs that border the headland (see Figure 92A). There is a stairway down the side of the rock cliff to a small sand beach (Figure 92B). In addition to a rather spacious parking area, picnic tables and restrooms are also available. The Bodega Head trail goes up high along the bluff. Beautiful place. Good place to spend some time.

If you happen to be there from late November to January, be sure to watch for the spouting of the gray whales as they migrate south along the longest migration path taken by any mammal (Figure 86). In the absence of any whales, the star of the show will probably be the 90+ million year old Cretaceous granodiorite in the cliffs discussed earlier (Figure 93), with the bonus of harbor seals and seabirds on the small sea stacks offshore. As is clear in the photograph in Figure 93, this granodiorite is greatly fragmented, possibly because it is located so near the San Andreas Fault.

As you walk around Bodega Head at the edge of the parking lot, the crumbly, granular material under your feet will be an accumulation of the rotten granodiorite and granite as it is breaking up under the influence of chemical weathering. The first step in the process occurs when water penetrates fractures in the granitic rocks and weathers part of it to soil. The grains of feldspar in the rock swell as they react with the water and some of the feldspar turns into clay. This expansion breaks the remainder of the rock into finer pieces composed mostly of quartz, weathered fragments of the feldspar, and possibly some individual flakes of mica, hornblende, and other minor minerals. Pick up a handful of the weathered material and you will be able see the individual quartz grains (clear to slightly opaque) and feldspar (opaque white minerals). This type of fine-grained weathered debris, called **grus** by geologists, is a common associate with granitic rocks worldwide, because of the susceptibility of feldspar to this type of chemical weathering and expansion (Alt and Hyndman, 2000).

After viewing the open ocean side of the headland, we recommend that you drive back the way you came a short distance and turn right on a gravel road and drive to the south at first until you come to another parking area. Drive carefully, because this area contains an abundance of California quail and black-tailed deer. From the parking area, there is a trail along which you can look down on the waters of the Harbor, the tidal flats with their fringing marsh, and the long, west-trending sand spit called Doran Beach, with its beautiful arcuate shoreline. On a clear day, from the trail you will have a great view of the entrance to Tomales Bay and parts of the Point Reyes area. In fact, this is a terrific view of most of Compartment 1 and part of Compartment 2 almost all the way to Point Reyes (Figure 90). If you continue walking to the end of the trail you will usually be able to observe and hear a group of harbor seals down on the rocks off the headland. This trail is definitely a place you don't want to miss, with its excellent scenic overview to the south, which includes a look at the rocky cliffs and mountains along the mainland shore.

As you leave the top of the headland, drive to the bottom of the hill and take an immediate right into the parking area for Campbell Cove. At this location, there is a moderately wide muddy sand tidal flat backed by a narrow sand beach along the side of the main channel into the Harbor. This is another fine birding spot at low tide.

Also, don't miss a short walk to the west along a path that leads to an observation deck overlooking a small pond, an important historic site. As it turns out, in the 1960s, PG&E selected this site for the location of a nuclear power plant, and construction was actually started on it. As noted by Wright (1996), *nuclear power plants need cooling water, and the most abundant supply of cold water is along the coast. Unfortunately, there are many active faults along the coast, and this site was right next to the San Andreas Fault.* At the beginning of the project, a huge pit was dug for the foundation of the power plant in which a fault was discovered. The pond by the observation deck occupies this pit.

Doris Sloan (pers. comm.) gave the following account of the discovery of the fault through the potential reactor pit:

A geophysicist came to look at the site at the request of the opponents of the proposed plant. I was living in Sebastopol at the time and working with the plant opponents. I took the geophysicist out to the site on the day he found the fault. He walked around the construction site and saw the fault in the granitic walls of the pit, The rock visible at the surface and into the pit was all granodiorite. His report triggered more careful investigation (of the geological setting of the site).

Wright (1996) commented that this finding *indicated that there had been an earthquake with offset along this fault less than 40,000 years ago.* As a result of this discovery and some public pressure, PG&E abandoned plans to build the power plant at this location, and it was eventually located at Diablo Canyon in Compartment 7.

To continue your tour of this area, go back all the way around the Harbor and just after you leave the side of the tidal flats turn right into the parking area for the Birdwalk Trail. Along the trail, you first pass by a couple of dredge spoil impoundments, but you eventually encounter a good-sized marsh after a few hundred yards, called Doran Park Marsh. According to a kiosk on the trail, Bodega Bay *was discovered in 1775 by Juan Francisco de la Bodega y Codra, a Spanish explorer on his way to Alaska.* The most notable birds cited on the kiosk include the short-eared owl, snowy egret, northern harrier, northern pintail, willet, and marbled godwit.

In 1993, a new wetland habitat and waterbird foraging pond was constructed near the Cheney Creek levee in this same area. Essentially, what happened is this used to be a mudflat and then, because of land-use practices of agriculture and grazing, the flat filled in with sediment and a marsh grew on top of it. Now, they've actually constructed some ponds to try to replicate the open water habitat that was originally there. This is one of those funny cases where the marsh is not considered to be a part of the natural ecosystem. This high marsh habitat is just about 100% common pickleweed (*Salicornia virginica*).

Once you leave this parking area, it is only a short distance to the turn to Doran Beach Regional Park, located on the arcuate spit at the shadow zone of the Bodega Bay crenulate bay (illustrated in Figures 90 and 91). This is a relatively unspoiled medium- to fine-grained sand beach. Because of the way waves normally refract/diffract into the Bodega Bay crenulate bay, the waves typically approach this beach straight on, which leads to the creation of **beach cusps** (illustrated in Figure 38). As noted earlier, beach cusps are extremely common features all along the sand beaches of the Central California Coast. They are the scalloped seaward margin of an intertidal sand berm consisting of a regularly spaced sequence of indented bays and protruding horns.

The Doran Beach Park, which has a long list of amenities, including picnic tables and camping facilities, also provides the opportunity for long beach walks. If you head toward the southeast on such a walk, you will eventually come to some interesting and spectacular rock cliffs and sea stacks in the Franciscan Complex.

CENTRAL UPLANDS

General Description

This area (Zone B on Figure 90) is the uninhabited, hilly shoreline of the northeast side of Bodega Bay (the water body) accessible only by foot trails and kayaks. Two moderate-sized creeks empty into linear, incised, relatively flat-bottomed and meandering estuaries called Estero Americano, at the boundary between Sonoma and Marin Counties, and Estero de San Antonio, located a little over two miles to the southeast of that county line. Both of these coastal water bodies, which do indeed meet the criteria of the classic definition of **estuaries** (i.e., drowned creek valleys with tides and freshwater influx), contain some marshes and tidal flats and, hence, some significant bird populations. The two inlets at the entrances to these estuaries conform to the class **stream-mouth/bar systems**, with both the sandy spit at the entrance to Estero Americano and the sand-and-gravel spit at Estero de San Antonio having been built to the north across the entrances for reasons discussed in Section I (Figure 94).

Much of the shoreline consists of high scarps in crumbly bedrock of the Franciscan Complex. A few gravel beaches and four short sand beaches are located in narrow embayments at the base of the scarps. The most striking aspect of this shoreline is the phenomenal number of **sea stacks** in the shallow waters off the eroding scarps, possibly the most profuse collection of these features anywhere along the Central California Coast.

Places to Visit

To get to this area from the town of Bodega Bay, drive east on Highway 1 for about 3.6 miles and turn right on the Valley Ford Cutoff. After rejoining Highway 1, you will soon be in the small town of Valley Ford. Once there, take a right on Franklin School Road. About a mile further along, you will cross the Estero Americano (as well as the border into Marin County). This bridge area is a popular place to launch kayaks for trips down the Estero to Bodega Bay. You should check the tide tables before attempting this. No doubt, this is a very enjoyable float trip.

Continuing to the south along this road for another 2.5 miles or so you will cross the bridge over Estero de San Antonio. We are sure you will find this drive through beautiful Marin County, with its rounded grassy hills, exposed isolated rock outcrops from the Franciscan

FIGURE 94. The Esteros of Bodega Bay (Sonoma/Marin Counties; Zone B, Figure 90). (A) Oblique aerial view of Estero Americano taken on 5 October 2005, courtesy of the California Coastal Records Project, Kenneth and Gabrielle Adelman. The sand spit across the mouth of the Estero projects to the north. (B) View looking north at the entrance to Estero de San Antonio at low tide on 10 November 1992. The sand and gravel spit that closes off this Estero also projects to the north.

mélange, and picturesque farms a most rewarding one. Classic Marin County, California, so unique in its own way.

DILLON BEACH AND DUNES

Introduction

This area (Zone C on Figure 90), unlike the one just discussed, is accessible by paved road at the end of which is the small town of Dillon Beach, where vacation homes and year-round residences line the narrow, winding streets. In addition to these permanent residences, the area is host to hundreds of mobile vacation homes and trailers, particularly on weekends. The mass exodus from this location on a late Sunday afternoon is a challenge for

the narrow roads. Dillon Beach is a beautiful little town in a majestic setting with views that look west across the open Bodega Bay and to the south across the entrance to Tomales Bay, with the craggy projection of Tomales Point as a backdrop.

Beaches and Dunes

A sandy, spit-like peninsula extends for a little over a mile and a half away from the eroding bedrock scarps to the north-northwest of the town of Dillon Beach. The beach itself, which is composed of medium- to fine-grained sand, faces directly into the dominant westerly waves that bend into Bodega Bay. Over the past several hundred years or so, some parallel beach ridges have slowly prograded to the west as sand eroded from the rocky shoreline to the north and possibly with some input from the two creeks (Esteros Americano and de San Antonio) has gradually accumulated on this peninsula. As is clear on the vertical image in Figure 95, much of this accumulated sand has been blown almost a mile to the east to form a large, active dune field. The foredunes along the beach are vegetated, but the large, transverse dunes in the main dune field (illustrated in Figure 76), created by the prevailing northwesterly winds, are not.

Beach Erosion

Savoy et al. (2005) mapped this outer coastline as having a *Caution: moderate risk* for beach erosion, with the exception of the fact that *homes situated on the cliff edge at Dillon Beach are threatened by erosion.* They are referring to about fifteen houses built fairly close to the rocky scarp a little ways beyond the northern edge of the sand beach. In this case, landslides are probably more of a threat than beach erosion *per se.* Most of the rest of the permanent houses in the town are set well back from the beach.

General Ecology

Dillon Beach is host to many shorebirds and gulls. Western snowy plovers (75-123 birds) nest on Dillon Beach March-September. It is clear why these shorebirds, one of which is pictured in Figure 96, are listed as threatened; their preferred habitats are the same as many people who enjoy visiting the coast. Snowy plover chicks are precocial (that is, they are covered with down and having eyes open; therefore, they are capable of leaving the nest within a few hours). These chicks were described as follows by the U.S. Fish and Wildlife Service:

Snowy plover chicks leave the nest within hours after hatching to search for food. They are not able to fly for approximately four weeks after hatching, during which time they are especially vulnerable to predation. Adult plovers do not feed their chicks, but lead them to suitable feeding areas. Adults use distraction displays to lure predators and people away from chicks. Adult plovers signal the chicks to crouch, with calls, as another way to protect them. Most chick mortality occurs within six days after hatching.

Places to Visit

To get to Dillon Beach, continue south on Highway 1 after you cross the bridge over Estero de San Antonio until you reach the town of Tomales. Once in town, turn to the right on Dillon Beach Road. It is about 3.5 miles from this intersection to the shore at Dillon Beach. You will be impressed with the wild topography you drive through on your way to the beach – big rounded grassy hills with no trees but littered with isolated mounds of rocks and huge boulders separated by deep valleys. These miscellaneous boulders and blocks are composed of resistant Franciscan rocks of different types (e.g., red radiolarian chert) that were mixed into a sheared mudstone matrix during the process of the formation of the **Franciscan mélange** discussed earlier (Cardwell and Detterman, 2003).

When you come into the town take a left down to the public beach, from which you get a great view of the multitude of sea stacks along the eroding rocky northeast shoreline of Bodega Bay. The tall, vegetated dunes just behind the beach are also interesting. A hike back to the barren sand dunes is also a unique opportunity for this part of the coast.

If you go there on a weekend, you will be amazed at the number of people who visit this somewhat remote beach, especially the teeming masses in the trailer park.

TOMALES BAY

Introduction

Tomales Bay (Zone D on Figure 90) is a fourteen mile long, narrow water body averaging about one mile in width that is situated within the eroded valley along the San Andreas Fault zone (Figure 23). The entrance to the bay is occupied by the large "half inlet" illustrated in Figure 73, with a sandy beach on the northeast side of the inlet and a relatively steep, rocky shore (with some

FIGURE 95. The Dillon Beach Area (Marin County; Zone C, Figure 90). The sand dunes in the unvegetated area are exceptionally large transverse dunes. Compare these dunes with the diagram in Figure 76. Like most coastal sand dunes in California, these dunes are formed primarily by winds from the northwest. Note how the crests of these dunes are oriented perpendicular to the northwest wind. NAIP 2005 imagery courtesy of the California Spatial Information Library (CaSIL).

FIGURE 96. Western snowy plover. Photograph courtesy of Will Elder, National Park Service.

narrow sand beaches) on the opposite side. The southeast end of the Bay contains the modest-sized delta of Olema Creek. The southwestern shore is located along the steep-sided limb of Inverness Ridge, and the northeast side adjoins the less-elevated, rolling hills of Marin County. The aquaculture of oysters and other fisheries activities take advantage of these pristine waters.

Ed Clifton and his associates conducted a study of the impacts of the great 1982 *El Niño* storms on sedimentation within the Bay (reported by Anima, Bick, and Clifton, 1988). This storm caused some big landslides along the west side of the Bay near Inverness and lots of water flowed down the creeks as a result of the heavy rainfall accompanying the storm. However, he (pers. comm.) concluded that *I suspected that we would see a layer of sediment added to the floor of the Bay, but it turned out that the only sedimentation occurred on the delta fronts.*

Deltas

River deltas are nonexistent on the open coast of California, where the large waves disperse sediment brought to the shore by streams. However, a few do occur where a stream empties into a sheltered bay, as is the case for two streams that empty into Tomales Bay. The **Walker Creek Delta** lies about 2.5 miles inside the entrance on the northeast side of the Bay (Figures 90 and 97). The **Olema Creek Delta** is located at the very head of the Bay (Figure 90). Being the most significant depositional features within the Bay, these deltas contain the largest extent of tidal flats and salt marshes, the most sensitive and ecologically significant of the coastal habitats in the region.

Beaches

Long depositional beaches are relatively scarce within Tomales Bay, but there are a number of short, pocket beaches with a wide range of compositions, especially along the southwest shoreline. The four pocket beaches in Tomales Bay State Park are discussed later. Note the prominent peninsula on the northeast shore located half way between the Walker Creek Delta and the end of the sandy shore at Dillon Beach (Figure 90). This peninsula, which on the ground can be seen to consist of sandy sediment (although salt marshes seem to dominant the shoreline at the present time), probably formed in a fashion very similar to the one that now extends away from the rocky shore north of Dillon Beach. According to this hypothesis, the backup

peninsula would have formed when the end of Tomales Point was opposite to it, similar to its present position relative to the Dillon Beach spit. How long ago was that? Assuming that Tomales Point is moving northwest at the rate of 2 inches/year, the time that the second spit form was active was sometime between 80 and 100 thousand years ago, depending upon the exact location of the end of the Tomales Peninsula at the time when the old spit was forming. Of course, sea level was at its lowest during the last ice age sometime around 20 thousand years ago, so, if our theory is correct, this southern spit/beach must have formed earlier than that, probably during the previous highstand around 100,000 years ago. No spit would have developed when sea level was out. Of course, the modern beach has formed during the present highstand within the past 5,000 years or so.

An interesting, rather small triangular-shaped projection of sand beaches called a **cuspate spit**, illustrated in Figure 98, occurs on the southwest shoreline of the Bay almost due south of the mouth of Walker Creek (Figures 90 and 98). Cuspate spits like this one are most common on shorelines that are parallel to two opposing wind directions. These two opposing wind directions generate waves that play a role in molding the sediments into the characteristic morphology, in this case, a triangular projection of sand offshore, such as the one shown in Figure 98. As proposed by the pioneering Russian coastal geomorphologist, V.P. Zenkovitch (1967), waves striking a shoreline at a large angle are capable of high rates of longshore sediment transport. As the sediment moves along shore, the longshore transport system becomes "saturated" (Zenkovitch's term). As a result of this "saturation," features such as rhythmic topography (discussed earlier; Figure 37 – Upper Right) make a bulge in the beach as the more exposed side of the sand bulge orients perpendicular to the wave approach direction (parallel with the approaching wave crests). Where that reorientation takes place, the rate of sediment transport slows down, because of the near-parallel wave approach direction; however, the bulge continues to grow slowly further out into deeper water and the resulting shoreline "rhythm" continues to migrate slowly downdrift. In the case of cuspate spits, however, when the wind blows from the opposite direction, the other side of the sand bulge responds in a similar fashion, creating an arrow-shaped tip of the bulge that builds out into even deeper water. Where the sandy shoreline is long, a number of equally spaced cuspate spits may form. They are called cuspate forelands, or capes, if they are large (e.g., the four major capes off the coast of North and South Carolina – Capes Hatteras, Lookout, Fear, and Romain).

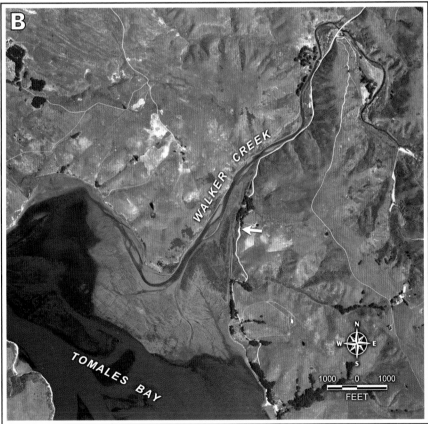

FIGURE 97. The Walker Creek Delta on the northeast shoreline of Tomales Bay (Marin County; Zone D, Figure 90). (A) Ground view looking west on 17 August 2008. The marsh grass is primarily *Salicornia* (pickleweed). (B) Image of the delta taken at low tide. The maximum width of the tidal flats, measured west/northwest by east/southeast, is close to one mile. Arrow points to spot from which the photograph in A was taken. NAIP 2005 imagery courtesy of the California Spatial Information Library (CaSIL).

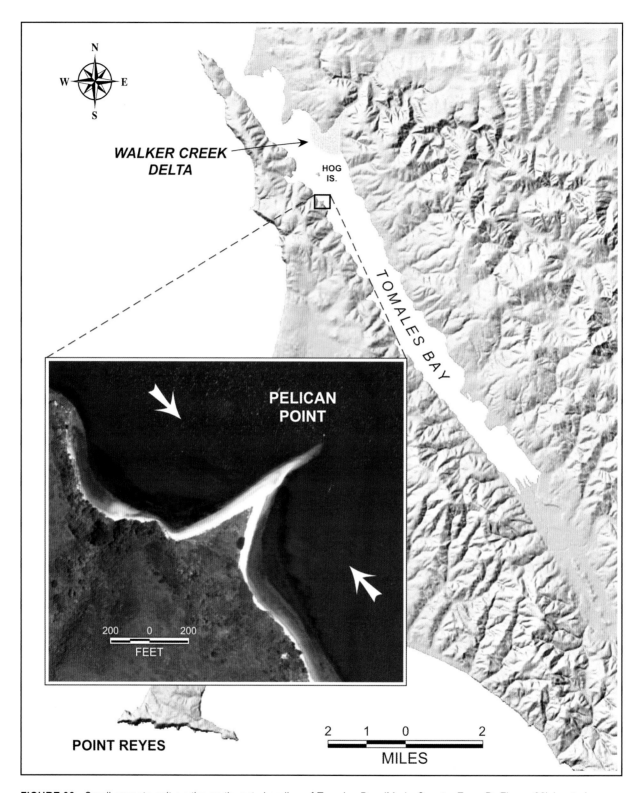

FIGURE 98. Small cuspate spit on the northwest shoreline of Tomales Bay (Marin County; Zone D, Figure 90) located across the Bay from the Walker Creek Delta that is illustrated in Figure 97. The spit is located within the square on the background map. Such spits are built by waves that approach it from opposing directions, in this case from the northwest and southeast, because of the long fetches in those two directions down the long axis of Tomales Bay. NAIP 2005 imagery courtesy of the California Spatial Information Library (CaSIL).

Even though cuspate spits and cuspate forelands are very common around the world's shoreline, this spit, and an even smaller one that projects in the same direction (northeast) out into Tomales Bay off nearby Hog Island, are the only ones on the Central California Coast. It is not unusual for just one of these features to form in long, linear water bodies like this one (usually closer to the middle of it), because the long axes of the water body provide the only effective fetch distances for the winds to generate waves. The spit form develops at the location where the energy expended by the two opposing wave systems is balanced. In this case, assuming this interpretation is correct, the more effective fetch is from the northwest, as is to be expected because that is the predominant wind and wave-approach direction on this coastline (Figures 40 and 41). However, as pointed out by Ed Clifton (pers. comm.), who expressed some doubt about our explanation for the origin for this particular cuspate spit, *winds from the southeast or east are very uncommon* in this area. Our defense of this idea is twofold: 1) the fetch from the southeast is twice as long as that from the northwest; and 2) maybe the southeast winds associated with the big *El Niño* storms, though infrequent, are effective enough to move the sand the other way, creating the cuspate spit. Anyway, guess we will stick with our hypothesis until proven otherwise.

General Ecology

The waters of Tomales Bay are part of the Gulf of the Farallones National Marine Sanctuary. The 483 acres of salt marsh and tidal flats at the head of the Bay are protected as the Tomales Bay Ecological Reserve managed by the California Department of Fish and Game. In winter, the Bay supports nearly 21,000 shorebirds (with dunlin and western sandpiper most abundant) and 22,000 waterbirds (surf scoter, bufflehead, and greater scaup most abundant). Tomales Bay provides particularly important winter habitat for red-throated loons, common loons, eared grebes, horned grebes, black brant, surf scoters, and black scoters (Kelly, 1998). The wetlands at Walker Creek and the head of the Bay support both the greatest concentrations of shorebirds and the highest densities for more shorebird species than other areas in the Bay (Kelly, 2001). Walker, Olema, and Lagunitas Creeks support threatened/endangered runs of coho and chinook salmon and steelhead, and there are major efforts under way to restore the watershed and spawning habitats in these streams.

The Bay is one of four commercial oyster-growing areas in the State. The commercial oyster farms grow the Pacific oyster (originally imported from Japan but now larvae are produced by U.S. hatcheries) in mesh bags attached to racks or lines because the native oyster was depleted in the late 1800s as siltation covered the rocky substrate on which they grew. The oysters grow as individuals instead of in clusters. There are efforts underway to restore native oysters to the Bay, but there are now new problems, including non-native predators such as green crab and oyster drills that eat the larvae and young oysters.

As you explore Tomales Bay, the differences in the vegetation on west and east sides of the Bay are striking: the dense bishop pine and Douglas fir forests on the west versus the open oak woodlands and grasslands with dairy and beef ranches on the east. The vegetation types reflect the differences in the bedrock and soils on either side of the fault zone, as well as the slope, precipitation, and other climatic factors.

Places to Visit

As you head down toward Tomales Bay along Highway 1, you will encounter the **Walker Creek Delta** with its wide marsh complex (Figure 97). Assuming you come at low tide, you can see gravel bars exposed in the riverbed as you drive along. At the point where you come to the main body of the Bay, there is a spacious pullover spot from which you can view the Delta and its wide tidal flats (Figure 97B). No doubt there is good birding at low tide here. From the pullover you can also see across Tomales Bay to the Inverness Ridge on the southwest side, which is steep and forbidding looking, being composed mostly of heavily forested steep slopes with a few randomly scattered minor beaches that are too small to see from this pullover without binoculars.

Driving further down the east side of the Bay, you come across some of the oyster farming activities out in the water, as well as numerous fishing boats, boat landings, etc. (During our trip there in August 2008, we had a great lunch of barbecued oysters at one of the roadside cafés). Also, there is some spectacular **salt pruning** of the vegetation along this side of the Bay. [NOTE: Another term for salt pruning is **flagged**, as the trees or shrubs bend away from a persistent prevailing wind, with the growth only occurring on the side of the tree protected from the salt spray so that the limbs extend downwind like flags blowing in the wind. This presumably has something to do with the dehydrating effects of salt spray, but there may be other reasons. Whatever the reason, this is a very common phenomenon all around the world.]

About three fourths of the way between the pullover and the head of the Bay, a salt marsh fringe borders the road. There also is a good-sized marsh where the small Millerton Creek Delta builds into the Bay.

Olema Creek has built a modest-sized delta at the head of the Bay (Figure 90). Fairly extensive salt marshes occur around the margin of the wide tidal flats on the delta surface. There is some *Spartina* in the marsh at this locale, lower in the channels than the *Salicornia* up on the flats. The birding should be excellent during low tide on these fairly extensive tidal flats.

Once you reach the town of Point Reyes Station, we recommend that you follow the signs to the **Point Reyes Seashore Bear Valley Visitor Center** (see location on Figure 99), where you can walk the famous earthquake trail that memorializes the 1906 earthquake that was discussed in Section I (under the topic of earthquakes). A couple of the highlights along the trail are the blue posts marking the trace of the fault line and the recreated 20 foot offset of a fence. This is a good walk to take, because it gives you a clear mental impression of the fault that is difficult to get in any other way.

In addition to musing about the big earthquake, this is probably a pretty good birding walk for chickadees and warblers. While we were last there, we saw an acorn woodpecker trying to drink from a water faucet. The Visitor's Center has an abundance of books and maps, as well as an informative display.

From the Center, go back into town in order to get on the Sir Francis Drake Boulevard and drive along the eastern side of Inverness Ridge. At the Tomales Bay Trailhead, which is located near the southeast end of the Bay, you can walk down a steep trail to the marsh.

As you continue driving down the west side of Tomales Bay through the town of Inverness, you will pass by some big mud flats that are hundreds of feet wide. There is some salt marsh at several spots along there as well. A little ways beyond the Golden Hinde Marina (on your right), the highway turns left up the hill toward the top of the Inverness Ridge. About 1.7 miles along the drive up the hill, take a right on Pierce Point Road. You will then be headed toward Tomales Bay State Park. As you drive along the top of the Ridge, you can see the abundant large trees down the slope on the right and the bare, grassy land on the tops of the rolling hills on the left. As best we can tell, the heavy forest on the east side of Inverness Ridge is there because of some kind of wind-shadow effect that produces more rain on these east-facing slopes of the Inverness Ridge than elsewhere in this region (presumably somehow related to the dominant northwesterly wind direction). The lack of trees on the fields to the left may also be due, at least in part, to agricultural activities. About 1.2 miles further along, the road takes an abrupt turn to the left. At that point, pull over to the parking area on the right and take a walk down the Jepson Trail, if you are interested in visiting a virgin bishop pine forest.

There should be a lot of birds in this area at the right time of the year. On the way up the hill, you passed by the boundary to the Point Reyes National Seashore, but the Jepson Trail itself is within the boundary of the Tomales Bay State Park.

The Jepson Trail is a pleasant one with a moderate slope that reaches a stand of virgin bishop pine (*Pinus muricata*) about half a mile in. There was a big fire there in 1995 that burned most of the trees on Inverness Ridge. According to Lage (2004), this grove of bishop pines (Jepson Memorial Grove) is one of the few remaining virgin bishop pine forests. These pines, which have a very thick bark, are not very long-lived trees. Unlike the lingering redwoods, they only live about 150 years. They are somewhat unique in that they are fire-obligate trees. That is, they cannot reproduce without a fire to open up the pine cones and free the seeds.

This is a great walk through these multi-shaped pines, most of which are highly distorted, with only a few straight ones being present.

Back on Pierce Point Road, continue west for about a quarter mile, turn right, and take the road down to **Tomales Bay State Park**. This drive down to the State Park, which passes through a large number of good-sized trees, will probably strike you as a bit unusual if you have spent much time driving through the relatively treeless hills in Marin County or further west into the grassy hills of the Point Reyes National Seashore. Trees present in this forest, in addition to the bishop pines, include madrones, California laurels, oaks, red and white alders, willows, and buckeyes.

The road down the hill leads to Hearts Desire Beach, which is composed of coarse-grained sand that looks like it has just recently weathered straight out of the granite/granodiorite source. At the south end of this beach, there are some outcrops of "very interesting" metamorphic rocks (Doris Sloan, pers. comm.). Three more of these sheltered beaches on the shore of Tomales Bay are located within the State Park – Indian Beach, composed of mixed sand and gravel, Pebble Beach, and Shell Beach. Trails within the Park lead to these other three, usually uncrowded and serene, beaches.

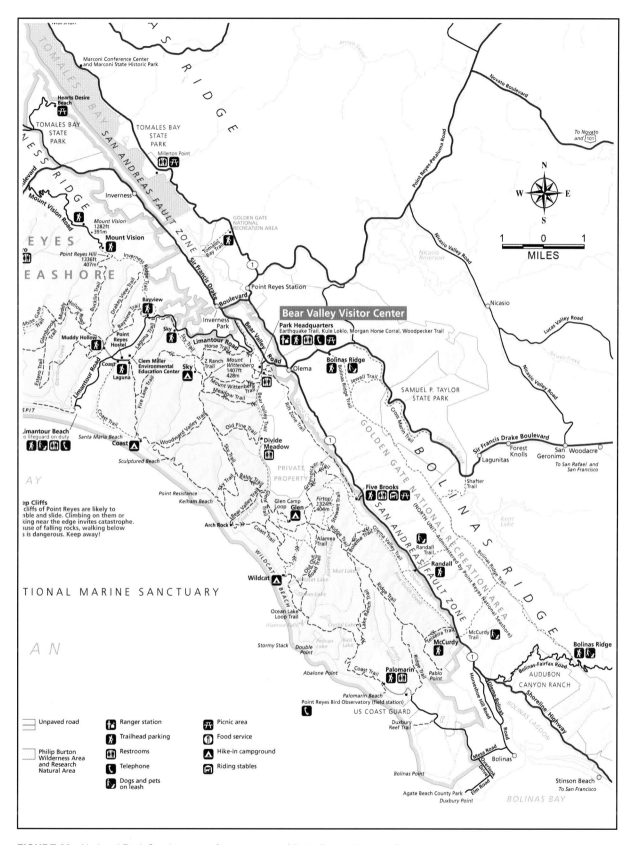

FIGURE 99. National Park Service map of eastern part of Point Reyes National Seashore.

14 COMPARTMENT 2 – POINT REYES

INTRODUCTION

All of the 70,000 acres of this naturalist's paradise is located within the **Point Reyes National Seashore**, probably the jewel of that entire seashore system (Figures 99, 100, and 101). It includes two areas with spectacular rocky cliffs, the western shoreline of Tomales Point and Point Reyes itself, both of which contain exposures of the Cretaceous granitic rocks that have been transported hundreds of miles to the north of their continental origin aboard the Pacific Plate (as it slides toward the northwest along the San Andreas Fault zone in Tomales Bay). The almost twelve mile long Point Reyes Beach to the north of Point Reyes, with its abundant beach cusps and rhythmic topography, faces directly into the dominant northwesterly waves. Another long sandy shoreline around Drakes Bay, located east of Point Reyes, circumscribes one of the largest crenulate bays in the state (Figure 67B). Possibly the most pristine estuary on the west coast, Drakes Estero, occupies center stage of this Compartment (Figure 101). Excellent birding abounds, most notably along the cliffs of Point Reyes (another famous whale watching spot).

Two items of information that you should have that will enhance your visit to this site are the detailed map and brief discussion of the Seashore presented in a fold-out information flyer published by the National Park Service, that you can get from any of the visitor centers (there are three within the Seashore's boundaries; see maps in Figures 99 and 100), and a guide to the Seashore written by Jessica Lage (2004). Lage's book will be especially valuable if you plan to do much hiking, because it has detailed information on all the major hiking trails.

It is difficult to pick the most outstanding attribute of this exceptional location, but the abundance of birds possibly heads the list. Information on one of the web sites of the National Seashore (2009) outlines the birds you might see during your visit. With nearly 490 species recorded (45% of the species of birds in North America), Point Reyes National Seashore easily claims the prize for the greatest avian diversity in any U.S. national park. Some of the factors responsible for attracting this amazing diversity are Point Reyes' location at an optimal latitude, its diverse habitats, its location along the Pacific Flyway, and the shape of the peninsula which acts as a geographic magnet.

In addition to the natural scenery, geology, and birds, the area also has an interesting **human history**. As noted in the NPS flyer, *the moderate climate and the fertility of the land and its nearness to the sea made this area attractive to humans from very early days. The Miwoks, a peaceful people, harvested acorns and berries, caught salmon and shellfish, and hunted deer and elk. In 1579, Francis Drake, an English adventurer serving Queen Elizabeth I, appeared offshore in his ship Golden Hind. Evidence indicates that Drake careened his ship (leaned it on its side to expose the hull) in Drakes Estero to make repairs, staying there some five weeks. The Miwoks supplemented the Englishmen's rations with boiled fish and meal ground from wild roots. ..... Before Drakes Golden Hind sailed west to cross the Pacific on it's round the world voyage, he named this land Nova Albion, meaning New England. He doubtless thought it resembled the Dover coast of the English Channel* (The White Cliffs of Dover).

Before you begin exploring the individual sections of the National Seashore, a good way to get a general overview of it is to drive up to the end of Mt. Vision Road (Figure 100). To get there, take Sir Francis Drake Boulevard through the town of Inverness and drive up over the crest of Inverness Ridge. About a mile past the intersection with Pierce Point Road, take a left on Mount Vision Road, which has several sharp switchbacks. As you make your way up the slope toward the mountain top, you will have a great overview of the whole south-central portion of the Seashore. As a bonus, it would be a rare instance that you could look up into the sky and not see a bird circling as you drive along, most commonly lots of turkey vultures and the occasional golden eagle will be in view. Hawks and ravens are abundant as well. Further up the hill, there are several great views to the

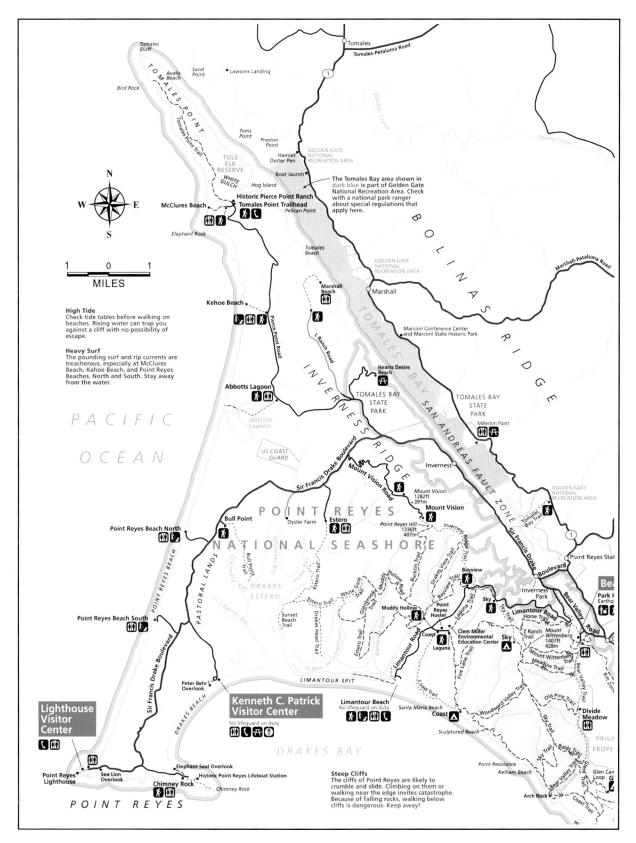

FIGURE 100. National Park Service map of western part of Point Reyes National Seashore.

163

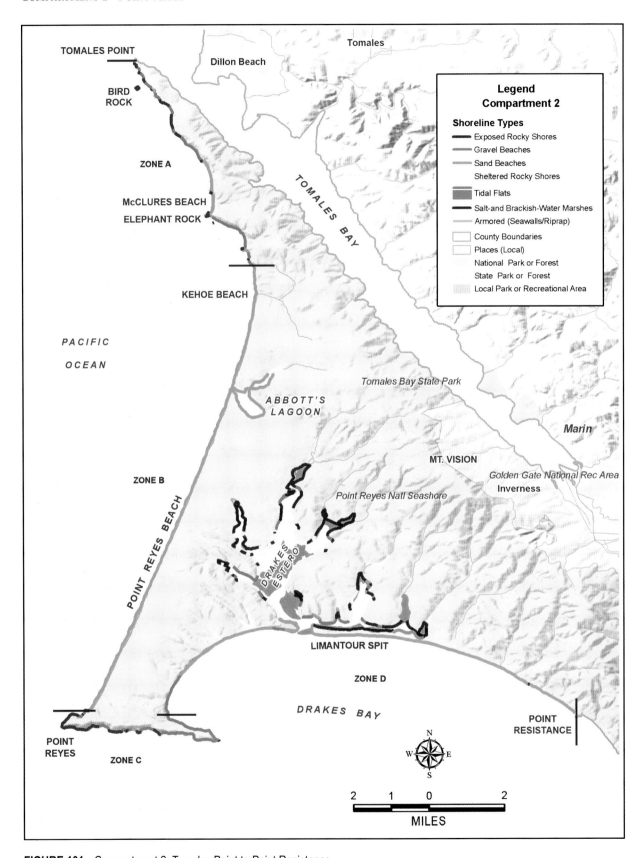

FIGURE 101. Compartment 2, Tomales Point to Point Resistance.

southwest of Drakes Estero. You will eventually reach the end of the road near the top of Inverness Ridge, which parallels the San Andreas Fault Valley (Figure 23) and approaches 1,400 feet in elevation in places. While there, you will also be treated to an excellent view of Tomales Bay.

Several thick growths of small pines are scattered around the ridge top. These are **bishop pines** that have grown up since a fire scorched the area in 1995, opening up the cones of the pines that were consumed by the fire and freeing their seeds. That was only thirteen years ago (in 2008), so these pines are pretty fast growers. Because of their uniform size, there is little doubt that all of these pines were seeded at about the same time

This drive is definitely worth the time for the vistas of both the southwest and northeast flanks of Inverness Ridge, as well as of the Drakes Bay shoreline. The changes in the vegetation on the way to the top are also quite interesting. It would be best to take this drive on a weekday, because the road can get pretty crowded on weekends.

For purposes of discussion, we have further subdivided this Compartment into four Zones (as mapped in Figure 101):

A) **Tomales Peninsula** is a narrow promontory oriented parallel to the San Andreas Fault Valley that has a core of the Mesozoic granitic rocks of the Salinian Block. Much of the shoreline is backed by high scarps cut into the rocks by the large waves that approach it from the northwest across the open Pacific Ocean. Large numbers of tule elk graze the grassy hills of the Peninsula.

B) **Point Reyes Beach** is one of the longer reaches (12 miles) of pure sand beach on the Central California Coast. The beach, which is backed by a wide zone of mostly vegetated sand dunes, faces directly into the approach direction of the dominant northwest waves.

C) **Point Reyes** is a three mile long eroding bedrock headland that projects far out into the Pacific Ocean. Like the Tomales Peninsula, it also has a core of Mesozoic granitic rocks. This granitic foundation is overlain by some deep-water conglomerates deposited in the Paleogene period (Table 1). This is also a great spot to observe a wide variety of birds and marine mammals.

D) **Drakes Bay Shoreline** is an open, arcuate shoreline with the classic crenulate bay shape. The refraction/diffraction pattern of waves entering this crenulate bay is illustrated in Figure 67B. A relatively flat and wide sand beach borders the margin of the bay. Much of the shoreline contains high, picturesque wave-cut scarps in the light-colored Purisima Formation, which is composed of fine-grained sedimentary rocks of upper Miocene/Pliocene

age. The pristine Drakes Estero, located near the center of the Bay's arc, is a classic estuary with well developed ebb- and flood-tidal deltas (Figure 68B).

TOMALES PENINSULA

Introduction

As shown by the map in Figure 101, the Tomales Peninsula (Zone A) is a narrow, mountainous, six mile long sliver of land that parallels the San Andreas rift zone. Its northeast shoreline is along the relatively quiet waters of Tomales Bay, but the huge waves of the open Pacific Ocean pound the southwest shore. Because of this wave action, over half the outermost shoreline on the ocean side of the Peninsula is made up of steep rocky cliffs with a few scattered pocket beaches of gravel and/or sand.

Geological Framework

As shown by the geological map of Point Reyes in Figure 102, the Peninsula is underlain by the granitic rocks of the **Salinian Block** which, like the rocks exposed at Bodega Head, have ridden to the north for possibly several hundred miles aboard the Pacific Plate. According to Cardwell and Detterman (2003), *the basement of the Salinian Block in the Point Reyes area is composed of three different granitic bodies: tonalite of Tomales Point, granodiorite and granite of Inverness Ridge, and porphyritic granodiorite of Point Reyes Promontory. The granodiorite is intruded by numerous light- and dark-colored dikes.* They also made the important observation that these Salinian Block igneous rocks are all Late Cretaceous in age (around 65-100 million years old) and that they are considered to be a piece of what they called a *continental plutonic arc* similar to those in the roots of the Sierra Nevada.

[NOTE: As discussed earlier, a **porphyritic** igneous rock is one that has exceptionally large, separate, and distinct individual crystals imbedded within a finer-grained matrix. The larger crystals were formed first, during which time they "floated around" within the molten magma as they slowly grew larger. The matrix, which is finer-grained, crystallized more abruptly when a cooler temperature of the molten magma was reached that promoted the more rapid crystallization of the matrix minerals. The texture of the different granitic rock types is illustrated in Figure 17A. An intrusive **dike** is a relatively narrow rock band (or zone) of distinctly different composition (usually) that has intruded into and cuts across an older rock unit.]

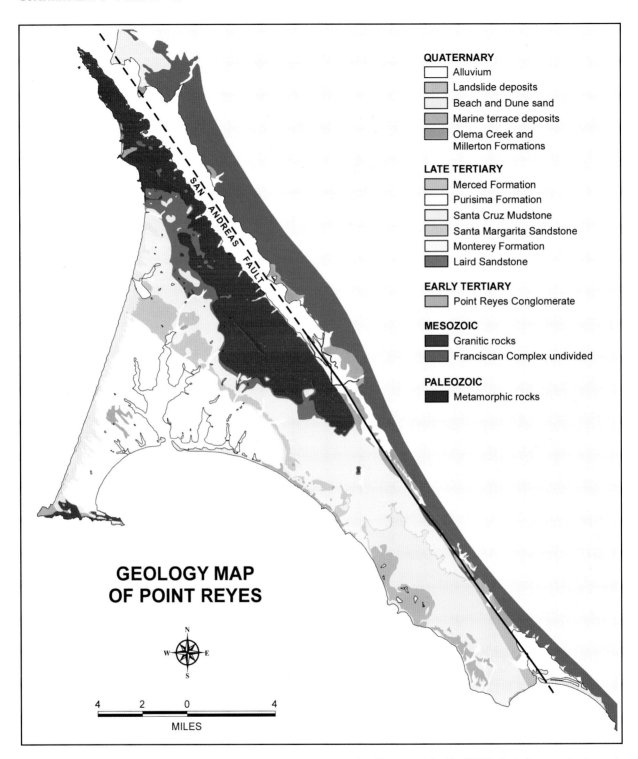

FIGURE 102. Geological map of Point Reyes. Modified only slightly after Clarke and Brabb (1997). A similar map is shown in Sloan (2006).

The Tomales Peninsula has the most continuous exposure of these Mesozoic granitic rocks anywhere along the shoreline of the Central California Coast north of San Francisco, but access to them is difficult, because of the steep scarps in the wave-cut rock cliffs. As the waves erode these granitic (tonalite?) bluffs, the residual sand grains created by the erosion are transported to the south, some of which ends up on the spectacular Point Reyes Beach.

Beaches

Elephant Rock and another rocky headland a few hundred yards to the north of it (Figure 101) form natural groins that have backed up about a mile of sandy beach called McClures Beach, a popular destination for visitors to the Seashore. This relatively remote beach is an outstanding one to visit because of its wide, sandy berms and dynamic waves, as well as the excellent outcrops of granitic rocks in the wave-cut rock cliffs and nearshore sea stacks.

All of the other beaches in this Zone are short pocket beaches composed of both sand and gravel perched at the base of steep, rocky cliffs.

General Ecology

Along the western shoreline, there are four harbor seal haulouts, a seabird nesting colony on Bird Rock, and a large patch of kelp forest off Tomales Point. The nearshore waters are important wintering areas for a long list of seabirds, including auklets, murrelets, grebes, loons, and shearwaters. Whales, dolphins, and sea otters are also present.

The vegetation on the Peninsula reflects a long history of human influence. By studying pollen and dating charcoal particles deposited in lakes, Anderson (2005) showed that the vegetation on Point Reyes consisted of conifer forests until about 10,200 years ago. Then, the climate became drier, with mixed forests and scrub, until about 6,500 years ago. Native Americans populated the area starting 5,000 years ago; the studies show evidence of increased fire activity and vegetation dominated by coastal scrub and grasslands. The interpretation is that native coastal grasslands were created and maintained through frequent burning by Native Americans, to provide grazing areas for deer and force oak trees to produce higher yields of acorns. When Europeans arrived and developed extensive dairy farms, non-native annual grasses took over, and grazing further damaged the native vegetation. In the parts of the Seashore where cattle are now excluded (and tule elk introduced), the vegetation is changing, with greater abundance of several native perennial grasses, along with other non-native perennial and annual grasses, and an increase in scrubs such as coyote brush and yellow bush lupine. Current vegetation maps show grasslands on the high central ridge and the western, windward slopes, and scrub communities dominated by native coyote brush covering the southern and eastern slopes. Without fire, grasslands will gradually become scrub, which will gradually become forests. In 1995, a fire

from an illegal campfire spread, burning 12,000 acres; you will see evidence of this fire on Inverness Ridge. The National Park Service is conducting research on the positive and negative impacts of fire suppression and targeted prescribed (intentional) burning of certain types of habitats. Burning of grasslands could promote native grasses; as discussed earlier, bishop pines require burning to reproduce in large numbers. Prescribed burning also reduces the amount of fuel and the risk of more intense fires.

Places to Visit

To get to the west side of the Tomales Peninsula, follow the Sir Francis Drake Boulevard toward the northwest through Inverness and drive up to the top of Inverness Ridge, where you turn right on Pierce Point Road (Figure 100). Follow this road, along which you will pass by the exit to Tomales Bay State Park and the parking areas for the trails to Abbots Lagoon and Kehoe Beach, to its end near Historic Pierce Point Ranch. About a mile and a half past the Kehoe Beach parking area, you will be driving along the crest of Inverness Ridge from which you will have a great view of Tomales Bay, with the Walker Creek Delta being very conspicuous. You will also enter the **Tule Elk Preserve**, within which the elk are contained with fences.

In an informative discussion of tule elk, Lage (2004) pointed out the following:

1) Large populations of tule elk occupied Point Reyes for thousands of years.

2) Habitat shifts from perennial bunchgrasses to European annuals devastated the herds.

3) Excessive hunting didn't help much either, and by the mid-1800s, they were considered to be extinct throughout California.

4) However, a Central Valley landowner discovered a small population of the elk on his land and protected them. The few that remained gained official protection by the state in 1971.

5) In 1978, the National Park Service reintroduced ten of the elk to this section of Tomales Point.

6) Eventually, the herd expanded to over 400, outgrowing the contained reserve.

7) Such intense pressure on the land threatens rare plants and butterflies, and with such a large population of animals, disease is more likely and forage is less available.

8) Some elk have been relocated to other parts of the Seashore.

9) And, finally, the National Park Service has started a tule elk birth-control program (immunocontraception

vaccine). The population is apparently now under control at a number of around 450.

Back to the drive, at the end of which is the start point for the 9.4 mile (round trip) **Tomales Point Trail**. Among other glowing comments about this trail, which she *calls one of the most dramatic trips in the Seashore,* Lage (2004) noted *that the wide, sandy trail meanders through coyote bush, low-growing blue lupine, and the sweet-smelling yellow flowers of coastal bush lupine (Lupinus arboreus).* The trail is also a place to view the conspicuous tule elk, abundant wild flowers in the spring, and a variety of birds, including an array of hawks and kestrels flying overhead during the migratory season (spring and fall). A little over half a mile from the end of the Peninsula, a large sea stack in the Pacific Ocean called Bird Rock should also be of interest to the birders on the hike. However, the best part of the hike is the phenomenal views along the trail, most notably: 1) Tomales Bay and its northern shoreline to the east; and 2) the striking cliffs along the Pacific Ocean's shore to the west.

Whether you take the Tomales Point Trail or not, you certainly don't want to miss the short walk down to **McClures Beach**. As you get close to the beach, right along the trail you can see the highly weathered and jointed granitic rocks of the Salinian Block. A sedimentary unit, presumably the Laird Sandstone of Miocene age (Figure 102), referred to as a basal transgressive sandstone unit by Cardwell and Detterman (2003), overlies the granodiorite (or is it tonalite?) at this location. The term **transgressive** refers to the condition where a rising sea floods the landscape, with the eroding shoreline leaving a sand deposit in its wake as it moves further landward. As this remaining sand is buried by sediments deposited on it later, it eventually converts to sandstone. The presence of such sandstone layers at this location raises a puzzling question with reference to their location on the geological map in Figure 102. It should be clear to the reader by now that the two red zones on the map (along Tomales Point and at Point Reyes) represent the older granitic rocks that have been transported to the north on the Pacific Plate since their original formation in magma chambers deep within the Earth's crust (somewhere far to the south).

However, a legitimate question would be how did the bands of Tertiary sedimentary rocks that separate the two granitic exposures get there? Apparently, the sediments that eventually became these rocks were deposited in a moderately deep basin filled with marine water during the time interval between middle Miocene and Pliocene time (probably somewhere on the order of 5-20 million years ago). According to this concept, the

Laird Sandstone was deposited as the basin first began to fill with marine waters. As the waters deepened, the sediments that eventually became the rocks that now make up a formation called the Monterey Formation were deposited within the basin, so that they now overly the Laird Sandstone. This is a somewhat confusing picture, even to geologists like ourselves. Just remember that the sedimentary rocks present in this location today make up only a preserved fragment of the rocks deposited within the whole original basin, which was not restrictively scrunched in between Tomales Point and Point Reyes. Later tectonic movements and erosional events have resulted in the loss of a significant part of the sedimentary rocks originally deposited within the basin, with individual fault-bounded and disparate pieces of them being scattered along the coast. The preserved sedimentary rocks in the middle zone of the Point Reyes National Seashore are one of those disparate pieces. Also keep in mind that the original basin was located far to the south of where these rocks are now, because of their position on the west side of the San Andreas transform fault.

Standing on the beach at the end of the trail at McClures Beach, you can see all the way to Bodega Head (also underlain by granitic rocks). This sand beach is located at the mouth of a little stream, which seems to be a consistent pattern for the location of sand beaches all along the Central California Coast. The beach itself is a beautiful, medium- to fine-grained sand beach with what appears to be a granitic source (Figure 103A). Against the rock cliff behind the beach is a zone of only moderately sorted, but rounded boulders composed mostly of the granitic rocks, which indicates that the sand normally found on this beach may be carried offshore during a major *El Niño* storm, allowing the large waves to round the boulders by smashing them together.

At the southeast end of the beach, a sharp, spear-like limb of the Elephant Rock sea stack is composed of granitic rocks (see photograph in Figure 103B). This is one of the most unweathered outcrops of pure granitic rocks found anywhere along the coast north of San Francisco. Other outcrops of relatively unweathered granitic rocks are present on the Farallon Islands and the Monterey Peninsula.

POINT REYES BEACH

Introduction

With Point Reyes itself acting as a natural groin, this twelve mile long segment of the shoreline, Point Reyes Beach (Zone B on Figure 101), which faces directly

FIGURE 103. McClures Beach in the Point Reyes National Seashore (Marin County; Zone A, Figure 101). (A) An 0.4 mile trail leads from a parking area to this beautiful, medium-grained sand beach located about four miles south/southeast of the end of Tomales Point (see details on location in Figure 100). This photograph, taken on 17 August 2008, looks toward the south end of the beach to where it abuts against some conspicuous outcrops of the Salinian Block granodiorite. (B) Close up view of the granodiorite.

into the approaching northwest wind and waves, is an effective sand trap. Figure 104 shows an overview of this sand beach taken from the northwest end of Point Reyes, and Figure 38B shows a close-up view of beach cusps near the Point. Being exposed to the open ocean, the waves are so large and chaotic that the National Park Service recommends that this beach not be used by swimmers, especially because rip currents are a common occurrence.

Beaches and Dunes

We do not have detailed information on the mineral composition of this medium- to fine-grained sand beach, but inspection with a hand lens indicates that the granitic rocks in the area have made significant contributions. This is not a big surprise, considering the presence of the granitic rocks along Tomales Point. Such an extensive sand beach at this location is, however, something of a puzzle, because the scarps along Tomales Point could not possibly have supplied this much sand. Another potential source for the sand is the continental shelf, which was exposed during the lowstands of sea level during the ice ages and quite possibly contained some extensive dune fields at that time, as was the situation off the area south of the Golden Gate (discussed later). These dune sands could have been transported toward the present beach by wave and wind action as sea level approached its present position. Also, it is not unusual for sand to be transported from the shelf to the beach during mild storms. At lowered sea levels, the streams that carved the valleys in which Estero Americano and Estero de San Antonio are now located could also have delivered a considerable amount of sand to the shelf.

Because of its orientation, with the dominant waves coming straight on, the longshore sediment transport rate should be rather slow and probably generally toward the south. However, strong waves from due west or a more southerly direction could move the sand to the north during those episodes. In terms of beach erosion, Savoy et al. (2005) classified the entire length of Point Reyes as *Stable: low risk*, which is to be expected, considering its continuing rate of sand deposition.

The occurrence of an extensive zone of **sand dunes** back of the beach provides further evidence that this beach is, in general, a sediment sink, or trap. The first major line of dunes back of the beach occurs just south of the stream outlet at Kehoe Beach, which is located south of the last of the eroding bedrock scarps (see Figure 101). These dunes form a continuous zone all the way to the northern side of Point Reyes itself (colored light yellow on the geological map in Figure 102). The landward

FIGURE 104. Oblique view along the 12 miles of Point Reyes Beach in the Point Reyes National Seashore (Marin County; Zone B, Figure 101). This beach faces directly into the dominant northwest waves. Beach cusps and some irregular rhythmic topography were present on the beach this day, 18 August 2008.

edge of the line of dunes is multi-lobate (digitate), with individual dune lobes extending landward for almost half a mile in a few places, but more generally on the order about 500-700 feet. Most of these dune lobes are now vegetated; however, barren sand sheets with some active dunes do occur at three localities: 1) just north of the outlet of Abbotts Lagoon; 2) just north of the parking area at Point Reyes Beach North; and 3) on eight active dune lobes on the southern end of the beach, as well as on a few bare spots a bit further to the north.

General Ecology

Point Reyes Beach is designated critical habitat for western snowy plover (Figure 96), a federally listed endangered species, typically supporting 30-35 adult breeding plovers. Protection efforts include roping off breeding habitat on upper sections of beaches, construction of exclosures around their nests immediately after an egg is laid, closing certain areas to dogs, and restoring dune habitat. According to the

National Park Service, snowy plovers used restored dune habitat in 2007 for the fifth consecutive year, building 4 of 28 nests in the restoration areas. Nesting occurs March-September. The beach supports many other shorebirds as well.

Between February and August, the offshore waters support up to 300,000 seabirds, 150,000 diving birds, and 20,000 gulls that nest on the Farallon Islands. After the nesting season, these birds disperse to feed on the waters from Point Reyes to the Golden Gate. Common species include common murres, marbled murrelets, rhinoceros auklets, Pacific loon, western grebe, surf scoters, white-winged scoters, and sooty shearwaters.

Freshwater Abbotts Lagoon is an important migratory stopover for ducks and geese. Hawks and osprey are also commonly seen there March-September.

Places to Visit

Driving south along Pierce Point Road after leaving the parking area for McClures Beach, in a couple of places you get good views of the long Point Reyes Beach. There is a pullover area where you can park and take the trail down to Kehoe Beach, located near the north end. Going down the Kehoe Beach trail, you pass along the side of a small valley floored by some fairly extensive wetland plants, including cattails and common tule (*Scirpus acutus*). This is no doubt a good birding spot at the right time of year. When we were there in August 2008, the wetland was full of birds, including a flock of about 100 red-winged blackbirds. Right before you get to the beach there is some water in this channel (see photograph in Figure 105A). A line of vegetated high dunes closes off much of the mouth of this little valley.

Also, the side of the bluff north of the trail is underlain by tilted layers of white sedimentary rocks of the famous Monterey Formation of Neogene age (Miocene; deposited 20-10 million years ago), noted for the fact that it is a major source rock, as well as a reservoir rock, for much of the petroleum in the sedimentary rocks of coastal California (as briefly discussed earlier; see Figure 105B). Fractures in the rocks provide the permeability necessary for removal of the oil. As noted by Sloan (2006), microscopic fossils of single-celled plants (diatoms) and animals (foraminifera) are common in these rocks.

In a discussion of the Monterey Formation, Cardwell and Detterman (2003) provided a possible explanation for the origin of the basin in which the Laird Sandstone and Monterey Formation were deposited. They reported that prior to about 5.5 million years ago, the motion along the San Andreas Fault was slightly tensional (i.e., opposite sides pulling away from each other) as well as

FIGURE 105. Kehoe Beach in the Point Reyes National Seashore (Marin County; Zone B, Figure 101). (A) Ponded wetland along the trail to Kehoe Beach located in the valley of a small stream inland from the line of foredunes straight ahead. The active beach is a short distance on the other side of the dunes. (B) An outcrop of the famous Monterey Formation of Miocene age located in the bluff landward of the beach. Both photographs taken on 18 August 2008.

parallel to each other as is expected along a transform, or strike-slip, fault. As the two sides moved perpendicular to and obliquely away from each other, deep basins resulted that were filled with sediments, with the total thickness of the Monterey Formation on the Point Reyes peninsula estimated to be about 3,500 to 5,000 feet.

Kehoe Beach is a medium- to fine-grained sand beach next to a small stream outlet. About a half mile north along the beach from where the trail comes out, there is some pegmatite dike-riddled granite (or granodiorite and/or diorite). From that spot on north toward Elephant Rock and McClures Beach, the beach gives way to some very steep, rocky scarps. There are also a number of major landslide areas along those scarps that provide another possible source for the sediments transported south along the shore. Kehoe Beach marks the beginning of the Point Reyes sand beach, and from out on the beach, you can see all the way along the straight beach front to Point Reyes. The sand dune area landward of the beach starts a little ways to the south of the stream outlet at this locality.

If you return to your car and drive south along Pierce Point Road for another two miles or so, you come to the parking area for the trail to Abbotts Lagoon, which is a 3.2 mile round trip to the beach and back (Figure 100). Along your walk around the 282-acre lagoon, the birding is diverse, as pointed out by Lage (2004) who recommended that you *skirt the lagoon, looking for shorebirds foraging along the water's edge. Caspian terns, coots, western and pied-billed grebes are among those you may see feasting on clams, snails, crabs, and worms, and diving for small fish.* Furthermore, this area is noted as a great place for birding in the fall, because it has a freshwater pond, a brackish pond, and also an open ocean shoreline, so there is a variety of habitats available for migratory birds.

Lage observed that this is a prime wildflower viewing area during the right season. In addition, this trail gives you the opportunity to view an active dune field that is building into the north side of the lagoon. And, of course, there is the spectacular beach itself, with its rhythmic topography, beach cusps, rip currents, and so on (refer once again to the oblique view of this beach in Figure 104).

There are two more beach access areas to the Point Reyes Beach that you can drive to. To do so, continue southeast on Pierce Point Road until you reach Sir Francis Drake Boulevard, where you turn right and head toward Point Reyes. About 5 miles along, you reach the turnoff to Point Reyes Beach North, and about a mile and a half from there you encounter the turnoff to Point Reyes Beach South (Figure 100). This drive is through some barren, treeless areas, but the multiple, striking colors of the invasive ice plant (*Carpobrotus edulis*) on the sides of some of the hills is remarkably appealing (in August anyway). With regard to these two beach areas, their morphology is clearly illustrated by the photograph in Figure 104.

POINT REYES

Introduction

This rocky headland (Zone C on Figure 101), composed mostly of stark and steep wave-cut cliffs with a multitude of offshore sea stacks, is a little over three miles long. There are a few small pocket sand beaches at

the base of the scarps, mostly on the eastern half of the headland, which are usually covered with sea lions and seals at low tide (refer to the Environmental Sensitivity Index map of this area in Figure 85). The largest of these beaches, a curving arc about 200 yards long located due south of the historic Point Reyes Lifeboat Station on Drakes Bay, may contain over a hundred hauled-out animals on any given day.

The historic lighthouse on the west end of the headland is a major attraction (Figure 106A). Like other places in the world where a rocky promontory sticks out into the ocean, historically there have been a significant

FIGURE 106. Point Reyes Lighthouse (Marin County; Zone C, Figure 101). (A) There is a lot of spectacular geology to see on the walk down these stairs. (B) A 50 million-year-old early Eocene conglomerate that is thought to be a turbidite deposit (scale is ~ 6 inches). The arrow points to a porphyritic volcanic cobble that, according to Sloan (2006), has its nearest known source in southern California or Mexico, another small piece of evidence supporting the idea that these rocks in the Salinian Block are riding north on the Pacific Plate. Both photographs taken on 18 August 2008.

number of ship wrecks on this headland, hence the need for the lighthouse. The fairly common foggy conditions along this shore is also a part of the problem, especially in the summertime.

Geological Framework

Possibly after the marine mammals and the birds, the geology of the headland is its most exciting attribute. As noted by Sloan (2006) in her book on the *Geology of the San Francisco Bay Region, when you cross the wide, straight valley of the San Andreas Fault and step onto Point Reyes Peninsula from mainland North America, you step into another world, geologically speaking* (see geological map in Figure 102). As mentioned in the discussion of Tomales Point, the rocks in the Seashore are aboard the Pacific Plate; therefore, they have traveled far to the north in the past 25 million years or so. Again, it is the Mesozoic granitic rocks, also discussed elsewhere in considerable detail, that form the core of the Point Reyes headland, which is composed of porphyritic granodiorite. At Point Reyes, the granodiorite is overlain by the Point Reyes Conglomerate of Eocene age (Sloan, 2006), a deep marine deposit. As noted by Ed Clifton (pers. comm.), *some geologists* (specifically Burnham, 2005) *believe the conglomerate at Point Reyes is the same unit that crops at Point Lobos* (in Monterey County; discussed in detail later).

A thick sequence (almost 15,000 feet) of sedimentary rocks, including the Laird Sandstone and Monterey Formation (Figure 102), occurs within the Seashore area. Doris Sloan (pers. comm.) shed more light on this story, explaining that *the Laird is a shallow marine deposit* (as noted in the discussion of McClures Beach area), *then comes the deeper Monterey, and then comes a change to more shallow water* (in the Late Miocene). *It's not clear whether the Laird ever was deposited on Point Reyes Conglomerate,* which overlies the porphyritic granodiorite at Point Reyes. Both of the other two granitic sequences in the area (Inverness Ridge granite and Tomales Point rocks) are not overlain by the Point Reyes Conglomerate, being instead overlain by the Laird Sandstone. She concluded by saying *some major faulting is going on, I think.* The younger of these sedimentary rocks are most perfectly displayed in the cliffs along the shoreline of Drakes Bay.

Sloan (2006) described the outcrop of the Eocene Point Reyes Conglomerate that is located below the lighthouse at Point Reyes (pictured in Figure 106A). *It includes quartzite, large cobbles of the underlying granite, an unusual purple volcanic rock, and others. The purple volcanic rocks, which contain small white crystals of the mineral feldspar* (a porphyritic rock with a fine-grained matrix, due to their volcanic origin), *are evidence for the long distance these rocks have traveled with the Pacific Plate. The nearest known source of these distinctive rocks is southern California or Mexico.* One of the gravel clasts composed of these purple volcanic rocks is shown in Figure 106B. According to Cardwell and Detterman (2003), this unit was deposited as *turbidites and debris flows in deep water.* Presumably, they *were deposited in a submarine channel that was cut into the underlying granodiorite.*

[NOTE: A **debris flow** is defined as the rapid movement of a partially consolidated mass of sediment and broken up rock fragments, some up to boulders in size, down a steep slope, usually in a marine setting. Debris flow deposits are commonly associated with turbidites, which are deposited by more fluid plumes of sediment-laden water (Figure 21A).]

Beach Erosion

Savoy et al. (2005) mapped the outer coastline of this headland as having a *Hazard: high risk* for beach erosion, which is made abundantly clear by the scalloped margin of the tops of the wave-cut rock cliffs. Some of the hiking trails lead fairly close to these cliffs, so hikers should be especially careful not to get too close to the edge, particularly along the margin of the more recent landslides.

General Ecology

The rugged isolation of Point Reyes makes it ideal for seabird nesting. There are two large colonies on the outer shore (shown on Figure 85). Based on the data in the 2006 coastal resource atlas (NOAA/RPI, 2006), the species and number of birds in the eastern colony are: common murre (15,155), Brandt's cormorant (1,522), pigeon guillemot (616), pelagic cormorant (266), western gull (178), ashy storm-petrel (15), black oystercatcher (6), tufted puffin (4), and rhinoceros auklet (unknown). The southeastern point hosts a rookery for northern elephant seals (2,000 individuals, pupping December-March) and harbor seals (292-319 individuals, pupping March-June), and a year-round haul-out area for California sea lions (ranging from 11-1,388 individuals) and the occasional Stellar sea lion.

Extending 10 miles into the Pacific Ocean, Point Reyes provides an excellent location for whale watching. For gray whales, the peak of the southern migration usually occurs in mid-January; the northern migration peaks in mid-March. Mothers and calves can be seen late April and early May. In addition, humpback whales can be seen March-November, and minke whales can be seen year round.

Places to Visit

The first place to go is the parking area for the lighthouse. If you walk out onto the headland a ways from the parking lot, you will be treated to the view of the Point Reyes beach shown in the photograph in Figure 104. From there, it is a walk of about 0.4 miles to the stairs that go down to the lighthouse (Figure 106A). As you go down those stairs, you will see several splendid exposures of the 50 million year old Eocene conglomerates (Figure 106B) that cap the Mesozoic granodiorite that forms the core of the headland. These conglomerates are thought to have been deposited as turbidites and debris flows, as evidenced by their graded bedding and other definitive sedimentary features.

Also along this walk, you get some outstanding views of the wave-cut cliffs, as well as a chance to see large numbers of birds. You may also see some marine mammals, and from November through February, this is one of the best spots along the California Coast to see the gray whales on their way south.

If you can tear yourself away from the lighthouse area, we recommend that you go back along the road you came in on for about 1.5 miles and then travel east on the one-way, paved Chimney Rock Road (Figure 100). Along this road, you will have some good views across Drakes Bay, but the ultimate goal is the parking lot for the Chimney Rock Trail. This trail across the crest of the east end of the Point Reyes headland provides phenomenal views of its eroded face and, if you are there at low tide, you may see a host of seals and sea lions hauled out on a couple of the sand beaches. Beautiful place! Close to being one of a kind.

DRAKES BAY SHORELINE

Introduction

This shoreline (Zone D on Figure 101) is a classic crenulate bay into which the waves refract/diffract in the pattern illustrated in Figure 67B. From the Point Reyes headland to the tidal inlet at the entrance to Drakes Estero (Figure 100), the wide and flat beach is backed by high eroded cliffs in light-colored sedimentary rocks (Drake's White Cliffs of Dover). East of the Drakes Estero Inlet, a long, sandy spit, called Limantour Spit, closes off several arms of the Estero. A few hundred yards east of the head of the spit, rocky cliffs once again dominate the

backbeach. A rocky promontory that projects a short distance out into the ocean, called Point Resistance, marks the eastern boundary of this Zone as well as of Compartment 2 – Point Reyes (Figures 1 and 101).

Geological Framework

The two photographs in Figure 107 illustrate the thick accumulation of the horizontal layers of the Purisima Formation, which form the cliffs around the margin of the shoreline all the way to the head of the Limantour Spit (Figures 101 and 102). This formation, of upper Miocene/Pliocene age, is the youngest of four Neogene (Table 1) formations that dip gently toward the southwest and occupy a trough between the granitic rocks of Point Reyes and Inverness Ridge. Clearly, the deposition of these formations took place somewhere a good distance to the south and the detailed accounting of their evolution is too complex to try to elucidate here (even if we could!).

[NOTE: We have three sources that have commented on the distance of travel of the Purisima Formation on the Pacific Plate:

1) Before delving too much into the literature, we concluded that if you assume Salinian granitic rocks have moved 300 miles from their place of origin, and you accept an age of the Purisima Formation of 4-5 million years ago, as well as that the movement on the San Andreas Fault started 25 million years ago and continued at a relatively constant rate of around 2 inches/year, these rocks that now border Drakes Bay were deposited about 94 miles to the south of where they are now (a little to the south of the present town of Monterey).

2) On the other hand, Sloan (2006) noted that these Drakes Bay sedimentary rocks are similar enough to rocks in the Santa Cruz Mountains and on the Monterey Peninsula that they may have been contiguous at one time. She concluded further that *these rocks have been offset about 90 miles by movement along the San Gregorio and San Andreas Faults,* which is in pretty good agreement with our numbers.

3) Clifton (pers. comm.) suggested that if we had paid more attention to the literature, we would not have needed to go to the trouble to make the calculation we gave in number 1. He said that *several USGS geologists working in this area in the 1980s noted a similarity between the stratigraphic section here* (Drakes Bay) *and in the area around Monterey and suggested that they indicated 94 miles of cumulative offset along the San Andreas and San Gregorio Faults (USGS, 1997). Burnham* (see Burnham, 2005) *followed with a dissertation at Stanford trying*

FIGURE 107. Drakes Bay shoreline in the Point Reyes National Seashore (Marin County; Zone D, Figure 101). (A) View looking north along the beach near the Kenneth C. Patrick Visitor Center (Figure 100). (B) The rocks of the Pliocene Purisima Formation are thought to have reminded Sir Francis Drake of the "White Cliffs of Dover" when he made his historic stopover at Point Reyes in the year 1579. Both photographs were taken on 17 August 2008.

to prove it with a detailed petrographic and chemical analysis of cobbles and pebbles in both conglomerates. I remain uncertain of the connection, but it is probably worth noting.]

The Purisima Formation was deposited in shallow, marine waters as the basin gradually filled up with sediments, and it is composed of diatomaceous mudstones and siltstones. It also contains spheroidal carbonate concretions that commonly contain cetacean (whales, dolphins, etc.) bones (Cardwell and Detterman, 2003).

Beaches

As shown by the photograph in Figure 107A, at low tide the beach along the northwest side of Drakes Bay is wide and flat and, as a general rule, the waves are of modest size. Along the shoreline in the vicinity of the Kenneth C. Patrick Visitor Center (Figure 100), the beach has been erosional for some time, such that eroded wave-cut rock platforms are exposed in the intertidal zone in places. Furthermore, the rock scarps landward of the beach are subjected to erosion during storms. However, this is not the case along the Limantour Spit beaches, which are generally accretional, or at least, stable. To the east of the east end of the spit, the beaches become erosional again, with eroded rock scarps located along the backbeach. According to Savoy et al. (2005), the entire shoreline within this Zone was mapped as *Caution: moderate risk*. However, the only man-made structure along the whole beach is the Kenneth C. Patrick Visitor Center, and it is set a fair distance back of the high-tide line.

Drakes Estero

This coastal water body is one of the most pristine on the California Coast. The waters of Drakes Estero were designated by Congress as potential wilderness by the 1976 Point Reyes Wilderness Act (Public Law 94-544). It designated 25,370 acres as wilderness, and 8,002 acres of potential wilderness. This is the only federal marine coastal wilderness from Washington State to the Mexican Border. Only eleven marine wilderness areas exist in the U.S.

This coastal water body is an **estuary** according to the definition of Pritchard (1967) because: a) it is a classic example of several stream valleys flooded during the last rise in sea level, reaching near its present size about 5,000 years ago, or so; b) it has freshwater input by several small streams; and c) it has a considerable tidal

range of around twelve feet during spring tides. Also, the tidal inlet at the entrance to the estuary has a classic morphology, with relatively equal-sized tidal deltas, as illustrated in Figure 68B. According to Anima (1990), because of the large tides and the relatively shallow water (maximum depth of around 25 feet), the waters of the estuary are vertically mixed from top to bottom within the 1,300 acres of the central arm of the Estero. Salinities range between 34 and 35 parts per thousand throughout the estuary, and relatively strong tidal currents flush it daily.

The following summary of the Estero's characteristics is based on the National Park Service website:

• Extensive eelgrass beds that support rare and specially protected species.

• Reduced presence of non-native species: Recent surveys show that many invasive species are only found where mariculture and oyster racks occur, but not in Limantour Estero (the eastern arm of Drakes Estero).

• One of the largest harbor seal populations in California, with numbers surpassing 1,800.

• Identified as significant area for the U.S. Shorebird Conservation Plan: 86 species of birds recorded in 2004, including osprey and black brant.

• There are 18 species of concern, including red-legged frog, western snowy plover, brown pelican, peregrine falcon, and marbled murrelet.

• Recent fish survey identified over 30 species of fish, including rare and endangered species such as coho salmon, steelhead trout, and three-spined stickleback.

• Rare plants occur along the shoreline of the estuary.

The coastal resource map (Figure 85) shows the extensive eelgrass beds, tidal flats, and the many animals that use this area. The western end of Limantour Beach is designated critical habitat for the federally threatened western snowy plover (see the discussion of protection actions under Point Reyes Beach). Drakes Estero was designated in 1976 as California's first estuary wilderness area. The eastern arm is designated as a National Wilderness, and the entire Estero will become a Wilderness in 2012. This designation has raised issues with the commercial oyster farm that has been operating in Drakes Estero since the last quarter of the nineteenth century, and whose permit expires in 2012.

The Wilderness Act states *"...there shall be no commercial enterprise and no permanent road within any wilderness area designated by this Act and, except as necessary to meet minimum requirements for the administration of the area for the purpose of this Act (including measures required in emergencies involving the health and safety of persons within the area), there shall be no temporary road, no use of motor vehicles, motorized equipment or motorboats, no landing of aircraft, no other form of mechanical transport, and no structure or installation within any such area…"* (16 U.S.C. 1131-1136, Section 4). The National Park Service commissioned a study by the National Academies to review the scientific information on the ecological impact of the commercial oyster farming activities. The study (NRC, 2009) concluded that *"there is a lack of strong scientific evidence that shellfish farming has major adverse ecological effects on Drakes Estero at the current (2008–2009) levels of production and under current (2008–2009) operational practices, including compliance with restrictions to protect eelgrass, seals, waterbirds, and other natural resources. Adaptive management could help address effects, if any, that emerge with additional scientific research and monitoring to more fully understand the Drakes Estero ecosystem and the effects of shellfish farming."* The National Park Service has until 2012 to make a decision about whether or not to permit shellfish farming in Drakes Estero after it becomes a National Wilderness.

Places to Visit

The first two places you might want to experience Drakes Bay as you enter the Seashore are hiking trails to the shoreline of the Drakes Estero. The first one, the Estero Trail, is reached by driving from the Tomales Bay shoreline up over Inverness Ridge along Sir Francis Drake Boulevard. About a mile past the Mount Vision Road intersection, turn left on a road marked by an Estero Trailhead sign (Figure 100). About a half mile or so down this road, you reach the parking area for the trail, which is a two mile round trip to a bridge over an arm of the Estero, and a 9.4 mile round trip to an overlook at Drakes Head. According to Lage (2004), this trail *offers excellent bird-watching in the Estero, the largest harbor seal breeding colony at Point Reyes, abundant iris in the spring, and outstanding views from Drakes Head, which makes this a great all-around hike or bike ride.*

Back on Sir Francis Drake Boulevard, continue toward Point Reyes for another half a mile or so, where you come close to the landward edge of the main arm of Drakes Estero (Schooner Creek), which is bordered by a beautiful marsh system. Continue along the main road for another two miles and pull into the parking area on the left to reach the trailhead for the Bull Point Trail (Figure 100). Lage (2004) mentioned that if you take this 3.8 mile round-trip trail, you are likely to *see many migratory birds in the Estero, and seals and sea lions often gather on the beach at the trail's end.*

Continuing on toward Point Reyes for another 2.5 miles, you reach the road to the Kenneth C. Patrick Visitor Center on Drakes Bay, a very popular spot with an excellent display on the human history of the Bay, as well as a collection of maps, books, and gifts. The beach at this spot is pictured in Figure 107. The spectacular eroding bluffs in the light-colored Purisima Formation are thought to have reminded the homesick sailors in Francis Drakes crew of the White Cliffs of Dover in the English Channel, which are composed of chalk (pure white calcium carbonate of Late Cretaceous age) and look remarkably similar to the cliffs around Drakes Bay, even with the same kind of undulating cliff tops (Figure 107A).

A great hike to take from here is a 1.5 mile trip along the beach to the tidal inlet at the entrance to Drakes Estero (pictured in Figure 68B). We have taken this walk several times and have always been impressed with the large number of birds we have seen. Witnessing the chaotic interaction of the dynamic waves and tidal currents around the margin of the ebb-tidal delta is also exciting. Along the walk you will no doubt notice that the wrack on the beach there is different from what you will have seen in most other places, because it commonly contains an abundance of bright green eelgrass blades transported out of Drakes Estero by the strong ebb currents.

The last place of relatively easy access to Drakes Bay is Limantour Beach, which can be reached by going back along Sir Francis Drake Boulevard to Inverness Park, and taking a right on Bear Valley Road (Figure 100). Within a few hundred yards beyond this turn take another right onto Limantour Road. The first half of this drive is through some of the forest on the east flank of Inverness Ridge, an area ravaged by a big forest fire in 1995.

At the end of Limantour Road, there are two good-sized parking areas with restroom and picnic facilities within a short walking distance of the beach. Walking toward the west either along the beach itself, or along a trail in the dunes back of the beach called the Limantour Spit Trail, you will eventually come to the tidal inlet into Drakes Estero. About half way down as you walk along the beach, you will encounter features we call **washover fans**, where waves wash all the way across the spit during storms or during especially high spring tides, creating a flattened-off surface along which sand is transported across the spit's surface. You can also see some active sand dunes toward the end of the spit along this walk. This is a great hike with lots of birds and even some solitude, if you walk far enough. Limantour Beach, where the waves are a bit larger than elsewhere in Drakes Bay, is a very popular beach during some parts of the year.

You can also take an equally enjoyable and interesting hike along the beach toward the east. Within a few hundred yards, you will begin to encounter wave-cut rock cliffs landward of the beach. The further you go, the more exotic the erosional patterns in the rocks become, with sea caves and sea arches, as well as eroded washboard textures (vertical circular pillars) in the cliff faces. The end of this Zone is marked where an irregular rocky headland projects out into the water beyond the low-tide line (called Point Resistance; see maps in Figures 100 and 101). There are also some wave-cut rock platforms in this area. If you can't get all the way down the beach, because of high water or the headlands are sticking too far out into the water, there is a well-traveled, and very scenic, trail along the upland behind the beach.

15 COMPARTMENT 3: SAN FRANCISCO'S OUTER SHORE

INTRODUCTION

The centerpiece for this Compartment is the huge ebb-tidal delta that projects offshore from the deep (300 feet) channel under the Golden Gate Bridge (Figures 108, 70, and 71). Much of this shoreline is located east of the San Andreas Fault zone, which passes offshore near Daly City in the south and comes ashore at the southern end of the San Andreas Fault Valley in which the Bolinas Lagoon is located (Figures 23 and 108). Because of its location east of the San Andreas Fault zone, excellent outcrops of the pillow basalts, ribbon cherts, and turbidites of the Franciscan Complex are present along the hillsides, rock cliffs, and road cuts in the region just north of the Golden Gate (Figure 20). The mudflats of Bolinas Lagoon have an abundance of shorebirds, and the tide pools at Duxbury Reef are some of the most accessible and popular ones in the state (Figure 46A).

For purposes of discussion, we have subdivided this Compartment into the five Zones mapped on Figure 108:

A) **Double-Point Uplands (Point Resistance to Bolinas Point)** is located on a southeast trending upland averaging about three miles wide that is bounded by the San Andreas Fault Valley to the northeast and the rugged shoreline of the Pacific Ocean on the southwest.

B) **Bolinas Lagoon and Vicinity** is composed of three very distinct geomorphological units: 1) the flat-topped headland between Bolinas and Duxbury Points flanked by a broad intertidal rock platform called Duxbury Reef that contains numerous fertile tide pools; 2) the biologically rich Bolinas Lagoon; and 3) an arcuate sand spit that marks the northern boundary of a classic crenulate bay.

C) **Muir Beach Uplands (Gull Rock to Point Bonita)** is a spectacularly beautiful mountainous coast with three short sand and gravel spits at the mouths of small streams. Steep, landslide-prone wave-cut scarps dominate the shoreline. The bedrock geology is entirely Franciscan Complex on the North American Plate, with the trace of

the San Andreas Fault being located offshore.

D) **Golden Gate Entrance** encompasses the two flanks of the entrance to San Francisco Bay. Both sides of the entrance are composed mostly of high, wave-cut rock cliffs scalloped by numerous landslides. The centerpiece of the Zone is the huge ebb-tidal delta off the entrance to the Bay.

E) **South Beaches (Lands End to San Pedro Rock)** consists of long sand beaches along the northern third of its shore (e.g., Ocean Beach) that are composed of sand derived from both the eroding cliffs along the coast and continental shelf sands that were carried down from the Sierra Nevada during the ice ages. The North American and Pacific Plates meet where the San Andreas Fault slices offshore at Mussel Rock.

DOUBLE POINT UPLANDS (POINT RESISTANCE TO BOLINAS POINT)

Introduction

Most of this segment of the shore, Zone A on Figure 108, is contained within the Point Reyes National Seashore (see maps in Figures 99 and 108), with hiking paths being its primary means of entry. This 8.5 mile stretch of shoreline is backed in most areas by high, steep, landslide-prone bluffs in the Miocene Santa Cruz Mudstone and some marine terrace deposits (Figure 102). Wave-cut rock platforms are present along parts of the shore and rocky protuberances occur in a few places, most noticeably at Point Resistance, Arch Rock, and the two headlands at Double Point. Because of the steep eroded scarps along much of the shoreline, beach access is limited.

Geological Framework

As shown on the geological map of the Point Reyes area in Figure 102, Tertiary sedimentary rocks of the Monterey Formation and Santa Margarita Sandstone

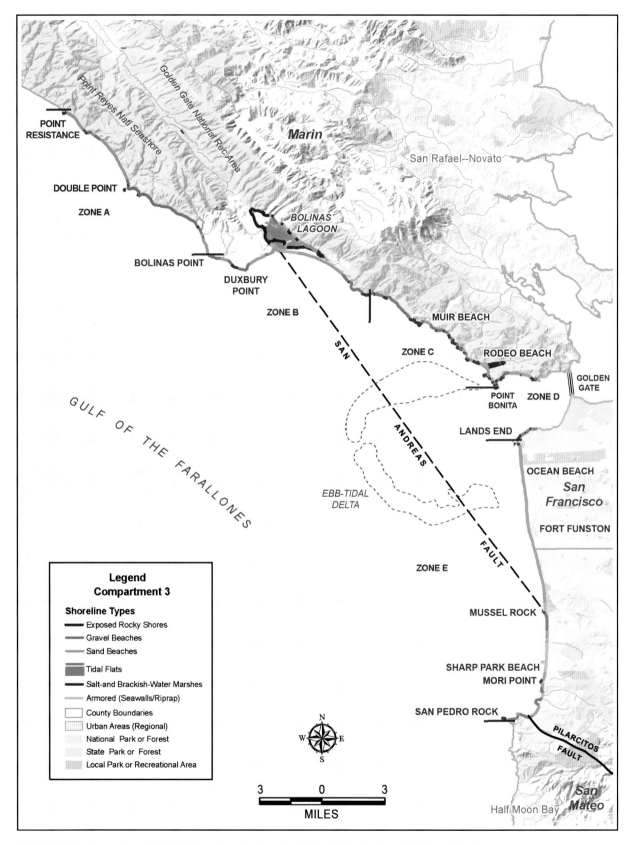

FIGURE 108. Compartment 3, Point Resistance to San Pedro Rock. Faults after USGS geological maps.

crop out in the hills and along the backbeach area in the northwestern one fourth of this Zone. Further to the southeast, the Santa Cruz Mudstone prevails. The characteristics of the Monterey Formation, a major petroleum reservoir, were discussed earlier (see photograph in Figure 105B). The Santa Margarita Sandstone is a poorly consolidated marine sandstone that overlies the Monterey Formation unconformably, which means that there is a time gap with a missing sedimentary record between the two formations, presumably due to an erosional episode. The Santa Cruz Mudstone, which is thought to have been deposited fairly far offshore, is composed of fine-grained silt and clay that contains an abundance of the siliceous shells of one-celled marine plants called diatoms.

In a discussion of the shoreline southeast of Double Point, Sloan (2006) observed that *several remnants of an uplifted marine terrace perch 150 feet or so above the waves,* and noted further *that these flat lying bits of land seem out of place along the rugged shore.* This marine terrace was carved (at sea level) during the highstand of sea level that preceded the last major glaciation and was eventually uplifted, following the pattern illustrated by the model given in Figure 48. There are some differences of opinion on the exact timing of the erosion of the terrace, with dates ranging from 80,000-125,000 years ago. According to Doris Sloan (pers. comm.), Karen Grove of San Francisco State University obtained a date of 80,000, based on luminescence dating.

Sloan (2006) pointed out another somewhat unique aspect of the geology of this Zone, noting that *toward Bolinas at the south end of the peninsula, a group of lakes marks yet another geological process in action, an area of extensive landslides that slid westward off Inverness Ridge and blocked the drainage, ponding the water into lakes.* This area of landslides is depicted clearly on the geological map in Figure 102.

Beaches

Medium- to fine-grained sand beaches are present along much of the northwestern half of this shoreline. Coarse-grained gravel beaches occur in the Double Point area, and sand and gravel beaches are common from Double Point to the southeast end of the Zone at Bolinas Point. With regard to beach erosion, Savoy et al. (2005) rank the shoreline of the northwestern one third of the Zone as *Caution: moderate risk,* similar to the rest of the beaches along Drakes Bay. For the southeast two thirds of the shore, however, the ranking is *Hazard: high risk.* They describe this high risk area as *steep cliffs with large scale landsliding.* However, because there are no

man-made structures along this section of the coast, the *high risk* applies mostly to hikers who may get too close to the edge of the cliffs.

General Ecology

This remote stretch of shoreline is host to very large numbers of seabirds and marine mammals. The seabird nesting colonies on the rocky headland at Point Resistance and Arch Rock, on either side of Kelham Beach, have over 4,000 common murres, 105 pigeon guillemots, Brandt's and pelagic cormorants, and black oystercatchers, with peak nesting March-July. Peregrine falcons also nest nearby; seabirds are a favorite prey. There are three seabird nesting colonies around Double Point. This is also a good area to see marbled murrelets, sooty shearwaters, and maybe leatherback sea turtles. Harbor seals haul out near Arch Rock, Double Point (nearly 1,000 individuals), and Bolinas Point with pupping March-June. Double Point is a state-designated Area of Special Biological Significance because of the importance of this area for seabird nesting and harbor seal pupping.

Places to Visit

One way to get to this Zone is to leave the Bear Valley Visitor Center and drive a short distance to the east along Bear Valley Road to the juncture with Highway 1, where you turn right and continue for about 3.5 miles until you reach the trailhead for the Five Brooks Trail, one of the two main entrances into this part of the Seashore (see map in Figure 99). There are several options for hikes beginning at this trailhead; see Lage (2004) for details. The most direct route to the shoreline is along Stewart Trail, a 12.8 mile round trip on a well-worn trail on which bikers and horses are allowed. Lage noted that *this trip offers great views on both the west and east sides of Inverness Ridge.* One of the more appealing aspects of this trail is its ascent up the east side of Inverness Ridge, along which it climbs gently *through a lush forest of Douglas-fir, bay laurel, big-leaf maple, hazelnut, and alder.* The ultimate destination of this trail is the shoreline at Wildcat Camp, located on a flat terrace landward of a high wave-cut cliff that is presently partially vegetated. A trail along side of a small stream leads to the moderately wide sand beach in front of the cliff, which extends for several hundred yards in both directions along the shore.

As you leave the Five Brooks trailhead and drive toward the southeast, you will be moving along the floor of the eroded valley located along the San Andreas Fault zone where the North American and Pacific Plates

181

slide past each other (see Figure 23). On your right, the northeast slope of Inverness Ridge, with its coat of very tall and straight Douglas firs, rises abruptly. On the left, the northeast side of the valley, the Bolinas Ridge, which is underlain along most of its length by the turbidity-current deposited sandstones of the Franciscan Complex, also rises to form the northeast limb of the San Andreas Fault Valley.

Continuing along Highway 1, this is a surprisingly nice drive through a very pretty area. As you approach the Bolinas Lagoon, the valley flattens out quite a bit, but then you go down through one more fairly steep incline through some big eucalyptus trees. After that you pass through the small hamlet of Dogtown, then through another growth of tall eucalyptus trees. These trees were introduced into California from Australia in the 1850s, with mixed results as is usually the case for introduced exotic plants.

As you approach the Lagoon, turn right on Horseshoe Hill Road (Figure 99). Continue on around the Lagoon, and at a little under two miles, turn right on Mesa Road. At about four miles on this somewhat winding road, you will reach the Point Reyes Bird Observatory on your left. According to their web site, visiting hours at the Observatory are: sunrise to sunset every day but Monday from May to Thanksgiving, and from sunrise to sunset Wednesday, Saturday, and Sunday from Thanksgiving to May.

Continuing west along the road, you will shortly reach the entrance to the Palomarin Trail, the southernmost trailhead in the Point Reyes National Seashore. This trail leads to the Coastal Trail which, in addition to numerous spectacular overlooks at the magnificent wave-cut rock cliffs and coarse-grained beaches, has at least three other notable features: 1) the trail passes by two good-sized lakes, Bass Lake and Pelican Lake, that formed as a result of the landslides down the southwest side of Inverness Ridge (mapped on Figure 102); 2) an overlook of the rocky headlands at Double Point; and 3) a beautiful waterfall called Alamere Falls that careens down the face of the wave-cut rock cliff. Wildcat Camp can also be reached on this trail. Some one-way distances involved include: 2.7 miles to Bass Lake; 4 miles to Double Point; 4 miles to Alamere Falls; and 5.7 miles to Wildcat Camp (Lage, 2004).

Once past the Bird Observatory, it is only a few hundred yards more to a pullover on the left at the head of a mile-long trail down to Palomarin Beach. We have not taken this trail, but it looks like some pretty rough going over a couple of landslides. A few hundred yards to the north of Bolinas Point, the boundary of this Zone, there are some truly unique perfectly spherical boulders

located on the landward boundary of a wave-cut rock platform. These boulders are probably concretions weathered out of the Santa Cruz Mudstone in the cliffs, because of their uniform size (averaging well over a foot in diameter), color, and composition. The thin-bedded sedimentary rocks in the platform at this location are tilted offshore at a high angle.

[NOTE: Concretions are defined as local concentrations of chemical compounds, such as calcium carbonate or iron oxide, within a sedimentary rock layer that usually take on the form of a spherical nodule.]

BOLINAS LAGOON AND VICINITY

Introduction

Positioned right across and along the San Andreas Fault zone in the same fashion, this area (Zone B on Figure 108) is a twin to the Bodega Bay complex (Compartment 1 on Figure 1; Figure 90), with a conspicuous headland that projects out into the Pacific Ocean and a curving sand spit that almost closes off a shallow bay/lagoon system. The high, wave-cut rock cliffs in the Santa Cruz Mudstone, with their associated wide wave-cut rock platform (Duxbury Reef), at Bolinas Point mark the western boundary of this area. The precipitous cliffs on the eastern edge of a protruding rocky headland in the Franciscan Complex near Gull Rock, with its associated huge boulders, marks the eastern boundary (Figure 108).

Geological Framework

The most compelling aspect of the geology of this Zone is the fact that it is split in half by the San Andreas Fault zone. Although the Bolinas/Duxbury headland sits astride the Pacific Plate, the granitic rocks of the Salinian Block lie out of sight at this location, buried beneath the outcropping Santa Cruz Mudstone of Late Miocene age (see geological map in Figure 102). Therefore, in a geological sense, this headland differs from the ones at Bodega Head and Point Reyes, where the Salinian Block granitic rocks are at the surface. The most striking, and ecologically significant, aspect of the headland is the broad intertidal wave-cut rock platform known as Duxbury Reef. The exceptional width of this wave-cut rock platform is presumably a result of the relatively high erodability of these sedimentary rocks, as well as their exposure to open ocean waves, compared to the rocks in other parts of the shoreline. A map of the "reef" area, which is noted for its extensive tide pools, is given in Figure 46A. This famous rock platform is shown in the ground photograph in Figure 109A. This rather extensive projection of intertidal and shallow subtidal rocks has historically been a hazard to marine navigation, being the site of seven major ship wrecks between 1853 and 1914.

The topography of the headland is also of interest. According to Sloan (2006), the village of Bolinas sits on the largest remaining piece of an uplifted marine terrace that extends to the northwest along the shore for several miles. This flat, raised tableland is called "the Mesa" by the local inhabitants. As noted earlier, this marine terrace was carved at sea level during the last highstand that preceded the present one (80,000-125,000 years ago), and has been uplifted by the process illustrated in Figure 48.

The rocks on the uplands along the eastern side of the lagoon and the San Andreas Fault, which are consequently located on the North American Plate, belong to the Franciscan Complex, a mix of volcanic, metamorphic, and sedimentary rocks incorporated into a major subduction zone between Late Jurassic and Late Cretaceous time (see Figure 14). The rocks that underlie these hills have been mapped as Franciscan mélange (Sloan, 2006). Therefore, many of the large boulders along the shoreline to the southeast of the village of Stinson Beach are most likely blocks of miscellaneous compositions that were incorporated into a sheared mudstone matrix during the formation of the mélange. Similar randomly distributed and conspicuous hunks of rock dot the beautiful rolling hillsides of Marin County.

Beaches

The primary beach in this area is the 2.5 mile long medium- to fine-grained sandy spit informally called "Sea Drift" that has been built in a northwest direction across the San Andreas Fault Valley, creating Bolinas Lagoon (see photograph in Figure 109B). This beach tends to have a lot of beach cusps because of the straight-on wave approach resulting from their refraction/diffraction into this crenulate bay. Where the sand came from to build such a long spit is not readily obvious, with no major riverine source and little but San Franciscan mélange in steep cliffs to the southeast as an obvious source. It seems likely that a considerable amount the sand came from the continental shelf, having been deposited there during the ice ages (derived from the Sierra Nevada), and later blown inland or moved onshore by wave action as sea level rose. On the other hand, the Bolinas Lagoon has an abundance of sand flats, so, perhaps, Pine Gulch Creek could have carried some sand out onto the shelf at times of lowered sea level as well.

The beaches around the base of the bluffs on the

Bolinas/Duxbury headland are composed for the most part of mixed sand and gravel, but those under the bluffs southeast of the village of Stinson Beach are composed mostly of gravel, some containing very large boulders.

Beach Erosion

This is the first Zone encountered with beach erosion problems, except for the relatively minor problem at Dillon Beach. According to Savoy et al. (2005), who classified most of the shoreline of this Zone as *Hazard: high risk,* the erosion rates of the ocean side of the Bolinas/Duxbury Bluff range between 6 and 24 inches per year. The western half of that side of the headland is not developed, so there is little concern about that area. However, they pointed out several problem areas in the developed parts of the headland, such as the area on the cliff edge directly to the northeast of Duxbury Point, observing that *since the area was initially subdivided in 1927, many of these ocean-front lots and those fronting Ocean Parkway have been either partially removed or damaged by cliff erosion.* There are several other problem areas along this side of the bluff related to landslides as well as wave-generated cliff erosion. For details, we recommend that you get a copy of *Living With The Changing California Coast.*

The sand spit that partially closes off Bolinas Lagoon (Figure 109B), where approximately 150 beach cottages are presently located a short distance back of the high-tide line, has also had some pretty serious erosion problems. During winter storms in 1977-78, nine beachfront homes were endangered as *10 to 90 feet of protective sand dunes were eroded* (Savoy et al., 2005). During the *El Niño* of 1982-83, 16 houses were destroyed and over 50 more were damaged. More damage was suffered during the *El Niño* of 1997-98. This spit, which is somewhat protected from storm waves approaching from the northwest, is significantly exposed to storm waves from the southwest, a fairly common occurrence during *El Niño* storms.

General Ecology

In 1998, Bolinas Lagoon was designated as "A Wetland of International Significance" for waterfowl habitat under the Convention on Wetlands of International Importance, because of its importance along the Pacific Flyway. The Point Reyes Bird Observatory has been monitoring winter waterbird numbers on Bolinas Lagoon since 1971, maintaining detailed charts for individual species (PRBO, 2009). The results indicate significant declines, sometimes to zero, in many species. One of the key reasons for these declines is the loss of intertidal and subtidal habitat from increased siltation. In the past 30

FIGURE 109. Vicinity of Bolinas Lagoon (Marin County; Zone B, Figure 108). (A) Intertidal rock platform at Duxbury Point, the acclaimed "Duxbury Reef" illustrated in Figure 46A. The ecology of the tidal pools in this area has fascinated thousands of visitors over the years. (B) Overview of a west trending sand spit in the shadow zone of the Bolinas crenulate bay. Both photographs taken on 19 August 2008.

years, the "upland" areas surrounding the Lagoon have increased by 67%. During this same period, the subtidal areas have decreased by 29% and the intertidal by 5%. Most of the Lagoon is accessible only by canoe or kayak, and even then only at high tide. The U.S. Army Corps of Engineers has developed a restoration plan that includes dredging and fill removal. Implementation awaits funding, as of 2008.

The Lagoon has a population of about 500 harbor seals that are commonly seen hauled out on the tidal flats at the lagoon entrance. Pupping occurs March-June.

Duxbury Reef is a state-designated Area of Special Biological Significance because of the richness and width of the intertidal rocky habitat (Figure 109A). It is a popular place for exploring tide-pool animals and plants, fishing, and wildlife viewing. One of the common intertidal biota you will see is the giant green anemone with an appropriate species name of *Anthopleura elegantissima*. The regulations for the area include:

> *No form of marine life may be taken from the ocean area within 1,000 feet of the high tide mark in the Duxbury Reef Reserve without a written permit. Recreational take of abalone, Dungeness crab and rock crabs, rockfish, lingcod, cabezon, surfperch, halibut, flounder, sole, turbot, salmon, kelp greenling, striped bass, steelhead, monkeyface-eel, wolf-eel, smelt and silversides is allowed. All recreational take of marine aquatic plants is prohibited.*

Places to Visit

As you approach Bolinas Lagoon from the north on Highway 1, turn right on Horseshoe Hill Road and drive around the west side of the Lagoon (Figure 99). If you come at low tide, you will see some wide tidal flats with a fringe of salt marsh, no doubt a great birding spot at the right time of the year. This Lagoon is very shallow, and unlike Bodega Bay, it has no port or harbor facilities. Keep driving around the edge of the Lagoon, eventually turning right into the village of Bolinas, and follow the signs to Agate Beach County Park. Along the way, you will notice that you are driving along the very flat surface of an elevated marine terrace. When you reach the parking area, first you should walk out to the overlook, where you can see the wide rock platform where, according to the kiosk at the site, you will be able to see *the nearly two miles of tide pools filled with beautiful and exciting marine life. The soft shell reef is the largest in California and is named after the sailing ship Duxbury that was ship wrecked there in 1849.* The access to these tide pools is exceptional, if you come during a low spring tide, of course.

There is a trail down to the rock platform, at the end of which you will see algae, many snails, and so on, all the things you would expect to see in tide pools. The photograph of the rock platform in Figure 109A is taken from this spot. The sedimentary rocks of the Santa Cruz Mudstone in the wave-cut rock cliff have been jumbled up a bit by tectonic processes, most likely those associated with the San Andreas Fault.

To leave the headland, drive back down to the shore of the Lagoon, eventually getting back on Highway 1 and driving around the northeast side. Toward the head of the Lagoon, there are a couple of little side embayments with pretty wide salt marshes (relatively speaking). There are some freshwater plants up along the most landward edges of these marshes. As you drive along, there are a couple of pullovers where you can stop and look for birds. This is a good birding spot if you come at low tide. Must be fabulous during the spring and fall migrations. The main channel impinges against the road for about the middle part of the northeastern side of the Lagoon. In August 2008, we saw a long-billed curlew, common egrets, some ducks far out, and about 40 harbor seals hauled out on the other side of the channel.

A few hundred yards after you leave the side of the Lagoon, you can pull into a large parking area with access to the fine-grained, moderately dark-colored sand beach (Figure 109B). When we were there in August 2008, the waves were plunging and had formed some well-developed cusps near the high-tide line. This is a very popular beach that is quite wide and relatively flat at low tide. If you are feeling adventuresome, you should walk the half mile or so to the southeast along the beach, so you can inspect up close the wave-cut rocks cliffs and boulders at the updrift end of the spit. Beautiful spot.

A bit south of the exit to the sand beach, Highway 1 winds up the side of the Franciscan rocks-based mountain. There are a couple of pullovers where you get an overview of the sand spit (Figure 109B), as well as the spectacular eroding wave-cut rock cliffs, which have an abundance of sea stacks located shortly offshore. Further along, you can look straight down on the gravel beaches and an incredible number of giant boulders along the base of the cliffs, no doubt eroded out of the Franciscan mélange.

MUIR BEACH UPLANDS (GULL ROCK TO POINT BONITA)

Introduction

This section (Zone C on Figure 108) is six miles of very scenic, steep rocky coast on the southwest flank of the Marin Peninsula. The base of the cliffs along the northwestern third or so of the shoreline contains gravel beaches, most of which incorporate an abundance of massive boulders. Three short spits composed of sand and gravel at Rodeo Lagoon (Rodeo/Cronkhite Beach), Muir Beach, and Tennessee Beach are located at the outlets of steep valleys. The high, landslide-prone, wave-cut cliffs are composed of the rocks of the Franciscan Complex. Nearshore sea stacks are populated by thousands of pelicans and cormorants.

Geological Framework

For this treatment of the geological framework, the area of discussion is extended all the way around the shoreline to the Golden Gate Bridge [includes a portion of Zone D (Figure 108), i.e., the Marin Headlands shore along to the Golden Gate] in order to take in the entire southeastern Marin Headlands geological province. The Franciscan Complex, securely aboard the North American Plate along this shoreline, has better exposures here in this part of the Marin Peninsula than anywhere in the Coast Ranges, according to Sloan (2006). This is where you can see clearly exposed the three basic rock types of the sequence, the origin of which is illustrated in Figure 22. Especially noteworthy are the exposures of the radiolarian chert, which were described as follows by Sloan (2006): *All along Conzlman Road, west from Hwy. 101 at the north end of the Golden Gate Bridge, are exposures of tightly folded and contorted layers of chert and shale that represent over 200 million years of deposition.* A photograph of one of the exposures of chert in this general area is given in Figure 20A. The basal pillow basalt is exposed along the walk to the Point Bonita lighthouse and near the Nicasio Reservoir, also in Marin County (Figure 19). For more details on the fascinating geology of this part of the Marin Peninsula (called the Marin Headlands) see Sloan's (2006) book on the *Geology of the San Francisco Bay Area*. Also for details on what you can see in the field, you should refer to the outstanding field-trip guidebook for the Golden Gate Headlands by Elder (2001), which is available on the internet.

The rocks in the northwestern half of this shoreline didn't fare so well in the subduction zone as those on the Marin Headlands, having been crunched mercilessly and turned into a massive mélange. Large, metamorphosed blocks of the original material are scattered over the hillsides, with many such massive boulders on the beach where the waves have eroded the hinterland.

Beaches

In addition to the gravel beaches with their overabundance of large boulders along the northwestern shoreline already alluded to, three sand and gravel spits are present at Muir Beach, Tennessee Beach, and Rodeo Beach (sometimes referred to as Cronkhite Beach). These beaches are all a part of minor tidal inlets, called stream-mouth/bar systems, the most common type of tidal inlet on the California Coast (Hayes and Montello, 1993a). In our description of these types of inlets in Section I, we noted that the stream mouth usually has a sand or sand and gravel spit built across it. The spit, which is typically a few tens of yards wide, is frequently overwashed during high spring tides and storms and it may be removed or highly modified during the larger floods on the stream. The inlet at Rodeo Beach was one of the type examples of this kind of inlet we used for constructing our tidal-inlet-classification scheme (see illustration of this particular spit in Figure 72). The spit that closes off Rodeo Lagoon is composed of sand and gravel (pebbles mostly), and is commonly cuspate, as the photograph in Figure 110A shows. The pebbles on the horns of the cusps are composed of a high percentage of ribbon chert (see photograph in Figure 111). Parts of the spit are almost pure sand and/or granules (Figure 110B).

General Ecology

This stretch of coast is under management of both the National Park Service's Golden Gate National Recreational Area (GGNRA) on the land and out to about 0.25 miles offshore and the National Oceanic and Atmospheric Administration's Monterey Bay National Marine Sanctuary, which extends an average of 30 miles offshore. The GGNRA contains 33 federally protected endangered and threatened species – more than any other unit of the National Park System in continental North America – more than Yosemite, Yellowstone, Sequoia, and Kings Canyon National Parks *combined*.

This area is a good place to discuss the Beach Watch program that is a joint effort of the Gulf of the Farallones and Monterey Bay National Marine Sanctuaries. Beach Watch, founded in 1993, is a year-round assessment of 42 beaches from Bodega Head to Año Nuevo that is conducted by dedicated volunteers who regularly survey an assigned beach. Volunteers collect data on live and dead species of birds and marine mammals, tarballs, beach use, etc. Birds and marine mammals die during both natural and man-made events, and the Beach

FIGURE 110. Rodeo Beach (Marin County; Zone C, Figure 108), which showed a distinct change from coarse on the southern end (with pebbles) to sand on the northern end the day we took these photographs (20 August 2008). (A) Beach cusps on the southern end of Rodeo Beach, which is located about a mile northwest of Point Bonita at the north entrance to the Golden Gate area (see Figures 72 and 108). The general model of beach cusps is illustrated in Figure 38A. The spacing between the two closest horns was 86 feet. (B) Sand beach at the northern end. This part of the beach is frequently composed of pure granules.

FIGURE 111. Close up of the pebbles on the horns of the cusps shown in Figure 110A. The orange colored pebbles are composed mostly of ribbon chert of the Franciscan Complex. Scale is ~ 6 inches.

Watch program provides baseline data and the ability to quickly detect significant events.

Figure 112A shows plots of the total numbers of beached birds found per kilometer of shoreline, and Figure 112B shows the number that were oiled for the period 1993-2008 from a summary by Roletto et al. (2003). Two patterns are very clear: 1) the top plot shows that high numbers of dead birds (especially local nesting species) wash ashore during the post-breeding season (July-October), and 2) the bottom plot shows that there is a higher rate of oiled birds during winter months. Figure 113 shows the number of oiled birds per kilometer on 40 beaches over the period 1993-2007 (Lyday et al., 2008). There are two sharp peaks, during the big *El Niño* season of 1997-98 and in 2002-2003, although there were no observations of oil slicks during these periods. In 2002, nearly 2,000 oiled birds were collected, instigating efforts to locate the source of these "mystery" spills. Chemical "fingerprint" analysis of oil samples showed a match with mystery spills starting in 1992, thus the oil was not likely from illegal discharges from passing vessels. The oil did not match natural seeps. Hindcast modeling, satellite imagery, and overflights were used to narrow the source area. Shipwreck databases contained information on over 700 shipwrecks in the region. After analysis, eight vessels were targeted for assessment; first on the list was the *SS Jacob Luckenbach*. Anecdotal information obtained from recreational divers confirmed that the *SS Jacob Luckenbach* was known to leak oil. During the initial assessment, oil was observed rising from the wreck. Oil collected from within the hold (by recreational divers) was a match to that on the oiled birds. The vessel, a C-3 freighter fully laden with 1,950

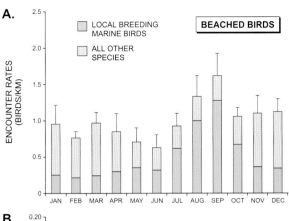

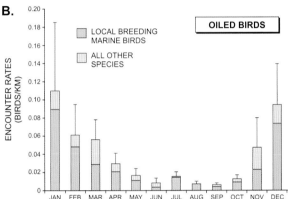

FIGURE 112. Summary of the Beach Watch Program where volunteers routinely survey the beaches from Bodega Head to Año Nuevo for dead birds and oiled birds. (A) The monthly average number of beached birds per kilometer for local breeding marine birds (blue) and for all other beached birds (yellow) over the period from September 1993 to August 2002. Note the high mortality of local breeding birds in July-October, representing die-off of newly fledged birds after the breeding season. (B) Monthly average number of oiled beached birds per kilometer of shoreline surveyed for the same period. Note the higher rates of oiling during the winter months (Roletto et al., 2003).

tons of fuel oil, sank in 185 feet of water on 24 July 1953 as a result of a collision. During the summer of 2002, 350 tons of heavy fuel oil were recovered from the sunken ship; residual oil continued to leak out until early 2003. Since then, the frequency of oiled beached birds and tarballs has dropped dramatically. On some beaches, the tarball stranding rate for 2005-2007 dropped to zero, that is, until the container ship *Cosco Busan* spilled 53,000 gallons of heavy fuel oil in San Francisco Bay in November 2007. Tarballs and oiled birds on the outer beaches suddenly increased! The Beach Watch data were invaluable in determining the difference between the background rate of dead and oiled birds and those that died as a result of the oil spill.

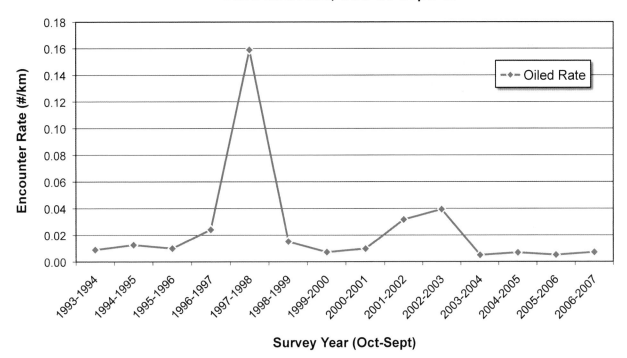

Oiled Bird Rate, Oct. '93-Sept. '07

FIGURE 113. The yearly average rate (number per kilometer) of oiled beached birds over the period 1993-2008. Note the peak during the big *El Niño* season of 1997-1998 and during 2002-2003, which was linked to oil being released from the sunken ship, *SS Jacob Luckenbach* (Lyday et al., 2008).

Places to Visit

If you are approaching this area from the north along Highway 1, your tour will begin with a breathtaking (scary?) drive along a high ridge above the shore along which, on a clear day, there are several pullovers from which you can take some superb photographs of this classic, tectonically active, rocky coast. Eventually, you will reach the village of Muir Beach, where you can take a right turn and drive down to the beach itself. The top of the spit, which trails away from a gravel beach on its southeastern side becoming more sandy to the northwest, is quite wide, and beachface cusps are common. A small stream outlet is located at the northwest end of the spit.

The area to the east of Muir Beach has beautiful rolling hills within the Golden Gate National Recreational Area. There are several hiking trails in this section of the coast where you can get good views of the rocky shore between Muir Beach and Rodeo Beach. The cliffs behind the beach in this area are very precipitous, with a number of large landslide scars, so don't get too close to the edge. One of the most popular hiking trails is the one to Pirates Cove. Tennessee Beach, another beautiful sand and gravel spit like the ones at Muir and Rodeo Beaches, can also be reached along one of these trails.

Another possibility after you leave Muir Beach would be to drive up the mountain on Muir Woods Road and visit the Muir Woods National Monument. This monument protects 554 acres of forested area populated by coast redwood (*Sequoia sempervirens*), one of the last remaining such stands in the immediate San Francisco Bay area.

To get to the rest of this coastal area, you need to follow Highway 1 up the mountain to Highway 101. From there, follow the signs to get back down to the beach at Rodeo Lagoon. When you get to the bottom of the hill on your way to Rodeo Beach, take a right and drive around the Lagoon to the parking area for the beach, where you will encounter a kiosk with the following information:

Much of the year this coastal lagoon is a stagnant body of water but it still supports a variety of wildlife and vegetation. This enriched environment sustains many small plants and animals providing a feeding ground for egrets, herons, ducks, and migrating birds. Rodeo Lagoon may be one of the last habitats for the tidewater gobi, a small fish.

The inlet into the Lagoon opens up sometimes during storms, and storm waves frequently wash over the berm and into the Lagoon.

When we visited Rodeo Beach on 20 August 2008, the beach had a higher berm composed mostly of sand, and a lower berm with mixed sand and gravel. We walked down to the southeast side of the spit where beach cusps were forming, as shown in the photograph in Figure 110A. The horns of the cusps were composed mostly of pebbles (Figure 111).

Co-author Michel has visited this beach many times, most notably during the response to the *Cosco Busan* oil spill beginning in November 2007. As we were walking along watching the cusps form, she said that "it seems like this is the way this beach has always been. It just doesn't seem to change that much. Now, along the south end you get mixed sand and gravel and it is flatter. The north end is sandier and then it becomes granule and is steeper, every time I come here."

The sediment on this beach has to be derived locally from the eroding wave-cut scarps on either side of it, because there are no streams of any size bringing in sediment. As shown by the photograph in Figure 111, the pebbles are composed mostly of the radiolarian chert that is so common in the Franciscan Complex rocks all along southeastern end of the Marin Headlands (see photograph in Figure 20A).

A strange small rectangular, steep-sided embayment with a short sand beach at its head can be observed from a trail that goes up the side of a steep hill on the northwest side of the Rodeo Beach. This feature, which is striking from the air (it is visible in the northwest corner of the image in Figure 72) and from the trail overlook, is no doubt the result of erosion along two parallel joints, somewhat like the one illustrated in Figure 43.

When you leave the beach, go back past the head of the Lagoon and take a right to reach two very worthwhile destinations; 1) an overlook of Bird Island; and 2) the walkway to the lighthouse at Point Bonita. Both the overlook and Bird Island are shown on the striking oblique aerial photograph in Figure 25A. The white color on Bird Island is caused by guano produced by large numbers of roosting birds over the years.

The walk out to the lighthouse will provide views of the stretch of wave-cut rock cliffs between Bird Island and Point Bonita shown on the photograph in Figure 25A. Such wave-cut rock scarps dominate most of the shoreline of this Zone. The lighthouse walk also provides the opportunity for a few glimpses back at the Golden Gate Bridge. This walk is additionally an important site from a geological perspective, inasmuch as the lighthouse is perched on an extensive outcrop of the basal pillow basalts of the Franciscan Complex, one of the best exposures of these basalts in all of Marin County.

GOLDEN GATE ENTRANCE

Introduction

Because this book deals only with the outer shoreline of the Central California Coast, omitting San Francisco Bay, the northern segment of the shoreline of this Zone extends from Point Bonita to the northern end of the Golden Gate Bridge. The southern segment extends from the southern end of the Golden Gate Bridge to a modest-sized rocky headland called Lands End. These two segments combined make up Zone D on Figure 108. The Zone is therefore bisected by the 300+ feet deep channel that runs through the Golden Gate into San Francisco Bay. The shoreline of the northern segment is mostly steep, landslide-prone wave-cut cliffs in Franciscan Complex rocks with scattered, relatively short sand and gravel beaches. The southern segment also has several stretches of wave-cut rock cliffs, but it additionally contains a mile or so of sand beach (Bakers Beach), as well as some sand and gravel beaches with large boulders.

The Ebb-Tidal Delta

The huge ebb-tidal delta outside the Golden Gate Bridge, illustrated in Figure 70, is probably the largest such feature in the whole world. It is there because of the strong ebb-tidal currents that flow under the Bridge during falling tides that are partially depicted in Figure 32. These strong currents generate the large sand waves on the bottom of the channel shown in Figure 71. Much of the sand in the tidal delta was delivered to the continental shelf by the ancestral stream that no doubt carved the deep entrance channel at the times of lowered sea level during the ice ages (discussed in more detail later).

Geological Framework

The Franciscan Complex rocks along the wave-cut scarps in the northern segment compose the triad of pillow basalt, radiolarian chert, and turbidite graywackes already discussed in the description of the geology of the southeastern portion of the Marin Peninsula. There are landslide deposits in a couple of places along the shore as well.

The rocky scarps of the southern segment are also Franciscan rocks on board the North American Plate, inasmuch as the trace of the San Andreas Fault is offshore of this area (Figure 108). The rock cliffs of the first third of this segment heading south from the Golden Gate

Bridge are composed of serpentinite mélange, which is clearly distinguished by the massive character of its matrix and the inclusion of large, randomly distributed blocks of solid rock. The base of the cliffs is lined with accumulations of the large boulders freed from the mélange, one of the most spectacular examples of such boulder beaches found anywhere along the Central California Coast. In this case, a sand blanket covers the beach seaward of the boulder berms. The cliffs in this area are also riddled with massive landslide scars, as might be expected in rocks with such a friable matrix.

The rest of the rocks along the shoreline to the south are composed mostly of graywackes of the so-called Marin Headlands Terrane, signifying its correlation with the rocks of the Franciscan Complex on the other side of the bridge. The rocky scarp at Lands End, the southern edge of the Zone, is fittingly composed of pillow basalt at the base of the "Marin Headlands" assemblage, which dips toward the northeast.

[NOTE: A **graywacke** is a type of sandstone that has been defined in different ways, principally to differentiate them from sandstones composed mostly of well-sorted quartz grains, called orthoquartzites or quartz arenites, that are usually whitish in color. Graywackes are usually dark-colored sandstones, made that way by either containing a lot of clay matrix between the sand grains or having abundant sand particles composed of dark-colored rock fragments from a variety of rock sources, such as metamorphic slates or schists. The sands that eventually became quartz arenites are thought to have been deposited mostly on beaches or in sand dunes where all but the resistant quartz grains were abraded away over a long period of time as the sand grains were repeatedly impacted against each other under the influence of strong wave and/or wind action. Graywackes are deposited more abruptly than quartz arenites, as might be expected in a turbidity current, for example, so that some of the accompanying silt and clay or fragile rock fragments will be preserved within the sand deposit. Deep ocean trenches within subduction zones, such as the one in which the Franciscan Complex sandstones were deposited, have ideal conditions for the formation of graywackes.]

Beaches

Within the northern segment, there are several stretches of sand and gravel beaches between Point Bonita and the Golden Gate. The source for these beach sediments is mostly the adjacent rock cliffs in the Franciscan Complex; therefore, the gravel typically contains abundant rounded fragments of radiolarian chert similar to that on Rodeo Beach (Figure 111).

There are three distinct types of beach zones in the southern segment. From the Golden Gate Bridge south to a low rocky area where the sand beach begins, gravel occupies the upper part of the beach, with boulders being common in some areas. However, sand does occur in the lower half of the beach in most places. Next are several hundred yards of relatively pure medium- and fine-grained sand beach (Bakers Beach). The south end of Bakers Beach terminates with a steep scarp backed by numerous residences just north of China Beach. Several, well-defined pocket gravel beaches are present between China Beach and Lands End.

Beach Erosion

The shoreline north of the Golden Gate is eroding fairly impressively in places, but no man-made structures are endangered there, because this is a part of the GGNRA. The same thing is true for most of this Zone south of the Golden Gate, except for a section at the southern end of Bakers Beach, about which Griggs, Fulton-Bennett, and Savoy (2005) observed that *houses have been built on the sandy bluffs and rock cliffs above Bakers Beach. This beach is somewhat protected from large surf* (by the shoals of the ebb-tidal delta), *but most of the oceanfront properties directly above the beach are now protected by low concrete seawalls.* They did, however, conclude with a word of caution about the possible *slumping of loose cliff materials* in the future.

Places to Visit

It is difficult to get down to the shore between Point Bonita and the Golden Gate Bridge except for the beach at Kirby Cove. To get there, park by the gate at Battery Spencer on Conzelman Road and walk down the gated road to the beach. Sloan (2006) waxed eloquently about this site, noting that the sand and pebbles on this beach, as well as at nearby Rodeo Beach, are unusually dark in color, *because the source rocks are Franciscan. The reddish brown of chert grains is the dominant color, along with dark grays and brown of the graywacke and pillow basalt. Scattered among these dark pebbles are the vibrant red, blue, green, and mustard yellow pebbles of altered chert, which tumble down from the ridge above Kirby Cove.*

If you didn't take enough pictures of the Golden Gate Bridge from down on the beach, there are several pullovers where you can stop and add your few photos to the tens of millions of others that have been taken of the Golden Gate Bridge from these overlooks along the end of the Marin Peninsula.

Not far below the southern end of the bridge, there are several trails along the cliffs in the serpentinite mélange. If you are feeling adventuresome, taking one of these trails down to the beach would be worth the trip, if for no other reason than to observe the grand scale of the boulder accumulations. Also, from there, and maybe further down the road at Bakers Beach, you could add a few more shots to the tens of millions of pictures taken of the Golden Gate Bridge from this side of the bridge as well. Bakers Beach is a typical sand beach with a well-developed, typically cuspate berm most of the time and, as already noted, the waves are usually relatively mild.

To get to the end of the southern segment at Lands End, go to historic Lincoln Park, a 100 acre Park dedicated to President Abraham Lincoln in 1909. The Park is part of the Golden Gate National Recreational Area. From the parking area, you can follow a trail for a few hundred yards to the overlook at the primary rock headland, composed of pillow basalt, called Lands End. From the overlook on a clear day, you can see all the way up the shoreline to Point Reyes, a magnificent view. One thing you don't want to miss here, however, is a visit to the gravel beach on the south side of the headland. This may be one of the few chances you will have during your visit to the Central California Coast to so easily access and examine such a classically formed gravel beach, although you will have to navigate quite a few stairs to get there.

SOUTH BEACHES (LANDS END TO SAN PEDRO ROCK)

Introduction

The relatively straight shoreline of this Zone occupies about twelve miles of the west side of the San Francisco area (Zone E on Figure 108). It begins at the Lands End rocky headland and ends at the rocky headland called San Pedro Rock (located on the vertical image in Figure 114). The San Andreas Fault trace crosses the shoreline at Mussel Rock. The northern half is dominated by over three miles of wide sand beach called Ocean Beach that has a variety of backbeach conditions, including two major seawalls, a narrow zone of vegetated dunes, an eroding low scarp in older sediments, and riprap.

Geological Framework

Because it is cut almost in half by the San Andreas Fault, the northern portion of this Zone rides on the North American Plate and the southern portion on the Pacific Plate. Between Lands End and the Cliff House,

the irregular wave-cut rock scarps are composed mostly of Franciscan graywackes. Landward of these scarps and of Ocean Beach to the south, more than half of the whole of the San Francisco area shown in Figure 114 is underlain by Quaternary **dune sand** (<2.6 million years old; Table 1).

The origin of the dune sand that now covers much of this area, which an 1857 geological report described as a desert with sand dunes 60 feet high at that time, involved the following steps, according to Sloan (2006):

1) Some of the sand comes from erosion of the local wave-cut rock cliffs, such as the eroding sandstones at Fort Funston. The longshore sediment transport direction along this beach is from south to north as a result of wave refraction around the giant ebb-tidal delta off the entrance to the Bay, and the beach's exposure to any southerly waves that may occur, which do not have any offshore obstruction to their approach.

2) However, the bulk of the sand was derived from the Sierra Nevada 150 miles away. During the ice ages, rivers draining the mountains, namely the ancestors of the present Sacramento and San Joaquin Rivers, delivered sediment eroded from the mountain sides by the glaciers through the Carquinez Strait and out into the ocean.

3) When sea level was lowered hundreds of feet during the glaciations, the shoreline was far out on the present continental shelf beyond the present Farallon Islands.

4) The rivers draining the mountains during these periods of lowered sea level deposited large volumes of sand along those shorelines located far offshore from the present one.

5) Then, as Sloan pointed out, much of this sand *was then picked up by the prevailing westerly winds and blown back across the exposed continental shelf* to eventually cover much of the area now known as San Francisco. Doris Sloan (pers. comm.) noted that *there are also sand dunes south of San Francisco on the peninsula (in San Mateo County)*. Therefore, much of the landscape onshore from Ocean Beach is underlain by dune sand.

The first place to see actual rock formations along the beach south of the Cliff House is the wave-cut rock cliffs that begin just as the main highway turns away from the beach a ways beyond the San Francisco Zoological Gardens. These sedimentary rocks, called the Merced Formation (pictured at Fort Funston in Figure 115), continue along the beach scarps to near Mussel Rock.

Ed Clifton, who has studied the Merced Formation in detail (see also Clifton and Leithold, 1991), described it as follows in a personal communication:

This unit records coastal deposition in cycles that alternate between outbuilding of the shore (prograda-

FIGURE 114. Image of Zone E (Figure 108) in San Francisco and San Mateo Counties (South Beaches - Lands End to San Pedro Rock). NAIP 2005 imagery courtesy of the California Spatial Information Library (CaSIL).

FIGURE 115. Wave-cut rock cliff at Fort Funston (Zone E, Figure 108). Oblique aerial view taken on 1 October 2008, courtesy of the California Coastal Records Project, Kenneth and Gabrielle Adelman. These dipping sedimentary rocks are a part of the Merced Formation, which was deposited during Late Pliocene to Early Pleistocene time. The arrow points to a layer of sediment called the Rockland Ash, which was erupted from a volcano near Lassen Peak about 570,000 years ago (Sloan, 2006).

tion) and incursion of the sea (transgression). The cycles in the lower part of the Merced are much thicker than those in the upper part. Either the lower cycles represent longer time periods or they accumulated during more rapid subsidence. The cycles include shelf, surf zone, beach, estuary, sand dune, and coastal marsh deposits. Because the rocks are dipping toward the northeast as can be seen in the photograph in Figure 115, the younger sediments are visible in the Cliffs under Fort Funston, with older strata exposed further south. The deposits lying above the mineralogical change noted in the following paragraphs are dominated by non-marine deposits, although beach and surf zone sand and gravel are not uncommon. The strata below the change consist mostly of shelf sandstone and mudstone, although beach, surf zone, and nonmarine deposits occur in most of the cycles.

Clifton added that the remarkable rate of tectonic uplift in this part of the coast should be pointed out. The

cliffs shown in Figure 115 are cut into a section that has tilted and uplifted in the last few hundred thousand years given the fact that the entire section in the cliffs formed just above (and below) sea level. I know of no other place on the Central California Coast that demonstrates such remarkable dynamics.

The arrow on the photograph of the Fort Funston cliffs in Figure 115 points to a layer of sediment called the Rockland Ash, which was erupted from a volcano near Lassen Peak about 570,000 years ago. As Sloan (2006) pointed out, at about 380 feet **below** the ash layer within this Merced Formation, where the sediments are about 620,000 years old, the mineral composition abruptly changes from a local source of sand to a non-local granitic source in the Sierra Nevada mentioned previously. Sands from this more distant source continue to be part of the sediments from that point on to the top of the formation. This abrupt change in the composition of the sand was explained by Sloan (2006) as follows:

1) Until about 3 million years ago, the rivers that drained the Central Valley area had a southern outlet.

2) About the time of the onset of the ice ages around 2.6 million years ago, this southern outlet was closed off by movement along the San Andreas Fault, creating a large lake in the Central Valley area.

3) After a major glaciation event about 620,000 years ago, the ponded lake spilled through a low area in the Coast Ranges, resulting in the scouring out of the deep valley (canyon?) now known as Carquinez Strait. This allowed the Sierra Nevada sediments to be carried out onto the continental shelf, providing a different type of source material from that of the sediments that formed the lower part of the Merced Formation. It is thought that the first outlet to the sea was a valley called Colma Strait, and it was located not too far from where the San Andreas Fault crosses the present shoreline. However, some time later, the present inlet into San Francisco Bay was carved. Actually, Sloan said the river drainage *may have* flowed to the ocean through "Colma Strait." She also noted that the *change in drainage from the Colma Strait to the Golden Gate probably occurred sometime in the past several hundred thousand years.* [She added in a later pers. comm. - *I think that's all the present evidence allows one to say* (i.e., *may* and *probably*)].

Ed Clifton (pers. comm.) added that the Rockland Ash lies about 650 feet below the top of this stratigraphic sequence, which *includes at least 3 more transgressive/ regressive cycles. So, the top of the section is probably no more than a couple of hundred thousand years old.*

The shoreline between Mussel Rock and the southern edge of San Pedro Beach in Pacifica is relatively low lying, except for two significant rocky headlands at Mori Point and the one that separates Rockaway and San Pedro Beaches. This segment of the shore is bordered by two major northwest/southeast trending faults, the San Andreas on the north side and the Pilarcitos on the south side. The surface rocks in this area are Franciscan Complex coherent rocks (Sloan, 2006), with conspicuous outcrops of them exposed in the two aforementioned rocky headlands.

Once across the Pilarcitos Fault (Figure 108) and onto the San Pedro Peninsula, which ends with tilted rock layers in sea stacks that project straight offshore (San Pedro Rock, the southern boundary of this Zone), the geology is significantly different, having been described as follows by Sloan (2006):

1) A thick sequence of Paleogene/Neogene marine sedimentary rocks – sandstone, shale, mudstone, and conglomerate – Paleocene to Pliocene in age (65-2.6 million years). The Paleocene rocks at the San Pedro

headland are *spectacular examples of turbidites* (Ed Clifton, pers. comm.).

2) They record a marine basin that existed after the Franciscan subduction ended.

3) Since their deposition as flat-lying layers of sediment on the sea floor, they *have been tilted up almost vertical by tectonic forces.*

Speaking of Sloan, she made the following statement in the beginning of her book on the geology of the Bay area, which would seem to be more than just appropriate at this point:

Millions of years of movement along the Bay Area's many faults have rearranged the rocks into a geologic complexity that defies order and reason.

Beaches

Going from north to south, the irregular medium-sized rocky cliffs between Lands End and the historic Cliff House have some fairly extensive gravel beaches between Lands End and Point Lobos. Boulder beaches line the shore along the more southerly part of this rocky peninsula. A short seawall/riprap zone and a small sand beach occur in front of the Sutro Baths structure, just to the north of Cliff House.

The northern part of the long, sandy strand of Ocean Beach is backed by a sturdy sea wall, called the O'Shaughnessy seawall, built in 1929 (Griggs, Fulton-Bennett, and Savoy, 2005). Some vegetated sand dunes occur back of the high-tide line in the north-central portion of the beach. Another, more recently deployed seawall bisects the vegetated sand dunes further to the south, and beyond the end of the seawall, there is another zone of vegetated dunes extending further to the south. Some riprap has been added to the backbeach along the most southerly portion of Ocean Beach. The sand on this beach was discussed at some length by Sloan (2006), who recorded that the sand comes from both the nearby eroding wave-cut rock cliffs in the Merced Formation and other sources, and the *older Sierra sand from the continental shelf.* This sand has a variety of compositions (and colors), from miscellaneous rock fragments to a variety of minerals, including: 1) colorless clear quartz; 2) opaque milky feldspar; 3) reddish chert; 4) greenish serpentine; 5) pink garnet; and 6) black magnetite. Magnetite (Fe_3O_4) is one of what sedimentologists call **heavy minerals**, because they are considerably denser than associated minerals like quartz and feldspar (light minerals). As you walk along the beach, you will no doubt occasionally encounter narrow swash lines, or traces, of thin layers of pure black minerals, which might also include heavy minerals other than magnetite (e.g., hornblende, ilmenite, etc.). As briefly discussed in Section I, at some point in their existence on the surface of the beach, a very shallow, but fast moving swash of a wave may pivot the larger grains and move them away, leaving behind a layer of the smaller, and less pivotable, heavy mineral grains, thus creating the narrow swash line of black minerals you see on the surface of the beach (refer to Figure 63B). Ed Clifton (pers. comm.) pointed out that *among the scattered pebbles on this beach there are some small flat round sandstone masses that contain a sand dollar, which have weathered out to the lower part of the Merced Formation.* Parts of the Merced Formation also contain numerous mollusk shells and echinoid tests.

A sand beach extends to the south from the end of Ocean Beach, passing below the wave-cut rock scarp at Fort Funston, and continuing on for around three more miles. On our visit to the Fort on 16 August 2008, the sand beach south of it was highly cuspate.

As shown by the two photographs in Figure 116, the sediment on the beach north of Mori Point, called Sharp Park Beach, is dark gray in color, similar to the beaches on the Marin Peninsula (e.g., Rodeo Beach; Figure 110). This darker color is no doubt due to the fact that their source is the rocks of the Franciscan Complex that are exposed in the adjacent rocky headlands, a similar type of source rocks to those that provide the sand and pebbles on the beaches north of the Golden Gate. Note the abundance of orange radiolarian chert and grey graywacke in the granules on the berm top in the photograph in Figure 116B. This typically wide and cuspate beach extends to the north from Mori Point for 0.9 miles to where the backbeach is heavily armored with riprap at the pier on Pacifica Beach.

The two sand and sand and gravel beaches at the southernmost end of this Zone, San Pedro and Rockaway Beaches, are very popular surfing beaches, especially San Pedro Beach, where large waves refract/diffract around the end of San Pedro Rock and into the embayment between two rocky headlands within which the beach is located. This beach contained gravel cusps with cobble-sized clasts on the horns during our visit in August 2008.

Beach Erosion

Griggs, Fulton-Bennett, and Savoy (2005) rank the rocky shoreline from Lands End to the Cliff House as *Hazard: high risk,* but there are no man-made structures close to the beach in this area, except for the Sutro Baths ruins, which are protected by a seawall, and at the Cliff

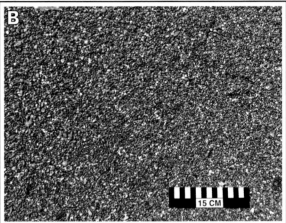

FIGURE 116. Sharp Park Beach (San Mateo County; Zone E, Figure 108 and Figure 114). (A) View looking north along beach from near Mori Point. The pier at Pacifica is visible in the distance. (B) Granules and coarse sand on the beach in the foreground of the photograph in A. Scale is ~ 6 inches. The source for these sediments is rocks of the Franciscan Complex. Both photographs taken on 17 August 2008.

House, which sits atop a resistant bluff of Franciscan graywacke. They rank the northern part of Ocean Beach, which is protected by a seawall as *Stable: low risk,* with the rest of this beach down to the Lake Merced area as *Caution: moderate risk.* In the southernmost part of the beach, just north of where the wave-cut rock cliffs behind the beach begin, riprap has been recently added to protect the highway. From where the bluffs start at Lake Merced all the way to Mussel Rock, the higher bluffs are subject to massive landslides. Two segments of the bluffs north of Mussel Rock have been protected with riprap. This segment of the shore was ranked as *Hazard: high risk* by Griggs, Fulton-Bennett, and Savoy (2005), who concluded that: *The primary coastal hazard in this area is movement and headward or landward expansion of the large landslides that threaten numerous homes along the crest of these high, unstable coastal bluffs. Wave erosion, heavy and sustained rainfall, and seismic shaking can all contribute to the movement of these slides.*

Ed Clifton (pers. comm.) pointed out several aspects of landsliding in this area, including the fact that results of an immense landslide (a low-lying landfill area) are visible on recent aerial photographs (taken by Kenneth & Gabrielle Adelman) of the location where the San Andreas Fault crosses into the sea just north of Mussel Rock. Numerous attempts have been made over the years to stabilize this part of the coast. Homes all along the cliff top are lost routinely as the sliding progresses. The riprap piled along the coast in this area will slow the wave erosion, but not the landsliding. *The Coastal Highway (Rte 1) used to run between Mussel Rock and Ocean Beach along a bench cut about halfway down the sea cliff. In 1957, an earthquake with a magnitude of 5.3 and its epicenter at Daly City (UCal-Berkeley, 2009) generated landslides that forced the permanent closure of the road.* In the first decade of the 21st Century, remnants of the road are still visible along parts of the sea cliff there, although it has been mostly lost to landsliding and subsequent wave erosion. *Another giant active landslide lies between Fort Funston and Thornton Beach. Attempts to create a park at Thornton Beach in the 1970's utterly failed as the slide, within a few years, destroyed both roads and facilities.*

The shoreline between Mussel Rock and San Pedro Rock, the corporate boundaries of Pacifica, has suffered damaging erosional episodes for decades except, of course, for the shorelines along the four main rocky headlands (Mussel Rock, Mori Point, the headland between San Pedro and Rockaway Beaches, and San Pedro Rock). Actually, there is a small sandy pocket beach on the San Pedro headland called Shelter Cove, where a small community of homes has been damaged by large waves in recent years. To add insult to injury, landslides during the 1983 *El Niño* winter cut off vehicular access to these homes permanently, several of which were eventually abandoned (Griggs, Fulton-Bennett, and Savoy, 2005).

The beaches from Mussel Rock to San Pedro Rock (between the rocky headlands) are backed by scarps in poorly consolidated sand and gravel generally less than 100 feet high that historically have eroded tens of feet during a single storm. After the *El Niño* storm of 1983, measures were taken to protect most of the shoreline in front of these scarps by the addition of riprap, seawalls, or a combination of the two, but erosion continues in some areas. The whole grim and uncensored story of the

mayhem caused by big storm waves and landslides in this area is presented in Chapter 13 of the book *Living With The Changing California Coast* (Griggs, Patsch, and Savoy, 2005).

General Ecology

The extensive sand beaches support numerous shorebirds including the federally threatened western snowy plover, which can be found almost year-round from Ocean Beach to Mussel Rock and at San Pedro Beach. The National Park Service biologists reported that unleashed dogs represent the most significant recreational threat to western snowy plovers on these high-use beaches, because of the prolonged and repeated disturbance created when they chase birds. Ocean Beach is a good place to spend some time searching the nearshore waters for the federally threatened marbled murrelet, present year-round. High vantage points such as Lands End, Cliff House, Fort Funston, and Mussel Rock are excellent sites for looking for migrating humpback whales, California sea lions, harbor seals, and the occasional sea otter. Also, there is a bank swallow colony at Fort Funston. An extended bird list for that area is presented on the web (Eaton, 2009).

The offshore rock islands and pinnacles along the coast, such as Point Lobos, Seal Rocks, Mussel Rock, and San Pedro Rock, are part of 20,000 such features along the California coast administered by the Bureau of Land Management. One of the last acts of President Clinton was to establish, by Presidential Proclamation number 7264 on 11 January 2000, the California Coastal National Monument, to elevate the protection of "all unappropriated or unreserved lands and interest in lands owned or controlled by the U.S. in the form of islands, rocks, exposed reefs, and pinnacles above mean high tide within 12 nautical miles of the shoreline of the State of California." These are among the most viewed but the least recognized of any of the Nation's national monuments, for sure.

Places to Visit

Assuming that you are traveling from north to south during your tour of this Zone, we recommend that you first stop at the Fort Miley parking area and walk out to the overlook at Point Lobos and other overlooks in that area. From there, on a clear day, you can see all the way to Point Reyes to the north and to the Montara Mountain headland to the south. The scenery is spectacular. Numerous sea stacks commonly covered with birds are present immediately offshore of the overlooks. There are also a number of cobble/boulder beaches in the small embayments between the rock promontories.

After leaving this rocky headland, go south on the Great Highway, where you will be driving landward of the large seawall on the north end of Ocean Beach. There are many places to park and walk out across this wide medium- to fine-grained sand beach. Pick up a hand full of the sand and note the wide variety of mineral types that make up the sand grains that were described earlier. Some kind of magnifying lens would be a big help, in this case. If you look offshore, you might see the waves breaking on the sand bars of the giant ebb-tidal delta. Some of the sand in those bars was formerly brought all the way to the continental shelf from the Sierra Nevada. However, the two major rivers that drain the high Sierra, the Sacramento and San Joaquin Rivers, are not bringing sand to the shelf at the present time, because the present highstand of sea level has created the monster sediment trap called San Francisco Bay. Not to mention the role that the dams on the Sierran streams play in trapping any potential sand that might go further down to the coast.

At about the south edge of the Golden Gate Park, the seawall ends and the beach is backed by vegetated sand dunes with linear scars, gaps, and minor sand lobes created by the dominant northwesterly winds. These dunes are continuous for six blocks, until they diminish significantly and another high seawall has been constructed along side of the road. This seawall continues for five blocks until the dunes and a mound of older sediments replace it. This next partly vegetated area continues for four more blocks to where the backshore becomes a low eroding scarp, which is presently protected (hopefully) with riprap. This erosion zone is just opposite the San Francisco Zoo.

If you continue south, the highway will join Highway 35 and you will drive around the west shoreline of Lake Merced. Continuing on for a short distance, you will reach the turn to Fort Funston, a place we highly recommend you visit. First visit the overlook, from which you could look down on the sand beach and also south along the high scarps in the Merced Formation all the way to Mussel Rock. The scenery up there is ridiculously good! When we were there on 16 August 2008, the wind was howling and there were hang gliders all over the place. We saw five in the air at one time. Also on that day, the beach to the south was very cuspate as the big, spilling waves were coming straight onto the beach.

There are a number of hiking trails along the top of the scarp near the overlook. On these hikes you will see some of the most spectacular **salt pruning** of the small trees and shrubs you will see anywhere.

One of the hikes you might take would be down to the beach to inspect the sedimentary rocks in the Merced Formation that were discussed earlier and are shown in the oblique aerial view in Figure 115. One thing to look for is the repeated cycles of sediment types from those deposited in deeper water to those laid down in shallower water.

The next place that we recommend you visit is the Mussel Rock overlook located about three miles south of the Fort Funston overlook (as the eagle flies). You reach it by driving south on Highway 35. When you reach the intersection with Highway 1, you will be entering the San Andreas Fault zone.

The Mussel Rock overlook is a little difficult to find. First, get off Highway 1 onto West Manor Drive. From there go down W. Manor to Esplanade, then turn right and continue to Palmetto Street to where you turn left and then turn left again onto Westline Drive. Drive to the end of Westline Drive to a parking area by the high overlook at Mussel Rock. There are some really high scarps all along this section of the shore, with very narrow cuspate sand beaches in places and gravel beaches in others. Looking to the south, the San Pedro Rock sea stacks stand out strikingly. This is an area that has lots of sea ducks both north and south of Mussel Rock, and lots of pelicans and cormorants on the rock itself, which is straight down under the overlook. Last time we were there, we saw a big golden eagle hovering over the crest of the bluff for a long time. This is a great place. If you have about three hours of time to spend, this would be a good place to do it. There is a long trail down to the beach as well as other inviting trails along the top of the high bluff.

Moving further south, you may want to take the time to walk out onto the Pacifica pier, where you can observe the waves close-up from above, pretty exciting when the waves are big.

We recommend that you next keep going south on Highway 1 until you reach the village of Sharp Park, where you can take a hike of a bit over half a mile to the base of Mori Point from which the photograph in Figure 116A was taken. This allows you to walk on Sharp Park Beach and examine the gray sediment illustrated in Figure 116B that is obviously eroded out of the Franciscan Complex in the adjacent rocky headlands. There is also a trail to the top of Mori Point for more breathtaking views.

Back on Highway 1 and driving south for a half mile, you reach the trailhead for the Reina Del Mar Trail, a paved trail with views of the ocean and headlands that goes along side of the hill and down along Calera Creek to Rockaway Beach.

There are two fairly short sand beaches, Rockaway (500 yards), and San Pedro (1,300 yards), at the southernmost end of this Zone. The two beaches are separated by an impressive rock headland as wide as a football field that has numerous sea stacks offshore of it. When we visited San Pedro Beach on a Saturday morning (16 August 2008), the water was full of surfers, dozens of them, taking advantage of the big waves. The sand on the north end of San Pedro Beach was fine-grained and contained a lot of heavy minerals. However, the southern end was cuspate with gravel up to cobbles in size on the horns. This is an interesting, complex beach with very large waves.

In these first three shoreline Compartments (1-3, Figure 1), the sediments on the different beaches have showed some remarkable contrasts. We collected sediment samples at seven of the beaches in the first three Compartments, displayed in Figure 117 and described below:

1) Duxbury Reef shoreline – Light brown granules and pebbles composed of sedimentary rock fragments.

2) Hearts Desire Beach in Tomales Bay State Park – Light colored granules and coarse sand derived from granitic rocks of the Salinian Block.

3) McClures Beach, Point Reyes Seashore – Gray medium-grained sand with a variety of sources. Abundant heavy minerals.

4) The sand spit that closes off Bolinas Lagoon – Dark colored fine-grained sand with a Franciscan Complex source (mostly).

5) Rodeo Beach – Brown, orange, and dark gray pebbles and granules with a Franciscan Complex source.

6) Sharp Park Beach – Dark gray granules with Franciscan Complex source.

7) San Pedro Beach – Dark gray, fine-grained sand with a mostly Franciscan Complex source.

These samples can be compared with those taken from beaches on the Salinian Block and Nacimiento Block to the south (shown in Figure 182).

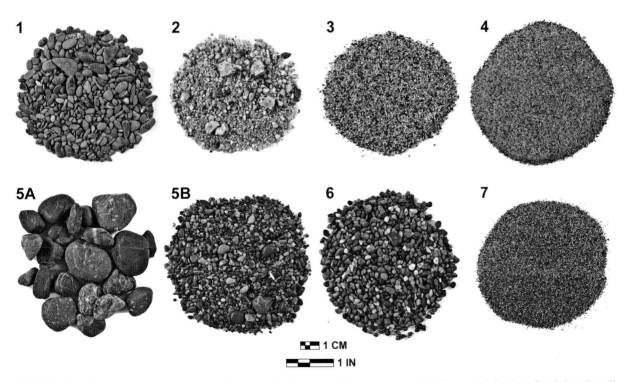

FIGURE 117. Suite of sediment samples collected on the beaches of Compartments 1-3; Figure 1. 1) Duxbury Reef shoreline; 2) Hearts Desire Beach in Tomales Bay State Park; 3) McClures Beach; 4) Sand spit that closes off Bolinas Lagoon; 5) Rodeo Beach (two samples; A is from a cusp horn); 6) Sharp Park Beach; and 7) San Pedro Beach.

16 COMPARTMENT 4: CENTRAL HEADLAND

INTRODUCTION

Going from north to south, this Compartment starts with some spectacular, high rock cliffs at Devils Slide where the rocks of the Salinian Block, including some weathered granites, are exposed (Figures 118 and 119). Not far to the south of Devils Slide, 30 acres of intertidal rock platform with its rich tide pools are exposed at low tide at the Fitzgerald Marine Reserve. Next, the classically formed crenulate bay at Half Moon Bay, with some extensive sand beaches, forms the shoreline. South of the mouth of Tunitas Creek, four wide sand beaches are fed by relatively small individual streams draining the adjacent mountains. The Año Nuevo State Preserve provides one of the best opportunities in the state to observe marine mammals, most notably elephant seals. From Santa Cruz for a few miles to the north, spectacular wave-cut rock cliffs, mostly in the Santa Cruz Mudstone, provide viewing of numerous sea caves, sea arches, and sea stacks (Figure 42A). As a bonus, some of the best examples of raised marine terraces in the state are present along this same shoreline.

For purposes of discussion, we have subdivided this Compartment into the six Zones mapped in Figure 118:

A) **Montara Uplands (San Pedro Rock to Pillar Point)** is a subtle north to south headland that projects out into the ocean between San Pedro Rock and the north end of a major crenulate bay called Half Moon Bay. It is composed of a complex mixture of precipitous wave-cut rock cliffs, protruding granitic headlands, beautiful tan-colored beaches, and the eroded margin of a major raised marine terrace.

B) **Half Moon Bay** is dominated by a classic crenulate bay. A rectangular-shaped chunk of raised marine terrace underlain by dipping sedimentary rocks of the Purisima Formation, called Pillar Point, marks the northern boundary of the Bay. Two long breakwaters protect a harbor in the shadow zone of the crenulate bay.

An arcuate, wide sand beach composes the shoreline of the southern half of the Bay.

C) **Scalloped Terrace (Miramontes Point to Tunitas Creek)** is an irregular-edged shoreline oriented northwest/southeast parallel to the San Gregorio Fault zone (Figure 118). The entire shoreline is backed by a flat, raised marine terrace that is bounded by a high wave-cut scarp in marine terrace sediments underlain by dark-colored sedimentary rocks imbedded with numerous spectacular sea caves.

D) **The Sandy Strand (Tunitas Creek to Pescadero Creek)** is oriented almost straight north and south. Much of the shoreline consists of a wide sandy beach flanked on its landward side by a high, near vertical wave-cut rock cliff composed of the tan-colored Purisima Formation of Pliocene age. Four moderate-sized, steep streams probably supply the bulk of the sand to the beaches, but the eroding Purisima Formation exposed in the wave-cut rock cliffs is possibly also a contributor.

E) **Pigeon Point Uplands (Pescadero Creek to Año Nuevo Creek)** are seaward-projecting headlands with a 14.5 mile long irregular shoreline, accented by three conspicuous promontories – Pigeon Point, Franklin Point, and Point Año Nuevo. Eroding rocky scarps of moderate heights, generally underlain by steeply dipping sandstones, conglomerates, and shales of the Late Cretaceous Pigeon Point Formation, are the most abundant shoreline type.

F) **Santa Cruz Uplands (Año Nuevo Creek to Capitola)** runs straight southeast for about nine miles before curving east to form the northern border of Monterey Bay. The shoreline is mostly composed of wave-cut rock cliffs in the Santa Cruz Mudstone of Late Miocene age, which reach a maximum height of near 150 feet at Waddell Bluff. Raised marine terraces, up to five in number, back the southern half of the shoreline in a most spectacular fashion. The sand beaches are typically quite short and are usually located off the mouths of streams.

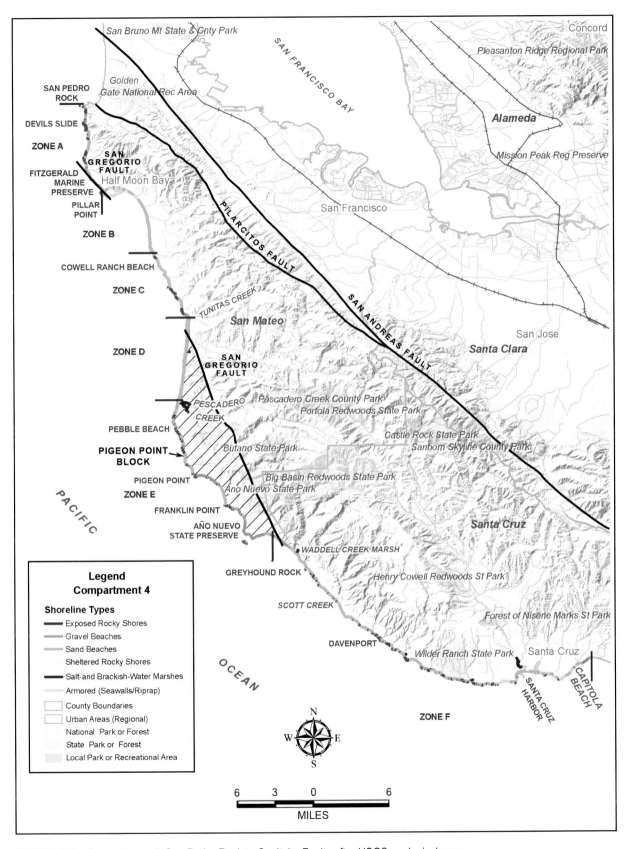

FIGURE 118. Compartment 4, San Pedro Rock to Capitola. Faults after USGS geological maps.

FIGURE 119. Oblique aerial view of Devils Slide (San Mateo County; Zone A, Figure 118) taken on 1 October 2008, courtesy of the California Coastal Records Project, Kenneth and Gabrielle Adelman. A tunnel being dug through the other side of the mountain will replace this segment of Highway 1 (for obvious reasons). The gray rocks in the foreground are granitic rocks of the Salinian Block.

MONTARA UPLANDS (SAN PEDRO ROCK TO PILLAR POINT)

Introduction

This projecting headland area (Zone A, Figure 118) begins on the north end at San Pedro Rock, a sea stack composed of steeply dipping Paleocene rocks, which were treated earlier in the discussion of the geology of the southernmost Zone of Compartment 3. A precipitous, high wave-cut rock cliff, which extends to the south for 1.4 miles in the friable sedimentary rocks, ends abruptly at a headland called Devils Slide, which is composed mostly of Salinian Block granitic rocks of Cretaceous age (Figure 119). These rocks are called the Montara granodiorite by some geologists. A section containing seven projecting headlands in the granitic rocks and two beautiful pocket sand beaches, including Gray Whale Cove State Beach (Figure 120A), extends for a straight line distance of another 1.4 miles. At the end of these headlands, a linear

sand beach called Montara Beach (Figure 120B) runs for 0.8 miles north to south seaward of a raised terrace underlain by Quaternary marine terrace deposits. Next is about 1.5 miles of irregular shoreline consisting of low wave-cut rock cliffs (some of which have granitic bases), including the most prominent one on which the Montara lighthouse is located, gravel and sand pocket beaches, armored areas, and so on around the populated shorelines of the towns of Montara and Moss Beach. Then there is the 0.6 mile boundary of the Fitzgerald Marine Reserve, with its wide intertidal rock platform and frequently visited tide pools. The last approximately two miles is the elevated marine terrace upon which the Half Moon Bay airport is located. Most of the shoreline of this segment is backed by a moderately high wave-cut cliff in Quaternary alluvium underlain by sedimentary rock layers. Sediment grain size of the beaches in this segment varies from large boulders at the base of some of the scarps to finer-grained gravel on wave-cut rock platforms to pure sand.

FIGURE 120. Beaches near the Devils Slide headland (San Mateo County; Zone A, Figure 118). A) Gray Whale Cove State Beach, located 0.7 miles south of Devils Slide, which is shown in Figure 119 and located in Figure 118. (B) Montara State Beach, located 1.4 miles south of Devils Slide. Both of these beaches contain at least some sand derived from the granitic rocks of the Salinian Block, which crop out at Devils Slide. Both photographs taken on 17 August 2008.

Geological Framework

Some of the steepest and most perilous scarps anywhere along the Central California Coast occur along the high cliffs in the 1.75 mile zone south of San Pedro Rock. The thin-layered and steeply dipping sedimentary rocks composed of sandstone, shale, mudstone, and conglomerate of Paleocene age are extremely friable. Highway 1 runs for 0.8 miles along the unstable scarp's edge, a route so hazardous that work is under way to replace it with a 1.1 mile long bridge and tunnel system, the tunnel passing through Montara Mountain (called San Pedro Mountain on some maps). This route has been closed periodically by landslides since it was first opened in the 1930s.

These friable cliffs end at Devils Slide, where a headland composed almost entirely of granitic rocks is spectacularly displayed (see oblique aerial view in Figure 119). As noted by Sloan (2006), these granitic rocks are similar to those already described for the granitic outcrops at Bodega Head and Point Reyes, all of which, having been transported from their place of origin far to the south, are presently riding to the north aboard the Pacific Plate. This surface exposure of granodiorite, which begins at Devils Slide, extends south in an eight mile long, three mile wide swath along the high Montara Mountain, with its southern extremity being close to where Highways 92 and 35 intersect. Outcrops of the weathered and fractured granodiorite are accessible at a couple of pullovers above Devils Slide along Highway 1. All of the six additional protruding rock headlands between Devils Slide and Montara State Beach are also

cored by granitic rocks.

South of these rocky headlands, a clearly defined raised marine terrace, the edge of which is shown in the photograph in Figure 120B, extends inland for at least 500 yards or so, on the landward side of Montara Beach. The material in the wave-cut scarp along this sand beach is horizontally layered material of a relatively young age (probably no older than Pleistocene).

The irregular, jagged rocky headlands between the south end of Montara Beach and the Fitzgerald Marine Reserve have a granitic core in places, but the granitic rocks are usually overlain by yellowish tan sediments that are probably Quaternary alluvium (stream) deposits.

The famous wave-cut rock platform, with outstanding tide pools, at the Fitzgerald Marine Reserve is composed of marine sandstones and shales of the Purisima Formation of Late Miocene/Pliocene age (4-6 million years ago, or so), earlier described as the same formation that makes up the light-colored cliffs at Drakes Bay (Figure 107). Sloan (2006) showed a low-tide picture of the rock platform at the Fitzgerald Reserve where a **syncline** in the folded layers of the Purisima Formation is exposed. She attributes the syncline's presence in the rock platform on the Preserve to the fact that the Seal Cove Fault runs along a northwest/southeast trace through this area. The Seal Cove Fault is now considered to be part of the San Gregorio Fault system (USGS/Seal Cove, 2009).

[NOTE: The terms syncline and anticline are used to described layers of folded rocks, similar to what you would get if you pushed one edge of a rug along the floor, folding it into parallel bands. Where the folded layers bend down, the fold is called a syncline, and *vice versa* for an **anticline**. The simple understanding of folded rocks has lead to the discovery of many, if not most, of the major oil fields in the world. Because oil is lighter than water, in a folded rock layer that is permeable enough to allow the fluids to flow within it, the oil will rise to the top of an anticline with the water staying at the bottom of the adjacent syncline.]

South of the Fitzgerald Reserve, you will be driving across an elevated marine terrace, this one considerably wider and also apparently higher than the one at Montara Beach. The fact that this scarp is near the trace of the Seal Cove Fault might explain the discrepancy in elevations of the terraces. This Zone ends at a peninsula with a rectangular end, with dimensions of 460/380 yards, called Pillar Point, a flat-topped remnant of a raised marine terrace. Pillar Point has a long breakwater attached to it that protects Pillar Point Harbor, which is located in the shadow zone of the large crenulate bay known as Half Moon Bay (Figures 46B and 118).

Beaches

Going from north to south, there are extensive gravel beaches, composed of pure boulders in a few places, between San Pedro Rock and Devils Slide. None are accessible by hiking trails, because the wave-cut scarps are so high and steep. Next are the five rocky headlands that don't appear to have any beaches. A beautiful 500-yard-long sand beach called Gray Whale Cove State Beach (Figure 120A) is located just beyond the end of the fifth headland. Compare the tan/pink color of the sand on this beach and that on Montara Beach (Figure 120B) with the gray sand at Sharp Park Beach (Figure 116). The sand on these beaches, reflecting their source from the granitic Salinian Block rocks, is composed of quartz, feldspar, small pieces of granitic rocks, plus a little mica and hornblende. The sand at Sharp Park Beach is derived from the graywackes and radiolarian cherts of the Franciscan Complex. This is a striking contrast, a common occurrence on the beaches of the Central California Coast (see Figure 117), reflecting the local source for the sand along much of this shoreline. A major exception to this trend is the sand carried down from the Sierra Nevada and dumped on the continental shelf off the entrance into San Francisco Bay.

A shorter sand beach, about 190 yards long and similar in appearance to the Gray Whale Cove Beach, is located between the last two of the granitic headlands. South of the last headland, the beautiful and wide-bermed Montara State Beach (Figure 120B) extends nearly a mile to the south.

A mile and a half of irregular rocky shore around the edge of the towns of Montara and Moss Beach extends from the south end of Montara Beach to the entrance to the Fitzgerald Marine Reserve Visitors Center. There are some scattered short pocket beaches of gravel and sand along this area, with some strange accumulations of rounded boulders on granitic ledges in some areas. Miscellaneous shore protection structures, mostly riprap, are also present.

The shoreline between the entrance to the Fitzgerald Marine Reserve Visitors Center and end of Pillar Point consists of the following components from north to south, most of which have a wide intertidal rock platform offshore of them:

1) About 300 yards of a moderately wide, tan medium-grained sand beach backed by a steep bluff composed of tan-colored sediments.

2) A moderately steep bluff about 200 yards long with some sedimentary rocks at the lower half joined at the base by the rock platform with little accompanying beach sediment.

3) A 400 yard sand beach with a couple of zones of riprap backed by a low vegetated bluff.

4) About 1.3 miles of steep bluff composed of sedimentary rocks with scattered sand and/or gravel along the base.

5) An arcuate sand beach about 600 yards long in the neck of the Pillar Point Peninsula.

6) A steep bluff composed of sedimentary rocks that extends for 500 yards to the southwest end of the Pillar Point Peninsula.

7) On the very southern end of the Peninsula, about 350 yards of sand beach composes the southern face of the Point. The eastern end of this beach is a famous, and very dangerous, surfing area called the Mavericks.

Beach Erosion

Beach erosion in the Devils Slide area was summarized succinctly by Griggs et al. (2005) as follows: *Devils Slide is an area of very high, steep, and unstable cliffs. Some segments of the Ocean Shore Railway bed can still be seen, and Highway 1 has a constant maintenance problem as it crosses the upper portion of the slide area.* The Ocean Shore Railroad Company went bankrupt in 1921, due at least in some measure to the costs involved in building and maintaining a useable track across the dangerous, steep wave-cut rock cliffs along this stretch of coast.

The next segment with beach erosion problems, as one might expect, is the developed shorelines of Montara, Moss Beach, and the small village of Seal Cove south of Moss Beach. A noteworthy example of erosion is at the restaurant on top of the bluff at the southern end of Montara Beach, where 33 feet of bluff erosion occurred during the 1983 storms. Later, a 130-yard-long revetment of riprap was installed to protect the restaurant and a parking lot.

As already noted, several measures, including riprap revetments and adding gunite to a slope on the bluff, have been installed to protect the structures built close to the bluff's edge in Montara. The granite at the base of the bluffs in this area is helpful, but the more friable sediments that overlie the granite are subject to both wave erosion and landslides in places. A large landslide historically has plagued part of the development at Seal Beach. See Griggs et al. (2005) for more details.

General Ecology

The Fitzgerald Marine Reserve is one of the most popular sites along the coast for exploring tide pools. A kiosk at the reserve states:

The Reserve has been described by some as one of the most diverse regions in California. This is easy to see given that up to 30 acres of reef may be exposed during a low tide with over 200 species of animals and 150 species of plants available to observe. There are 126 species of gastropods alone, which account for 77% of the state's species, nudibranchs, which account for 17% of the state's species, and chitons, which account for 37% of the American species. The Pacific coast of America supports a wealth of marine life. Its species abundance and diversity is matched by few other coastlines in the world.

However, the impact of 135,000 visitors annually has resulted in severe trampling and disturbance of the fauna and flora. "It's got to the point," said Bob Breen, supervising naturalist, "where many plants and animals have disappeared and may not return for many years." Plans are underway to close off some areas and limit the number of visitors per day. The Reserve is open during daylight hours but the life in the tide pools can only be seen at low tides, below +1.0 feet, so it is important to check the tides prior to your visit.

The rocks offshore Pillar Point Lighthouse are used as haulouts for about 100 California sea lions and 200 harbor seals. This location is also good for watching for migrating gray, humpback, and minke whales.

Places to Visit

Approaching this Zone from the north on Highway 1 and after leaving Pacifica, first you drive up the side of Montara Mountain through a thick grove of eucalyptus

trees. In the road outcrops near the top of the climb, there are some excellent exposures of thin-bedded turbidites and shales of Paleocene age, the same general age and rock types as the dipping layers exposed at San Pedro Rock. As observed by Sloan (2006), these are the rocks *causing so much trouble at Devils Slide.* By that she meant the loss of the railroad, the need to construct the new tunnel, and other erosional nightmares.

Going through the gap at Devils Slide, that is about as steep and high a rocky shore as you will see almost anywhere. The face of the granitic outcrop of Devils Slide, pictured in the oblique aerial view in Figure 119, is scary just to look at as you drive along the road. However, this thrill will no longer be possible (while driving along this road) as soon as the tunnel is finished. On the other hand, if you do come after that project is completed, hiking and biking trails along these slopes are supposed to be available.

When you come out of the tunnel going south at some time in the future, you will be looking straight down on a beautiful Gray Whale Cove State Beach (Figure 120A), as well as have a spectacular view of the rock cliffs. There is a parking area on the east side of the road for access to the very steep trail and stairs down to the sand beach, the mineral composition of which was discussed earlier. The primary source for the beautiful medium- to coarse-grained sand is the granitic rocks of Cretaceous age in the adjacent cliffs that arrived here in transit on the Pacific Plate. Massive exposures of the granodiorite are also present in the road cuts along Highway 1, which may have to be reached in the future along hiking and biking trails.

As you come down the road out of the granitic uplands, you will drive out onto a flat valley floored by a raised marine terrace that is bordered on the west by the nearly mile long, exquisite tan-colored sand beach called Montara State Beach (Figure 120B). Even though the beach is quite wide, there is a wave-cut scarp in the valley sediments, shown clearly in Figure 120B, that is located just shy of the road. Three relatively spacious parking areas are located along the west side of the road that provide ready access to this beach.

Once you enter the towns of Montara and Moss Beach, the primary place for a naturalist to visit is the James Fitzgerald Marine Reserve, preserving three miles of intertidal wave-cut rock platforms from Montara Point, which has an historic lighthouse you may want to visit, to Pillar Point. The tide pools on the rock platform, described in some detail earlier in the section on general ecology, are the main attraction, of course. However, there are some pleasant hiking trails along the bluff to the south, which include a walk through a dense Monterey

Cypress stand. The beach sand itself near the visitors center is pretty coarse-grained, being mixed with some gravel. A massive riprap revetment intended to preserve the bluff from further erosion begins a little to the north of where the path from the visitors center exits onto the beach.

The unique intertidal rock platform in this area was given protection in 1969, when it became a designated marine reserve. The visitors center has a number of excellent printed guides to the animals and plants living in the tide pools, and other items of interest.

In a personal communication, Ed Clifton mentioned the famous surfing Mecca, called the Mavericks, which lies about a half mile offshore from Pillar Point. An unusual bottom configuration there creates breakers that, on a big day, can exceed 40 feet high and commonly reach 20 feet. The annual Mavericks Surfing Contest has been called the "Super Bowl of big-wave surfing" by Sports Illustrated. See Mavericks (2007) for more information.

HALF MOON BAY

Introduction

Half Moon Bay (Zone B, Figure 118) is a classic example of a **crenulate bay**, a shoreline type illustrated by the general model in Figure 67A. As the model implies, the original shape of most of the crenulate bays on the Central California Coast, including Half Moon Bay, which has a shoreline length of 6.75 miles, is the result of the refraction/diffraction of the predominate northwesterly waves around a rock headland to the north of the Bay, Pillar Point in this case. Figure 67B shows this classical wave pattern modeled for another Central California crenulate bay, Drakes Bay, which is an exception to the general rule, being oriented east-west.

In 1959, two overlapping breakwaters were built to create Pillar Point Harbor in the shadow zone of the crenulate bay (see Figure 46B). These are long breakwaters, the east one measuring 0.8 miles and the west one 0.7 miles, which provide a shelter for the only small boat harbor between San Francisco and Santa Cruz.

Geological Framework

The entire landscape surrounding Half Moon Bay is composed of a raised marine terrace at an elevation of around 80 feet. This flat terrace extends landward for over a mile in places. The surface geology of this level area is mapped as either Quaternary (Pleistocene and Holocene; <2.6 million years in age) marine terrace deposits or Quaternary alluvium (Sloan, 2006). These flat-lying deposits of poorly consolidated gravel, sand, and silt are exposed in the modest-sized eroded scarps that line the landward side of the wide sand beaches in the southern half of the shoreline arc. The Purisima Formation of Late Miocene/Pliocene age (4-6 million years ago, or so), composed of layers of sandstone, mudstone, and siltstone deposited in marine waters, are exposed along the base of the eroded cliffs along Pillar Point. These rocks are sheared and folded both at Pillar Point and in the Fitzgerald Marine Reserve further north as a result of movement on the Seal Cove Fault (Sloan, 2006; Alt and Hyndman, 2000). The Seal Cove Fault is a segment of the San Gregorio Fault, named before it was known of the connection.

Beaches and Beach Erosion

Before the breakwaters were constructed in 1959, sand beaches existed around most of the shoreline of Half Moon Bay. As noted in the discussion of crenulate bays (Section I), studies by two researchers (Krumbein, 1944; Bascom, 1951) found that the grain size of the sand and the beachface slope of this bay increased from the shadow zone to the tangential end, a common trend for crenulate bays. It is also probably safe to assume that the sand beaches were in a more-or-less equilibrium condition at that time. The construction of the harbor disrupted that equilibrium, interfering with the normal sediment transport patterns, such that erosion of the shoreline south of the east breakwater increased from a few inches to as much as 80 inches per year (Griggs et al., 2005). After construction of the west breakwater, *a county road and the underlying sewer lines were gradually undermined and destroyed, and State Highway 1 was threatened* (Griggs et al., 2005). A picture of the remaining remnants of this road taken in 1972 graces the cover of the book *Living With the Changing California Coast*. Riprap now lines this part of the shore. About 1,000 yards south of the end of the east breakwater a number of homes and businesses in the community of Miramar are threatened by erosion. Around 550 yards of riprap have been placed in front of these structures to protect them (hopefully). Most of the southern half of this arcuate beach is not eroding at the present time, and developments are set back a good distance from the beach.

We visited this area on 16 August 2008, stopping at Blufftop Coastal Park about three fourths of the way down to the south end. On that day, the beach had a long, moderate-sized berm with a simple west-coast type profile, like the one illustrated in Figure 50A, a classic California profile, we guess you could say. The medium-grained sand, composed of quartz, feldspar, and some heavy minerals, was the same tan color we had seen on the beaches up around Devils Slide. There were a few unspectacular rip currents in the surf zone. We estimated the long and consistent bluff landward of the beach to be about 50 feet high.

General Ecology

About 60 western snowy plovers spend the winter and a few breed during the summer on the beaches of Half Moon Bay State Beach. The five-acre nesting area is fenced off, and there is an active Plover Watch volunteer program that erects protective enclosures when eggs are laid. To protect these federally threatened birds, dogs are prohibited on these beaches.

Marbled murrelets are another federally threatened species in peril in this area. In a 2008 study for the California State Parks, researchers estimated the Central California population to be 174 birds offshore of breeding habitat between Half Moon Bay and Santa Cruz (Peery et al., 2008). These estimates represent 54-55% declines since 2007 and 71-80% declines since 2003. They did not find any juveniles during the 2008 surveys. The dire prediction was that marbled murrelets in Central California will almost certainly become locally extirpated when the current cohort of adults dies.

Places to Visit

Going from north to south, we recommend that you first go west on the road to Pillar Point and stop at a parking lot on the left about 0.25 miles after you leave the edge of the town of Princeton-by-the-Sea. From the parking lot, it is about 230 yards down a trail to a small sand beach along the harbor inside the breakwater. After walking about 200 yards along the beach to the northeast, you will come to a tidal inlet that drains a small, seemingly pristine salt marsh complex. The birding should be good there at low tide, not only on the beach and in the salt marsh, but also on the tidal flats exposed out in the harbor. While you are in that general vicinity, you might want to check out the fishing from a long pier that juts out into the harbor about another 200 yards to the northeast.

The next place we recommend that you visit is the beaches along the southern half of the arcuate shore. There are at least eight major parking areas that provide good access to the beach. This is a very pleasant beach with lots of room to roam. There is a paved bike trail all along the edge of the bluff in some areas, should you be in the frame of mind to ride, jog, or hike along it.

On the drive out, you will be impressed with the width and flatness of the top of the raised marine terrace.

SCALLOPED TERRACE (MIRAMONTES POINT TO TUNITAS CREEK)

Introduction

Except for the intense scalloping, this shoreline (Zone C, Figure 118) follows a relatively straight course for about 5.5 miles in a northwest/southeast direction, parallel to the San Gregorio Fault, the trace of which is located about a mile offshore (Alt and Hyndman, 2000; Sloan, 2006). The entire length of the coast is backed by a raised marine terrace, generally around a half mile wide, at average heights ranging from 70 to 160 feet (Griggs et al., 2005). Except for the huge Ritz Carlton Hotel and grounds at Miramontes Point on the northern edge of the Zone, and a small private village at Martins Beach about three quarters of the way to the south end, there is no human development close to the beach along this coast, which is mostly farmland that extends out to the edge of the scarp in some places. This Zone ends where the valley of Tunitas Creek intercepts the coast.

Geological Framework

The geological map for this area indicates that the hinterland close to the coast is everywhere underlain by Quaternary marine terrace deposits of sand and gravel up to 65 feet thick, which are in evidence in the upper portions of the wave-cut scarps all along the coast. Quaternary alluvium is also present in a couple of stream valleys. The lower parts of the erosional scarps commonly are composed of darker-colored sedimentary rocks (the Purisima Formation ?) that are steeply dipping, as well as jointed and fractured in places, probably as a result of their proximity to the San Gregorio Fault. These wave-cut scarps are eroded into spectacular sea caves in a number of places (see photograph of one of them in Figure 44). The wave-cut scarp in these sediments and sedimentary rocks is truly a thing of beauty (Figure 121A).

Beaches and Beach Erosion

There are seven tan-colored, commonly cuspate, medium-grained sand beaches measuring between 230 and 1,000 yards in length along the base of the wave-cut scarps (see example in Figure 121B). There are also

FIGURE 121. Wave-cut rock cliff and sand beach at the Cowell Ranch Beach public access (San Mateo County; Zone C, Figure 118). (A) Oblique aerial view taken on 1 October 2008, courtesy of the California Coastal Records Project, Kenneth and Gabrielle Adelman. The upper horizontal layers in the cliff are Quaternary marine terrace deposits, but the lower section consists of sandstones, mudstones, and siltstones of a sedimentary rock unit thought (by us) to be the Pliocene Purisima Formation, which is thought (by others) to have been deposited in a shallow marine basin around 2-6 million years ago. (B) The sand beach adjacent to the cliff (left side of the photograph in A). It was highly cuspate on the date this photograph was taken, 16 August 2008.

a number of shorter pocket sand beaches, but gravel beaches are present in only a few localities, probably the result of the fact that their potential source, the wave-cut cliffs, are mostly soft sediments. There are a number of tombolos in the lee of small sea stacks and rock ledges on the narrow wave-cut rock platforms that are randomly distributed along the coast. Some of the rock ledges form natural groins that trap sand on their northern sides.

With a couple of exceptions, these wave-cut rock cliffs have been relatively stable over the past 80-120 years (Griggs et al., 2005). Storms have occasionally threatened the beach homes at Martins Beach, and a large section of cliff face collapsed in front of the Ritz Carlton Hotel, resulting in the addition of riprap to protect that portion of the golf course. This riprap, which was on the beach as late October 2005, was mysteriously missing by October 2008. The formerly "protected" portion of the golf course was still there. Griggs et al. (2005) concluded their analysis of beach erosion in this area by stating that *these two examples notwithstanding, the coastal hazards and risks in this coastal segment are very low, due primarily to sparse development and limited access to the shoreline.*

General Ecology

There are four harbor seal pupping areas along this coast, at Miramontes Point, Eel Rock, Seal Rock, and just north of Martins Beach. There are five seabird nesting colonies, with relatively low numbers of birds present. The offshore waters have moderate occurrences of Clark's grebes, common murres, cormorants, Pacific loons, brown pelicans, rhinoceros auklets, surf scoters, white-winged scoters, and sooty shearwaters. Lobitos and Tunitas Creeks support runs of steelhead trout.

Places to Visit

It is possible to gain access to the beach in the Ritz Carlton area by driving to a parking area west of Miramontes Point Road about 500 yards short of the entrance to the hotel. Stairs lead down to a 230-yard-long beach called Arroyo Cañada Verde Beach. Riprap protects the base of the stairs.

Back on Highway 1, which runs along the landward margin of the raised marine terrace, go south for 0.6 miles and turn right into the parking area for Cowell Ranch Beach, a place you will not want to miss, as well as the only remaining public access to the shoreline in this Zone.

It is about a 0.5 mile walk from the parking area to the bluff overlook. This walk is over a flat surface along a road across farm land, including some big artichoke fields. On our visit there on 16 August 2008, the shrubs and fields along this walk were full of sparrows. So, if you are looking for seed-eating birds at the right time of the year, this would probably be a pretty good place to visit.

At the overlook, there are stairs down to the beautiful, tan-colored sand beach pictured in Figure 121B, which had some perfectly formed cusps when we were there.

Looking to the south from the overlook, we saw several seals resting on a pair of twin tombolos. That overlook is visible in Figure 121A. The ledge that projects offshore from the base of the bluff is composed of very steeply dipping sedimentary rocks.

While we were there, a big golden eagle was hanging in the stiff breeze coming off the bluff, and some pelicans were doing the same thing. This, the best place to access the shoreline in this Zone, is truly a special spot!

Going further south along this scarp there are a number of sea caves carved into it, one of which is shown in Figure 44.

THE SANDY STRAND (TUNITAS CREEK TO PESCADERO CREEK)

Introduction

This straight, 6.5 mile long, almost north/south oriented shoreline (Zone D, Figure 118) is a sand beach backed by wave-cut rock cliffs in the Pliocene Purisima Formation almost all the way except for a few areas where the rock cliffs project out a ways and the adjacent intertidal zone is a wave-cut rock platform. The sand is fed to the coast by four modest-sized, steep-gradient streams that drain the western flank of the Santa Cruz Mountains – the Tunitas, San Gregorio, Pomponio, and Pescadero Creeks. These streams are located on the vertical image in Figure 122.

Geological Framework

According to Alt and Hyndman (2000), the rocks that compose the bluffs along this shoreline consist entirely of the sandstones, mudstones, and siltstones of the Pliocene Purisima Formation, which is thought to have been deposited in a marine basin around 2-6 million years ago. Griggs et al. (2005) observed that the rocks in the *400 to 500 foot high irregular coastal bluffs in the northern part of this area, between Tunitas and San Gregorio Creeks, are sheared and fractured by the San Gregorio Fault, leaving these bluffs unstable and scarred by deep gullies and large landslides.* In the southern half of the area, however, the bluffs in this Pliocene formation continue to be very high in places, and it's amazing how vertical these scarps are, with few landslides in this more stable area. The distinct layers in the formation are horizontal in some areas and dip slightly to the north in others. In addition, white beds of volcanic ash are spectacularly displayed in the cliffs near the mouth of San Gregorio Creek (Ed Clifton, pers. comm.).

Beaches and Beach Erosion

These beaches along this Zone, which are composed of medium- to fine-grained sand, are typically quite wide and commonly cuspate. The beach at the mouth of Pomponio Creek is shown in Figure 123B. In a few places, particularly near the stream mouths, a high, vegetated foredune ridge is present landward of the beach. Another common occurrence is lots of drift wood both in the intertidal zone and high in the dunes.

When we visited this area on 16 August 2008, we kept remarking to each other how the widest and most extensive sand beaches seemed to be near the mouths of the four streams. This is generally true, but the strong north to south longshore transport direction driven by the prevailing northwest winds that strike the north/south oriented shore is able to move the sand along the coast effectively. Therefore, when the beach is wide during a long interval between storms, this is essentially a sand beach all the way except for a few areas of wave-cut rock platforms.

Griggs et al. (2005) classified this entire stretch, which has no endangered man-made structures, as *Stable: low risk* with regard to beach erosion issues. They did express some concern for the fate of Highway 1 in the future in the event of a significant rise in sea level.

General Ecology

In winter, the wide, driftwood-strewn San Gregorio State Beach is host to up to 75 western snowy plovers, so no dogs are allowed. At more than 500 acres, the Pescadero Marsh Natural Preserve is the largest coastal marsh between San Francisco Bay and Elkhorn Slough. Over 160 species of birds have been sighted there; the best birding is in late fall and early spring. Six species are listed as threatened or endangered in Pescadero Marsh: steelhead trout, coho salmon, California red-legged frog, San Francisco garter snake, tidewater goby, and California brackish water snail. Like many coastal estuaries in California, the marsh habitat has been severely degraded by sedimentation from the watershed, building of levees, and construction of the bridge over the stream mouth that prevents its natural migration. In fall, when the stream flows wane and the waves increase, the mouth is closed off by the buildup of sand. Then the marsh is flooded by increasingly fresh water. Winter rains cause the water level to rise, eventually so high that it breaches the beach and the marsh drains very rapidly. Since 1995, there has been a large fish kill during the breaching event; with three endangered or threatened fish species being present, the causes of these fish kills

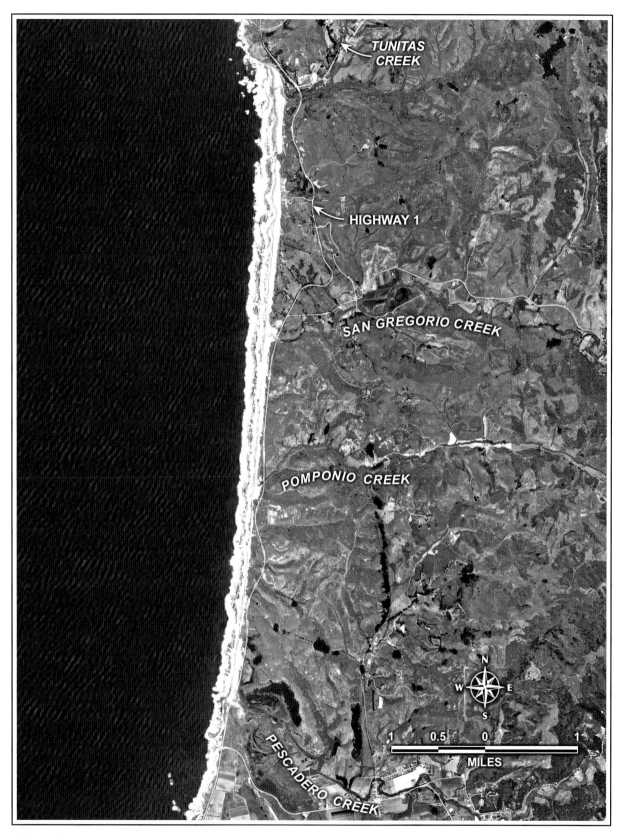

FIGURE 122. Image of the area between Tunitas Creek and Pescadero Creek (San Mateo County; Zone D, Figure 118). NAIP 2005 imagery courtesy of the California Spatial Information Library (CaSIL).

FIGURE 123. Two ground views in San Mateo County (Zone D, Figure 118). (A) View looking north of the raised marine terrace between Martins Beach and the mouth of Tunitas Creek. This 1978 photograph taken from a short distance south of the mouth of Tunitas Creek. (B) Medium-grained sand beach at the mouth of Pomponio Creek on 16 August 2008.

are of great concern, but not understood as of 2008.

Places to Visit

As you are traveling south into this area on Highway 1, find a pullover early and take a look back to the north at the spectacular eroded bluffs along the edge of the raised marine terrace that was described in the discussion of the next Zone to the north (the Scalloped Terrace). Our picture of this much-photographed raised marine terrace is given in Figure 123A.

Each of the three state beaches in this Zone, San Gregorio, Pomponio, and Pescadero, have spacious parking facilities. From north to south, the first public access you come to is the San Gregorio State Beach, which has a great overlook on top of the wave-cut rock cliff from which you can see a long ways up and down the beach. Where the stream outlet crosses the beach is a great place to observe antidunes forming at low tide (see Figure 60 for illustrations of antidunes). There are also opportunities to inspect the geology of the rock cliffs in several spots. This is a very sizable and generally uncrowded beach.

Heading south from San Gregorio Beach, Highway 1 is right on top of the bluff, and passengers on the right side can look directly down on the ocean, a pretty spectacular ride along these high and vertical scarps.

Next, you will drive down into the valley of Pomponio Creek, where another parking area is located on the north side of the mouth of the creek (Figure 123B). If you go down to the south end of the beach, you will encounter a big vertical bluff in the Purisima rocks at that point. If the tide is real low, you may be able to get around the edge of the cliff and visit some spectacular

sea caves a short distance down the bluff. If you do try this, be careful and keep your eye on the tide level.

Continuing south for another 1.8 miles, you will reach another parking area on your right. This stop provides an excellent opportunity for some close-up inspection of the rock cliffs, which stand out boldly a little ways north along the beach.

The last parking area in this Zone is reached by going south another 0.7 miles and crossing the bridge over Pescadero Creek. The main attractions there are another opportunity to have a long view north along the beach from an overlook on top of the scarp, and a wetland with some channels and tidal flats associated with Pescadero Creek. This marsh has a big, open tidal flat and no doubt there are lots of birds on it during migrations.

PIGEON POINT UPLANDS (PESCADERO CREEK TO AÑO NEUVO CREEK)

Doc walked on the beach beyond the lighthouse. The waves splashed white beside him and sometimes basted his ankles. The sandpipers ran ahead of him as though on little wheels. The golden afternoon moved on toward China, and on the horizon's edge a lumber schooner balanced.

John Steinbeck – SWEET THURSDAY

Introduction

This 14.5 mile long protrusion of the shoreline (Zone E, Figure 118), which is backed by a raised marine terrace for the most part, is complex and irregular. It contains both gravel and sand beaches, a lengthy intertidal rock platform with numerous tide pools in the northern section, and a wave-cut bluff of modest height all along the shore that is underlain by marine terrace deposits on the surface and some spectacular steeply dipping sedimentary rocks of Late Cretaceous age at depth. Three prominent headlands, Pigeon Point, Franklin Point, and Point Año Nuevo, are associated with arcuate, primarily erosional, shorelines on their downdrift (southern) sides. Numerous sea stacks contain thousands of sea birds, and the tide pools are equally rich with a variety of marine life.

Geological Framework

The bedrock is dominated by relatively old sedimentary rocks that are riding to the north on the Pacific Plate as a part of what is called the Pigeon Point Block (Sloan, 2006; Figures 16A and 118). This block is bounded on the east by the San Gregorio Fault that runs to the northwest at a slight angle to the San Andreas Fault, which is located about twelve miles to the northeast of it at this point. The surf-resistant point at Año Nuevo is composed of rocks of the Monterey Formation, the only outcrop of this famous Miocene formation along this part of the coast.

The star of the geological show here, however, is the Late Cretaceous sedimentary rocks, pictured in Figure 124, that are said to have the following characteristics by Alt and Hyndman (2000), Sloan (2006), and Leech (2001):

1) These rocks, which are usually tilted and folded in this area (Figure 124A) and were deposited during Late Cretaceous time (70-80 million years ago), make up a sequence about 8,000 feet thick.

2) They are not part of the Salinian Block, which contains the granitic rocks at Devils Slide, Point Reyes, and elsewhere. They are part of the Pigeon Point Block. The San Gregorio Fault system, which separates these two blocks (see Figure 118), is a young fault, that might have *split off a chunk of the Salinian Block and displaced it to the northwest* (Ed Clifton, pers. comm.). Therefore, these rocks lie west of the San Gregorio Fault, which crosses the shoreline just south of Año Nuevo and lies inshore until it goes out to sea again just north of San Gregorio State Beach (Figure 118). These Cretaceous rocks may have originated in a setting like that shown for the Great Valley Sequence in Figure 14 (and some geologists even refer to them as part of that Sequence). The Great Valley Sequence was formed in a deep forearc basin (during the Franciscan event) on top of basal, deep

oceanic igneous rocks, and they generally crop out more to the east side of the Coast Ranges. Therefore, they are usually located to the east of the rocks of the Franciscan Complex, which are more highly deformed (e.g., those along the shoreline of the Marin Headlands north of the Golden Gate). Obviously, they are located to the west of the Franciscan rocks at this location because of their being transported to the north as part of a fault bloc. Their origin was many, many miles to the south, perhaps in the southern end and probably on the eastern side of the Franciscan trench shown in Figure 12 (deposited on top of the granitic rocks; Ed Clifton, pers. comm.). These rocks correlate with a formation of similar age located 100 miles to the south in Monterey County (to be discussed later). Yes, the faults associated with the colliding North American and Pacific Tectonic Plates do make this geology complex and exciting.

[NOTE: This geology presents many intricate questions that geologists are still trying to solve. For example, Ed Clifton (pers. comm.) stated that it is likely *that the Pigeon Point Formation in San Mateo County* (the area under discussion at this point in the text) *lies on granitic crust (Alt and Hyndman, p. 215, flatly say so). Both the Pigeon Point Formation and the granitic rocks that compose the Farallon Islands lie on the east side of the San Gregorio fault and are not that far apart. The granite is Middle Cretaceous (94-95 million years ago; Kistler and Chanpion, 2001) and the sedimentary rocks are Late Cretaceous (<84 million years ago; Hall, Jones, and Brooks, 1959). Unless there is a large unknown fault offshore, Alt and Hyndman are probably right.* An important point being that the Great Valley Sequence present on the east side of the Coast Ranges, to which these rocks have been assigned by some geologists, was deposited on top of the Coast Range Ophiolite (Figure 14).]

3) The original sediments of the Pigeon Point Formation were deposited mostly by **turbidity currents** in a deep submarine canyon something like the one that now occupies the center of Monterey Bay, but probably not nearly that big. The evidence for this interpretation is the graded bedding of the sandstones and conglomerates with coarse-grained gravel layers occurring at the base of the individual depositional units, which range in thickness from 3-20 feet. The gravel at the base of these deposits may be overlain by parallel-bedded, medium-grained sand topped by rippled sands and silts. In addition to turbidity currents, some of the thicker and more massive of these sedimentary rock layers may have been deposited as debris flows, which were described in Section I.

FIGURE 124. The rocks at Pebble Beach, a State Beach located about one mile south of the mouth of Pescadero Creek in San Mateo County (Zone E, Figure 118). (A) Strongly dipping sedimentary rocks, including numerous turbidites, of Late Cretaceous age (75-65 million years ago). These rocks, which occur within the Pigeon Point Block (located on Figure 118), are said to be part of the "Great Valley Sequence" by some geologists. They were deposited far to the south of this locality. (B) Highly weathered surface of a sandstone layer (called *tafoni*). Both photographs taken on 16 August 2008.

Beaches and Dunes

Gravel beaches are sparse in this area, but one with easy access is located near the parking area 0.4 miles south of the bridge across Pescadero Creek on Highway 1. The gravel on this beach ranges up to boulders two feet or more in diameter (see photograph in Figure 26B).

Sand beaches are also relatively scarce, being located mostly downdrift from the mouths of several of the small streams that have outlets on this low, flat-topped headland. Examples include: a) 300 yard beach south of Arroyo de los Frijoles; b) 900 yard beach south of a stream north of Pigeon Point Lighthouse; c) 1,500 yard beach south of Gazos Creek; and d) 800 yard beach north of Año Nuevo Creek, the mouth of which is in the lee of the projecting Point Año Nuevo (i.e., it is sheltered from northwest waves, hence the northerly longshore sediment transport direction).

There are two areas of coastal sand dunes, both on the northerly flank of a projecting rocky point. At Franklin Point, a wedge of vegetated dunes over 400 yards wide with a maximum length of over 600 yards is fed by sand delivered to the coast by Gazos Creek. The beach runs north/south, so the dunes, formed by winds blowing straight out of the northwest, strike the shoreline obliquely. At Point Año Nuevo, the dunes, which are also mostly vegetated and have the same northwest/southeast orientation as those at Franklin Point, have a maximum width of 700 yards and maximum length of 1,700 yards. Figure 125 shows photographs of the dunes at Point Año Nuevo.

Beach Erosion

Griggs et al. (2005) ranked the hazard level of the shoreline of this Zone as *Stable: low risk*, because the sandstones and conglomerates of the Pigeon Point Formation are so highly resistant to wave erosion. An exception occurs just south of the mouth of Pescadero Creek, where large storms during the winter of 2002-2003 damaged several portions of Highway 1. In early 2003, *the roadbed for Highway 1 was moved inland and riprap revetments were constructed at three points to protect the highway from further damage* (Griggs et al., 2005). Erosion also threatened a couple of houses to the south of there. The shoreline at Point Año Nuevo has a very interesting erosional history, but that is presented later in the more detailed discussion of that area.

General Ecology

Along the rocky platforms between Pescadero and Pigeon Points, there are three harbor seal haulouts used by more than 500 animals; pupping occurs March-June.

FIGURE 125. Vegetated coastal dunes at Año Nuevo State Reserve, San Mateo County (Zone E, Figure 118). (A) Oblique aerial view taken on 1 October 2008, courtesy of the California Coastal Records Project, Kenneth and Gabrielle Adelman. Arrow points to near area from which the photograph in B was taken. (B) Ground view of the dunes shown in A, taken on 16 August 2008.

Harbor seals congregate in groups only during breeding and molting; otherwise, adults are solitary. In fact, when hauled out, they generally do not touch each other, in contrast to California sea lions (Figure 126). Molting, which is the regeneration and shedding of their hair, occurs after every breeding season and takes 1-2 months. Harbor seals do not migrate and tend to have favorite haulout sites that vary seasonally, probably due to prey availability.

Año Nuevo Island and the adjacent mainland beaches make up one of the most important pinniped rookery and resting areas in Central and Northern California (Figure 126). Año Nuevo State Reserve is the site of the largest northern elephant seal breeding colony on the mainland in the world, with up to 10,000 animals present. Amazingly, they were first sighted using this site only in 1955. According to the Marine Mammal Center (2008):

FIGURE 126. Aerial views of marine mammals hauled out on a sand beach on Año Nuevo Island (San Mateo County; Zone E, Figure 118) in May 1992. This is the largest California sea lion haulout in Central California; they are very social and like to pack closely together at favorite spots.

The northern elephant seal is a conservation success story. They were hunted to the brink of extinction, primarily for their blubber, which was used for lamp oil. By 1910, it is estimated that there were less than 100 elephant seals, all found on Guadalupe Island off Baja California, Mexico. Today, the northern elephant seal population is over 150,000 and is probably near the size it was before they were over-hunted.

A good time to see the elephant seals is during the breeding season, December through early March; the weaned pups remain through April. They return to molt April-July, when they shed not only their hair but also the upper layer of their skin as well. When not ashore, they live up to 5,000 miles offshore and dive to 5,000 feet to feed. Males feed near the eastern Aleutian Islands and in the Gulf of Alaska; females feed a little more to the south.

Año Nuevo Island also has the largest California sea lion population in Central California, with a population range of 2,300-9,440. It is also the southernmost breeding site for Steller sea lions.

Birds that nest on the island include rhinoceros auklet, Cassin's auklet, pigeon guillemot, cormorants, black oystercatcher, and western gull. This island is one of only three main colonies in California of rhinoceros auklets, which nest in burrows. The island became denuded of vegetation by 1997, leading to declines in nesting. Researchers are trying to restore the central part of the island not used by seals and sea lions by planting native shrub and grass assemblages and using temporary ground-cover to reduce erosion until natural vegetation is re-established.

Access to Año Nuevo Island is limited to approved researchers. The structures visible from land are the remains of a fog signal, light tower, keeper's dwelling, tramway, dock, boathouse, and various other support facilities of the Año Nuevo Island Light Station that operated between 1872 and 1948. Some of these abandoned structures are visible in the photographs in Figure 126.

The Bean Hollow Beach trail is a good place to observe rocky intertidal communities; be sure to plan your visit during low tides and bring a field guide to identify the plants and animals. You will know you are there at the best time when you see the bright green surf grass that occurs only in the lower intertidal zone.

Places to Visit

Starting at the north end of the Zone and driving south on Highway 1, you can visit a well-developed gravel beach if you pull over to the parking area on the right at 0.4 miles south of the bridge over Pescadero Creek. The parking area is directly across the road from where Pescadero Road joins Highway 1. If you go down the short stairs to the right, you will be able to walk across a beautiful high-tide gravel beach with a little bit of sand on the lower part. There are some boulders up high on the beach and the gravel on the lower part of the beachface shows some subtle imbrication (Figure 26B). At least that was the case when we visited this beach on 15 August 2008. At that time, the beach did not have a well-defined storm berm. Note how well rounded the cobbles and boulders are, a result of frequent reworking by large waves. A sea stack with a diameter of between 150-200 feet located about a hundred yards offshore was covered with pelicans and cormorants.

The next place you shouldn't miss is the parking area at Pebble Beach, located 1.7 miles further south on Highway 1. This is one of the most exciting geological stops you can make anywhere between Santa Cruz and the Golden Gate. The steeply dipping rocks you will see are the deep-water turbidites and debris flows deposited around 70-80 million years ago in water possibly as much as over a mile deep. As you walk down to the beach you will cross over the flat-lying sand layers of the marine terrace deposits. Note the contact between these sediments and the steeply dipping sandstones and conglomerates of the Pigeon Point Formation that underlie them. That contact between the lower steeply dipping beds and the overlying flat beds is called an **angular unconformity**. Inasmuch as the marine terrace deposits are probably considerably less than two million years old, the unconformity represents a time gap with a missing geological record of something near 70 million years!!

The tan, relatively thick sandstone beds of the Pigeon Point Formation have gravel at their base, which supports the turbidite or debris-flow hypothesis. Probably more amazing are the holes and crinkles weathered out of the rock's surfaces, known as **tafoni**, or honeycomb weathering (shown in Figure 124B). According to Sloan (2006), *two mechanisms, not mutually exclusive, have been proposed for its formation: chemical interaction of salt spray with the sandstone and differential erosion of more weakly cemented portions of the rock.*

Between Pebble Beach and Bean Hollow Beach there is a trail along the top of the wave-cut bluff that extends for 0.8 miles, and there is a moderately wide wave-cut rock platform the whole way. We took this walk at low tide on 16 August 2008, noting that there is some complex and bewildering geology exposed on the platform – folded rocks, steeply dipping rocks, and beautiful thick sandstone layers – amazing. Also, there are a few narrow gravel beaches along the base of the scarp. The rock platform is a little tough to get to, but there are a lot of tide pools exposed at low tide.

On this walk, we saw some ruddy turnstones, a curlew, numerous cormorants, brown pelicans, and

Herrmann's Gulls. This is probably one of the best places for birding along this part of the coast, because of not only the abundance of different birds, but also because of the convenient walking path that allows you to cover such a lengthy segment of the shore. We also observed a good number of harbor seals hauled out on the higher ledges.

Driving south again for a little less than a mile, you will come to the parking area for Bean Hollow State Beach. This is a shapely, 300-yard-long, wide and arcuate sand beach inside a break in the offshore rock platform, presumably the result of a small valley having been cut through it by the stream now called Arroyo de los Frijoles during a time of lowered sea level. The big waves come right through this little valley and arc onto the beach as large, curving breakers. This beach no doubt receives its sand from the adjacent stream during times of flood. This is a beautiful sand beach that is unusually steep for reasons we could not figure out during our visit there.

Continuing south along Highway 1 for a little over three miles, you will come to the turn to the Pigeon Point Lighthouse, a worthwhile stop. Along the way you will be passing wide agricultural fields on the flat, elevated marine terrace that lies adjacent to most of the shoreline in this area. The Pigeon Point Lighthouse is tall and stately, being the subject of almost as many photographs as the Golden Gate. As the oblique aerial photograph in Figure 127 shows, the lighthouse is located on a projecting point in the resistant rocks of the Pigeon Point Formation.

Once you reach the parking area, if you take the side trail to the south you can get good views of the sea stacks

FIGURE 127. Oblique aerial view of Pigeon Point Lighthouse (San Mateo County; Zone E, Figure 118). The former lighthouse keeper's housing is operated as a hostel. Photograph taken on 1 October, 2008, courtesy of the California Coastal Records Project, Kenneth and Gabrielle Adelman.

offshore. Also notice that there are some massive, dark-colored conglomerates exposed in the scarp with clasts as big as your fist. These are the largest clasts we have seen in the conglomerates in this 70-80 million year old deep-water deposit that was presumably laid down in a massive submarine canyon. More debris flows?

Continuing on south you will reach the Gazos Creek access at about 2.3 miles. A short walk from the parking lot allows passage out onto a 1,500-yard-long sand beach, fed by Gazos Creek, the stream with the largest watershed in this Zone. The stream comes out about a third of the way south along a 1.7 mile-long flattened out crenulate bay, initiated in the north at the southern end of the extended intertidal rock platform discussed earlier, and ending at the rocky extension of Franklin Point. There are two of these flattened crenulate bays in this Zone, the other one being to the north of Point Año Nuevo. Another one extends to the south into the next Subcompartment (Zone F, Figure 118) away from the south end of Point Año Nuevo.

Walking from this access point is an enjoyable way to reach Franklin Point, a walk of about a mile. This is a pleasant, wide sandy beach unless you happen to go there shortly after a major storm. About half the way down, you have to navigate over a narrow wave-cut, rock platform. Should you prefer a shorter walk to Franklin Point, get back on Highway 1 and drive south for 0.8 miles and pull over on the right for another access trail to the Point. This trail is 0.6 miles long, more than half of it being through an open field. After leaving the field, you will be walking along the front of some vegetated coastal dunes, a remote natural area. At the very end of the walk, the rocky headland provides some lengthy views in both directions along the coast.

The last place we recommend you visit in this Zone is the Año Nuevo State Reserve. According to the California Coastal Access Guide, this Reserve has the following appealing characteristics:

1) Probably most notably, it is the protected breeding ground for northern elephant seals.
2) It contains 4,000 acres, covering a wide range of coastal and terrestrial habitats.
3) It is the site of one of the "remaining active dune fields."
4) California gray whale sightings are possible in January and March.
5) Excellent spring and fall migratory bird watching.

A three mile hiking trail leads from the parking area to Point Año Nuevo. One of the first things you come to is a beautiful little pond with a big stand of wetland plants all around it. This is a beautiful place with views of the background mountains, the pond, and rip currents off the beach along the way. That little freshwater pond is full of activity, because birds such as brown pelicans and Herrmann's gulls land on the pond and splash around washing the salt off their feathers.

Eventually, the trail out goes over one fair-sized, barren sand dune but, as shown by the photographs in Figure 125, most of the dunes are vegetated. There are some large, blow-out type dunes (Figure 76 and 125A).

When we visited there on 16 August 2008, the sea stacks offshore of the point were covered with sea lions, cormorants, and pelicans. It looked like every square foot was taken. We saw numerous **elephant seals** lounging on the shore and frolicking in the water; this is the place to see these marine mammals.

According to historical record, Point Año Nuevo was named for the new year on 3 January 1603 by the Spanish explorer Sebastian Vizcaino. An important element missing from his log was Año Nuevo Island, a barren rocky island about 400 yards long and 75 yards wide that now exists 2,300 feet offshore of the southern tip of Point Año Nuevo. Several abandoned houses are still present on the island (Figure 126). The fascinating evolution of this headland was explained in detail by Griggs et al. (2005). The evolutionary sequence follows:

1) 1603 – A broad dune field extended all the way across the area now occupied by water between the island and the point.
2) Early 1700s – Surface faults and/or subsidence during a large earthquake changed conditions sufficiently enough to initiate erosion of a channel between the present-day Point and the island.
3) Early 1800s – The channel was over 1,000 feet wide and a low-tide sand spit, over which wagons could be driven, still connected the island to the Point.
4) 1900 – The channel was over 1,800 feet wide and large portions of the Point had been eroded away.
5) 2003 – Active dune formation had essentially stopped along the northern shore, and the dunes that were left had become mostly vegetated, as shown in Figure 125. Erosion of the shoreline is occurring in several places.

This is one of the better documented cases on the California Coast of dramatic shoreline erosion, sand loss, and the evolution (or should we say death) of dune fields. But, don't worry, the elephant seals still have many years left to call this place home. That is, unless sea level rises abruptly or another fault causes significant subsidence. Come to think of it, maybe you should go out to visit this place as soon as possible, just in case.

SANTA CRUZ UPLANDS (AÑO NUEVO CREEK TO CAPITOLA)

Introduction

This 26 miles of the shoreline (Zone F, Figure 118) form a south to east arc that terminates as the northern boundary of Monterey Bay. Except for a one mile extension of the Point Año Nuevo raised marine terrace at the northern end of the Zone, the northern one third (8 miles) of the shoreline is backed generally by high wave-cut scarps in the Santa Cruz Mudstone. The remaining length of the shoreline from the mouth of Scott Creek south is backed by at least one, but up to five, raised marine terraces, one of the most striking examples of this phenomenon in the world. See Figure 48 for the model of how such terraces form.

Geological Framework

Except for rocks of the Purisima Formation in the wave-cut rock cliff at Santa Cruz, the lower portions of the cliffs that border the bulk of this shoreline are composed of the Santa Cruz Mudstone of Late Miocene age (around 10 million years ago). This formation was described by Clark, Brabb, and Addicott (1979) as a thin- to medium-bedded, pale yellow-brown siliceous mudstone that locally grades to siltstone. They further concluded that the fine-grained siliceous nature of the formation implies that it is the result of organic deposition, presumably containing abundant radiolaria, in a quiet-water, marine basin. Many of the bluffs are backed by raised marine terraces, and sediments deposited on these terraces, several feet thick in some cases, commonly overlie the mudstone. Along most of the scarps, the mudstone layers are near horizontal.

From the geomorphological perspective, the usually near-vertical rock cliffs in the Santa Cruz Mudstone have been eroded into a smorgasbord of artistic features, including untold numbers of sea caves, arcuate scallops into the cliffs, numerous sea stacks (see Figure 42A), and wave-cut rock platforms of various shapes and sizes. The longevity of these features, as well as the abundance of vertical slopes along the scarps, is no doubt the result of the siliceous cement that holds the fine-grained material together.

As shown on Figure 118, a branch of the San Gregorio Fault intersects the coast a short distance north of Waddell Creek. According to the State geological map, this fault separates the middle/lower Pliocene rocks to the north from middle Miocene strata to the south. There are also some sand and gravel Late Miocene deposits that crop out in this area and north and east of Santa Cruz. *These appear to represent sand and gravel deposited in a tide-dominated seaway into the central valley area* (Ed Clifton, pers. comm.).

Weber and Allwardt (2001) published a fieldtrip guidebook that gives details on the topographic levels and rates of uplift of the world-class, raised marine terraces that exist in this area. They also described Late Miocene bituminous sandstone outcrops up Majors Creek that have been mined for asphaltic paving material.

Beaches and Beach Erosion

There are eighteen sand beaches in this Zone, none of which are particularly long and most of which are located off stream mouths. Two of the more notable ones are located off the mouths of Scott and Waddell Creeks. Almost three miles of segmented sand beach fed by at least three streams, most notably the San Lorenzo River, are located along the crenulate bay known as Santa Cruz Harbor. Gravel beaches are also present along the base of some of the wave-cut rock cliffs.

Beach erosion along this section of coast borders on the sensational, partly because it has been studied in such intricate detail by Dr. Gary Griggs and his associates at the UC Santa Cruz Marine Laboratory. The shoreline from the mouth of Año Nuevo Creek to the western border of Santa Cruz was classified as *Caution: moderate risk* by Griggs and Patsch (2005). Areas with erosion serious enough to require the addition of riprap include: 1) along the high eroding scarp at Waddell Bluff; and 2) at the mouth of Scott Creek, where a combination of high creek flow and wave erosion during exceptionally high tides led to armoring of the road fill along the backbeach. All of the shoreline of Santa Cruz and Capitola, except for the sand beach a few hundred yards long off the mouth of the San Lorenzo River, was classified as *Hazard: high risk* by Griggs and Patsch (2005). Over half of this shoreline has been armored with riprap and other means of armoring. More of the details of the shoreline erosion in this area are presented under the discussion of the individual coastal sites.

General Ecology

Theodore J. Hoover Natural Preserve near the mouth of Waddell Creek is an undisturbed and rare freshwater marsh along the coast (Figure 128). It is home to many threatened or endangered species including California black rail, brown pelican, red-legged frog, San Francisco garter snake, western snowy plover, tidewater goby, steelhead trout, and coho salmon. It is also a good

FIGURE 128. Waddell Creek marsh (Santa Cruz County; Zone F, Figure 118) in the Theodore J. Hoover Natural Preserve, a rare, undisturbed freshwater marsh. Photograph taken on 16 August 2008.

birding spot.

From 2000-2005, Scott Creek Beach had up to 106 wintering western snowy plovers and 12 breeding birds in the summer. Laguna Creek Beach, Wilder Ranch State Park, and Seabright Beach also have snowy plovers in both summer and winter. The Kingfisher Flat Fish Hatchery is located on a Scott Creek tributary to produce juvenile steelhead and coho salmon for release to streams where they were historically present.

There are six seabird nesting colonies in this area where black oystercatcher, pelagic and Brandt's cormorants, pigeon guillemot, and western gull nest between February-September. Numerous seabirds, diving birds, and gulls winter in the coastal waters.

Several hundred harbor seals haul out on offshore rocks between Waddell and Scott Creeks. The kelp beds from Scott Creek to Santa Cruz are important sea otter habitat, though it can be difficult to see the otters among the kelp floats. They eat mostly abalones, sea urchins, crabs, and clams; sea otters consume about 25% of their body weight daily. Pupping occurs January-March; they give birth in the water rather than on land like other otters. Santa Cruz Harbor is a good location for going on a whale-watching cruise into the Monterey Bay National Marine Sanctuary. The sighting reports most often list humpback whale (March-November), followed by gray whale (November-May), blue whale (June-November), and killer whale (year-round). They also list sea lion, sea otter, dolphin, and porpoise. If you do not have the time or inclination to take a cruise, you still have a chance to see these animals from the many shore viewing points.

Places to Visit

Heading south on Highway 1 from the Año Nuevo State Preserve you will be driving across the flat topography of the raised marine terrace that makes up most of the surface of the Preserve. After going one mile or so, the flat ends and the road goes up an incline by a steep wave-cut rock scarp called Waddell Bluff. The sedimentary rocks that compose the bluff are the flat-lying layers of the Santa Cruz Mudstone, which was described in the earlier discussion of the geological framework of this Zone.

A detailed analysis of the activities of the California Department of Transportation in dealing with the erosion of Waddell Bluff, which reaches heights of almost 150 feet, was part of the discussion of rock cliffs in Section I. Briefly, the talus that accumulates at the bottom of the cliff on the east side of the road is periodically removed to the beach on the other side of the road so that it can be dispersed by the waves. Much of the high-tide area along the beach at the base of the bluff is armored with riprap to protect the road from eventual erosion.

About one mile beyond the north end of the high rock scarp (still going south), there is a parking area on the right that allows access to Waddell Creek Beach. This wide beach is a famous kite surfing location. It is to wind-surfers what Mavericks (near Pillar Point at the head of Half Moon Bay) is to body surfers – a very popular spot.

Of course, when Waddell Creek is in flood, it cuts across the beach to be closed off only after the flood has passed. The stream was completely closed off when we were there on 15 August 2008, but in an aerial photograph taken on 1 October 2008 by the California Coastal Records Project (Kenneth & Gabrielle Adelman), a large pond seaward of the bridge was closed off by a narrow sand bar. Therefore, the stream had been flowing through the beach and into the sea not long before the picture was taken. This wide beach extends to the south for about 900 yards, at which point a narrow sand and gravel beach flanked by a wave-cut rock platform begins.

Across the road, the Theodore J. Hoover Natural Preserve harbors the picturesque wetland shown in Figure 128.

Back on Highway 1 and driving south, it is 1.4 miles to the parking area for Greyhound Rock, a north/south oriented sea stack composed of barren rock that measures 435 by 100 feet. It is commonly covered by sea birds. As shown by the vertical image in Figure 129A, there is a classically shaped crenulate bay located to the south of the rock. The bay has a straight-line length of

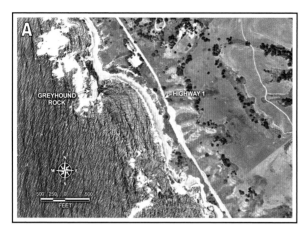

FIGURE 129. Greyhound Rock area (Santa Cruz County; Zone F, Figure 118). (A) Image of the Greyhound Rock crenulate bay. NAIP 2005 imagery courtesy of the California Spatial Information Library (CaSIL). (B) Oblique aerial view of Greyhound Rock that shows a subtle tombolo in the lee of the rock. Photograph taken on 1 October 2008, courtesy of the California Coastal Records Project, Kenneth and Gabrielle Adelman.

760 yards. Also, a subtle **tombolo** has developed in the lee of the rock, as is clearly shown by the oblique aerial photograph in Figure 129B. A steep path leads down to this appealing, medium-grained sand beach, another spot favored by surfers. Greyhound Rock is also a public fishing access.

South of Greyhound Rock, Highway 1 runs for 3.4 miles in a straight line fairly close to the wave-cut rock cliff, which has spectacular scallops with sea caves and sea stacks. There are some gravel beaches and wave-cut rock platforms along the base of these moderately high scarps in the Santa Cruz Mudstone, which also is exposed in a number of road cuts along this section of the road. This uniform stretch ends at the mouth of Scott Creek, where it is possible to pull over into a roadside parking area. There have been some erosion problems there, so riprap has been added to the side of the road where it crosses

the Scott Creek valley. A beautiful wetland, pictured in Figure 27B, is present along the valley on the north side of the road.

A wide sand beach 1,100 yards long extends to the south of Scott Creek, the mouth of which had been deflected to the north before it was closed off in August 2008. Review of the historical oblique aerial photographs of the California Coastal Records Project revealed that the mouth of the stream was deflected to the south on 30 September 2002; therefore, the stream mouth orientation probably responds simply to the direction the waves were coming from at the time immediately before the flood level had receded to the point that the waves could close the stream mouth again. Some beaches like this stay open during the rainy winter season if the rains have been strong enough to get the streams into flood stage. When the streams diminish significantly in flow during the drier, summer months, the big waves can close off the stream mouth by building a spit across it.

Scott and Waddell Creeks have both carried enough sand to the shore to create the largest sand beaches north of Santa Cruz. Both streams have large watersheds that extend into the Big Basin Redwoods State Park high up in the mountains. According to the California Coastal Access Guide, this, the oldest California State Park, was established in 1902. It has over 18,000 acres of old growth redwoods, 80 miles of hiking trails, numerous waterfalls, and abundant wildlife. A side trip to visit this Park would certainly seem to be a good idea.

From the Scott Creek area heading south on Highway 1, you will be driving over raised marine terraces all the way to Santa Cruz. A little over a mile along you will reach the turn to Davenport Landing Beach, a pocket sand beach a few hundred yards north of the town of Davenport. This pocket beach, which is only 350 yards long, probably derives most of its sand from three small streams that have cut a shallow valley through the raised marine terrace.

Before you reach the main part of Davenport, you will pass by a large cement plant on the east side of the road that has been producing cement for many decades, including supplying materials used in World War II, in the construction Golden Gate Bridge, and other important projects over the years. The rocks of an ancient pre-Franciscan formation of Paleozoic age (probably Mississippian or Pennsylvanian – Table 1) called Gavilan Limestone, derived from mines up in the hills northeast of the plant and north of Santa Cruz, are the source materials for the cement. Although they were originally named a limestone, in actuality they are composed of **marble** (metamorphosed limestone). These rocks, which are intruded in places by the Cretaceous-age granodiorites of the Salinian Block, are also a part of that suite of rocks riding north on the Pacific Plate. Each time we drove by the plant during our surveys of the coast, co-author Hayes was always puzzled as to what rocks were used to make the cement, considering the fact that the omnipresent 8,000+ feet thick Santa Cruz Mudstone contains a lot of SiO_2 and not so much $CaCO_3$. The marble is almost pure $CaCO_3$. He did some research before we wrote this, and now the puzzle is solved.

There is public access to Davenport Beach, located opposite of the historic town. You park in a dirt parking area, walk across the railroad and take one of a number of trails along the top of the bluff [good for whale watching January through March (California Coastal Commission, 2003)]. You can also work your way down to this 400-yard-long pocket sand beach, which will allow you to observe sea caves and a conspicuous sea stack up close. This sea stack is called Sentinel Rock and it has been a landmark in this area for many years. Griggs and Ross (2006), in their book titled *Santa Cruz Coast*, showed pictures 50 years apart of this beach, which showed almost no recognizable change during that time. The reason they gave for this surprising lack of change was that *the cliffs and Sentinel Rock consist of a geological formation known as the Santa Cruz Mudstone, a rock that is generally quite resistant to erosion.* Presumably, it is resistant to erosion because of the strong silica cement that holds it together.

Back on the road traveling south, you by all means should stop at the pullover 0.4 miles south of the center of the town of Davenport for an inspection of the spectacular rectangular embayment illustrated in Figures 42A and 43. This is a great place to see a variety of erosional features in the wave-cut rock cliffs, including sea caves and a large nearshore sea stack (Figure 42A). Also, as illustrated by the sketch map in Figure 43, and discussed at some length in the earlier presentation on rock cliffs in Section I, the effects of the joint patterns in the rock cliff are very striking and account for the straight sides of the embayment, the location of the beach, and other features.

In another half mile to the south, you will reach a pullover from which you can access Bonny Doon Beach, a wide, 350-yard-long sand beach in an indention in the wave-cut rock cliff. You might be interested to know that this beach is currently listed as *clothing optional.* The pullover is located just opposite to where Bonny Doon Road joins Highway 1. At this site, you can also walk out to the edge of the top of the scarp for great views up and down the shoreline. If you look to the south, you can see a wave-cut rock platform up to 70 feet wide in places that has been uplifted several feet.

While driving south from there on to Santa Cruz, you will be driving over the lowest of the famous raised marine terraces in this district. Two examples of raised marine terraces are illustrated in Figure 47, and their mode of formation is explained by the model in Figure 48. As you ride along, the youngest one, with its extensive flat agricultural fields, will be easy to recognize; however, if you look up the side of the hills to the left, you may also be able to pick out some of the older ones.

Yellowbank Beach is about another 0.6 miles south of the pullover for Bonny Doon Beach. This beach, composed of a short and a long segment, extends a little over 500 yards. This is another place to get close-up views of sea caves in the wave-cut rock cliffs in the Santa Cruz Mudstone.

Heading south for another 0.6 miles you come to the parking area for Laguna Creek Beach. This 500-yard-long pocket sand beach has a small wetland on its landward side between the beach and the road. This is a pleasant natural setting that is no doubt a great birding spot during migrations.

One mile further southeast along Highway 1, you come to the turn to the Red, White and Blue Beach, a pretty big one for this area, 1,000 yards long (at low tide anyway). It is private and clothes are optional (it was apparently closed in 2007).

It is 1.1 miles from the exit road at the Red, White and Blue Beach to the first parking area for Wilder Ranch State Park. There are extensive trails along the crest of the bluff in this Park, from which you can view a variety of sea caves, sea stacks, and so on, as well as a plethora of seabirds and shorebirds. The trails lead to three very serene-looking sand beaches.

If you continue east on Highway 1 over the youngest of the flat, raised marine terraces for about three more miles, you will be in the city of Santa Cruz. The first place we recommend you visit in the city is Natural Bridges State Beach. To get there, soon after you enter the city turn right onto Western Street (there is a sign for the State Beach). You will next come to Mission Street, which you take for one block, and then take a left and you are on Natural Bridges Drive that takes you for one block right to the parking area for the Natural Bridges State Beach. As recently as 1978, there were two **sea arches** in the ridge of rock that projects offshore from this spot. A photograph we took of the natural bridge in 1978 (Figure 45A), shows the most landward of the two arches (the offshore one is somewhat obscured by a shadow). The inner arch had long since collapsed before we took a photograph of it in August 2008 (Figure 45B).

When you finish there, we recommend that you drive to the east along Westcliff Drive for 1.8 miles until you get to Lighthouse Point. All along this road, you will be driving on top of the wave-cut rock cliff, a drive definitely worth doing. A variety of erosional features are present along this route, including several sea stacks. Figure 42B is a photograph of one of the sea stacks in this area covered with brown pelicans and cormorants. If possible, it would be a good thing if you had a copy of *Santa Cruz Coast* by Griggs and Ross (2006) with you during this drive, because this book has a plethora of historic pictures of the Santa Cruz shoreline, along with intriguing tidbits about the coastal geology and the human history of this cliff-bound coast.

Somewhere along this drive the geology changes, with the Purisima Formation of Pliocene age (3-2 million years of age), which lies unconformably on the Santa Cruz Mudstone, forming the cliffs the rest of the way to Capitola. We have encountered this formation in this discussion several times before, most notably along the shoreline of Drakes Bay (Drake's "White Cliffs of Dover;" Figure 107). The Purisima Formation at this location was described by Cummings, Touring, and Brabb (1962) as a tan-yellow to light brown and gray, fine-grained sandstone and siltstone. The evidence would seem to indicate that it is more susceptible to erosion than the Santa Cruz Mudstone.

Leaving the lighthouse and going east, you will then encounter the sizeable crenulate bay know as Santa Cruz Harbor. This bay has a sand beach shoreline 2.8 miles long that is interrupted by: a) a 350-foot-long and 50-foot-wide sliver of the Purisima Formation that projects straight across the beach immediately to the east of the mouth of the San Lorenzo River; b) a set of jetties that provide an entrance into the Santa Cruz Small Craft Harbor (Figure 82A); and c) a short rock headland. These beaches are frequented by large numbers of people.

Another mile and a half to the northeast you will encounter the last sand beach in this Zone, the 500-yard-long Capitola Beach, another extremely popular bathing beach. This beach has had some erosion problems in the past that were described as follows by Griggs and Ross (2006). *In the late 1960s, the beach at Capitola narrowed as sand moving down coast was trapped by the construction of the jetties at the Santa Cruz Harbor (1962-1964). In an effort to regain the beach, which Capitola relied on for summer tourists, a 250-foot-long groin was constructed in 1969 and about 2,000 truck loads of local quarry sand were imported.* The beach is now 220 feet wide and the groin is filled to its end (Figure 82B).

The shoreline 1,000 yards east of Capitola Beach is backed by a modest-sized, wave-cut rock cliff in the Purisima Formation. This formation in this area is said to contain an abundance of Pliocene-age marine fossils that range from common mollusks to seal bones and even whale bones (Geology News Blog, 2008). The beach itself contains an abundance of cobble-sized blocks of rock that have fallen off the rock scarp, mostly along the high-tide line.

17 COMPARTMENT 5: MONTEREY BAY AND VICINITY

INTRODUCTION

The western boundary of Monterey Bay has the largest stretch of sand beaches on the Central California Coast (Figure 130). The central, dividing point of these long beaches is the head of the massive Monterey Submarine Canyon, which starts a short distance off the Moss Landing jetties. The biologically rich Elkhorn Slough is located immediately to the east of the jetties. The sand on the beaches of the south limb is fed by the Salinas River, made famous in the writings of John Steinbeck. A large field of sand dunes, created by the predominant northwest winds, is located landward of the southern shoreline. The Monterey Peninsula contains beautiful, white sand beaches, but the dominant feature is the massive outcrops around the shoreline of the Porphyritic Granodiorite of Monterey (also called the Hobnail Granite in some sources) of Cretaceous age, a part of the Salinian Block. The southern margin of the Compartment is along Carmel Bay, where the fine-grained, white sand of Carmel City Beach stands out in strong contrast to the granules that compose the steep beachface of Monastery Beach.

For purposes of discussion, we have subdivided this Compartment into the five Zones mapped in Figure 130:

A) **The North Beaches (Capitola to Moss Landing Inlet)** form the northern flank of the sand-beach shoreline of eastern Monterey Bay. This area has eleven public-access beaches, including six state beaches.

B) **Moss Landing and Elkhorn Slough** contain one of the more extensive and biologically rich wetlands along the Central California Coast. A jettied inlet, constructed in 1947, is located a short distance from the head of the Monterey Submarine Canyon.

C) **The South Beaches (Moss Landing to Monterey Harbor)** are the southern mirror image of the North Beaches. The Salinas River, which is subject to dramatic seasonal changes, flooding some winters and dry most summers, is the primary source for the sand on the beaches.

D) **Monterey Headland (Monterey Harbor to Pescadero Point)** is cored by the resistant Porphyritic Granodiorite of Monterey. The best and most readily examined exposures of the Mesozoic granitic rocks of the Salinian Block that are moving north on the Pacific Plate occur on this peninsula. A variety of exquisite shoreline types, from snow-white fine-grained sand beaches to armored boulder beaches abound around this headland.

E) **Carmel Bay (Pescadero Point to San Jose Creek)** consists mostly of long sand beaches that exhibit a variety of grain size and composition. Monastery Beach, located near the head of the Carmel Submarine Canyon, a stable, steep beach, composed of granules and coarse sand, is of special interest.

THE NORTH BEACHES (CAPITOLA TO MOSS LANDING INLET)

Introduction

Sand beaches compose the entire intertidal shoreline (Zone A, Figure 130), with armoring being abundant along the backbeach in the shadow zone of the crenulate bay (north arm of the sand beach at the head of Monterey Bay), most notably along the 6.4 miles of developed shoreline from the parking area at New Brighton State Beach to the southern parking area for Manresa State Beach. Another mile and a half of riprap has been installed to protect structures built on the sand spit to the north of the mouth of the Pajaro River. Public access to these wide sand beaches is possible at numerous locations.

Geological Framework

Approximately the northern half of the shoreline, from New Brighton State Beach to La Selva Beach, is backed by a near vertical bluff eroded in a 100-foot-high

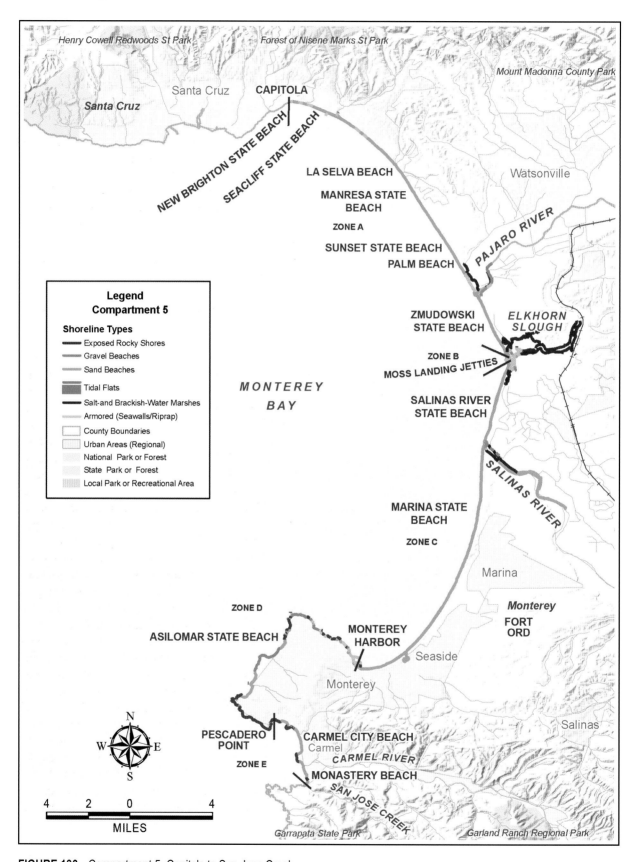

FIGURE 130. Compartment 5, Capitola to San Jose Creek.

marine terrace underlain by the *moderately resistant sandstones and siltstones of the Purisima Formation* (Griggs and Patsch, 2005), the same formation that underlies the bluffs in Capitola. Between La Selva Beach and the beginning of the sand spit at the mouth of the Pajaro River, the bluff is composed of much less resistant dune sand of Pleistocene age (Aromas Sand). This raised marine terrace is highly populated in the north and heavily farmed in the south.

The other major geomorphological feature of significance in this area is the wide valley and flood plain of the Pajaro River, which has an average width of around 2 miles in its lower reaches. This flat area, composed of very rich soil that is intensively farmed, was created by the infilling of a valley carved by the ancestral Pajaro River during the last major fall in sea level. As sea level rose and eventually stabilized around 5,000 years ago, the sediments carried by the stream gradually filled the valley to its present level as a result of the rising of the lowest level to which the river flowed (know as its **base level** which, in this case, was sea level). This continual rise of the sea, or base level, slowed the downcutting of the valley by the stream to the point that sediment that under earlier conditions would have been deposited far out into the ocean was, instead, deposited in the valley itself, creating the wide flood plain.

A question of primary importance with regard to the coastal geomorphology of this Compartment is why is there such a big shoreline embayment here? As it turns out, the Monterey Lowlands, which back most of Monterey Bay, occupy an area that has subsided (sunk) during most of the last half million years or so (based on observations made by Bill Dupré when he was with the U.S. Geological Survey; e.g., Dupré, 1990). This is in contrast to the area north of Santa Cruz just described, which contains an abundance of raised marine terraces, and the presently elevating Big Sur Country to the south. Ed Clifton (pers. comm.) remarked that he thinks this downwarp is *probably the primary reason for the existence of Monterey Bay.* However, he also noted that he doesn't think the rate of downwarp was rapid enough to create the embayment, *but it provided the site of continued deposition of young, readily eroded sediments, in contrast to the older, more resistant rocks that front the coast north and south of the Bay.* Therefore, his theory is that it was the continued assault of the waves on these less resistant materials that created the Bay.

Beaches and Beach Erosion

During the fifteen year project by the U.S. Geological Survey to measure the profiles of sand beaches on the

eastern shoreline of Monterey Bay, summarized by Dingler and Reiss (2002) and discussed at some length in Section I, four beaches in this Zone were surveyed – Seacliff State Beach, Manresa State Beach, Sunset State Beach, and Moss Landing State Beach.

At the start of that project in early 1983, these four beaches had been severely eroded during the strong *El Niño* season of 1982-83. Figure 55, which plots beach width versus the dates of the surveys, gives the results for the profiles at Sunset State Beach. As noted earlier, the curve in Figure 55 aptly demonstrates the fact that as soon as the major storms of 1982-83 ended, the beach at Sunset State Beach began to rebuild to significant widths, an average of about 164 feet, and it stayed fairly wide, with some sizable fluctuations, until the next major *El Niño* season of 1997-98. The plots for Seacliff and Manresa State Beaches showed a very similar trend to the one for Sunset State Beach. The sand beach at Moss Landing State Beach showed somewhat different results, presumably because of wave refraction over the adjacent submarine canyon causing waves to approach the beach at an angle. Accordingly, rhythmic topography (illustrated in Figure 37 – Upper Right) commonly occurs on that beach, apparently more so than on the other beaches where beach cusps are more usual as a result of the waves more typically coming straight on.

A study by Moore and Griggs (2002) showed that the bluff in this section of the shore retreats at rates of 3-6 inches per year, which is to say that the bluffs are not eroding much. The photograph of the bluff at Sunset State Beach (Figure 26A) shows that the bluff slope is vegetated, further evidence that it is pretty stable.

Sand is abundant on these beaches and, due to their sheltered position relative to the dominant northwest waves, the beach would normally be wide enough to protect the bluff from waves approaching from that direction. As a result, presumably, numerous houses have been built along the base of the bluff in the more highly developed areas in the northern part of the Zone. As discussed in Section I, during *El Niños*, the waves usually approach the shore from the west and southwest, exposing the shadow zone of the crenulate bay (northern end of this Zone) to extreme waves causing erosion problems there. Griggs and Patsch (2005) rank most of the northern area down to La Selva Beach as *Hazard: high risk* with regard to beach erosion, with much of the shoreline being "protected" by sea walls and riprap. Also, 1.5 miles of the sand spit on the north side of the Pajaro River mouth has the same ranking.

Griggs and Patsch (2005) tell the gruesome history of erosion of these beaches in gory detail, once again causing us to remind you to read *Living With the*

Changing California Coast to learn more specifics. One of the problem areas discussed in the book, Seacliff State Beach, is briefly summarized as follows:

1) Ten times in 58 years, about every 6 years on the average, seawalls and bulkheads on the beach have been destroyed or heavily damaged.

2) After extensive damage in 1939 and 1940, the timber seawall was rebuilt.

3) Storms in the winter of 1941 destroyed it again.

4) After more damage in 1978 and again in 1980, a new bulkhead 2,600 feet long was built.

5) *In late January 1983, within two months of its completion and dedication, El Niño storm waves combined with very high tides overtopped the bulkhead. Large logs battered and loosened the timbers and pilings, and about 700 feet of the bulkhead were destroyed.*

General Ecology

There are many shorebirds along the extensive sand beaches in this Zone. Western snowy plovers nest and winter along Manresa, Sunset, and Zmudowski State Beaches. The Watsonville Slough at the mouth of the Pajaro River hosts many birds, including wading birds, ducks, shorebirds, phalaropes, terns, gulls, and raptors. The high, vegetated dunes along this section of shoreline are very impressive; in many areas, park managers have made a concerted effort to remove exotic species (ice plant, European beach grass, and pampas grass are considered to be among the most widespread and of greatest ecological concern in coastal habitats) and replant with native grasses, such as native beach bur, sand verbena, lupine, and beach sagewort, which have much greater habitat value for songbirds, amphibians, reptiles, mammals, and butterflies than the exotic plants.

From June-September, sooty shearwaters may be seen feeding offshore in large numbers; hundreds of thousands of them come to feed on fish, squid, and krill during the upwelling season, making the sooty shearwater the most abundant seabird off the coast of California from May to September. A group of researchers (Adams et al., 2008) have been using satellite telemetry to track the annual migration of sooty shearwaters in California to their nesting sites in the southern hemisphere off New Zealand and Chile. They have found that the sooties traveling from California average about 15 miles per hour and can arrive at nesting sites in New Zealand nearly 6,400 miles away in about 17 days. Before these studies were conducted, the migration path was only hypothesized based on sightings from ships at sea. Now, researchers can track the hour-to-hour position of individual tagged birds, and couple their behavior with

physical oceanographic patterns to better understand how they are inter-related. Tagged sooties (and lots of other tagging studies of sea turtles, birds, marine mammals, etc.) can be viewed in real time on the web at Seaturtle.org (2009) (you have to register, but it is free).

Places to Visit

Assuming you travel from north to south, the first 6.4 miles of the shoreline are developed, from the parking area at New Brighton State Beach to the southern parking area for Manresa State Beach. If you stop at one of the more northerly beaches, you will see lots of riprap and seawalls, as well as experience the excitement of the sunken tanker at the end of the pier at Seacliff State Beach. The typically wide sand beach along much of this more northerly area is backed by a mostly vegetated sea cliff held up by horizontal rock layers of the Purisima Formation. A railroad track, which approaches the shore at La Selva Beach, runs along the face of the cliff to the south for about 800 yards before turning again landward at the northern Manresa State Beach parking area. This north parking area has an easy access down to the beach, as opposed to the southern parking area, located one mile to the south, which features a fairly extensive stairway for access to the beach. From that access point south to the northern entrance to Sunset State Beach, farm land occupies the surface of the raised marine terrace. Sunset State Beach, pictured in Figure 26A, is a beautiful secluded beach that is definitely worth a visit. Getting there is another story.

If you start your trip to Sunset State Beach by driving out of the northern parking area for Manresa State Beach, you will immediately encounter San Andreas Road, a name that should be easy to remember. Turn right on this road and go along it for about three miles, at which point you turn southwest on Sunset Beach Road. This intensely farmed, relatively flat land you will drive over is the surface of the raised marine terrace. The entrance to the State Beach is about three quarters of a mile from the turn.

As you can tell by the photograph in Figure 26A, this is a beautiful, wide medium-grained sand beach flanked by a steep, vegetated bluff. There is a walkway that slants down to the beach. Within the confines of the State Beach, there is a paved road that runs 1.3 miles to the south to another parking area near the southern boundary. This is apparently a popular fishing area. It is precisely at that point where the main bluff recedes landward and the new dunes and sand spit that build across the mouth of the Pajaro River begin to flare away from it. Within 0.8 miles, the abandoned bluff turns to the northeast becoming the valley wall of the Pajaro River Valley, with its rich, extremely flat farm lands.

Actually, there are seven miles of public access to beachfront in this area, if you include the Palm Beach area, which is located on the sand spit north of the Pajaro River mouth. The end of the spit is highly developed, with the structures on the last mile or so being protected with riprap. The river mouth, which was closed off by the sand spit on 23 September 2008, as determined from the oblique aerial photography of Kenneth & Gabrielle Adelman, is open after heavy rains.

If you should desire to visit the Palm Beach area by driving, go back out to San Andreas Road and go south. At 1.4 miles, you will pass down the valley wall and onto the flood plain of the Pajaro River. At about 0.6 more miles turn right on Beach Road and drive southwest. After another mile and a half you will be at Palm Beach.

The Pajaro River has a wide flood plain with phenomenal farm land, where they were harvesting brussels sprouts during our visit in December 2008. Big corporate farms abound across the flood plain of the Pajaro River, as well as the raised marine terrace in places, with multitudes of farm laborers in the fields.

To leave Palm Beach and continue your trip to the south, go back out on Beach Road for about 3 miles. When you get to within about 200 yards of Highway 1, turn right for one block at which point you can enter Highway 1 heading south. The city of Watsonville, the "strawberry capital of the world," is located not far from the place where you turned onto Highway 1.

A continuous sand beach extends for 2.5 miles from the mouth of the Pajaro River to the north jetty at the Moss Landing Inlet. Two State Beaches, Zmudowski and Moss Landing, are located in this section of the shore. The high-tide line in both of these State Beaches is backed by well-developed, vegetated modern sand dunes.

To get to Zmudowski State Beach, take Stuve Road off Highway 1 and turn on Giberson Road, which is laid out in a rectangular pattern. The parking area for Zmudowski Beach is located at the south end of an arcuate slough, that is presumably an old abandoned channel of the Pajaro River. The mouth of the Pajaro River was positioned about 750 yards north of the parking area in December 2008. When we did the protection strategy (from oil spills) for this tidal inlet in 1992 (Hayes and Montello, 1993b), the aerial photograph we used, which was dated 7 September 1982, showed the inlet to be about 1,000 feet north of where it was in December 2008, because the river had recently cut directly across the spit during a major flood. When we did our field survey of the site on 8 November 1992, the river mouth was almost

completely closed off, being washed over a little at high tide, and the channel was about 500 feet south of where it was in September 1982. This demonstrates the general pattern for tidal inlets on long sandy beaches exposed to a northwest fetch. That is, they are commonly closed off by southerly migrating sand spits during the dry season, as discussed in Section I.

This is an isolated area with beautiful vegetated dunes, and a popular fishing and birding spot as well. There is no high bluff there because the land has been eroded by migrating river channels forming a low flood plain. There are a number of abandoned sloughs nearby that have created a very complex coastal geomorphology, indicating that both the Pajaro River and the entrance to Elkhorn Slough have no doubt migrated around some in the past few thousand years (discussed in more detail in the treatment of the following Zone).

Moss Landing State Beach is considerably easier to find than Zmudowski. As you are heading south on Highway 1 and begin to cross the wetlands and channels of the Elkhorn Slough area, shortly after the road turns directly south, you will see a road that leads across the marsh/channel system, eventually reaching the dunes behind the beach. The turn is located 0.4 miles north of where the Highway 1 bridge crosses Elkhorn Slough. To the north of where you first reach the beach, there are some linear vegetated sand dunes created by the prevailing northwest winds, the best place to see these types of dunes in this Zone. There are also a few poorly developed dunes further north in the Zmudowski State Beach area. You will also be able to drive up close to the north jetty of the Moss Landing Inlet, which is a great birding area.

MOSS LANDING AND ELKHORN SLOUGH

Introduction

From an ecological perspective, this area (Zone B on Figure 130) is one of the more sensitive areas on the Central California Coast, being the home of the Elkhorn Slough National Estuarine Research Reserve. A significant harbor/marina system was created along with the construction of the Moss Landing jetties in 1947, which also provided ready access to a water intake for the massive Moss Landing Power Plant. Strong tidal currents that measure up to several knots flow in and out of the jetties as the tide changes. The head of the Monterey Canyon is located a mere few hundred yards or so offshore of the entrance to the jetties. Figure 131 shows a sketch of the jetty system, and Figure 132 shows the environmental sensitivity map of the entire Zone.

Geological Framework

A hypothesis regarding the origin of this over 20 mile wide indention of the shoreline called Monterey Bay was presented in the discussion of the geological framework of the Subcompartment located just to the north of this one (Zone A on Figure 130 - Capitola to Moss Landing). Observations of two retired USGS geologists, Bill Dupré and Ed Clifton, led to the conclusion that the Bay is there for two major reasons: 1) The land around the eastern margin of the Bay has been sinking (subsiding) during the last half million years or so, and relatively friable sediments have been deposited in the region during that time; and 2) These weaker sediments and rocks have been easily eroded, leading to the existence of the Bay. We have no reason to seriously doubt their hypothesis; however, at least one notable factor, the existence of a huge submarine canyon down the center of the Bay, is surely a significant stumbling block that begs for an explanation of its origin as well.

Some geologists think the canyon originated at least

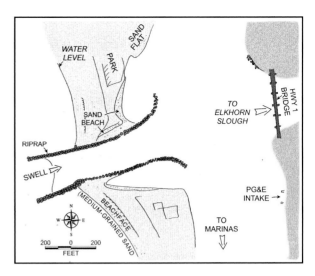

FIGURE 131. Sketch of Moss Landing jetties (Monterey County; Zone B, Figure 130) made on 8 November 1992.

30 million years ago far to the south, possibly off the location of the present city of Santa Barbara (around 150 miles south of where the canyon is today). That being the case, it began its growth even before the present

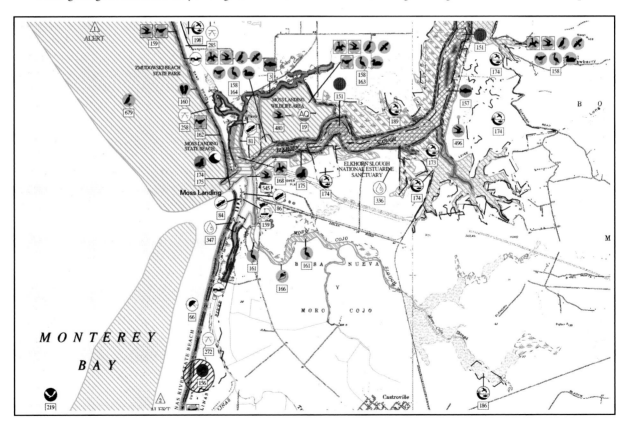

FIGURE 132. Environmental Sensitivity (ESI) Atlas of the Elkhorn Slough and vicinity (Monterey County; Zone B, Figure 130). From NOAA/RPI (2006). This area has a high diversity of birds. Note all of the green icons and the green hatch patterns showing concentration areas. Refer to Figure 85 for shoreline legend and page 132 for biological symbols.

Transverse Ranges were formed.

Inasmuch as the San Andreas Fault is located to the east of the Bay, though only a mere nine miles northeast of the shoreline at Sunset State Beach, the entire system must be riding to the north aboard the Pacific Plate. Therefore, it seems reasonable that the present Bay may have been inherited from further south (at least for some distance). A major river system that may have brought the sediment to the shore that was used by the submarine turbidity currents to carve this massive canyon in the first place is thought by some geologists to have been one that drained California's ancestral Central Valley many millions of years ago (Griggs and Patsch, 2005). That would have been when the canyon was located far to the south. There can be no doubt, however, that the long northwest/southeast trending 60+ mile-long valley of the Salinas River and the much shorter valley of the Pajaro River have contributed much sediment to the task, at least during the lowstands of the major glacial episodes of the past 2.6 million years. The sediments provided by the Salinas and Pajaro Rivers, as well as some minor ones, have also provided the rich soils of the flood plains and marine terraces that compose the low-lying band of land 5-9 miles wide that borders the eastern margin of the Bay from Capitola to Monterey.

In its present configuration, the submarine canyon is the major sink for the approximately 300,000 cubic yards of sediment that moves down the shoreline from the north, as well as possibly some from the south. These sediments are carried far offshore into very deep water by turbidity currents, as discussed in Section I. To repeat, the present thinking is that the canyon was basically cut by turbidity currents like the one illustrated in Figure 21A (e.g., Alt and Hyndman, 2000).

To complicate matters quite a bit, Ed Clifton (pers. comm.) raised the issue of a paradox that geologists are presently struggling with: *The age of Monterey Submarine Canyon and the Cenozoic offset of the San Gregorio Fault, which Carmel Canyon follows as a tributary of the larger canyon.* The trace of the San Gregorio Fault across the floor of Monterey Bay is shown on Figure 81. Weber and Allwardt (2001) *list current estimates of displacement across the San Gregorio Fault as being 19 kilometers* (11.8 miles) *in the last 3 million years. Nearly everyone agrees that the San Gregorio Fault is an active right lateral component of the San Andreas Fault system* (i.e., the land west of the fault is moving to the right of the land on the east side, if you are standing on the east side of the fault looking across the fault to the west). Clifton added that *at the estimated rate of offset (6 mm/yr), there should have been 6 kilometers* (3.7 miles) *of offset across Monterey Canyon in the last million years. One possibility is that the*

Monterey Submarine Canyon is, geologically, a very young feature. This is supported by the work of Bill Normark (another former USGS geologist), *who told me shortly before he passed away that he had found microfossils that indicated an age of 500,000 years or less in the strata that form the walls to the Canyon.*

Clearly, there is some serious disagreement among geologists regarding the timing of the origin of the Monterey Submarine Canyon, with Clifton getting the last word in this discussion - *In any event it is difficult to reconcile an ancient age for the Canyon with the conclusions of some that the Miocene and Pliocene rocks in the Point Reyes area were displaced 90 or so miles north along a fault* (discussed in some detail in our treatment of the Point Reyes area) *that does not displace the canyon where it crosses it.*

Sounds to us like this issue could be a topic for several dissertation studies. In fact, maybe some are already under way that we don't know about.

A detailed summary of the evolution of Elkhorn Slough and surrounding territory was given by Woolfolk (2005) who, using the sea-level data of Lambeck and Chappell (2001), noted that the freshwater streams in the area carved deep valleys when sea level was down as much as 400 feet 18,000 years ago. After sea level rose to near its present level around 5,000 years ago, the central part of the eastern margin of the Bay was composed of a number of estuarine systems containing abundant marshes and tidal flats with the usual complement of major tidal channels. These estuaries were the ancestors of today's Elkhorn, Moro Cojo, Bennett, Tembladero and McClusky Sloughs (located on Figure 133), as well as of the lower reaches of the Salinas and Pajaro Rivers. Now, many of these former estuaries are mere shadows of their former selves, due to land reclamation for agricultural purposes. Thus, these abundant ancient estuarine complexes with their amazing wealth of biological activity are no longer with us, hence the concerted effort to preserve what remains of Elkhorn Slough.

Studies of the historical ecology of Elkhorn Slough over the past 150 years by Van Dyke and Wasson (2005) showed that historically *more than half of the marshlands were diked, and more than two thirds have either degraded or been converted to other habitat types.* Construction of the new jetties in 1947 caused significant erosion that resulted in conversion of much of the original marsh into intertidal flats or open water. They concluded further that *degraded former marshland and unvegetated mudflat are now the dominant habitat types at Elkhorn Slough.* Don't be too discouraged, however, because the Slough is a wonderful place to visit and, hopefully, further efforts will aid in more restoration of the area.

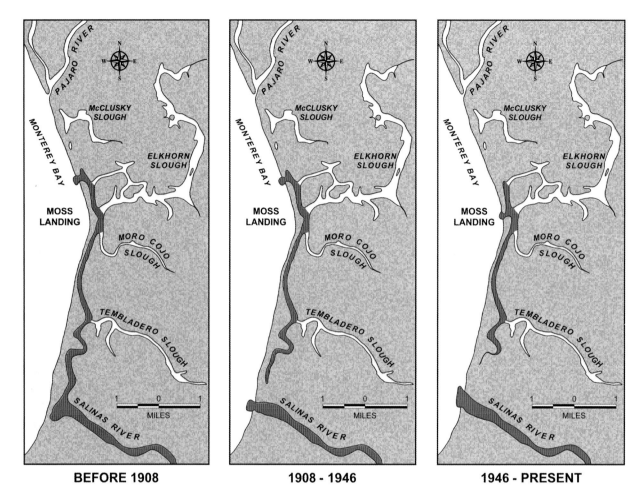

FIGURE 133. Historical changes of the mouths of the Salinas River and Elkhorn Slough (Monterey County; Zones B and C, Figure 130). Modified after Browning (1972). Several of these old channels are still sloughs that contain significant wetlands.

For other details on the history of this unique system refer to the discussions by Woolfolk (2005) and Van Dyke and Wasson (2005).

[NOTE: As further data become available, the exact numbers on times of the maximum lowstand of sea level, as well as how low the minimum was during the different glaciations will no doubt change somewhat, hence the inconsistency of values quoted in this book. This general pattern of change, however, is not in question. Furthermore, the present highstand having been achieved around 5,000 years ago seems to be fairly well accepted.]

Additionally, the mouths of the two major rivers have been anything but stable with regard to their location over the past 5,000 years, as the historical maps of the location of the mouth of the Salinas River in Figure 133 show. According to these maps, presented by Browning (1972), for some period before 1908, the river mouth was located to the west of the present Moss Landing Inlet, over 5 miles north of where it is now. Griggs and Patsch

(2005) observed that *the river actually discharged into Elkhorn slough until about 1910, when it broke through the dunes at approximately its present location.*

General Ecology

This area has a very rich avifauna with more than 340 species of birds recorded; the American Bird Conservancy designated Elkhorn Slough as a Globally Important Bird Area in 2000. Caspian terns nest on an island in the slough April-August, and relatively high numbers of western snowy plovers nest and winter along the mudflats and old salt ponds of the Moss Landing Wildlife Area on the north side of Elkhorn Slough. Seabirds present in large numbers seasonally include sooty shearwater and common murre; common shorebirds include western sandpiper and phalarope. Waterfowl, diving birds, such as grebes and loons, herons, egrets, gulls, terns, and brown pelicans are common.

With a population of nearly 300, harbor seals can

be seen hauled out on the tidal flats at low tide (pupping March-May). The sea otter population is about 100, feeding in the slough on the abundant clams and mussels. Elkhorn Slough is probably one of the best places to view sea otters up close.

A good introduction to the ecology of the Slough is presented in the book *Elkhorn Slough* by Silberstein and Campbell (1989) that was published by the Monterey Bay Aquarium.

Places to Visit

The two jetties at the entrance to the Moss Landing Inlet are good birding spots, but always be mindful of the strong tidal currents. The Moss Landing Wildlife Area is also a popular bird observation area. Otherwise, the main place to visit is the Elkhorn Slough National Estuarine Research Reserve.

The 1,400-acre Elkhorn Slough National Estuarine Research Reserve was the first of now 27 National Estuarine Research Reserves established nationwide as field laboratories for scientific research and estuarine education. You should plan a trip to the outstanding visitor center, which has great exhibits for all ages; it is open Wednesday-Sunday, 9 a.m. to 5 p.m. To get there, go east on Dolan Road (just south of the power plant) for 3.5 miles to Elkhorn Road, then it is 1.9 miles along a winding road through oak woodlands and farms to the gates of the Reserve. There are good road signs. Docent-led tours of the Reserve are held every Saturday and Sunday at 10 a.m. and 1 p.m. There are five miles of trails through beautiful oak woodlands, and along tidal creeks and freshwater marshes that provide excellent birding opportunities.

THE SOUTH BEACHES (MOSS LANDING TO MONTEREY HARBOR)

Introduction

This area (Zone C, Figure 130) is a flattened out crenulate bay shoreline on the southeastern margin of Monterey Bay that extends for sixteen miles from the jettied tidal inlet at Moss Landing in the north to Monterey Harbor in the south. The major differences between this flattened out crenulate bay to the south from the one on the north side of the eastern flank of Monterey Bay are that: 1) it contains the outlet of the Salinas River, no doubt the major supplier of sand to the beaches on the southeastern side of Monterey Bay; 2) the sand dunes are much better developed than on the northern flank of the bay; and 3) the waves are bigger,

producing significant erosion of the bluff in places.

Geological Framework

The two most prominent geological features in this area are the 6-8 mile wide flood plain of the Salinas River, and the 100-foot-high line of sand dunes on the landward side of the beach from south of the mouth of the Salinas River to Sand City. Like the flood plain of the Pajaro River to the north, the rich soils of the flood plain of the Salinas River are heavily farmed. The seasonality of the rains, which occur mostly in the winter time, exerts a major control over farming practices. Rainfall also dictates when the Salinas River floods, at which time it usually breaches a sand spit that tends to close off its mouth in the summertime, as is illustrated by the oblique aerial photograph in Figure 134A.

Based on their 15 year study of the Monterey sand beaches, Dingler and Reiss (2002) observed that this seasonal pattern of river flow also controls the grain size of sand deposited on the beach at the mouth of the river. They noted that *in September, the median sand size was 0.39 mm, while in April it was 0.99 mm.* They felt that the reason for this change was that during the strong flow of the river as a result of the winter rains, the finer sand is carried into the nearshore zone. Later, after the river is through flooding, the finer-grained sand is moved back onto the beach by wave action.

The line of sand dunes immediately back of the beach illustrated in Figure 135 were called **Flandrian dunes** by Cooper (1967), which means that they were formed sometime during the present interglacial period. The word Holocene is often used interchangeably with Flandrian, which roughly means less than 12,000 years old. This line of dunes is perched on top of a much larger dune field that stretches an average of 5.5 miles landward from the beach. This larger dune field was called **pre-Flandrian dunes** by Cooper, which presumably means that they formed earlier than the Flandrian dunes, sometime during the Pleistocene epoch. Cooper admitted that *the history of the existing pre-Flandrian dunes is obscure,* and that, indeed, at least one of these dune fields on the California Coast *may extend back through more than one glacio-eustatic cycle.* The sand in the pre-Flandrian dune field in this area reaches thicknesses of over 100 feet. According to Cooper, four raised marine terraces underlie the dune field, with the tops of the highest dunes located at an elevation of 790 feet. Much of this field lies within the historic Fort Ord military reservation.

[NOTE. The word **eustatic** means relating to or characterized by worldwide changes of sea level. Therefore,

FIGURE 134. The southeastern shoreline of Monterey Bay (Monterey County; Zone C, Figure 130). (A) Oblique aerial view of the mouth of Salinas River on 23 September 2008, courtesy of the California Coastal Records Project, Kenneth and Gabrielle Adelman. When the river floods during winter rains, the sand that blocks the river's outlet will be added to the longshore sediment transport system. (B) A ground view looking southwest along the shoreline at Marina State Beach, showing some well-developed rhythmic topography on this day (8 December 2008). See the location of this beach on Figures 130 and 135. The people in the picture are standing on one of the protuberances of the rhythms. Compare with photograph in Figure 37 (Upper Right).

FIGURE 135. Image showing coastal sand dunes on the southeastern flank of Monterey Bay just south of the mouth of the Salinas River (Monterey County; Zone C, Figure 130). NAIP 2005 imagery courtesy of the California Spatial Information Library (CaSIL).

a glacio-eustatic cycle is one in which sea level drops dramatically and then rises to near its original level as a result of global glaciation (and melting), as happened four times during the Pleistocene epoch.]

Beaches and Dunes

The average size of the waves is a key factor controlling the grain size on the sand beaches of Monterey Bay, with the waves being considerably larger along the beaches on this side of the Bay than those on the northeast side, because of their more open exposure to the northwest winds and the offshore deeper water of the submarine canyon being located within the dominant line of fetch. According to Dingler and Reiss (2002), the largest waves on the beaches of Monterey Bay occur at Fort Ord, (Figure 130), where breakers are commonly greater than 10 feet high and occasionally reach 35 feet. The sand on the Fort Ord beach is a bit coarser than elsewhere because the large breaking waves keep the smaller sand grains in suspension, moving them off the beach.

According to Thornton et al. (2006), there are two littoral cells in this Zone that join at the mouth of the Salinas River. These two cells, the result of *the refraction of waves over the Monterey submarine canyon and delta offshore of the Salinas River*, were described as follows:

1) Longshore sediment transport to the north between the river mouth and Moss Landing, *most of which eventually empties down the submarine canyon*; and

2) Transport to the south away from the mouth of the river, presumably the result of the dominant northwesterly wind direction.

During the fifteen year project by the U.S. Geological Survey to measure the profiles of sand beaches on the eastern shoreline of Monterey Bay, summarized by Dingler and Reiss (2002), five beaches in this area were surveyed. At the start of that project in early 1983, these five beaches had been severely eroded during the strong *El Niño* of 1982-83, and they recovered to relatively wide beach profiles over the succeeding fourteen years. They were eroded severely again during the *El Niño* season of 1997-98. The pattern illustrated in Figure 55 for Sunset Beach along the northern shore demonstrates a trend similar to that of three of the beaches in the south. Two of the beaches, located in the shadow of the crenulate bay close to Monterey Harbor, where beach erosion is generally severe, showed a similar but less striking trend with the berms being narrower during the recovery period. These results emphasize the importance of the influence of *El Niño* storms on the long-term cycle of beach changes in Monterey Bay.

The photograph in Figure 134B, taken at Marina State Beach, illustrates the character of the sand beaches in that area. A set of spilling waves was approaching the beach at a small angle out of the southwest, creating a series of rhythmic topography on that particular day. The high sand dunes that characterize this area are visible behind the beach.

The sand dunes behind the beach in this general area (Figure 135) are a part of the second largest dune system on the Central California Coast, the largest being further south in San Luis Obispo and Santa Barbara Counties (Figure 77). As shown on the image, these dunes are partly vegetated and are oriented in the northwest/southeast direction, being formed by the dominant northwest winds that occur in this area. Cooper (1967) referred to these dunes as *parabolic dunes*, and described them as being *half-canoe-shaped, usually 2-4 times as long as wide*. In places, they are cut off cleanly at the bluff edge, with the truncated marginal ridges appearing as *triangular facets topping the bluff*. These dunes have curving downwind slopes that can rise 60 feet or more, with a steeply sloping face in the landward direction (see also Figure 76).

The most natural, unmodified dunes of this type occur between Marina State Beach and the mouth of the Salinas River (Figure 135). As noted by Griggs and Patsch (2005), these large dunes are there for the following primary reasons:

1) A large supply of sand provided originally by the Pajaro and Salinas Rivers;

2) A broad low-lying area landward of the beach; and

3) A persistent and dominant onshore wind direction.

As the discussions of Compartments 6 and 7 show, these three factors account for all of the much larger dune fields to the south (illustrated in Figure 77). An additional factor is the location of the big dunes in a southerly position on a relatively straight reach of coastline that takes advantage of the general north to south longshore sediment transport direction along the beaches in littoral cells (Figure 66), as well as the long, open and unobstructed area the wind can blow across. Clearly, the presence of the huge submarine canyon complicates the picture somewhat in this area.

This general conclusion about coastal sand dunes on the Central California Coast applies all along the shoreline. However, the complete story of why such massive, over 100-foot-high dunes form the core and surface of the bluff in this area is a little more complicated than that. As noted by Chin, Clifton, and Mullins (1988), the core of the bluff is dune sand deposited during periods of lowered sea level, probably during both the last and previous glaciations. Similar to what we described in the discussion regarding the sand dunes that underlie the City of San Francisco, when sea level was lowered during the last major glaciation, sand dunes extended all the way across the exposed continental shelf. The distance across the shelf was approximately eight miles for the dunes on the southeast side of Monterey Bay. As sea level rose, the shoreline retreated at a rate of approximately two feet per year, ultimately resulting in the high scarp behind the parts of the shoreline away from the major river mouths. In the segment between Marina and the mouth of the Salinas River, the linear, vegetated Flandrian (Holocene) sand dunes extend on an average of about 2,000 feet landward from the edge of the bluff.

The abundance of such sand near large centers of population has led to **mining of the sand** in the surf zone and the sand dunes over the years, making Monterey Bay the most intensively mined shoreline in the U.S. between 1906 and 1990 (with regard to sand being removed directly from the surf zone). Possibly as much as 300,000 to 400,000 cubic yards of sand per year were removed from these areas over that period of time (Griggs and Patsch, 2005). Mining of the surf zone, which accounted for around 128,000 cubic meters per year between 1940 and 1948, was terminated in 1990 when *it was finally hypothesized that sand mining was a significant contributor to shoreline erosion* (Thornton et al., 2006). However, mining of the backbeach and dunes continues today in the Marina area. Because of its coarseness and composition, the sand delivered to the shore by the Salinas River is highly valued.

An attempt by Thornton et al. (2006) to determine the effect of sand mining on erosion of the sand beaches in this part of the Bay was rendered so complex by the accursed *El Niños* as to make the study almost inconclusive. However, such mining is surely a bad idea in the long run, considering the general decrease in the volume of sand presently being supplied by the rivers.

Beach Erosion

Griggs and Patsch (2005) ranked the shoreline from the jetties at Moss Landing to Fort Ord as *Caution: moderate risk* with regard to beach erosion. The only loss of man-made structures in the area was on the southern spit at Moss Landing, where liquefaction of the sand due to seismic shaking *destroyed the Moss Landing Marine Laboratories during the 1989 earthquake.* The average rate of erosion is around two feet per year, which is pretty high, but this part of the shoreline has only one development, the Monterey Dunes Colony, which is set back a ways from the high-tide line.

The major problems with beach erosion occur from Fort Ord to the Monterey Harbor, and this entire segment is ranked as *Hazard: high risk*. The sandy bluffs at Fort Ord historically have eroded faster than any other area around Monterey Bay, at the phenomenal rate of eight feet per year. Four sites along this northern half of the Zone have required either riprap or seawalls to preserve property. This is no doubt due to the larger waves that occur in this general area relative to other areas along the eastern shoreline of the Bay, plus a relatively slow rate of sediment replenishment.

General Ecology

The Salinas River National Wildlife Refuge is small (367 acres), but has a wide diversity of habitats, including coastal dunes, beaches, salt marshes, saline ponds, grasslands, and riparian habitats. Nesting species in spring and summer include American avocet, black-necked stilt, killdeer, Caspian tern, and western snowy plover. Wintering waterfowl populations vary from 500 to 3,000 depending on the availability of water.

The extensive sand beaches are good places to find wintering shorebirds. The offshore waters have moderate numbers of seabirds and diving birds, such as auklet, guillemot, grebe, loon, scoter, and cormorant. The Dune Nature Trail at Marina State Beach passes through a restored dune and provides good information on the animals and plants of the dunes.

The dominant vegetation you see in the dunes and along the roads is the exotic species called ice plant (*Carpobrotus edulis*); in some areas it is the only plant species present! Ice plant is native to South Africa and was introduced in California in the early 20th century as an ornamental and extensively planted along highways for erosion control. It is a robust, flat-growing, trailing perennial, rooting at nodes and forming dense mats. Ice plant grows aggressively, displacing native perennial species, thus eliminating habitat for native annual species. It also lowers soil pH and increases the organic content of the soil because the plant tissue is slow to decompose. The extensive ice plant mats provide only marginal wildlife habitat because they provide little forage value. Land managers have declared war on the ice plant, and many dune restoration projects focus on removal of this pest and planting of native plants. Many native dune plants are dependent on blowing sand and disturbances such as erosion for germination. And the loss of these native dune species affects the animals that have also adapted to this harsh environment. An example is Smith's blue butterfly, a federally endangered species that occurs only in the Central California coastal region, with the Monterey Bay dunes being an important habitat. They are at risk because all life stages are dependent on two species of buckwheat – seacliff buckwheat and seaside buckwheat. Emergence of the butterfly is timed with peak flowering of their host buckwheat plants: adults feed on the nectar and deposit eggs in the flowers; larvae feed on the flowers and seeds; and they overwinter as pupae on or below the plants. Individual butterflies are thought to stay within 200 feet of the plant on which they emerged. At Fort Ord Natural Reserve, part of the University of California Natural Reserve System, researchers conduct field experiments on methods to remove invasive plants, such as ice plant and, therefore, hopefully increase the survival of planted native species.

Places to Visit

The Salinas was only a part-time river. The summer drove it underground. It was not a fine river at all, but it was the only one we had so we boasted about it – how dangerous it was in the wet winter and how dry it was in a dry summer. You can boast about anything if it's all you have.

John Steinbeck – EAST OF EDEN

Two places we definitely recommend that you visit in this area are the Salinas River State Beach and Marina State Beach. The entrance road to the Salinas River State Beach is located on Highway 1 about a mile south of the bridge over Elkhorn Slough in Moss Landing. It is only 0.44 miles from the turn to the parking area for the State Beach. Just before reaching the parking area, you will drive across a slough that occupies the old channel of the Salinas River, which flowed through it as recently as 1910. A trail follows the west bank of this slough all the way north to the parking area for the Moss Landing State Beach, a distance of 0.6 miles.

A *Salicornia* salt marsh and an associated mud flat are present in this slough, an excellent birding spot. On 8 December 2008, we saw willets, marbled godwits, numerous gulls, ruddy ducks, great blue herons, coots, and sandpipers.

Out on the beach, there are long reaches of undeveloped shore in both directions. The low dunes, which have surficial heavy mineral layers in places, are mostly vegetated. This is a good place to take a long, uncrowded walk along the beach.

Once you are back on Highway 1, it is 1.8 miles south to the turn to Castroville and another 3 miles to the bridge over the presently active channel of the

Salinas River. All along this drive you will be crossing the flat topography of the flood plain of the Salinas River with its rich, productive soil.

Castroville is the start point for the Monterey Bay Sanctuary Scenic Trail, a 29 mile multi-use trail that stretches from there to Lover's Point in Pacific Grove. Walkers, skaters, joggers, and cyclists are permitted to use this trail, which is separated from the road most of the way.

For the next mile or so after you cross the river heading south, you can look across the fields on your right and see the landward side of the outstanding, high parabolic sand dunes that are shown on the vertical image in Figure 135. At 1.55 miles from the bridge, you can see one of the last sand mining operations on your right (also visible in Figure 135), and after one more mile you will reach the turn to Marina State Beach.

The photograph in Figure 134B illustrates the rhythmic topography on the beach on 8 December 2008. There were some large spilling waves coming ashore, and the berm top was covered with gulls. We took a sample of the medium-grained, yellow-tan beach sand, which contained a little bit of coarse-grained sand. Using a hand lens we estimated the sand contained 60% quartz, 10-15% feldspar, some translucent orange grains, and a few heavy minerals, which indicates that the Salinas River drains some pretty complex source material in its watershed. Looking south from the beach, we could see the high, eroded dune scarps located further to the south. This is another beautiful beach definitely worth the visit.

From Marina State Beach, it is 9.5 miles or so to Monterey Harbor, passing through the towns of Marina, Seaside, and Sand City. Driving along Highway 1 outside Sand City, you can see the enormous sand dunes behind the beach there. If you are of a mind to see some of the examples of efforts to curtail beach erosion, you may want to stop at a couple of the beach access points, such as del Monte Beach or Monterey State Beach. As you approach the Harbor, the beach becomes flatter as the grain size of the sand decreases.

MONTEREY HEADLAND (MONTEREY HARBOR TO PESCADERO POINT)

Introduction

Without question, the drive along the shoreline of the Monterey Peninsula (Zone D, Figure 130) has so many attractions, such as unbelievably white sand beaches, extensive outcrops of granitic rocks, spectacular trees, multitudes of birds, boulder beaches, and the sea lions

on Bird Rock, that a naturalist could spend a good part of a vacation in this one area and not be disappointed. And then there is the possibility of visiting the world's best sea aquarium, as well as an historic site to warm the heart of any Steinbeck fan.

Geological Framework

The geology of this Zone is pretty simple; it is a granitic rock mass, formally called the Porphyritic Granodiorite of Monterey (Doris Sloan, pers. comm.), that occurs almost everywhere along this shoreline. This unit has been "informally" called the Hobnail Granite. The exposure of these granitic rocks along the Monterey Peninsula are the most widespread you will find anywhere along the Central California Coast (see photograph in Figure 136A). Perhaps the most definitive characteristic of this granodiorite, besides its Mesozoic age (94-95 million years ago) and the fact that it is a part of the Salinian Block riding on the Pacific Plate, is its porphyritic texture with larger blocky crystals of *orthoclase feldspar that resist weathering better than the other minerals, and so tend to stand out in relief on weathered outcrops* (Alt and Hyndman, 2000). You can even spot these larger crystals from the car (see example of this porphyry in Figure 136B). Being a **granodiorite**, it is composed of 20-60% quartz, with the remaining mostly feldspar component being between 65 and 90% plagioclase (Figure 17B; obviously with some orthoclase squeezed in there somewhere!). It may also contain a few dark-colored minerals, such as hornblende and biotite.

Other geological features on the peninsula include outcrops of Paleogene sandstones and conglomerates that are elaborated upon further in the discussion of Point Lobos in Compartment 6, where the exposures are better and more accessible. There are also some exposures of the famous Miocene Monterey Formation. Two dune areas on the northwest face of the peninsula have been highly modified by human activities in recent years.

Beaches and Dunes

There are about a dozen small pocket sand beaches scattered along this shoreline. The longest and widest one is at Asilomar State Beach, which is very popular. The most appealing characteristic of these beaches is their snow-white color. This color, which contrasts strongly with the brownish color of the beaches along the bay shoreline to the north, shown in the ground photograph at Marina State Beach (Figure 134B). This white color is no doubt due to the source of the sand, presumably the

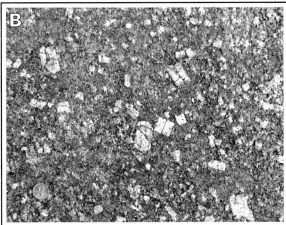

FIGURE 136. The Porphyritic Granodiorite of Monterey at Pescadero Point (Monterey County; Zone D, Figure 130). (A) The granodiorite at Pescadero Point on the 17-Mile Drive on the Monterey Peninsula, one of the best exposures of the Salinian Block granitic rocks anywhere along the Central California Coast. (B) This granodiorite has a prophyritic texture, as can be clearly seen in this photograph taken at Pescadero Point. The large crystals (phenocrysts; see Figure 17A) are composed of K-feldspar (Alt and Hyndman, 2000). Both photographs taken on 8 December 2008.

granodiorite itself, as well as a significant amount that was blown in from some sand dunes on the continental shelf when sea level was lowered during the last major glaciation (Ed Clifton, pers. comm.). The fine-grained sand is composed mostly of quartz with some feldspar and little else. It contains no rock fragments and just a few heavy minerals.

Beach erosion is a relatively minor problem in this area, because of the resistance of the granodiorite to erosion. Seawalls and riprap have been installed in a few places, but most of the shoreline is ranked as either *Stable: no risk* or *Caution: moderate risk* by Hapke (2005).

Cooper, in his 1967 report on the dunes in this area, lamented that the *purity of the sand has resulted in mining so extensive that whole dunes have been removed; residential development has added to the destruction.* There were originally two good-sized lobes of these Flandrian (Holocene) dunes on the Peninsula, one directly south of Point Pinos, and another just north of Point Cypress. What's left of a portion of the southern half of the Point Pinos dunes is within the confines of the Asilomar State Beach. What's left of the Cypress Point dunes is intermingled with two golf courses in the 17-Mile Drive area of the Peninsula.

General Ecology

The Monterey Peninsula is one of the best places in Central California to see, up close, a wide range of marine animals and plants. Large numbers of California sea lions can be seen at Monterey Harbor, Bird Rock, and Cypress Point. Harbor seals haul out on numerous sites. Hundreds of sea otters are present in nearshore waters. California brown pelicans are most common July-November when breeding adults disperse north from nesting colonies in the Channel Islands and Mexico. There are five colonies where black oystercatcher, Brandt's and pelagic cormorant, pigeon guillemot, and western gull nest February-September. Up to 1,300 pairs of Brandt's cormorants nest on Bird Rock; according to the kiosk at the parking area for the Rock, *they create nests out of seaweeds from the sea and twigs and leaves from land. Even when they are not nesting, cormorants spend a lot of time roosting on rocks; because they lack oil glands to keep their feathers dry, they must return to land after diving to dry their feathers. Therefore, they typically feed within a few miles of land.*

The many waterfront parks and trails provide ready access to rocky tide pools to search for intertidal animals and plants. Black abalone are (or were) common along the rocky shores of Asilomar State Beach. Abalone, a sea snail with a flat, coiled shell with a very large flat opening, live on intertidal and shallow subtidal rocks along exposed outer coasts. Because of over harvests and spread of a lethal disease, withering syndrome, in the 1980s, most forms of legal harvest in California have been suspended since 1993. The spread and impact of withering syndrome has been so severe that, in 2007, scientists from the National Marine Fisheries Service predicted that black abalone are likely to become effectively extinct within 30 years. On 13 January 2009, black abalone were designated as endangered by the federal government. However, red abalone can still be taken by free diving north of San Francisco.

Places to Visit

On the rocks off Hopkins Marine Station the sea lions barked with a houndish quaver. The silver canneries were silent under the street lights. And from somewhere on the beach Cacahuete Rivas's trumpet softly mourned the Memphis Blues.

John Steinbeck – SWEET THURSDAY

From Monterey Harbor it is about 0.9 miles south to the Aquarium as you drive through the heart of historic Monterey, no doubt remembering Mack and the boys and Doc as you move along. Staying close to the shore, it is about one mile from the Aquarium to Lovers Point Park. There are sand beaches and scattered coastal access in this stretch, which contains segments with riprap as well as some sea walls. However, from beyond Lover's Point for almost 3 miles, there is easy access to the shore with numerous parking spots and a trail that runs almost the whole way between the road and the high-tide line. This is a very beautiful area with numerous outcrops of the Monterey Granodiorite, some nice sandy beaches (mostly white fine-grained sand), sea stacks, many of which contain abundant birds with several being almost completely covered with cormorants, big waves at times, pebble and cobble beaches, as well as some armored boulder beaches. Needless to say, this is a good place for a walk with spectacular sightseeing.

This road you will be driving on appears to have several different names in different places – Ocean View Boulevard, Sunset Drive, and Highway 68. Just stay on the road closest to the edge of the shore and you can't go wrong.

As we were taking this ride on 8 December 2008, we happened to see a couple of fishermen climbing up onto some rocks, which brought back to mind Steinbeck's line in TORTILLA FLAT – *A little gold entered into the sunlight. The bay became bluer and dimpled with shore-wind ripples. Those lonely fishermen who believe that the fish bite at high tide left their rocks, and their places were taken by others, who were convinced that the fish bite at low tide.*

As a point of reference, it is 2.7 miles from Lover's Point, a worthwhile stop because of the granodiorite outcrops and the white sand beach, to Asilomar State Beach. When you reach the State Beach, you should pull over in the parking zone along the side of the road. There are two excellent reasons to stop there, the first one being the appealing white sand and other aspects of one of the most beautiful beaches anywhere. The other reason

is that if you walk across the road, there is a wooden boardwalk trail that goes for several hundred yards through the remnants of the coastal dunes, all within the boundaries of the State Beach. This is the only place on the Peninsula where you will have such a free reign to explore the dunes, whose beauty is also highlighted by the white sand.

The web page for **Asilomar State Beach** states that it is *a narrow one mile strip of sandy beach and rocky coves. A 3/4 mile coast-walking trail is open to pedestrians. Bicycle riding is allowed on the paved road bike lane in close proximity to the trail. There are no restrooms or picnicking facilities at the State Beach. Parking is available on Sunset Drive (Highway 68), the city street that runs parallel to the State Beach.*

After spending some time at this excellent site, we suggest that you continue south on Highway 68 for 0.6 miles and turn right to enter the 17-Mile Drive at Pacific Grove Gate. From there, continue south, hugging the shoreline the whole way to your exit at Carmel Gate. Of course, you may want to spend hours traveling along other parts of the drive, but we will just discuss the part that runs along the coast. In our opinion, this drive is definitely worth the $9.25 fee (as of 2008) from a naturalist's perspective, with the rest of the experience, including the eclectic architecture and a lot of upscale gardening design, being a bonus.

From the entrance at Pacific Grove Gate, it is 1.4 miles to a parking area by a gravel beach on Spanish Bay. In places, this cobble and boulder beach is pretty coarse. Unfortunately, humans typically amuse themselves by creating "art work" out of this gravel. In any event, this is a nice place for a stroll where you can watch the waves and the birds.

Continuing south, it is about 1.9 miles to Bird Rock, a site you most certainly do not want to miss. This may be the best location on the whole Central California Coast to get close enough to an offshore sea stack to easily inspect the multitude of residents that occupy the top of it. When we were there on 8 December 2008, there were around 200 sea lions on that rock, as well as a hundred Brandt's cormorants and several pelicans. The sea lions are especially noteworthy (see photograph in Figure 137). Seal Rock Picnic Area, located on another white fine-grained sand beach, is located a short distance to the south of the Rock.

From Bird Rock, it is about 0.9 miles to Fanshell Overlook, another location with good views. The famous *lone cypress tree*, the photographs of which have appeared in so many publications, is located a little over a mile beyond the Fanshell Overlook. The lone cypress seems to be sprouting right out of the granodiorite. You

FIGURE 137. Bird Rock, located beside the 17-Mile Drive on the Monterey Peninsula (Monterey County; Zone D, Figure 130). On the day this picture was taken, 8 December 2008, there were around 200 sea lions, a hundred Brandt's cormorants, and several pelicans on the rock. This is one of the few places along the Central California Coast where you can get this close to a sea stack that is so heavily utilized by both sea birds and marine mammals.

will also pass by Crocker Grove, a 1.3 acre nature reserve that contains numerous native pines and Monterey cypress, including the "granddaddy" cypress of them all. It is a very impressive stand and an interesting change of scene.

The last stop we recommend on the drive is at Pescadero Point, the best place to examine the famous granodiorite resident of the Pacific Plate that we have seen anywhere along the entire Central California Coast (see again Figure 136A).

From Pescadero Point, it is a couple of more miles on the 17-Mile Drive around the Pebble Beach Golf Course to the exit from the drive at Carmel Gate.

If you take this trip the way we recommended, you will have seen more granite (oops! granodiorite) than you might have imagined, beaches ranging from boulders to fine sand, with the sand dominated by the exquisite, white fine-grained variety, lots of birds, and a number of marine mammals.

We have taught several training courses for oil-spill response teams in Monterey over the years, but it had been quite a while since we had taken the 17-Mile Drive. We finally did it again on 8 December 2008, at the end of which co-author Hayes was heard to exclaim, "Wow, that far exceeded my expectations!" On this trip, maybe we were just seeing the place through new eyes, considering the fact that we were planning to tell you about it.

CARMEL BAY (PESCADERO POINT TO SAN JOSE CREEK)

Introduction

The southern three fourths of this area (Zone E, Figure 130), which includes the shoreline of Carmel-

239

by-the-Sea, is composed of three major sand beaches hundreds of yards in length separated by two rocky headlands in the Porphyritic Granodiorite of Monterey. The northern section contains two small bays and several small rocky headlands, much of which borders the Pebble Beach Golf Course.

Geological Framework

This part of the Monterey Peninsula is underlain by granodiorite, which was discussed in some detail in the previous section on the outer Peninsula. Two streams deliver sediment to the beaches, the Carmel River and San Jose Creek, but at least some of the sand appears to have been both: 1) eroded out of the granodiorite itself; and 2) transported onshore from the continental shelf.

It would seem to be imprudent for us to omit a mention of the Miocene Monterey Formation in this discussion of the Monterey area, inasmuch as this area is what geologists call the **type locality** of the Monterey Formation. The type locality, or type section, is the particular site where either the most typical rock types of the formation are located, or the location where the formation's name was given when it was first described. However, according to Bowen (1965), *the choice of the Monterey vicinity as the type locality of the Monterey Formation was unfortunate because the section there is not typical of the Monterey Formation in well-known localities elsewhere in California, either in age, thickness, or completeness.* In most areas, this formation, which was deposited in marine waters around 5-15 million years ago, is normally composed of light-gray, fine-grained sediments rich in organic matter, phosphate, and other unique ingredients (see photograph in Figure 105B). The Monterey Formation has already been discussed, and will be many times again, throughout this book, not the least reason for doing so being that it is thought to be the source for the petroleum in several of the marine basins along the Central California Coast.

Beaches and Beach Erosion

The three main beaches in the area are spacious; Carmel City Beach is 2,000 yards long, the part of Carmel River State Beach that closes off the mouth of Carmel River is 1,000 yards long, and an associated segment to the south, sometimes called Meadows Beach, separated from the northern unit by a couple of rocky intertidal zones, is 300 yards long. The southernmost beach, Monastery Beach, is 700 yards long. There is a wide range in sand composition among the beaches; Carmel City

Beach is composed of quartz-rich, white fine-grained sand, whereas Monastery Beach is composed mostly of granules and coarse sand that contain an abundance of feldspar and rock fragments. The reason for this contrast is possibly a significant difference in the source areas of the sediment for the beaches. On the other hand, the dissimilarity could be related simply to the difference in grain size, the quartz being finer grained when it came out of the source rock, with the bigger waves at Monastery Beach keeping these finer sand grains in suspension. Ed Clifton (pers. comm.) noted that sand on Carmel City Beach is *largely composed of well-rounded, frosted quartz grains. I think it is very likely that the present day beach sand there is derived from older dunes that existed in what is now below sea level during the last lowstand.*

[NOTE: **Rounded** means that the angular edges of the quartz sand grains have been abraded away, leaving a smooth, spherical grain. **Frosting** means that the surface is coated with tiny pock marks created by repeated impacts by other sand grains. These two conditions can only form when the abrasion and contacts take place in air. When the sand grains are in water, such as on a beach, the water cushions the blows. Therefore, both of these characteristics of the sand grains indicate that they were at one time in a dune field in the open air, something that we have already noted a couple of times with respect to the sand dunes that underlie San Francisco and elsewhere. These particular dune fields were located on the continental shelf at times of lowered sea level.]

Although riprap and seawalls have been installed in a few areas, the presence of the granodiorite makes this an area have little serious beach erosion problems. In fact, Hapke (2005) ranked the entire shoreline as *Stable: low risk.* However, Ed Clifton (pers. comm.) pointed out that *the mouth of the Carmel River has been shifting back and forth during the winter months with significant rainfall, causing some erosion at the margins of the beach.*

General Ecology

The 300 acre Carmel River Lagoon and Wetlands Natural Reserve was established in 1995. In summer, when the mouth of the river is closed off, a lagoon forms and supports abundant birds. It is also an important habitat for two federally listed species (threatened) – steelhead trout and California red-legged frog. The winter run of adult steelhead at the San Clemente Dam (18.6 miles upstream of the lagoon) has ranged from about 300 to almost 900 fish since 1995, which makes it one of the most significant runs within the south-Central California Coast.

Places to Visit

The Carmel is a lovely little river. It isn't very long but in its course it has everything a river should have.

John Steinbeck – CANNERY ROW

Pescadero Point, the northernmost promontory on Carmel Bay and Stillwater Cove, is the northern margin of this area (Zone E, Figure 130). If, in fact, you would want to start your visit to this Zone there, you would have to be on the 17-Mile Drive, which passes near the Point. From the Point, drive north on the 17-Mile Drive. Along parts of this drive, you can see outcrops of the Monterey Formation, which you probably haven't seen much on your drives around the Monterey Peninsula, although, as mentioned earlier, the Miocene Formation obviously derives its name from this area.

Go around the Pebble Beach Golf Course and exit the drive at the Carmel Gate. Once you pass through this Gate, go straight ahead to Ocean Avenue and turn right and you will soon be at Carmel City Beach. From the parking area there is a sandy bluff you walk down to get to the beach (Figure 138). Some of the bluff is vegetated and some has sand blown up it. The beach itself is composed of white, fine-grained, quartz-dominated sand that contains some heavy minerals, the origin of which was discussed earlier.

The next beach to the south is Carmel River State Beach. To get there, go into Carmel-by-the-Sea and drive to the parking area just to the north of the outlet of the Carmel River, the mouth of which is closed off during the non-rainy season. This is another wide and beautiful beach, similar in many respects to the Carmel City Beach to the north. According to Dingler (1981), this beach *commonly alternates between swell and storm profiles*. As noted earlier, these are terms suggested by Komar (1976), which mean a wide depositional berm under **swell conditions** (between storms) and a flattened out profile after **storms**.

Back on Highway 1, within a little over a mile, you can pull over on the right for the access to Monastery Beach, a beach you wouldn't want to miss. When we were there, on 8 December 2008, the beachface was pretty steep, as usual. We could see **surging waves** breaking right at the toe of the beach and surging up the beachface. The beach is typically composed of granules (diameter of 2-4 millimeters) or very coarse sand. On this day, there were lots of granules and some fine pebbles on the berm top. Figure 139 shows beach profiles and a field sketch of this site. As our surveyed profiles indicate, as well as our observations over the years, this beach is very stable in its configuration. Dingler (1981) reached a similar

FIGURE 138. Early morning view of Carmel City Beach (Monterey County; Zone E, Figure 130), looking south on 8 December 2008.

conclusion based on a three year study.

Why the sediment on this beach is relatively pure granules is an interesting question. Possibly, the waves are consistently so big they carry away the finer sand in suspension to be eventually lost down a major submarine canyon (Carmel Submarine Canyon), the head of which is located only a little over 700 feet offshore in water depths of 60 feet (Dingler and Anima, 1989).

The extraordinarily big waves can probably be explained by the fact that the head of the submarine canyon is so near shore. This means that the waves do not "feel the bottom" until they are very close to shore, which no doubt accents their large size and surging character (refer to Figure 34B, an illustration of the process of waves approaching shallow water). Shepard and Dill (1966) concluded that the submarine canyon is an extension of the adjacent on-land canyon presumably carved by an ancestral Carmel River, probably with a little help from an ancestral San Jose Creek. The long, unfettered fetch in the northwest direction away from the beach is another factor in the generation of the large waves.

Even though the beach profile remains relatively stable, apparently a considerable amount of the coarse-grained sediment eventually finds its way down the nearby Carmel Submarine Canyon by a process called grain flows (or sand avalanches). This process was studied by SCUBA divers in this Canyon, where the sand avalanches periodically move down a slope of 34 degrees at the head of the Canyon (Dingler and Anima, 1989).

All and all, this beach is a pretty spectacular location, from both the geological and oceanographic perspectives. The birding is also excellent. One last word of caution, this is a very dangerous beach, with the aforementioned big waves and a sharp drop-off at the base of the beach. *Drownings are not at all uncommon* (Ed Clifton, pers. comm.).

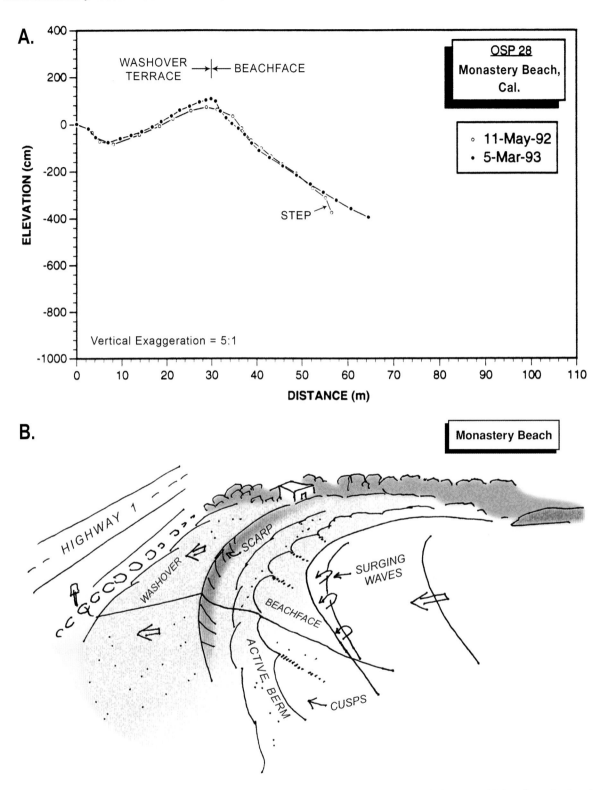

FIGURE 139. Monastery Beach (Monterey County; Zone E, Figure 130), located only a little over 700 feet from the head of the Carmel Submarine Canyon. (A) Topographic beach profiles measured on 11 May 1992 and 6 March 1993, which show very little change, typical for this particular beach, which usually contains an abundance of granules and some pebbles. (B) Field sketch (5 March 1993). At the time of this survey, the berm crest had recently been washed over and beach cusps were present along the high portions of the beachface. Surging waves were present, another common phenomenon on this beach.

18 COMPARTMENT 6: BIG SUR

INTRODUCTION

One of the most beautiful auto drives in the world, along Highway 1, follows closely around this classic, steep margin of the North American Continent (Figure 140). Starting with the preserved submarine-canyon conglomerates of the Carmelo Formation (of Paleogene age; Table 1) in Point Lobos State Reserve, the geology of the Big Sur Country shows an amazing amount of variability as the highway meanders through both the Salinian and the Nacimiento Blocks. The spherical boulders and cobbles of the gravel beaches and abundant sea stacks carved from the highly jointed granitic rocks in Garrapata State Park set the stage for more of the same to come as you drive south. The black, volcanic rocks of the Franciscan Sequence form the core of Point Sur which, itself, acts as a natural groin for a wide sand beach to the north. The steep shore contains numerous landslide scars, especially in the areas where the Franciscan mélanges occur. The sand beaches south of Point Piedras Blancas offer an excellent opportunity to observe hundreds of elephant seals.

We have subdivided this Compartment into the four Zones mapped on Figure 140:

A) **Carmel Uplands (San Jose Creek to Point Sur)** start in the north with the geological wonderland called Point Lobos State Reserve, home to ancient submarine canyon deposits. The rest of the area features two frequently photographed, classic Big Sur bridges, three long sand beaches, steep rocky shorelines with abundant sea stacks covered by thousands of sea birds, and numerous pocket beaches containing highly rounded cobbles and boulders.

B) **Santa Lucia Uplands (Point Sur to Lopez Point)** contain one of the most interesting and beautiful sand beaches on the Central California Coast, Pfeiffer Beach, as well as two spectacular State Parks – Julia Pfeiffer Burns and Pfeiffer Big Sur. The Franciscan volcanic rocks at Point Sur, the crossings of the shore by two major

faults, the Salinian Block granite rocks in the vicinity of the McWay Falls, and the Franciscan mélange chaos to the south make for a highly variable and interesting geological setting.

C) **Ragged Point Uplands (Lopez Point to Point Piedras Blancas)** are composed mostly of steep-sided mountains with abundant wave-cut rock platforms and gravel beaches. Raised marine terraces are sparse, except in the southernmost area. Sand beaches are few and far between. No major stream systems are present, except near the southern border. The rocks are composed of mélange deposits of the Franciscan Complex in many areas; however, in the south, exposures of the Coast Range Ophiolite are of considerable interest.

D) **San Simeon Strand (Point Piedras Blancas to Cayucos Creek)** is dominated by lowlands, in contrast to the more northerly areas. The shoreline has relatively low-lying, raised marine terraces most of the way, except for the area between Cambria and Estero Bay. Sea stacks are numerous off the low, wave-cut scarps. Several moderate-sized streams produce enough sand and fine gravel to supply several beaches. Marine mammals, especially elephant seals, are a main attraction, and birds are abundant all along the shore.

CLIMATE OF THE BIG SUR COUNTRY

The climate of the Santa Lucia Mountains and the Big Sur Country is classified as Dry Summer Subtropical, or Mediterranean. The mountains are close enough to the Pacific Ocean and high enough to allow the rising air off the ocean to generate some rainfall, which varies from 16 to 60 inches per year and occurs mostly in the wintertime. This precipitation allows conifers to grow on the western slopes, most notably the coast redwood (*Sequoia sempervirens*), the tallest tree in the world that has an average life span of 500 years. Fog is common in the summer, but nothing can detract from the beauty and tranquility of this place.

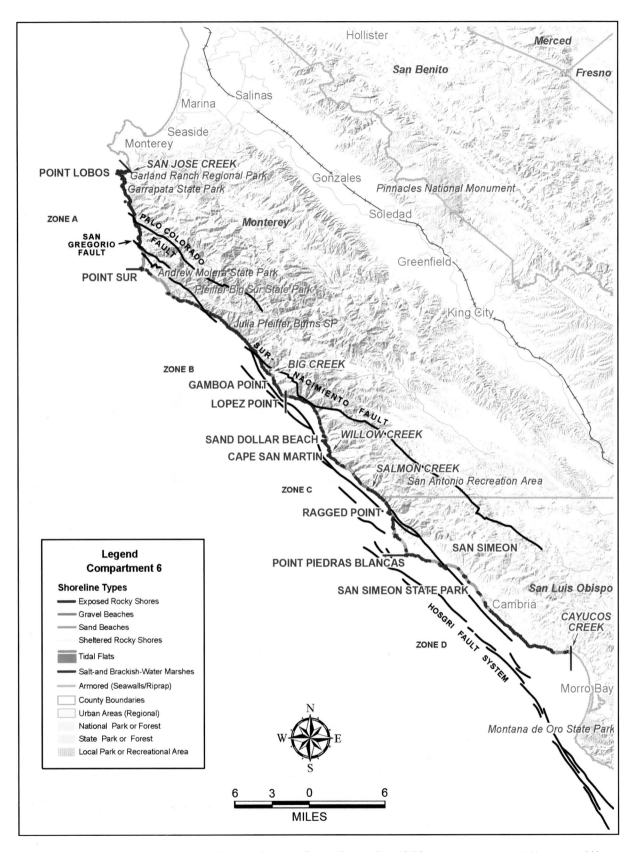

FIGURE 140. Compartment 6, San Jose Creek to Cayucos Creek. Faults after USGS geological maps and Henson and Usner (1993).

244

GEOLOGY OF THE BIG SUR COUNTRY

Because of major faulting episodes, the geology of the Big Sur Compartment is the most complex of the seven Compartments discussed in this book (Figure 1). As illustrated in Figures 16A and 18, all of the rocks along the shoreline in this area are located west of the San Andreas Fault, which means that they are all superimposed upon the Pacific Plate. The San Andreas Fault itself is located an average of about 45 miles inland of the shoreline.

A brief review of a typical cross-section of the subduction complex and adjacent environments along the California coast during the Late Mesozoic era is in order (see Figure 14). The different geographic/geologic elements of the California coastal area during that last major subduction episode, going from east to west, included:

1) An **ancestral Sierra Nevada mountain range** in which an older **Sur Series** composed mostly of metamorphic rocks, including a defining rock type called **marble** (metamorphosed limestone called the Gavilan Limestone), was intruded by a large series of magmas from which volcanoes emanated, forming a land-based **volcanic arc** similar to the one that now exists in western Washington and Oregon. The granitic magmas were eventually exposed at the surface as the original mountains were eroded away. These rocks are known as **Sierra Nevada granitic rocks** and they were formed mostly during the Cretaceous period (between 140-70 million years ago). Most of the original volcanic-arc material was eventually eroded away.

2) A **forearc basin** of unknown depths located between the ancestral Sierra Nevada volcanic arc and an accretionary wedge of lava flows and miscellaneous sedimentary materials (Figure 14). Both 1 and 2 were a part of the North American Plate. The area of the so-called forearc basin eventually filled up with sediment and geomorphologically became known as the Great Valley.

3) A **deep oceanic trench** into which turbidites were flowing in great abundance. Also, materials carried to the east aboard the rapidly sinking Farallon Plate were either scraped off the top of the oceanic crust and piled up against the accretionary wedge or carried deeper into the trench.

4) Initiation and continuation of the **subduction** process, during which the seaward edge of some of the sediments and volcanic deposits that were accumulating in the trench were crumbled and scraped downward into the subduction zone. These sediments, which were at least partly metamorphosed, are now known as the **Franciscan Complex**, with ages ranging from Late Jurassic up to Early Cenozoic. Some of the sedimentary layers beyond the eastern boundary of the trench were merely tilted to a steep angle or vertical position, maintaining some of their original qualities, such as distinct sedimentary layers with only slightly metamorphosed sediments. These rocks eventually became known as the **Great Valley Sequence**. These rocks were originally deposited in the forearc basin on top of an oceanic crust (the **Coast Range Ophiolite**). The Coast Range Ophiolite *is everywhere separated from Franciscan rocks by a fault – the Coast Range Fault* (Doris Sloan, pers. comm.)

5) Subducting oceanic crust, in this case, the **Farallon Plate**. This oceanic crustal material came from as far as a thousand miles away. The surface of this crust started out as pillow basalts as the lava flowed out from the mid-ocean trench onto the ocean floor (Figure 19). As the pillow basalts moved eastward, pulled by convection currents toward the trench along the west side of the North American Plate, untold numbers of radiolaria rained down on the basalt, accumulating in layers measured in inches. Some fine-grained rock fragments (silt and clay) were added as thin bands that separated the radiolaria deposits into distinct layers, which are now known as **ribbon cherts** (Figure 20). In the process of subduction, the pillow basalts, ribbon chert, and turbidite deposits all were incorporated into that complex assemblage of rocks called the Franciscan Complex, one of the most diverse and complicated sedimentary and metamorphic rock suites known to occur anywhere on Earth.

6) In some places, during the earlier parts of the Cenozoic era, steep submarine canyons were carved along the edge of the continental shelves of that time and coarse-grained, conglomeratic sedimentary beds were deposited in the canyons by turbidity currents and gravity slides. The **Carmelo Formation** at Point Lobos is a product of this process (see Figure 141).

7) After the Farallon Plate and the mid-ocean ridge were completely subducted, around 30 million years ago, the subduction process ceased, after which the **San Andreas Fault** was born, and the Pacific Plate started sliding north along side the North American Plate.

During the remaining time from 30 million years ago up to around 2 or 3 million years ago, marine basins developed on this complex mixture of rocks, in which thick sequences of sediments accumulated, including the oil-rich Monterey Formation. During the ice ages, marine terraces were carved during times of high sea levels and uplifted slowly, but relatively continuously, forming raised marine terraces, most of which contain a

FIGURE 141. The Carmelo Formation of Paleogene age (between 50-60 million years old) in the Point Lobos State Reserve (Monterey County; Zone A, Figure 140). These sedimentary rocks were deposited in a submarine canyon carved into the underlying granodiorite (Porphyritic Granodiorite of Monterey). Photograph taken on 7 December 2008.

variety of sediments on them, from alluvial fan deposits and debris slides, to sand and gravel beach deposits, to wind-blown sand dunes. Within the past 5,000 years or so, during which time the sea level has been near its present position, beach deposits, including sand spits and tombolos, and coastal sand dunes have continued to accumulate.

During your visit to the Big Sur Compartment, you will be able to see examples of sediments and rocks created in each of the episodes just described. The oldest rocks you will see in your travels are the rocks of the Sur Series, to which some geologists attach the age of pre-Cambrian (older than 542 million years), but most favor an early to middle Paleozoic age (around 500-400 million years old). As you have already seen in the discussion of the limestone plant at Davenport, the marbles of this series have had an important economic impact on the Central California Coast. The ophiolites are on the order of 165 million years old and the granitic rocks

range around 140-70 million years old. The Franciscan Complex, which is generally younger than the ophiolites and granites, also flourished during the Late Cretaceous, terminating in the early Cenozoic (60-50 million years ago). The early Cenozoic conglomerates at Point Reyes and Point Lobos are Paleogene in age (also 60-50 million years old). The time of deposition of the thick Mid- to Late-Neogene sedimentary rocks, including the Monterey and Purisima Formations, was on the order of 15-2 million years ago. Some Pleistocene terrace deposits a few hundred thousand years old may be encountered, as well as an abundance of sand beaches and sand dunes formed in the past 5,000 years.

It is a rare place where you can see such a wide range of ages of sediments and rocks on such a short drive. The reason for this is that these diverse units have been juxtaposed by some massive adjustments along three major fault systems. As discussed previously, the San Andreas Fault is a relatively vertical transform fault along which the two major plates slide by each other (with a few exceptions where there are bends in the fault trend and so on). This motion has brought the Salinian Block, composed of granitic rocks of Cretaceous age as well as some of the Sur Series rocks, from as much as 300 miles away, where these rocks are thought to have been a part of the ancient Sierra Nevada Range. The best exposures of these granitic rocks anywhere within the Central California Coast occur at Point Lobos and on the Monterey Peninsula (Figure 136A).

As illustrated in Figure 18, the Sur-Nacimiento Fault runs in a northwest/southeast direction roughly parallel to the San Andreas Fault, forming the southwestern boundary of the Salinian Block. As noted in California Coastal Commission (2007), *movement along the Sur-Nacimiento Fault brought northward the Nacimiento Block, consisting mostly of the Franciscan Complex and emplaced it west of the Salinian Block. These two blocks now lie far north of the location where the rocks in them were created.* Unlike the San Andreas Fault, the Sur-Nacimiento Fault is a thrust fault (Figure 10 – Middle) that has displaced the older rocks of the Salinian Block and deposited them on top of the younger rocks of the Franciscan Complex in the Nacimiento Block, which some geologists refer to as an *accretionary prism*, with reference to their being deposited in a subducting trench. Furthermore, this thrust fault is slanting at a low angle (discussed further in the sections on places to visit).

According to Trasko (2007), the Sur-Nacimiento Fault, which is inactive at present, is of particular interest because, inasmuch as the Salinian Block, a displaced piece of granitic rocks similar to the Sierra Nevada batholith, is juxtaposed with the Franciscan Complex, somehow the Great Valley Sequence and the Coast Range Ophiolite of the forearc basin have been omitted in the process. The point being, if you refer to Figure 14, you will notice that in this general cross-section, the Sierra Nevada batholith is separated from the Franciscan trench material by the Great Valley Sequence and the oceanic crust from which the Coast Range Ophiolite presumably came. This is just one more fact that emphasizes the complexity of the geology in this area. Hopefully, with a better understanding of the general setting illustrated in Figure 14, you will recognize the pieces of the puzzle when you see them, utilizing color, texture, composition, and so on as clues based on our descriptions as you ride along or inspect on foot the individual rock exposures, just as you would in reconstructing a jigsaw puzzle.

As shown in Figure 18, the Nacimiento Block is bounded on the west by the San Gregorio-Hosgri Fault which, according to Trasko (2007), *is an active strand of the San Andreas Fault system.* This means that, unlike the

Sur-Nacimiento Fault, this fault zone is still active. There are many other more minor faults in the Big Sur area in addition to the three major ones shown in Figure 18. For example, as the Franciscan Complex was raised during the formation of the Coast Ranges, this uplifting process was accommodated by normal faulting (Figure 10 – Top) *subparallel to and along the Sur-Nacimiento Fault during Late Cenozoic time* (Trasko, 2007).

CARMEL UPLANDS (SAN JOSE CREEK TO POINT SUR)

Introduction

This section of the Big Sur Coast (Zone A, Figure 140) is all about getting introduced to the beginning of one of the most stunning coastal areas on the North American Continent. Point Lobos, with its kelp beds, clear waters, almost unsurpassed geological features, and calming influence, is a great place to start. From there the steep shoreline at Garrapata State Park has all the defining features of rocky coasts, in the extreme. Other features include the classic Big Sur bridges at Bixby and Rocky Creeks. Surprisingly, there are three long sand beaches on this rocky shore, with the Little Sur River providing enough sand to allow for the development of a mile-long, pure white sand beach, as well as supplementing the accumulation of sand in the lee of Point Sur. The geology is complicated by the presence of a major fault zone, the San Gregorio-Hosgri Fault; hopefully, our explanations will keep your curiosity aroused and, once the general picture becomes clear, you will become as totally captivated by the workings of the great collisions of the North American, Farallon, and Pacific Plates as we have.

Geological Framework

As noted in the discussion on the geology of the Big Sur Country, this section of the coast contains examples of all of the geological components of the Central California Coast, from the oldest, the marble exposures of the Sur Series, to the youngest, the modern-day beaches and sand dunes. Perhaps the most noteworthy exposures are in the Point Lobos State Reserve, which contains spectacular outcrops of coarse-grained sedimentary rocks of Paleogene age that were deposited within a submarine canyon. For about the northern half of the area, granitic rocks of the Salinian Block are abundant in the wave-cut rock cliffs and road cuts along Highway 1, but the crossing of the coast by the San Gregorio-Hosgri Fault (Figure 140) complicates matters considerably by introducing a mix of other rock types. The southern extremity of the area is anchored by the dramatic headland at Point Sur, which is composed of layers of black volcanic rocks of the Franciscan Complex.

Beaches and Dunes

There are spectacular gravel beaches in Garrapata State Park. A sand beach a half mile long is present off the mouth of Garrapata Creek, one twice that long occurs off Little Sur River, and one only slightly longer than that is present on the north flank of a huge, low-lying triangle of land in the lee of the rock headland at Point Sur. Cooper (1976) referred to this triangular wedge as a low raised terrace, but others have called it a tombolo. Cooper was right, because the south flank of the peninsula is underlain by a rocky, flat terrace. There is, however, a well-developed sand beach on the north side of the peninsula, presumably the direction from which the sand comes. Otherwise, sand beaches are scarce in this part of the Big Sur coast.

A half mile wide band of low, linear sand dunes occupies the depressed land in the lee of the Point Sur headland, and some of the sand appears to blow all the way across it, for a maximum distance of around 0.8 miles. There are also some minor dunes to the south of the mouth of the Little Sur River.

Beach Erosion and Landslides

As observed by Hapke (2005), beach erosion *per se* is not much of a problem in this stretch of coastline. Point Lobos is mostly a rock coast and for the remainder of the shoreline down to Point Sur, *erosion hazards are fairly low*, with an average cliff erosion rate of *approximately 6 inches per year*. Hapke also observed that the granitic rocks, which occupy more than half of the shoreline of this area, are resistant to erosion, and the segment south of Carmel Highlands has little development. However, she also pointed out that *marine terraces have development potential, but also have a history of landsliding, which poses a moderate hazard*.

Landslides have definitely proven to be a problem in some areas. A massive slide just to the south of Hurricane Point resulted in the closure of Highway 1 for three months during the winter of 1999 (Hapke, 2005). This slide site occurs within the portion of the coast between the Little Sur River and Rocky Creek, about which the California Department of Conservation (2001) made the following comment:

The most rugged topography and the greatest landslide potential in the northern part of the Big Sur coast is concentrated in this short area. Northward from the Little Sur River, the highway ramps upward across the steep southwest flank of Sierra Hill. This side of the Hill is underlain by very weak rocks of the Tertiary Pismo Formation at the base, in fault contact with older Cretaceous rocks, which are in turn faulted against older Sur complex metamorphic rocks. Wave erosion of the weak rocks at the base tends to undermine the stronger rocks above.

This would certainly explain the big slide at Hurricane Point, which left a major scarp in the mountain side south of the Point that you will be able to see clearly if you pull over at Hurricane Point on any of your drives along Highway 1.

General Ecology

This rugged and largely inaccessible stretch of coast is rich in marine mammals and birds. There may be more than 2,000 California sea lions at Sea Lion Rocks at Point Lobos and up to 500 on Lobos Rocks off Soberanes Point. The Carmel River is the start of the California Sea Otter State Game Refuge, established in 1941 to help protect the sea otter; it extends to San Luis Obispo County. Thirteen seabird nesting colonies occur on offshore islets or remote cliffs, with the biggest ones at Bird Island, Point Lobos, with over 6,000 Brandt's cormorants, and at Castle Rock, with up to 2,000 nesting birds, mostly common murre and Brandt's cormorant.

Common murres are a dominant seabird in California; there are 32 breeding colonies from the Oregon border to Hurricane Point Rocks, just north of Point Sur (Manuwal et al., 2001). The annual cycle of colony attendance is divided into three distinct periods – the breeding season, the at-sea chick-rearing period, and the nonbreeding or winter season. Common murres start breeding at ages 5-7. Birds begin visiting breeding colonies in December and staying overnight at the colonies in April, when one egg per pair is laid on bare rock without a nest. The eggs have a pointed shape, so they roll in a tight circle, an adaptation necessary for birds that breed on narrow cliffs! Both parents incubate the egg over the 32 day incubation period. Chicks hatch in June and are fed 3-5 fish a day for three weeks, mostly anchovy, sardines, and rockfish. One parent stays with the chick, brooding it until it can thermoregulate, then protecting it from predators (mostly gulls, ravens, crows, and raptors). Chicks leave the colony after about 23 days. The male usually guards and raises the chick at sea for 1-2 months; during this period, the colonies are abandoned. Most birds forage within about 20 miles of their breeding colony.

As summarized by Manuwal et al. (2001), the Central California common murre population was estimated at 194,000-224,000 breeding birds in 1980-92. There was a steep decline (10% per year) between 1979 and 1989 due to mortality from gill netting, oil spills, and reduced productivity during the severe *El Niño* in 1982-83. The 1989 population was estimated at 90,200 breeding birds. When studies showed that more than 75,000 murres died in 1979–87 in central California as a result of gill-net fisheries, the State enacted gill-net closures, which has reduced this man-made source of mortality to common murres. Now, cumulative mortality from oil spills is probably the greatest source of human-related mortality affecting the central California murre population; over 9,000 common murres were estimated to have been killed over the winter of 1997-98 during a mystery spill that was eventually traced to the sunken vessel SS *Jacob Luckenbach* (Nevins and Carter, 2003). In 2006, counts using aerial photographs of the Central California common murre breeding colonies totaled 234,386 birds, with the largest colony on the South Farallon Islands with 115,864 birds (McChesney et al., 2006). A correction factor of 1.67 can be applied to whole-colony counts to estimate the number of breeding adults at colonies. It appears that the common murre has made significant recovery through both protection and restoration efforts.

It was the survival of a few sea otters near Big Sur that kept the southern sea otter from extinction. Through rigorous protection, this small population has expanded to over 2,800 animals by 2008.

Places to Visit

The truck crawled steadily and slowly but backward up Carmel Hill.

John Steinbeck – CANNERY ROW

As you leave the parking area at Monastery Beach and head south on Highway 1, you will be starting perhaps the greatest scenic auto tour in the world. It has been our good fortune to do many such tours, and the ones we would rank the highest include: 1) Going to the Sun Highway in Glacier National Park; 2) Trail Ridge Road in Rocky Mountain National Park; 3) the highway between Salalah, Oman and the Yemen border; 4) the drive between Anchorage and Homer, Alaska; and 5) the road from the Red Sea Coast up the side of the Red Sea scarp to Abha, Saudi Arabia, from barely above sea level

to 6,836 feet within a relatively short horizontal distance. And we would not want to omit the Blue Ridge Parkway between the Shenandoah National Park, Virginia and the Great Smoky Mountain National Park, North Carolina that runs within a few miles of our mountain home in western North Carolina, and one which we enjoy driving along several times a year. However, none of these, in our opinion, can match this ride through the Big Sur Country.

Going south for a little less than a mile, you will come on your right to the turn into the Point Lobos State Reserve (entry fee of $5 in 2008). This Reserve contains a large number of excellent, scenic hiking trails. The clear waters around the Point seem to contain an unusually large amount of extensive kelp forests. SCUBA diving is a popular way to explore the biologically rich nearshore waters, and the birding is great, as usual. But, to our minds, its geology is the greatest attraction of this Reserve.

After you enter the Reserve, we recommend that you turn right at 0.2 miles and drive to the parking area at Whaler's Cove. There are good exposures of the Porphyritic Granodiorite of Monterey with its *large twinned k-feldspar phenocrysts* in this area (see example of these phenocrysts in Figure 136B, a photograph of the granodiorite at Pescadero Point, Monterey Peninsula). Also, you should inspect the **nonconformity** between the underlying granitic rocks of Cretaceous age (around 94-95 million years old) and an overlying sedimentary rock formation of Paleogene age (between 65-50 million years old), called the Carmelo Formation (Jessey et al., 2003). Both of these rock units are a part of the Salinian Block and, obviously, they originated far to the south of Point Lobos and are presently en route to the north aboard the Pacific Plate. There are numerous sea stacks usually inhabited by birds off the headland by the parking area, as well as all around the whole peninsula.
[NOTE: A nonconformity is an unconformity along which older igneous or metamorphic rocks are overlain by considerably younger sedimentary rocks. An unconformity represents a missing interval of time between the origin of the two units.]

After finding the nonconformity, we recommend that you drive back to the main road and turn right and go for a little over half a mile to where the road circles to the left and goes south near the western shoreline of the Point Lobos Peninsula, along which there are seven major parking areas. From these parking areas, you can explore the fabulous outcrops of the Carmelo Formation, which are illustrated in Figures 21C, 141, and 142.

These sedimentary rocks, an amazing sequence of easterly dipping beds, are composed of coarse-grained

FIGURE 142. One of the conglomerates in the Carmelo Formation, some of the outcrops of which are shown in Figure 141. The scale is ~ 6 inches.

conglomerates (Figures 141 and 142), sandstones of different thicknesses, and some mudstones. This is one of the better locations to view sedimentary rocks along the whole Central California Coast. Based on 40 years of study, Ed Clifton, Emeritus Scientist of the U.S. Geological Survey in Menlo Park and fellow coastal geologist and long-term acquaintance and friend of co-author Hayes, has concluded that these sedimentary rocks were deposited in a submarine canyon that was carved into the underlying granodiorite. These sediments were deposited by both turbidity currents for the finer-grained sandstones, as indicated by the graded bedding shown in Figure 21C and other evidence, and submarine gravity flows for some of the conglomerates (see discussion of such flows in the present submarine canyon off Monastery Beach).

Clifton (1981; 2007) divided these rocks into six different *associations*: conglomeratic, thick-bedded sandstone, thin-bedded sandstone and mudstone, thick-bedded sandstone and mudstone, conglomerate/mudstone breccia, and conglomerate/pebbly mudstone. Therefore, as you can see, a variety of conditions prevailed as these coarse-to-fine sediments were transported down the steep slopes of this ancient submarine canyon. More details than this are beyond the scope of this book, but if you are a resident of the area or plan to spend a significant amount of time at Point Lobos, we recommend that you read the geological field trip guidebook written by Clifton (2007), which gives details on the best sites to visit to see the different rock types.

[NOTE: The term **breccia** usually means a conglomeratic assemblage of sedimentary material in which the individual clasts have sharp, unrounded edges (called angular), which indicates that they were not transported in such a way as to bump into each other many times, resulting in their losing their angular edges and becoming round, like those in most conglomerates (see example of such rounded and subrounded clasts in Figure 142). In the case of these submarine canyon deposits on Point Lobos, the breccias were probably deposited either as landslides or somehow associated with a faulting process, not as loose individual fragments in a turbidity current or a grain flow in which the individual particles came repeatedly in contact with each other.]

After indulging in this geological feast, we recommend that you go back to Highway 1 and head south. First, you will pass through the settlement of Carmel Highlands. From the south end of this town, it is 1.4 miles to Garrapata State Park. Along the way to the Park, you will pass by eight or so low scarps by the road that are composed of orange-colored Quaternary sediments that have weathered into intricate vertical chambers or skinny vertical spires similar in some respects to the **hoodoos** that occur in the High Plateaus region of the Colorado Plateau and in the Badlands of North and South Dakota. They are probably formed as a result of the combination of a clay content that holds the material together and rainfall runoff down the face of the scarps. For whatever reason, the composition of these scarps seems to be consistent for some distance along the coast, because such "hoodoos" are quite common in this area, not only in the road cuts but also in the scarps at the tops of the wave-cut rock cliffs along the shore.

Garrapata State Park is another place you don't want to miss. Its entrance is a little inconspicuous. Be on the lookout for the first place south along the road from Carmel Highlands where somewhat dense Monterey Cypress trees are present on both sides of the road. The rock cliffs, which are composed of the Porphyritic Granodiorite of Monterey, are very irregular in shape and sea stacks are common (see photograph in Figure 42C). In places, the granodiorite is eroded into rectangular coves, because of the presence of vertical joints, and the head of each slot has a perfectly rounded cobble or boulder beach which, curiously, do not have well-developed storm berms (Figure 143A). We speculate that the beach shown in Figure 143A does not have a storm berm, such as the one given in the general model in Figure 61, due to the absence of smaller cobbles and pebbles. This flat beach is composed mostly of boulder armor.

This is a great geology stop, spectacular really. There are several fantastic rounded, huge boulders on the beaches (Figure 143B). These granitic rock cliffs contain numerous sea caves. This is also an excellent place to see the granodiorite up close, if you haven't had that

FIGURE 143. Gravel beaches at Garrapata State Park (Monterey County; Zone A, Figure 140). (A) Highly rounded boulders and cobbles on a gravel beach in a jointed, rectangular embayment in the Porphytitic Granodiorite of Monterey (Cretaceous age; around 80-90 million years ago). A view looking north along the shoreline from near the same spot from which this photograph was taken, on 7 December 2008, is shown in Figure 42C. (B) Well-rounded boulders at the base of an irregularly eroded wave-cut scarp in the granitic rocks.

opportunity yet.

The park features a relatively narrow raised marine terrace, which has a thick layer of gravel on the surface that overlies the granodiorite unconformably (a nonconformity; see photograph in Figure 144). This is one of the better examples of a nonconformity you will see anywhere along the Central California Coast.

From the Park, it is 2.2 miles to the mouth of Garrapata Creek, another interesting location. The shoreline is very irregular in this interval, jagged with a lot of sea stacks. There is a narrow, somewhat tilted, raised marine terrace all along that area as well. The road then drops down a bit before you come to the 900-yard-long Garrapata Beach, which has an access trail with a stairway down to the wide sand beach. At the southern end of this beach, there is a bridge across the Garrapata Creek which, no doubt, supplies the beach with its sand.

About a mile beyond the bridge over Garrapata Creek, you will encounter a right turn to the Rocky Point Restaurant. Jessey et al. (2003) made note of the fact that, in the vicinity of the restaurant, there are some *NE dipping Upper Cretaceous, brown weathering coarse sandstone, shale, and siltstone* (of the "Great Valley Sequence"). They observed further that these *well-bedded sedimentary strata are derived from the Sierran plutonic-volcanic arc*. This means that somewhere in this one mile stretch between the mouth of Garrapata Creek and the restaurant, a major fault, called the Palo Colorado Fault (Hensen and Usner, 1993) has passed obliquely offshore, with granitic rocks of the Salinian

Block being located northeast of the fault and a sliver of Cretaceous sedimentary rocks being located on the southwest side. As noted by Ed Cliffon (pers. comm.), *the Palo Colorado Fault can be clearly seen on the beach separating the granitic rocks from the sedimentary section.* As you probably noticed, along the whole area between Point Lobos and the mouth of Garrapata Creek and a little beyond, the rocks in the wave-cut rock cliffs and road cuts, for that matter, have been composed entirely of the granitic rocks of the Salinian Block.

[NOTE: Although Jessey et al. (2003) and others refer to this sequence of sedimentary rocks at the Rocky Point Restaurant as Great Valley Sequence, Ed Clifton (pers. comm.) suggested that this term should only be applied to the rocks that are actually in the Great Valley. The rocks called "Great Valley Sequence" here on the coast *formed at a similar time and tectonic setting as those in the Great Valley, but their origins were hundreds of miles to the south* (near the south end of the trench shown in Figure 12). Clarifying the matter further, Clifton stated that: *Here, and in San Mateo County* (in this case, he is referring to the Pigeon Point Formation that was discussed in the description of Compartment 4), *the Cretaceous "Great Valley Sequence" was deposited on granitic or (as along this stretch of coast) on metamorphic basement that is locally intruded by granite and is part of the Salinian Block. In the present day Central Valley, the Central Valley Sequence is a vast wedge of deepwater clastics derived from the continent. They lie on oceanic crust in the western part of the valley and on granitic crust on the east side.*]

FIGURE 144. Unconformity, or more specifically, a nonconformity, at Garrapata State Park (Monterey County; Zone A, Figure 140). Marine terrace deposits containing abundant cobbles and a few boulders overlie the Cretaceous Porphyritic Granodiorite of Monterey. Arrows point to the nonconformity. Photograph taken on 7 December 2008.

Not long after you leave the sliver of Cretaceous rocks at the Rocky Point Restaurant you will be back in Salinian Block granitic rocks, because the steep slopes in the wave-cut rock cliffs you will be driving by are *dark gray, coarse-grained, biotite-rich charnockitic tonalite*, according to Jessey et al. (2003). *The tonalite contains medium-grained inclusions or metadikes (chiefly biotite + quartz + plagioclase). Float blocks of very coarse, white marble suggest proximity to the Sur Series. Presumably, the tonalite intruded the* (older) *carbonate strata, and then both were metamorphosed*. Lots of action went on down there beneath that original land-based volcanic arc far to the south of this location.

[NOTE: As discussed earlier, tonalite is one of the series of granitic rocks that happens to be rich in plagioclase (Figure 17B). Clearly then, though considerably jumbled up, these rocks are a part of the Salinian Block, so the main line of the Sur-Nacimiento Fault must be offshore at this point (Figures 18 and 140). Charnockite

is defined as *any of various faintly foliated, nearly massive varieties of quartzofeldspathic rocks containing hypersthene*. Thus, a key component of the tonalite in this area is **hypersthene**, which is *a green, brown, or black splintery, cleavable pyroxene mineral, essentially* $(Fe,Mg)_2Si_2O_6)$. A quick review of the literature on the complex metamorphic process of charnockite formation in igneous/metamorphic rocks from studies carried out around the world (e.g., India, Norway, and Antarctica) indicates that the influx of CO_2 from deep sources is usually an important factor. Therefore, deep faults that allow this influx are necessary. Some research indicates that its presence is apparently evidence for relatively rapid decompression. This is a complicated subject, well beyond the scope of this book, but clearly, these rocks were at one time subjected to some intense metamorphic processes at great depths.]

From the turn to the Rocky Point Restaurant, it is another 1.7 miles south to the famous, much photographed, Rocky Creek Bridge. Almost as soon as you leave the Restaurant turn, the road runs along the landward side of an irregular raised marine terrace that averages around 300 yards in width. At the mouth of Rocky Creek, with its picturesque bridge, another sand beach is present.

Shortly after you leave the bridge going south, you will be looking straight down the wave-cut scarp as the road clings to the side of the mountain. The next bridge is the even more famous and more photographed bridge over Bixby Creek. Just north of this bridge, a house sits on a projecting narrow limb of a raised marine terrace high above the water. There is a small, white sand beach at the mouth of Bixby Creek.

After you cross over Bixby Bridge, it is one mile to the pullover at Hurricane Point, another important stop at which you might get blown away by the strong wind, because the Point is aptly named. On the way there, you will be driving on the landward side of a tilted, raised marine terrace that averages about 250 yards wide. On this part of the drive, there are few rock exposures along the road until you get to Hurricane Point, but the rocks in the scarp below the road have a range of features, including long bands of dark brown country rock in the tonalite, an intruded zone of marble, and tilted layers, as well as an underlying unit of the easily erodable Tertiary Pismo Formation, according to the California Department of Conservation (2001). Needless to say, this is a region of very complex geology.

There are all kinds of things to see at Hurricane Point, not the least being the phenomenal views back along the shore to the north. Just to the south of the pullover, a large scallop created by a major landslide in 1999 has created a sheer slope all the way down to the water. There are numerous scattered blocks of white rocks up the slope above the road. These whitish chunks of marble are a part of the Sur Series, which is at least as old as the Paleozoic (some geologists estimate around 300 million years ago, but others think it is older than that). Therefore, Hurricane Point is the site that contains the oldest rocks you can see and touch anywhere along the Central California Coast. Associated metasedimentary rocks of different types can also be seen down the slope to the beach.

From Hurricane Point, it is 2.2 miles to the crossing of Little Sur River. Before you get to the river, you will have crossed the San Gregorio-Hosgri Fault and passed out of the Salinian Block and into the Franciscan Complex (Figures 18 and 140). The Little Sur River, which has a watershed mostly in the Salinian Block, has delivered enough beautiful white sand to the shore to create a sand beach one mile long.

[NOTE: At this location, the fault systems are quite complex. The San Gregorio Fault crosses the shoreline from offshore, running in a southeasterly direction more-or-less parallel to what appears to be the northwestern extension of the Sur-Nacimiento Fault [called the Sur Hill Thrust at this point by Hensen and Usner (1993).]

After crossing the river, it is 0.3 miles to where the road once again parallels the shore. Along this section, an 0.4 mile stretch of sand dunes have been deposited up the hill on the landward side of the road, another testament to the high sediment productivity of the Little Sur River.

You next come to the wide dune field on the triangle of low land behind the rock headland of Point Sur. This pronounced headland is shown on the oblique aerial view in Figure 145. The black rocks that make up the huge headland are volcanic lava flows of the Franciscan Complex (McWilliams and Howell, 1982; Alt and Hyndman, 2000).

During our 1992-93 studies, we surveyed the sand beach on the north side of Point Sur (Figure 52A). As shown by the profile plots in Figure 52B, the depositional berm on this beach was nearly 100 yards wide during both surveys, which emphasizes further the role of the rock headland as a sediment trap. It is probable that the Little Sur River has also contributed a significant amount of sand to this beach, in addition to the wide beach off its mouth.

The gated entrance to the headland/lighthouse area is located on Highway 1 directly in the lee of the headland. Guided tours to the lighthouse area are possible, but they are usually only given a couple of times a week. For specific information on the tours consult the website for Point Sur (2009).

SANTA LUCIA UPLANDS (POINT SUR TO LOPEZ POINT)

Introduction

This part of the shoreline (Zone B, Figure 140) is dominated by steep mountains with two short segments of raised marine terraces. The redwood forests and the blue-green waters of the Big Sur River in Pfeiffer Big Sur State Park make for a restful and peaceful atmosphere. Sand beaches are few and far between, thus the shoreline is dominated by wave-cut rock cliffs, sea stacks, and an occasional gravel beach. Rocks of both the Salinian Block and the Franciscan Complex are well-exposed in several places. As usual, the birding is superb, and the striking scenery with the steep mountains dipping into the sea is beyond compare.

FIGURE 145. Oblique aerial view of the black volcanic rocks of the Franciscan Complex (Jurassic to Cretaceous age) that form the core of the headland at Point Sur (Monterey County; Zone A, Figure 140). Arrow points to wave-cut rock scarp along the margin of a low raised marine terrace. Photograph taken on 23 September 2008, courtesy of the California Coastal Records Project, Kenneth and Gabrielle Adelman.

Geological Framework

Going from north to south, the black, volcanic lavas of the Franciscan Complex form the core of the headland at Point Sur. The Franciscan Complex rocks extend south for several miles but, ultimately, the **San Gregorio-Hosgri Fault** crosses the coast a few miles to the south of Pfeiffer Big Sur State Park, and Salinian granitic rocks (charnockitic tonalite is common) are present in the road cuts and wave-cut scarps all the way to a couple of miles beyond McWay Falls, where the **Sur-Nacimiento Fault** crosses the shoreline (Figures 18 and 140). From there to Point Lopez, Franciscan mélange dominates the shore, leading to high eroded scarps and many landslides.

[NOTE: In naming these major faults, we have used the general designation given by Hensen and Usner (1993; Figure 18). However, more detailed maps published in their book, *The Natural History of Big Sur*, shows the faults to be much more complicated than that. If you are interested in the details, we suggest that you refer to that book or to more in-depth geological maps published by the U.S. Geological Survey.]

Beaches

Sand beaches are scarce in this area, the most accessible one being Pfeiffer Beach, a tombolo composed of beautiful white sand with a generous supply of garnet-rich heavy minerals. There are several gravel beaches, but only the one at the mouth of the Big Sur River is readily accessible.

Beach Erosion and Landslides

Generally speaking, beach erosion *per se* is not as much of a problem in this Zone as it is in many other areas because of limited development. The biggest problem is landslides. Hapke (2005) listed the following damaging landslides that have occurred in this area in relatively recent times:

1) The Sycamore Draw slide closed Highway 1 for

approximately eight months in 1983.

2) J.P. Burns landslide closed Highway 1 for over a year in the winter of 1983.

3) Landslides during the *El Niño* of 1998 destroyed hot tubs at the Esalen Institute.

4) Chronic landslides at Hot Spring Creek caused the highway to be closed for short periods several times.

5) Slide at "Cow Cliffs" closed Highway 1 in January 1983.

6) Highway 1 closed at Big Creek because of slides for a week in March 2000.

The California Department of Conservation (2001) mapped fewer landslides in the central part of this segment, where the charnockitic tonalite prevails. They concluded that the tonalite is a *very hard, massive, coarse-grained igneous rock with relatively few fractures*, as opposed to the Franciscan Complex rocks, which border it to the north and south. They observed further that *this segment of the coast* (where the tonalite crops out) *has almost uniformly steep slopes of 50% to 65% from sea level to elevations of 2,000 feet or more. Although the slopes are very steep, they are apparently relatively stable.* They concluded further that *this area has moderate potential for landslides*, which occur only under *exceptional circumstances*. On the other hand, they concluded that *the Franciscan mélange contains the greatest density of landslides on the Big Sur Coast*, when speaking of one specific area north of Lopez Point. The mélange is composed of a loosely held together matrix of muddy sediments that contains randomly scattered huge blocks of miscellaneous rock types, whereas the tonalite is a hard, compact crystalline rock, except where fractured by faulting.

General Ecology

There are seventeen seabird breeding colonies along this section of coastline, with a total population of about 1,500 birds dominated by Brandt's cormorant and pigeon guillemot. The largest colony is off Burns Creek Bridge. The breeding season for most seabirds is February-September. In summer and fall, there will be many species of seabirds in their post-breeding dispersal, including Xantus' murrelet, which is both federally and state listed as threatened. California's entire population of Xantus' murrelets, estimated at 3,500, nests on the Channel Islands in southern California. The total number of Xantus' murrelets in North America during the nonbreeding season (1975–2003) is estimated at 39,700 birds, with 17,900 breeding birds and 21,800 nonbreeders (Karnovsky et al., 2005). They will be hard to see from land, though, because they are most common 15-100 miles offshore (and they are small–only 8 inches in length).

The kelp forest can form an almost continuous band offshore, and the sea otter population in this area was estimated at 268 adults and 57 pups in the 2003-2005 spring censuses. Just south of Grimes Canyon, there is a haulout where up to 722 California sea lions can be present, with pupping May-July; up to 72 California sea lions haul out off Lopez Point as well. For gray whales, the peak of the southern migration usually occurs in mid-January; the northern migration peaks in mid-March. Mothers and calves, which can be seen in late April and early May, often hug the shore to avoid the packs of hunting transient killer whales that can frequent these waters. In addition, humpback whales can be seen March-November, and minke whales can be seen year-round.

Big Sur River below the bridge at Highway 1 is open to steelhead trout fishing December-March, but only on Saturdays, Sundays, Wednesdays, legal holidays, and the opening and closing days. All fish must be returned to the water unharmed. Anglers should carefully check the rules and boundaries with the ranger staff to keep in line with the current regulations.

Places to Visit

"I've been to the ridge and looked down," Thomas said. "It's wild over there, redwoods taller than anything you ever saw, and thick undergrowth, and you can see a thousand miles out on the ocean. I saw a little ship going by, half-way up the ocean."

John Steinbeck – TO A GOD UNKNOWN

Driving south from Point Sur, you will be on the landward side of a 900-yard-wide raised marine terrace for the next 2.5 miles or so. This relatively low-elevation raised marine terrace has a white sand beach along the base of a low scarp much of the way.

After you leave the terrace and turn inland, within about a half mile you will come to the entrance to **Andrew Molero State Park**. The beautiful, blue-green waters of the Big Sur River run through this Park. If you have time, we recommend that you take the trail from the parking area to the mouth of the River, a walk of about one mile. As shown by the field sketch in Figure 146A, a sand spit usually builds from the south across the mouth of the river during the dry season. A gravel beach composed of well-rounded cobbles and boulders made up mostly of granitic rocks extends along the high part of the beach

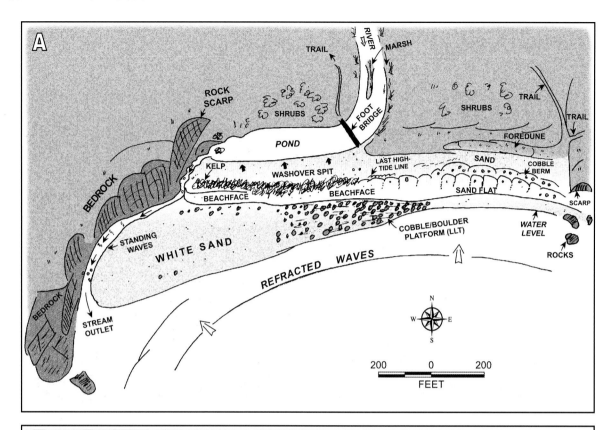

FIGURE 146. The mouth of the Big Sur River (Monterey County; Zone B, Figure 140). (A) Field sketch made on 7 November 1992. Sediment is a curious mix of fine-grained white sand and cobbles and boulders. (B) Close up view of the cobbles and boulders, many of which are granitic rocks. Scale is ~ 6 inches.

to the south (Figure 146B). This is despite the fact that the rocks in the surf zone appear to be of the Franciscan Complex. Of course, the watershed of the Big Sur River extends well into the core of the Salinian Block.

From the mouth of the Big Sur River, it is about 3.7 miles as the eagle flies to Pfeiffer Beach, which is accessible by driving. The shoreline in between the two sites is pretty rugged with two steep rocky headlands, but a sand beach is present much of the way, probably a result of the input of sand from the Big Sur River.

Continuing south along Highway 1, you will soon come to **Pfeiffer Big Sur State Park**, which covers *821 acres of redwood forests and the Big Sur River,* where swimming is possible. In addition to the redwoods, the Park contains conifers, oaks, sycamores, cottonwoods, maples, alders, and willows, plus open meadows. Wildlife is abundant throughout the Park. Popular spots along the trails include Pfeiffer Falls and the Gorge.

Before leaving the area, you should turn right on Sycamore Canyon Road and drive down to Pfeiffer Beach, an exquisite white sand beach (Figure 147A), part of a shallow tombolo located landward of a large sea stack that contains a couple of major sea arches. The stack is composed of dark-brown-colored turbidites of the Point Sur subterrane of the Franciscan Complex that were deposited in the deep Franciscan trench (Underwood et al., 1995).

A notable attribute of the white sand is its large percentage of heavy minerals, composed mostly of the reddish mineral called **garnet** (Figures 147B and 148). Under the right light conditions, some of the other sand beaches in this area take on a purple tint, because the heavy minerals are so abundant. Both the schists in the Franciscan Complex and the granitic rocks, particularly the diorite, of the Salinian Block contain significant quantities of garnet. The watershed of the Big Sur River, probably the source for most of the sand on Pfeiffer Beach, contains both of these rock types. The abundance of white sand on the beach would appear to favor the granitic rocks as the primary source for the garnet.

South of Pfeiffer Beach, the shoreline becomes quite steep with high scarps in the dark-colored rocks. Some landslides are present as well. The beaches alternate in composition between sand and gravel.

If you go back to Highway 1 and head south, you will at first be driving through redwoods that are growing in soils on Franciscan Complex rocks, including sheared serpentinite and dark-colored shales (Jessey et al., 2003). At 3.4 miles further along, you will be able to see outcrops of the Sur series marble and associated metamorphic rocks. As noted earlier, the Sur series represents the basement rocks of the Salinian Block, originally assigned

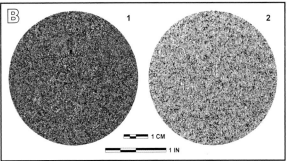

FIGURE 147. Pfeiffer Beach (Monterey County; Zone B, Figure 140). (A) The fine-grained, white sand beach. A tombolo has formed in the lee of the large sea stack in the upper left. Photograph taken on 7 December 2008. (B) Two sand samples from this beach. The white sand, which contains abundant quartz, is the dominant type of sediment on the beach, and the red sample was scraped from a thin surface layer of garnet, such as the one shown in Figure 148.

a Paleozoic age, but some now say Proterozoic (Table 1). The appearance of these rocks means that you will have crossed, or left behind, the San Gregorio-Hosgri fault some distance back (Figure 140). Actually, according to a report by the California Department of Conservation (2001), the road had been following the fault for some distance until it veered away from the coast starting a mile or so to the north of this location. From this point south all the way to the end of the Central California Coast at Point Conception, the fault is located offshore (see Figure 18).

Another 1.2 miles further along you will see well exposed outcrops of the Salinian **charnockitic tonalite**. Such outcrops are typical of the granitic rocks along the road for a number of the upcoming miles. Jessey et al. (2003) pointed out that *charnockites are relatively rare in the Sierra Nevada,* so they "guessed" that *the*

FIGURE 148. Surface layer of garnet at Pfeiffer Beach (Monterey County; Zone B, Figure 140). The photographs in Figure 147B compare this garnet-rich sand with the white, fine-grained, quartz-dominant sand that is the prevailing sediment type on this beach. Scale is ~ 6 inches.

more anhydrous nature of this assemblage suggests a higher T/P (temperature and pressure) *metamorphic overprint (deeper burial or proximity to the point of plate convergence?).* This conclusion seems to be in agreement with our earlier discussion of the charnockites.

The Julia Pfeiffer Burns State Park, with its world famous McWay Falls, is a place you would not want to miss. There are many exposures of the Salinian granitic rocks along the road north of the Park, with some massive granitic outcrops, said to be very similar to the Late Jurassic/Cretaceous rocks in the Sierra Nevada. These rocks are severely faulted and weathered in places, but it is consistently the same rock type. About a half mile north of the entrance to Julia Pfeiffer Burns State Park, there is a series of clearly exposed outcrops of the granitic rocks that extend for a few hundred yards. Sea stacks are very numerous off the bases of the wave-cut scarps in the granitic rocks.

The shoreline between Pfeiffer Beach and McWay Falls is composed mostly of high vertical scarps with no raised marine terraces whatsoever. This absence

of a raised marine terrace appears to be a common phenomenon where the adjacent shoreline is composed of the erosion-resistant, granitic rocks.

You would be the rare exception if you didn't walk down the trail in the Park to take a photograph of McWay Falls. Along the trail, you pass by a large piece of the granitic rock (tonalite?), that contains some big phenocrysts of feldspar, much larger than the groundmass. It also includes some streaks of country rock (material the granite intruded into), a brown sandstone. As pointed out by Alt and Hyndman (2000), the waterfall passes over the Salinian rocks at the top of the falls which sit on top of the major low-angle thrust fault, the Sur-Nacimiento Fault. In the more detailed map it is labeled as the McWay Thrust (Hensen and Usner, 1993).

After leaving the park, at 1.9 miles to the south you will cross the Sur-Nacimiento Fault, passing from the Salinian Block into the Nacimiento Block (also known as the Sur/Obispo Block). Up to this point, you will have passed through some relatively poor exposures of the rocks of the Salinian Block. The rocks are highly faulted and the geology is very complicated in this section.

About a tenth of a mile further along, the road outcrops include some conglomerates of the "Great Valley Sequence," originally deposited on the east side of the Franciscan subduction zone (Figure 14). According to Jessey et al. (2003), *where most fully developed, such as along the west side of the northern Great Valley, the aggregate stratigraphic thickness of the Great Valley Sequence is at least 7.4 miles* (i.e., a pile of sediments nearly 40,000 feet thick). The sediments forming these rocks were derived from the ancient continent and were apparently deposited mostly in a deep oceanic trench (perhaps more appropriately in the forearc basin; Figure 14).

[NOTE: Presumably these rocks are a part of the same misnamed "Great Valley Sequence" that was present further to the north at Rocky Point Restaurant and even further north in San Mateo County (e.g., at Pebble Beach; Figure 124). Because they don't have an accepted formal name on this side of the San Andreas Fault, we will continue to put the name in quotes. As noted earlier, the "Great Valley Sequence" here was formed at a similar time and tectonic setting as those in the Great Valley itself, but their origin was hundreds of miles to the south (near the south end of the trench shown in Figure 12).]

Along most of the coastline between the fault crossing and Big Creek Bridge, Highway 1 hangs precipitously along steep scarps in the adjacent mountains, with landslide scars being quite common. Sea stacks are numerous offshore of the eroded scarps and gravel beaches are abundant.

At four miles beyond the "Great Valley Sequence" conglomerates, you are right in the middle of typical **Franciscan mélange** of the Nacimiento Block, where disrupted graywacke sandstones, siltstones, and dark-colored shales are in abundance. These dark rocks are in strong contrast with the light-colored granitic rocks back at the Julia Pfeiffer Burns State Park and beyond.

Two or three hundred yards before you come to the rock catcher wires north of the Big Creek Bridge, there is a zone of large lenses of **Franciscan greenstones** that extend for almost a mile along the road. These greenstones are fragments of oceanic crust, with mid-ocean-ridge basalt being the most common type (Shervais, 1990). Stop to think about that as you look at this good outcrop of greenstone. This substance came from a deep mid-ocean ridge hundreds if not thousands of miles to the west of where you are standing!

At about 1.75 miles beyond the Big Creek Bridge, one of the most photogenic of them all, there is an excellent pullover/parking area at Gamboa Point, a great spot to take a picture of the Big Creek Bridge (see example in Figure 149). This is another very scenic drive through the mélange with the road barely hanging on in a few places. There are three beaches, two pure sand and one with a storm berm of cobbles and boulders, with the sand probably being derived from Big Creek.

The parking area at Gamboa Point is encircled by about 25 large blocks of rock that have been collected out of the mélange, providing a rare opportunity to see up close the range of rock types that occur in the mélange. The blocks are composed of serpentinite, greenstone, graywacke, marble, and so on.

Most of the geology from Gamboa Point to Lopez Point is similar to that leading up to Gamboa Point, except for the occurrence of a distinct, high, raised marine terrace that starts 1.6 miles beyond Gamboa Point. This terrace is rather narrow, averaging around only 130 yards in width. Except for the one south of Point Sur, this is the only clearly conspicuous raised marine terrace in this whole Zone. It is probably significant that both of these raised terraces occur in the Franciscan Complex.

RAGGED POINT UPLANDS (LOPEZ POINT TO POINT PIEDRAS BLANCAS)

Introduction

Highway 1 in much of this area (Zone C, Figure 140) typically runs along either the edge of a steep scarp or across steeply sloping former landscape scars. Great views are common everywhere, most notably at the Ragged Point Inn, where views to the north of the steep slopes of the Santa Lucia Range and to the south of the wide, raised marine terraces between Ragged Point and Point Piedras Blancas are some of the best in the Big Sur Country. The innards of the Earth are exposed in many places, especially where the serpentinite rocks from the mantle are exposed in the wave-cut rock cliffs, most notably at Jade Beach and Sand Dollar Beach. This Zone composes the southern core of the Santa Lucia/Big Sur Country.

Geological Framework

This entire shoreline is within the Nacimiento Block, which is located to the west of the Sur-Nacimiento Fault. It is composed mostly of rocks of the Franciscan Complex, with mélange deposits being common throughout. Outcrops of the Coast Range Ophiolite, one of the most extensive tracts of oceanic crust in North America, are clearly seen in the wave-cut rock cliffs between Ragged Point and Point Piedras Blancas.

Beaches

Sand beaches, especially ones that are accessible, are quite rare in this area. Gravel beaches are considerably more common, showing classic features such as armoring and imbrication. Boulder beaches are probably more numerous here than anywhere else on the Central California Coast, because of the exceptionally steep slopes and abundant mélange. Sand Dollar Beach, which is readily accessible by means of a staircase down to the beach, is the textbook example of a sand and gravel beach. It has a storm berm composed mostly of cobbles in the upper part of the beach and a wide sand berm and low-tide terrace in the mid- to low-tide portions.

Beach Erosion and Landslides

Beach erosion *per se* is not much of a problem in this section of the coast, because of limited development. However, there have been many problems with landslides. An exception occurs along the low, raised marine terrace between San Carpoforo Creek and Point Piedras Blancas, which is not very resistant to erosion. In some places where Highway 1 has been built close to the shore, it has been necessary to add riprap to protect it.

Hapke (2005) listed the following damaging landslides that have occurred in this area in relatively recent times:

1) The shoreline south of Lopez Point has numerous problems because of unstable slopes. A slide in one area displaced 20 million cubic yards of material in the winter

FIGURE 149. Big Creek Bridge from Gamboa Point (Monterey County; Zone B, Figure 140). View looks north on 7 December 2008.

of 2000.

2) The shoreline near Rockland Landing has had many slides. In March 2000, the highway was closed for three and a half months.

3) A slide at Cape San Martin caused the road to be closed for 10 days in 1965.

4) Near the town of Gorda, which *sits on an ancient landslide*, Highway 1 was closed for a month in January 1997.

5) At Redwood Gulch, located north of Salmon Creek, the road was closed for over two months because of a landslide in March 1986.

The California Department of Conservation (2001) indicated that the area between Lopez Point and Willow Creek is relatively free of landslides. However, with respect to the area between Willow Creek and Salmon Creek, they observed that *the abundance of very large slides suggest that this part of the mélange bedrock is notably weaker than the same unit immediately to the*

south. They also noted that the abundance of slick serpentinite along shear zones may contribute to the problem in that region. And, as you might expect if you have ever driven along the road there, the area between Salmon Creek and Ragged Point Inn *has a moderate to high potential for landslides due to the steep topography.*

General Ecology

With the numerous sea stacks along this coast, there are many seabird breeding colonies–eighteen, in fact. The dominant bird in these colonies is Brandt's cormorant. Large colonies can be seen at Point Piedras Blancas, Cape Saint Martin, and just south of Sand Dollar Beach. There are eight California sea lion haulouts and a population of up to 3,864 animals along this stretch of coastline. The largest haulout is at Point Piedras Blancas, with up to 2,816 animals present; pupping occurs May-July. There are fourteen harbor seal haulouts and hundreds of harbor seals in the area. The extensive kelp forest and its associated sea otter population continue offshore. When resting or sleeping in the kelp, sea otters will drape strands of kelp over their body to prevent them from drifting away.

Places to Visit

South from Lopez Point on Highway 1, it is about a mile to the settlement of Lucia. From there it is another mile until you reach a half mile long major landslide scar in the Franciscan mélange. There are huge numbers of sea stacks all along this section of the shore, where the big chunks of large, resistant blocks of rock from the mélange have accumulated along the base of the wave-cut scarp, ideal conditions for the creation of numerous relatively small sea stacks. It is one mile from the south end of the landslide scar to the entrance to Limekiln State Park. A few hundred yards before you get there, you will be driving along a zone of protective wire mesh draped over Rain Rocks, a steep scarp composed of **Franciscan greenstone**. A report by the California Department of Conservation (2001) concluded that Rain Rocks *originally formed very steep cliffs that rose hundreds of feet above sea level. Construction of the highway across these cliffs resulted in hundreds of feet of very steep slopes above the Highway. At Rain Rocks, these slopes are nearly vertical and composed of hard rocks that are prone to rockfalls* (hence the wire mesh). This is another great location to view the greenstone.

The historic activities at Limekiln State Park were outlined as follows by Jessey et al. (2003): *The manufacture of lime at this locality dates back to the 1880s. …Four stone and steel furnaces were built at the base of a natural limestone landslide. The kilns were loaded with limestone (marble) then stoked with redwood cut from nearby slopes.*

The marble they used surely came from a formation called the Gavilan Limestone of the Sur Series, which crops out on the east side of the Sur-Nacimiento Fault. The Fault runs more-or-less parallel with the shoreline and is located a ways up the side of the mountain. At the shoreline, the extensive outcrops of the Franciscan greenstone are located on the west side of the Fault.

When we visited this area in December 2008, the Park was closed because there had been a damaging fire, called the Chalk Fire, which burned several thousand acres of forest in Los Padres National Forest in the fall of 2008.

Another major, landslide-scared slope is located a short distance to the south of the entrance to the Park. It continues for about another mile along which the mountain is very steep and whole sides of it have slid

into the ocean some time in the not-too-distant past. At this point, you are back into the heart of the mélange. Jumbled material continues to crumble down, with some rocks on the road. From this slide-prone area, it is about another mile to the Kirk Creek Campground, which is located on the flat top of a raised marine terrace about 150 yards wide.

A very short distance past the Kirk Creek Bridge, the Nacimiento-Ferguson Road starts its winding path up the side of the mountain. This drive provides some terrific views along the shoreline to the south. Also, if you are so inclined, you can reach the Sur-Nacimiento Fault by driving 2.7 miles up this road. The fault, which runs northwest/southeast, is a thrust fault (Figure 10 - Middle) that dips approximately 59 degrees to the east, according to Jessey et al. (2003), who also made the following interesting speculation: *The juxtaposition of oceanic crust (Franciscan greenstones) to the west and Sierran volcanic arc rocks to the east* (Salinian Block) *suggests this fault may have marked the Cretaceous convergent boundary between the continental North American and the Pacific/Farallon Plates.* As we said, an interesting speculation, presumably based on the original paper by Page (1970). At this location, the fault line is located about 0.9 miles as the eagle flies to the northeast of Highway 1.

The Mill Creek Picnic Area, another good place to stop, is located 0.4 miles south of the Kirk Creek Bridge. This is one of the few places along the Central California Coast where it is possible for you to walk right up to a pure boulder beach (Figure 150). As you can tell in the photograph, these boulders have a wide range of compositions, including sandstones, metamorphosed mudstones, greenstone, and a variety of others. Another enigma at this site is a large block of vertical brown

FIGURE 150. Boulder beach at the Mill Creek Picnic Area (Monterey County; Zone C, Figure 140), which shows a wide range in composition of the individual boulders.

sandstone. Owing to its location, it almost has to be from the Franciscan Complex. A extra large block in the mélange? Also keep in mind that the Sur-Nacimiento Fault is only a short distance east of the highway at this spot.

Looking to the north along the beach straight ahead from the Mill Creek area for 4 or 5 miles, you can see the very high mountains of the Santa Lucia Range, with peaks over 4,500 feet.

From Mill Creek on south for a little under a mile the shoreline is a steep, badland-like, gullied topography in what appears to be Quaternary material with an orange-reddish color. The base of the scarp is some kind of resistant bedrock and the beach is mostly gravel, being composed of pure boulders in some segments. The scarp is steep and the road hugs close to the mountain side above the water.

For the next 3.4 miles beyond these narrow, steep scarps, Highway 1 runs along the landward side of a flat, raised marine terrace of highly variable widths, with a maximum width of a little over 500 yards. The irregular eroded margin of the terrace has numerous headlands with abundant sea stacks offshore. The widest part of the terrace is located just to the north of the Sand Dollar Beach embayment (Figure 151A). Figure 47A shows another view of this raised marine terrace.

When you reach the Sand Dollar Beach Picnic Area, we highly recommend that you pull into the parking area, go down the long stairs to the beach, and spend some time observing a number of interesting features along the shore. To get an overview of the shoreline, see the photograph in Figure 151B that illustrates a line of spilling waves coming into the embayment on 7 December 2008. As shown on the vertical image in Figure 151A, Sand Dollar Beach is located at the head of an 0.8 mile long arcuate embayment separated by two projecting rock headlands. The embayment has been eroded into the wide, raised marine terrace, which can be clearly seen in the upper right in Figure 151B. The beach there is a classic example of a mixed sand and gravel beach that has a high storm berm composed of gravel (mostly pebbles and some small cobbles; Figure 152A/B) and a lower sand berm and sandy low-tide terrace, as illustrated by the beach profiles given in Figure 152C.

The geology of the rocks in the wave-cut scarp behind the beach is possibly just as interesting as the beach morphology itself. Reeg (1999) stated that the Sand Dollar Beach area is in the Sur-Nacimiento Fault "zone," but it is our understanding the main fault line is located a bit further up the mountain to the east; if so, the rocks in this area are within the Nacimiento Block and hence in the Franciscan Complex (Figures 16A and

FIGURE 151. Sand Dollar Beach (Monterey County; Zone C, Figure 140). (A) Image of the Sand Dollar Beach embayment and vicinity. NAIP 2005 imagery courtesy of the California Spatial Information Library (CaSIL). (B) Overview of Sand Dollar Beach from the top of the wave-cut scarp. Spilling waves are the dominant wave type. Arrow points to the raised marine terrace to the north of the beach, which is also illustrated in Figure 47A. Photograph taken on 7 December 2008.

18). To see this geology in all its glory, walk to the north end of the beach where some **serpentinite** rocks that originated as mantle **peridotites** are clearly displayed (to do so, it is a good idea to visit this beach at low tide).

[NOTE: At first, we thought there may be some kind of

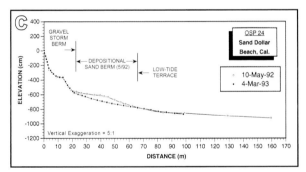

FIGURE 152. Sand Dollar Beach (Monterey County; Zone C, Figure 140). (A) View of the beach from trail at the top of the bluff, showing a gravel storm berm at the high-tide area (arrow) and sand across the lower part of the beach. (B) Gravel sediment (pebbles and cobbles) of the storm berm (scale ~ 6 inches). Photographs in A and B taken on 10 May 1992. C) Topographic beach profiles measured on 10 May 1992 and 4 March 1993. Labels of morphology apply to the May 1992 survey. The sand berm had been eroded away at some time before the March 1993 profile was measured.

a connection between the Franciscan Complex and the Coast Range Ophiolite. One of Doris Sloan's strongest critiques of our original manuscript (based on her detailed review of the text) was on this topic. She said

– The Coast Range Ophiolite is not part of the Franciscan Complex. It is the oceanic basement on which the Great Valley Sequence was deposited. The Coast Range Ophiolite is everywhere separated from the Franciscan rocks by a fault, the Coast Range Thrust – leaving very little room for doubt about her opinion on this topic, which is based on her detailed studies of the rocks in the San Francisco Bay Area and elsewhere.]

Needless to say, then, the geology of the Franciscan Complex relative to the Coast Range Ophiolite should be understood, so let us review it again as it was outlined by Reeg (1999):

1) The Ophiolite was formed between 165-153 million years ago in the mid-Jurassic (Hopson, Mattinson, and Pessagno, 1981). There doesn't appear to be much disagreement on this conclusion.

2) The initial serpentization of the peridotite occurred approximately 3-5 kilometers (2-3 miles) beneath the ocean floor in the Late Jurassic period. The serpentization process, which involved hot waters forced downward through fractures to convert the peridotite in the mantle to serpentinite, was discussed in some detail in Section 1.

3) During the Late Cretaceous, the Great Valley Sequence, consisting of sandstones, conglomerates, etc., was deposited mostly on top of the Coast Range Ophiolite (Bailey, Blake, and Jones, 1970), but along the east side of the Great Valley, they onlap the Nevadan and older basement terranes of the Klamath Mountains and Sierra Nevada (Martin, 2009). At approximately the same time, the Franciscan Complex of oceanic basalt, chert, sandstones, and conglomerates were accumulating in a deep oceanic trench located further offshore (but not on the Ophiolite, as was also explained in some detail in Section I; see Figure 14).

4) The Ophiolite was in the forearc basin of an ancient subduction zone and, thus, was in the hanging wall of the (new) subduction complex. In other words, the new subduction zone was thrust **under** the forearc basin rocks, as is generalized in Figure 14. Okay! They are separated by a **fault**!

5) Inasmuch as the Franciscan Complex was thrust beneath and now underlies the Coast Range Ophiolite (Dickinson, 1983), its jumbled, metamorphosed nature suggests (is further evidence that) it was part of the subduction complex that was accreted onto the underside of the Ophiolite during the (more recent) subduction process (Jayko, Blake, and Brother, 1986).

6) The last episode of thrusting along this boundary occurred in the Late Cretaceous/Early Paleocene epoch (69-65 million years ago) (Bogdanov and Dobretsov, 1987*).

Many thanks to Reeg (1999) for spelling this out so clearly. Being a graduate student, perhaps she felt the need to explain it in words so plainly that even coastal geomorphologists like ourselves could understand it. Sad to say, however, that the explanation she gave for the origin of the Ophiolite is still a matter of some dispute, as is mentioned later.

To clarify these points a little more, the serpentinite at the north end of the beach has been said by some geologists to be an exotic block bounded by graywacke within the Franciscan Complex. In other words, the serpentinite may be a broken off piece of the Ophiolite complex. And, again, inasmuch as these rocks are serpentized peridotite, they were derived originally from the mantle (Reinen and Davidson, 1999).

Another point to make here is that these Franciscan Complex rocks in the vicinity of Sand Dollar Beach are erodable enough such that, over time, a relatively

wide, raised marine terrace has been carved into them by the waves. One could speculate that the protruding, individual headlands are composed of some of the more resistant "exotic blocks" within the Franciscan subduction complex. Looking at these headlands, using Kenneth & Gabrielle Adelman's oblique aerial photographs, this does, indeed, seem to be the case (see again the spectacular example in Figure 47A).

From Sand Dollar Beach, it is about 0.6 miles south to the path that leads down to Jade Beach, a location that has fascinated people for decades. The pullover for the path is located about 30 yards from a sign that reads *Jade Cove Beach Recreation Area – Los Padres National Forest*. The trail crosses about 50 yards of raised marine terrace before reaching the top of the wave-cut rock scarp. Figure 153 is a view from the top of the scarp.

A kiosk by the trail warns that *prospecting, mining, or removal of any rock mineral or material is prohibited*

FIGURE 153. View looking north from the top of the bluff at the shoreline of Jade Cove, located a short distance south of Sand Dollar Beach (Monterey County; Zone C, Figure 140). The trail down to the beach is visible in the lower right. This is a popular spot to search for jade specimens, but they have to be collected from areas below mean high tide. These Franciscan Complex rocks were subjected to high pressure/low temperature metamorphism, according to Jessey et al. (2003). Photograph taken on 7 December 2008

above mean high tide. Jessey et al. (2003) commented that the jade can be seen quite well under water, because of its luminescence.

Jade is a semiprecious gemstone that takes a high polish. It is usually green, but sometimes whitish, and it typically consists of either **jadeite**, a pyroxene mineral with composition $NaAlSi_2O_6$, or nephrite. According to Jessey et al. (2003), however, the "jade" at Jade Cove is not jadeite, but is rather **nephrite**, which technically is a variety in a series of amphiboles, common minerals composed of silicates of magnesium and calcium (usually with aluminum and iron). They also reported that the rocks at Jade Cove, which are pictured in Figure 153, were subjected to high pressure/low temperature metamorphism (called the blueschist facies), often associated with plate subduction and accretion along a convergent margin. Therefore, rocks and minerals at this site were originally incorporated inside the subduction zone that produced the Franciscan Complex, a part of the Nacimiento Block, that has moved far north of its locale of formation. The association of the words subduction zone, upper mantle, and serpentinite with the "jade" at this locality fits well with the discussion of the serpentinite at Sand Dollar Beach, the previous recommended stop, which is only a half mile or so north of Jade Cove.

The raised terrace is high, around 120 feet, the scarp is steep, and the trail down the scarp is pretty rugged. The beach is composed of an assemblage of boulders that have fallen off of this eroded scarp. These boulders are composed of a little bit of everything, including *blue and green schist* (metamorphic rocks*), fibrous to asbestiform serpentine, graywackes, and nephrite jade, from which the name of the cove is derived* (Jessey et al., 2003). This is an amazing collection of boulders that you may want to check out, particularly any of you that may have rockhound blood.

After you leave Jade Cove, it is 1.6 miles to the next recommended stop, Willow Creek Picnic Ground. After about 350 yards or so, the flat, raised marine terrace will be left behind and the road will follow fairly close to the edge of the wave-cut rock scarp. About a quarter of a mile before the Willow Creek Bridge, you will encounter the edge of a large, fairly fresh landslide scar that extends to within a few hundred yards of the bridge. The cleaned off scarp has some excellent exposures of serpentinite that look similar to the rocks in the wave-cut rock scarps at Jade Beach and Sand Dollar Beach.

After you cross the bridge over Willow Creek and turn right, you can follow a paved road down to the beach. Before you go down, however, you may want to pull into the higher parking area to take some photographs of this very scenic shoreline (Figure 154A). The sediments on

FIGURE 154. Willow Creek Picnic Ground (Monterey County; Zone C, Figure 140). (A) View looking north along the coast. The bedrock in the right foreground and middle distance is an intricate mix of rock types in the Franciscan Complex. An armored boulder beach in visible in the lower left. (B) Close-up view of the armored boulder beach (compare with Figures 61 and 65). Both photographs taken on 6 December 2008.

the beach along this area are mostly gravel, which shows imbrication in the boulders along the low-tide line as well as some distinct **armoring** of the boulders (Figure 154B; also, you may want to refer back to the general model of gravel beaches given in Figure 61).

Continuing south on Highway 1 for 0.4 miles, you will come to a pullover spot on the right where you can view a striking sea stack, called San Martin Rock (Figure 155). This large stack, which is covered with white guano, is a roosting spot for numerous gulls, pelicans, and cormorants. There are a lot of gravel beaches along the shoreline in this area.

Immediately after leaving the pullover and for another half mile you will be driving along the landward margin of a well-displayed, 150 yard wide raised marine

FIGURE 155. Large sea stack called San Martin Rock located a little over 2 miles south of Sand Dollar Beach (Monterey County; Zone C, Figure 140). This area is still in the heart of the Franciscan Complex, a part of the Nacimiento Block (Figure 18). Photograph taken on 6 December 2008.

terrace. There are a lot of conspicuous sea stacks in the surf off the shoreline of the terrace. These stacks are large, resistant blocks of solid rock that eroded out of the finer, though somewhat bedded, groundmass of the Franciscan material. There is a steep, high scarp close to the road at Gorda, which was built on an ancient landslide.

South of Gorda, the road is close to the shore and large landslide scars are present in places. At 1.9 miles from Gorda, the barren slope on the left contains well-recognizable layers of radiolarian chert for about 0.4 miles. As discussed in some detail in Section I, this chert formed far out into the Pacific Ocean as the individual

radiolaria rained down incessantly on the basaltic sea floor as it moved eastward to eventually meet the subduction zone where the North American and Farallon Plates were colliding during Cretaceous times, plus or minus 100 million years ago. This chert, also known as **ribbon chert,** and the underlying **pillow basalts** form the base of the Franciscan Complex (see general model in Figure 22). To our knowledge, this is the most expansive and easily viewed outcrop of the ribbon chert exposed anywhere in the road cuts of Highway 1 from this location to the Golden Gate Bridge.

About another mile further along you will come to the area shown in Figure 64. The road is low near the water at this spot with some seawalls protecting it. Looking south from the pullover, you can see the bare scarps in the Franciscan mélange. The gravel beach at this spot shows both imbrication and armoring, as discussed in Section I (Figure 61).

About a half mile further along you will be driving through classic Franciscan mélange. As we were driving along there in December 2008, we made a note that we were "driving through the bowels of the Earth!" considering that we were so deep into the rocks of the subduction zone trench.

Soon you will make a sharp turn to the left and cross Salmon Creek, which has a frequently photographed waterfall a ways upstream. At 0.2 miles south of the bridge are good exposures of the Franciscan rocks shown in Figure 156, including a close-up of the serpentinite (Figure 156B). This is a curious jumble of material, and we could not help but speculate that the red material shown in Figure 156A was pure ophiolite. Co-author Hayes surveyed the shoreline in Oman in 1991-92,

FIGURE 156. Rock outcrops along Highway 1 near Salmon Creek (Monterey County; Zone C, Figure 140). (A) Long view of the outcrop of Franciscan Complex rocks, a part of the Nacimiento Block. The arrow points to a band of serpentinite. (B) Close-up view of the serpentinite shown in A. Scale is ~ 6 inches. Both photographs taken on 6 December 2008.

which has some world-famous ophiolite in the wave-cut rock cliffs there. These reddish rocks look very similar to the ones in Oman.

About 0.3 miles after the road once more parallels the shoreline, there are some pillow basalts exposed in the road cuts. For the few miles behind to the north and for the next 2.5 miles or so to Ragged Point Inn, the road winds its way through many exposures of the Franciscan mélange. On this ride, you will have driven past pillow basalt, bedded chert, serpentinite, as well as the classic mélange. They are all there, kind of jumbled up together. This general area provides *the choicest views of Franciscan rocks in Central California*, according to Alt and Hyndman (2005), whose book covers the geology to be seen along the highways throughout the state.

The views along this segment are also some of the best in the region. And for the last few miles to the Ragged Point settlement, the mountain side is very steep, sloping precipitously down to the sea, as shown in Figure

157. Large boulders have accumulated at the bottom of the slope.

The Ragged Point Inn area is one of the best places on the drive to stop for photographs, where you can view the steep slopes to the north (Figure 157), and the lower-level raised marine terraces, as well as Ragged Point itself, to the south. The photograph in Figure 158A was taken from beyond Ragged Point, looking back toward the mountains.

After leaving the Ragged Point Inn, the road runs along a steep scarp that flanks fragments of what appears to be a very high, raised marine terrace along the east side of the road that averages about 200 yards in width and around 300 feet high. It is 1.4 miles down the mountain to the bridge over San Carpoforo Creek, located at the northern margin of a wide lowland.

One would think that such a striking break in the topography would somehow be related to a fault zone, which it is, according to Jessey et al. (2003). The San

FIGURE 157. The Franciscan Complex near Ragged Point Inn (San Luis Obispo County; Zone C, Figure 140). Northerly view from the top of the bluff on 6 December 2008, one of the better views of the exposures of the Franciscan mélange along the Central California Coast.

FIGURE 158. Ragged Point and vicinity (San Luis Obispo County; Zone C, Figure 140). (A) View looking north across a raised marine terrace near Ragged Point (arrow). The Santa Lucia Range stands out conspicuously in the background. Photograph taken on 6 December 2008. (B) Arrow points to an exposure of the Coast Range Ophiolite that was recently described by Mehring and Cloos (2008). It is located approximately 4 miles south of Ragged Point. Photograph taken on 7 November 1992.

Simeon Fault, runs northwest/southeast along the trend of the shoreline across San Carpoforo Creek and along the base of the mountain front. This fault may be an onshore segment of the Hosgri Fault, which is located offshore parallel to it (Figure 18).

A 400-yard-long, north-oriented sand and gravel spit closes off the mouth of San Carpoforo Creek during the dry season. This stream has a moderately large watershed that drains the southern reaches of the Santa Lucia Range. The stream bed contains abundant gravel. Ragged Point itself protrudes out into the ocean about a half mile south of the stream mouth.

It is around four miles from Ragged Point to another major geomorphological feature, Arroyo de la Cruz (Figure 74), and about three miles beyond that to

Point Piedras Blancas. The shoreline in this whole area is backed by a low-elevation (around 100 feet), raised marine terrace with an average width of about 500 yards that has a number of irregular scallops in it. The raised terrace is illustrated by the photograph in Figure 158A. This part of Highway 1 is another very scenic drive, especially if you are driving north looking across the flat terrace to the abruptly elevated and imposing Santa Lucia Mountains in the background.

There are a number of well-developed sea stacks off the headlands and the scallops in some places have come so close to the road that riprap has been added to protect it. The 100 yard wide sea stack off Point Piedras Blancas, which has been covered with white guano produced by thousands of roosting sea birds over the years, is the Point's namesake.

There are sand beaches in places, which are mostly inaccessible along this whole stretch, and dune fields are located on the northern sides of both the rocky headland at Arroyo de la Cruz and at Point Piedras Blancas. The Arroyo de la Cruz shoreline, illustrated by the image in Figure 74, has several classic geomorphological features, including: 1) dune field on the northern flank made up of linear parabolic type dunes oriented northwest/southeast; 2) a north projecting sand and gravel spit that closes off the stream mouth during the dry season; and 3) a large sea stack, Cruz Rock, that has a strong influence in producing refracted/diffracted waves that move sediment on the spit back toward the north. There is a similar sand dune field north of the Piedras Blancas headland, with the sand blowing off a 500-yard-long sand beach.

This seven mile stretch of coastline is a geological hotspot at the present time. In a recent presentation, Mehring and Cloos (2008) stated that *a remnant of the middle Jurassic (165 million years ago) Coast Range Ophiolite is present near San Simeon, San Luis Obispo County, California. The Ophiolite is exposed along 5.5 kilometers (3.4 miles) of fresh, wave-cut coastline between Ragged Point and Piedras Blancas Point.* Figure 158B is a photograph of an exposure of this Ophiolite near Arroyo de la Cruz. The Ophiolite in this area has abundant **ultramafic rocks**, including gabbro and diabase, plus a variety of serpentinized minerals and other complexities.

[NOTE: Ultramafic rocks have the lowest content of silica of all the igneous rocks (<45%), and they are composed of minerals with high concentrations of magnesium and iron. **Gabbro** is a coarse-grained igneous rock composed of calcic plagioclase feldspar and pyroxene. **Diabase** is a igneous rock finer-grained than gabbro, but with a similar mineral composition. These rock types and basalts have a composition similar to the oceanic crust and mantle components of the Earth's interior. Igneous rocks composed of an abundance of lighter minerals, such as quartz and orthoclase feldspar (e.g., granite and rhyolite), are usually created in magmas derived from rocks of the continental crust or from the melting of sediments composed predominantly of the lighter minerals. Therefore, knowing the composition of the Ophiolite is highly relevant, because its composition clearly shows that it was derived from deep within the oceanic crust or the mantle and, therefore, it is significantly different in origin from other igneous rocks, such as those originally formed in the granitic batholiths of the Sierra Nevada.]

There is another remnant of this Ophiolite located to the south at Point Sal, the two remnants postulated to have been part of a single, contiguous remnant prior to offset along the San Gregorio-San Simeon-Hosgri fault system.

It is important to point out these ophiolites are only smeared out remnants of a much larger sequence of rocks called the Coast Range Ophiolite, the complete thickness of which is only preserved in a few places and nowhere along the Central California Coast. It is over two miles thick where it is totally preserved (Hopson, Mattinson, and Pessagno, 1981).

To review one more time, **ophiolites** in general are distinct igneous rocks that contain abundant iron and magnesium elements as well as plagioclase feldspars that are thought to represent oceanic crust and island arc systems of rocks (Shervais, 2003). It is a little confusing, but geologists have defined and commonly refer to a related group of rocks called the Coast Range Ophiolite sequence, which typically consists of three components: 1) layered ultramafics and gabbro (defined earlier); 2) pillow basalts; and 3) deep-water sediments (Jessey et al., 2003). The accretion against a moving continental plate by rock groups that include subduction zone wedges of sediments and island arc materials, including ophiolite sequences, has been the primary mechanism for continental growth back into pre-Cambrian times.

Research on the Coast Range Ophiolite sequence continues unabated, but geologists do not completely agree on its origin. Clearly, the Ophiolite itself is at least partly composed of oceanic crust and even some mantle material. Also, these rocks are typically in fault contact with the early ribbon cherts of the Franciscan Complex in some places and underlie the Great Valley Sequence in others. The major origins proposed for the Ophiolite sequence include: 1) originated at mid-ocean ridge and gradually moved to the continent margin; 2) on the ocean floor in a back-arc basin that was eventually covered with thick sediment piles; and 3) part of an exotic volcanic arc system that collided with North America during Late Jurassic time (Shervais, 2003). Debate about which, if any, of these origins are the correct one, even if we were qualified to do so, is beyond the scope of this book. But it is a fascinating topic, for sure.

SAN SIMEON STRAND (POINT PIEDRAS BLANCAS TO CAYUCOS CREEK)

Introduction

Having left the high mountains of the Big Sur Country behind, this area (Zone D, Figure 140) is quite subdued topographically. Much of the shoreline is a raised marine terrace of a modest elevation (40-60 feet). A major crenulate bay system dominates the northern third of this area; however, it is an unusual example in that most of its shoreline is erosional, except in the most northerly parts of the shadow zone, which contains a fine-grained sand beach. Perhaps one of the most interesting aspects of this shoreline is the large number of elephant seals that haul out and mate on the segmented sand beaches to the southeast of Point Piedras Blancas.

Geological Framework

The bedrock geology of this area is about half Franciscan Complex, made up of the usual package of ribbon chert, pillow basalts, and graywacke sandstones, and the other half is Late Cretaceous and Paleocene sandstones and conglomerates. However, outcrops one can inspect up close are few and far between. Quaternary deposits are present in some areas, but the relatively flat topography makes them fairly inconspicuous.

Beaches and Beach Erosion

Most of the major stream mouths have northerly oriented spits across their outlets that are composed of sand with auxiliary cusps composed of pebbles and granules. Pure gravel beaches are not common, especially ones that are easily accessible, an exception being at the south end of Cambria. Several fine-grained sand beaches in the northern part of this area provide haulout spots for marine mammals, including elephant seals.

The raised marine terrace that fronts much of this shoreline is underlain in many areas by relatively weak Franciscan rocks and Quaternary deposits and, therefore, is prone to erosion. For example, problems areas have developed north of San Simeon where Highway 1 runs close to the edge of the terrace. Several other developed areas have erosion problems, a prime example being in Cambria, where *all developed coastal areas are in high hazard zones and nearly every individual parcel has either a seawall or riprap for protection at the cliff base* (Hapke, 2005).

General Ecology

The beaches south of Point Piedras Blancas support a large elephant seal colony. The elephant seals were first observed at this site in 1990. The colony has expanded significantly; in 2005, there were about 14,000 adults in the colony, with about 3,500 pups born. A complete population count of elephant seals is not possible, because all age classes are not ashore at the same time; however, the most recent estimate of the California breeding stock was approximately 124,000 individuals (NOAA, 2008). There are public viewing areas adjacent to Highway 1, where you can get very close to the animals. Harbor seals also use this site; the pupping period is from mid-March through April. This is a great time to see pups and mothers and, if you are very lucky, the birth of a harbor seal. The Friends of the Elephant Seal has volunteer docents at the viewpoints to answer questions and make sure that visitors do not disturb the seals.

Harbor seals haul out at many sites along this coastline; there are seventeen mapped haulouts on the ESI maps between Point Piedras Blancas and Cayucos Creek. In comparison, there were 224 harbor seal haulouts in Central California in 2004 (Lowry, Carretta, and Forney, 2005). The California population has been generally rising since the 1980s and is thought to be stabilizing at about 34,000 seals.

You can get map information on seal and sea lions haulouts on the internet at NOAA (2009).

Kelp forests form an almost continuous band offshore, and sea otters number in the hundreds. The southern boundary of both the California Sea Otter State Game Refuge and the Monterey Bay National Marine Sanctuary is near Moonstone Beach.

There are only four seabird nesting colonies with relatively low numbers of nesting birds along this shoreline: near the elephant seal viewing areas near Point Piedras Blancas, on a small rock just west of San Simeon Point, and two sites between Little Pico Creek and Pico Creek. San Simeon Beach is a good place to potentially see wintering snowy plovers. Offshore, the following birds are present in moderate numbers at different times of the year: sooty shearwater, Cassin's auklet, pigeon guillemot, rhinoceros auklet, common murre, cormorants, Pacific loon, Clark's grebe, western grebe, white-winged scoter, and surf scoter.

See discussions of birds for Shamel County Park and San Simeon State Park in the next section.

Places to Visit

It is approximately six miles from Point Piedras Blancas, which has no public access, to the turn to Hearst Castle. Along the first segment of the drive, there are three very popular parking areas for viewing elephant seals and harbor seals. These marine mammals haul out on segmented fine-grained sand beaches. One of the parking areas is located a mile southeast of Point Piedras Blancas, another one 0.4 miles further along, where the photograph in Figure 159 was taken, and the last one another mile further.

The highway in this area is located on a relatively low-level, raised marine terrace 40-60 feet high that has a very irregular, scalloped margin. The wave-cut rock platform offshore is also irregular, with numerous scattered knobs of rock. Sea stacks are common, and there are two small sand dune fields where the terrace edge has jutted out to the south a sufficient distance for the northwest winds to create the dunes. According to the geological map, the rocks in the low scarps in this section are sandstones with smaller amounts of other sedimentary rock types of the Franciscan Complex (Jurassic to Cretaceous in age). The layered rocks are tilted in many areas.

As shown by the image in Figure 160, the William R. Hearst Memorial State Beach and its associated pier are located in the shadow zone of a three mile long crenulate bay. This bay is erosional everywhere except in the shadow zone, where a fine-grained sand beach

FIGURE 159. Fine-grained sand beach that serves as a haulout area for marine mammals on 6 December 2008. It is located beside Highway 1 about 1.4 miles southeast of Point Piedras Blancas (San Luis Obispo County; Zone D, Figure 140). Arrow points to a male elephant seal.

FIGURE 160. Image of an eroding crenulate bay in the San Simeon area (San Luis Obispo County; Zone D, Figure 140). The William R. Hearst Memorial State Beach is located in the shadow zone of the crenulate bay (arrow). NAIP 2005 imagery courtesy of the California Spatial Information Library (CaSIL).

approximately 1,000 yards long is located. The crenulate bay is created by waves from the west that refract/diffract around a low rocky headland, San Simeon Point, that projects about 700 yards out into the ocean.

From the road to the Hearst Castle south for a ways there are two pullovers where you can view sandy intertidal areas. These are good places to stop for pictures. There are lots of sea stacks offshore of the low bluffs with their rough-hewn, wave-cut rock platforms. Highway 1 is located close to the bluff edge along this section. A bridge over Little Pico Creek is located 1.4 miles south of the Hearst Castle road. There is a small sand and gravel beach at the creek's mouth, but there is no obvious public access to it.

It is about 1.5 miles from Little Pico Creek to the bridge over Pico Creek. The road is very close to the low, wave-cut scarp in this stretch, which has two pullover parking areas for more viewing of the short sandy beaches, sea stacks, birds, and possibly some marine mammals. Just before you get to the bridge over Pico Creek, there are two flat-topped headlands, fragments of the raised marine terrace, that project out into the ocean about 200 yards.

As soon as you cross Pico Creek turn right, opposite San Simeon Lodge, and drive a short distance to a parking area from which a stairway leads down to the 1,000 yard sand and gravel beach fed by Pico Creek. This is a narrow beach in places, so it is best to visit it at low tide. This beach has some pebble cusps with some boulders and cobbles scattered along the high-tide line that appear to have been eroded out of raised marine terrace deposits that overlie gray/green Franciscan rocks. This is not a well-defined, pure gravel beach, at least not near the stream mouth. When we visited this site in December 2008, there were about a thousand cormorants offshore in the kelp. This is a peaceful beach with lots of birds.

271

Assuming you are traveling south, it is a little under two miles from Pico Creek to the turn into San Simeon State Park, a place you shouldn't miss. Once you turn into the first parking area, there is a boardwalk that goes under the highway to a spot where you can walk out onto a sand and gravel beach by another creek outlet, this one for San Simeon Creek. There is usually some ponded water behind the beach where the stream comes out, except during floods, of course, when the spit across the mouth is washed away. This beach has a wide, cuspate berm with small pebbles and granules on the horns of the cusps (Figure 161A). As you can see in Figure 161B, the pebbles and granules are virtually free of any quartz, presumably having been derived from the radiolarian cherts and other rock types in the Franciscan Complex, although the rocks near the shore have been mapped as Late Cretaceous and Paleocene sandstones, conglomerates, and shales. The abundance of ribbon chert pebbles has resulted in some of these local beaches around Cambria being called "Moonstone Beach." One tourist brochure we read advertised that the Moonstone Beach contains jasper, jade, and moonstone agates. As best we can tell, these pebbles are mostly ribbon chert, but jade may be a possibility. Eroded rock scarps and associated sea stacks occur both north and south of the Creek's mouth. A large sea stack to the north is typically covered by birds (Figure 162).

On our trip to this beach in December 2008, we saw the following birds at this location: four surf scoters, lots of gulls, a mallard in the lagoon, some loons, and several western sandpipers. About 50 cormorants, as well as several brown pelicans, were diving into the waves as they participated in a feeding frenzy.

It is about two miles from San Simeon State Park to the heart of *Shangri-la*, otherwise known as **Cambria**. In between, Highway 1 stays close to the beach. There are several access points to Moonstone Beach where you can look for "moonstone agate," whatever that is. While you are trying to discover some of the mysterious pebbles, you may want to look north up the shoreline to where the lofty mountains of the Big Sur Country are still in view.

Once you get to near the middle of Cambria, follow the signs to Shamel County Park. The medium- to coarse-grained sand beach in the Park, which has highly elevated pebble/granule cusps, closes off the mouth of Santa Rosa Creek, except when it is in flood, of course. The sediments are composed of chert and rock fragments, with little to no quartz, no doubt with a Franciscan source.

When we visited this beach in December 2008, there was a huge flight of cormorants offshore, and a

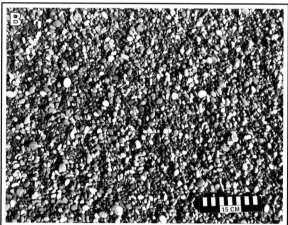

FIGURE 161. San Simeon State Park (San Luis Obispo County; Zone D, Figure 140). (A) View looking north along this beach, which has a wide, cuspate berm with small pebbles and granules on the horns of the cusps. (B) The typical sediment on the beachface, which is virtually free of any quartz, being composed mostly of rock fragments (e.g., radiolarian chert). Scale is ~ 6 inches. Both photographs were taken on 6 December 2008.

big batch of gulls on the water. Some vultures circling overhead were joined by hundreds of gulls as soon as we walked out onto the beach. There were also a number of loons and sea ducks on the water offshore. In fact, there were hundreds of birds out there riding on the surf, including brown pelicans. Being surrounded by all this, we remarked to ourselves that this was "the best birding spot we had experienced in our recent travels on the Central California Coast."

There is some great salt pruning of the vegetation along the edge of the stream mouth. Numerous sea stacks offshore add to the diversity of the place. In short, this beach "has a little bit of everything."

If you leave Cambria heading south, it is about eleven miles before you come back to the shoreline, bypassing

FIGURE 162. Sea stack north of the mouth of San Simeon Creek at San Simeon State Park (San Luis Obispo County; Zone D, Figure 140). The stack was covered with brown pelicans and cormorants on this day (6 December 2008).

a pretty rugged and inaccessible coast all the way. Not long after you come into view of the shore again, you will have arrived at a new State Park called Estero Bluffs (located about 1.5 miles east of the bridge over Cayucos Creek; Figure 140). This Park is located along the edge of a flat, raised marine terrace about 300 yards wide, which is thought by some to have formed about 120,000 years ago. The rocks in the terrace are Franciscan mélange, composed of a mixture of sandstones, chert, serpentinite, and so on.

The beach is mixed sand and gravel and sea stacks are common. The Park has four miles of trails, many of which are along the edge of the wave-cut scarp (California Coastal Commission, 2003). This is an outstanding place to take a walk.

19 COMPARTMENT 7: THE SOUTH BAYS

INTRODUCTION

Four large, arcuate shorelines with an abundance of sand beaches offer a strong contrast to the Big Sur County to the north (Figure 163). Morro Rock, a preserved volcanic neck of Miocene age, is a striking landmark off the entrance to Morro Bay, one of the largest estuaries on the outer coast of Central California. Montaña de Oro State Park is a geological wonderland with every conceivably type of eroded rocky shore in the thin-bedded sedimentary rocks of the Miguelito Member (basal member) of the Pismo Formation (Miocene age). Three crenulate bays to the south, the tangential ends of which face directly into the dominant northwest winds, contain major sand dune fields off their central and southern shores, by far the most sand dunes found anywhere on the Central California Coast (Figure 77). The two wide sand beaches originally fed by the Santa Maria and Santa Ynez Rivers contain rhythmic topography, constructional berms, and tons of birds! The southern end of this Compartment, as well as that of the Central California Coast, terminates at Point Conception along the western end of the Santa Ynez Mountains of the Transverse Range, with its strongly folded rock layers.

We have subdivided this Compartment into six Zones that are mapped on Figure 163:

A) **Estero Bay (Cayucos Creek to Islay Creek)** is dominated by the arcuate shoreline around Estero Bay. The most dominant feature is a volcanic neck on the order of 22-million years old, called Morro Rock, that has been eroded down to an isolated, circular core that stands 576 feet above the water surface just off the main shoreline. A sand spit covered with linear, parabolic sand dunes has extended for five miles away from a headland to the south to close off Morro Bay. Some of the most diverse birding opportunities on the Central California Coast are available in this area.

B) **Montaña De Oro Headland (Islay Creek to Point San Luis)** contains Montaña de Oro State Park, one of the most spectacular and interesting places on the Central

California Coast, particularly with respect to its coastal geomorphology. It is a rocky coast with only a few sand and gravel beaches that extends most of the way around the west and southwest margin of the Irish Hills. Much of the shoreline is backed by an exceptionally wide, raised marine terrace.

C) **San Luis Obispo Bay (Avila Beach to Point Sal)** is, in effect, a large crenulate bay. The bay's irregular, east/west oriented, northern shoreline contains some fine-grained sand beaches, a complex assortment of Paleogene/Neogene sedimentary and volcanic rocks, and some of the most picturesque, eroded rocky shores anywhere (see photograph on cover). The southern shoreline contains the usual arcuate, sandy strand. The largest area of coastal sand dunes on the whole Central California Coast is an added attraction.

D) **San Antonio Valley (Point Sal to Purisima Point)** is a smaller scale, mirror image of the Zone immediately to the north. Of special note is the occurrence of Coast Range Ophiolite sequence rocks at Point Sal, which are widely discussed in the geological literature. Two other features of special interest include major landslides south of Point Sal, and large fields of sand dunes landward of a long sand beach in the southern two-thirds of the shoreline. Vandenberg Air Force Base occupies most of this shoreline; therefore, access is limited.

E) **Santa Ynez Valley (Purisima Point to Point Pedernales)** has a central area composed of a long sand beach backed by sand dunes, with a zone of low wave-cut scarps in mostly Quaternary deposits to the north and steeper wave-cut rock cliffs in bedrock at the extreme southern end. The Santa Ynez River is the primary source for the sand on the beaches and in the dunes, which are vegetated for the most part. The wide, medium-grained sand beach at the mouth of the river is the most remote and underused sand beach in the area that is easily accessible.

F) **Santa Ynez Headlands (Point Pedernales to Point Conception)** cover a remote area with only one location that has ready access to the beach, Jalama Beach County

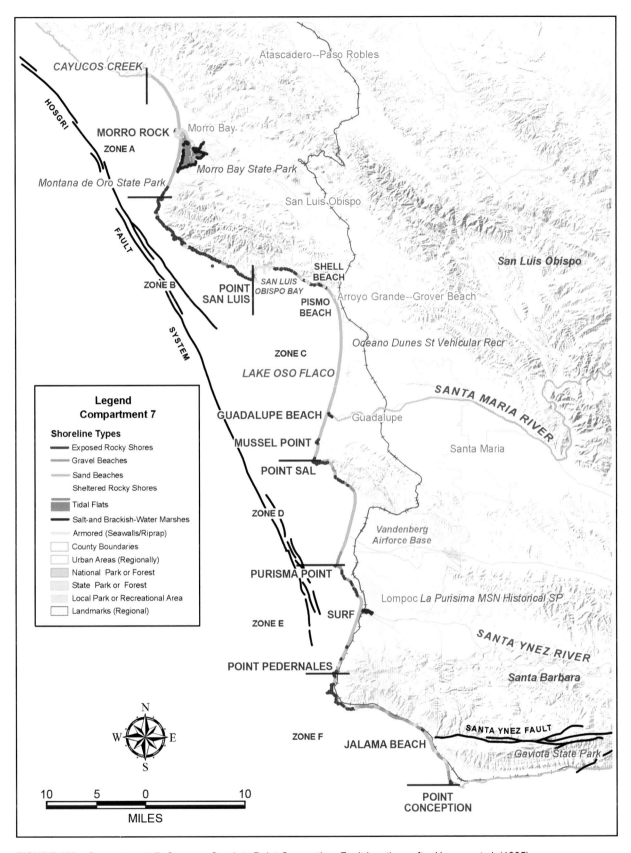

FIGURE 163. Compartment 7, Cayucos Creek to Point Conception. Fault locations after Hanson et al. (1995).

Park. Some of the best examples of raised marine terraces on the Central California occur along a significant length of this coastline. Three bedrock headlands, Point Pedernales, Point Arguello, and Point Conception, are being consistently pounded by huge waves. The sand and gravel beach at Jalama exhibits classic beach cusps, as well as a wave-cut rock platform dotted with tide pools in the mid to lower intertidal zone.

Geographically and geomorphologically, this Compartment is a dramatic departure from the Big Sur Country, being composed of conspicuous headlands formed by the northwest/southeast trending San Luis Range (in Zone B; Figure 163), the Casmalia Hills (northernmost Zone D), and the east-west trending Santa Ynez Mountains (an extension of the east/west trending Transverse Ranges) in the north part of Zone F. These headlands are separated by four wide, very fertile valleys (see general geomorphology diagram in Figure 77). Several moderately long rivers (compared to those in the Big Sur Country) have delivered an abundance of sand to the coast, resulting in the development of four major dune fields.

As clarified by Orme (2005), geologically speaking, this area is part of the **Santa Maria Basin**, *a broad triangular structure constrained between the Coast Ranges to the northeast, the Transverse Ranges to the south, and the active Hosgri fault zone offshore to the west* (Figure 163). Since about 12 million years ago, this basin has been receiving large amounts of sedimentary materials in miscellaneous marine embayments, creating some petroleum-rich formations that have produced many millions of barrels of oil. As expected, these oil-producing sediments overlie a downwarped complex of Franciscan, Ophiolite, and "Great Valley Sequence" rocks that are so prevalent in the mountains of Big Sur Country. A variety of Paleogene/Neogene formations with different origins occur within the basin, perhaps the most noteworthy one being the fine-grained, silica-rich Monterey Formation of Miocene Age. Some of these formations, as well as older Mesozoic rocks and some volcanic rocks, are exposed along the shore of the headlands at Montaña de Oro, Point Sal, and the Santa Ynez Mountains.

Orme (2005) also pointed out that in the past million years or so, tectonic forces have produced episodes of uplift in the area, particularly of the linear bedrock ridges. Evidence for this is a sequence of marine terraces that rise between 650 and 1,000 feet above sea level in different areas, the maximum uplift occurring to the east of Point Conception. In more recent times, a fairly wide wave-cut rock platform was carved in many areas during the last highstand of the sea, about 125,000 years ago. With continuing uplift of the land during the ensuing lowstand of sea level and a small amount during the stable stand of the past 5,000 years, that wave-cut rock platform, now called a **raised marine terrace**, is between 40 and over 100 feet above the present sea level in different places along the coast (Orme, 2005).

Sand beaches dominate throughout the area, except off the eroding rocky headlands. The prevailing longshore sediment transport direction is north to south, as has been true for the other areas to the north. There seems to be an **overabundance of sand** south of Morro Rock, a trend that continues increasingly all the way to near the south end of the Compartment. Orme (2005), in addressing this issue, observed that there are two major sources for the sand, Santa Maria and Santa Ynez Rivers, both of which have sizeable watersheds that drain a variety of rock and sediment types. He also noted that both of these rivers now have dams on them, so they presently *deliver less sediment than in historic times*. Other streams, including San Luis Obispo Creek, Arroyo Grande, San Antonio Creek, and Jalama Creek, also supply sand to some local beaches.

Even so, it seems unlikely that these streams under the current conditions, even before the dams were built, could have supplied enough sand to create the huge dune fields of both modern (Holocene) and Pleistocene dunes shown on the map in Figure 77. Orme suggested that the answer to this enigma is that *wetter climates and floods of Pleistocene time certainly yielded far more sediment than today's regimes*.

ESTERO BAY (CAYUCOS CREEK TO ISLAY CREEK)

Introduction

This area (Zone A in Figure 163) is a twelve mile long, half-moon-shaped arc along the shoreline of Estero Bay. In plan view, it looks like a miniature version of Monterey Bay. Unlike Monterey Bay, however, the occurrence of this embayed shoreline is not related in any way to the presence of a submarine canyon. Instead, the low-lying Los Osos Valley to the east has provided a topographically low area for the shoreline to arc into. Also, the Quaternary sediments in the valley are more easily eroded than the bedrock headlands to the north and south.

[NOTE: In fact, the suggested origin of this embayment just proposed matches fairly closely the origin suggested by Ed Clifton (and Bill Dupré) for Monterey Bay, in which they give little credence to the idea that the presence of the Monterey Submarine Canyon had anything to do with it.]

In comparison with Compartment 6 to the north, this Zone is highly developed, with three modest sized towns located along the shore, Cayucos to the north, Morro Bay in the middle, and Baywood-Los Osos in the south. However, there are still miles of beautiful natural beaches that are relatively free of development.

Three striking geomorphological features dominate the shoreline: 1) Morro Rock, the centerpiece of the shoreline arc, is a remnant volcanic neck that conspicuously "guards" the entrance to the largest estuary on the outer Central California Coast (with the obvious exception of San Francisco Bay); 2) The estuary itself, which contains a harbor/marina complex and the largest river delta on the Central California Coast (again with the exception of the much larger one in San Francisco Bay); and 3) a five-mile-long barrier spit that extends north from the headlands across the ocean side of the estuary, ending at the present engineered entrance to Morro Bay (Figure 164).

Geological Framework

Clearly, the most striking geological feature in this area is the huge, preserved volcanic neck called Morro Rock. Several more peaks with a similar origin extend in a straight line away from Morro Rock toward the southeast. There are some bedrock exposures, called Mesozoic volcanic rocks on the geological map, along the base of the low wave-cut scarp in the northern area. However, in most places, the bedrock, if present, is masked by overlying Quaternary sediments.

Orme (2005) listed the following steps for the origin of the estuary called Morro Bay, which is illustrated by the Environmental Sensitivity Index (ESI) Atlas of the area in Figure 165:

1) The estuary was born shortly after the present highstand of sea level was achieved around 5,000 years ago.

2) As sea level rose, waters *flooded the seaward end of a subsiding structural trough* that had been created during the evolution of the Santa Maria Basin and was by then occupied by Chorro and Los Osos Creeks.

3) The new coastal water body became separated from the open ocean by the development of sand bars south of Morro Rock.

4) Eventually, as the bars evolved, a barrier spit grew to the north, closing off the water body. This barrier spit migrated landward some at first, but it continued to grow as sand was added to it by longshore sediment transport, as well as by the growth of wind-blown sand dunes on top of the accreting beach sediments. This barrier spit

is now 1,000-2,000 feet wide and contains a continuous succession linear, parabolic dunes on its surface (see image in Figure 164).

5) Once stabilized, the estuary began to fill up with sediment, including the growth of a fairly large river delta at the mouth of Chorro Creek.

Beaches and Dunes

Sand beaches are the dominant shoreline type in this area. The six miles of shoreline between Cayucos Creek and Morro Rock is composed of a very wide, fine-grained sand beach with well-developed antidunes and other physical sedimentary structures. A modest line of vegetated sand dunes is located back of the high-tide line in the southern one third of this beach. The sand beach along the five mile long spit to the south of Morro Rock is a bit coarser-grained and has spectacular linear parabolic dunes all along the top of it. Most of the dunes are at least partly vegetated, but there are a number of blowouts that contain barren sand dunes. The dunes are oriented in a northwest/southeast direction in agreement with the prevailing wind direction, and there are no offshore obstacles anywhere south of Morro Rock to block the wind coming from that direction.

Beach Erosion

Hapke (2005) reported that *seawalls and riprap protect nearly each individual parcel along developed sections of Cayucos except where small promontories of more resistant rock occur.* Orme (2005) pointed out that riprap, breakwaters, and groins protect the channel at the entrance to Morrow Bay, and that dredging is necessary at times to keep it usable. He also classified the spit to the south as *Caution: moderate risk*, but that shoreline presently has no manmade structures on it.

General Ecology

The sand beaches of Estero Bay and Morro Bay are the northern extent of significant numbers of the California grunion, a 5-6 inch fish that spawns at night from March-August on sand beaches. Early Spanish settlers called this fish *grunion*, which means grunter, because they are known to make a faint squeaking noise while spawning. These beaches are also the northern extent of significant populations of the pismo clam, which was once an important fishery (the commercial fishery was prohibited in 1947 and all recreational fishing in Central California has been prohibited since 1993). It is thought that the extension of the range of the sea otter,

FIGURE 164. Morro Bay and vicinity (San Luis Obispo County; Zone A, Figure 163). NAIP 2005 imagery courtesy of the California Spatial Information Library (CaSIL).

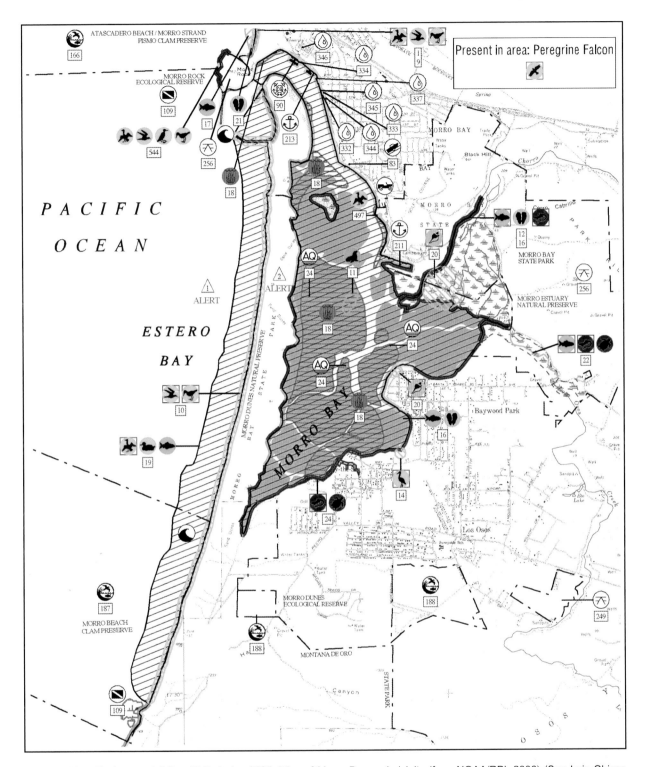

FIGURE 165. Environmental Sensitivity Index (ESI) Atlas of Morro Bay and vicinity (from NOAA/RPI, 2006) (San Luis Obispo County; Zone A, Figure 163). The bay is mostly filled with extensive muddy tidal flats (orange) and has large areas of eelgrass (purple hatch areas) that are important to migrating geese. Morro Rock is a seabird nesting colony. Refer to Figure 85 for shoreline legend and page 132 for biological symbols.

a major predator of pismo clams, has been a factor in the decline of this once abundant species.

The sand beaches are also home to large numbers of birds including snowy plover (Morro Bay had 87-205 nesting adults March-September over the period 2000-2005), California least tern (nesting May-August), and many shorebird species in large numbers. The Morro Bay estuary, composed of 2,300 acres of mud flats, marshes, eelgrass, and open water (Figure 165), was added to the National Estuary Program in 1995. Morro Bay is an important stop on the Pacific flyway; over 200 species of birds were recorded in the 2008 Christmas Bird Count in Morro Bay, including the California condor. However, sedimentation is filling the Morro Bay estuary, reducing the ecological functioning of this important habitat. Of particular concern is the decline in the quantity and quality of the eelgrass habitat, because of its importance to migrating black brant, for which eelgrass is their primary food. Morro Bay is a key migratory bird staging area that provides a food source that is geographically isolated from other large staging areas.

Places to Visit

The north end of this shoreline arc is made up of a fine-grained sand beach along the shore of the town of Cayucos. Cayucos State Beach, which contains a long fishing pier positioned a little over a hundred yards south of the mouth of Cayucos Creek, is located in this same area. Looking south from the pier, you can see the imposing headland of the Irish Hills south of Morro Bay, even though they are only 400 feet high, with Morro Rock boldly apparent in the foreground.

As noted earlier, many of the houses crowded along the beach in Cayucos are protected by seawalls and riprap, indicating that this beach, located in what is, in effect, the shadow zone of a crenulate bay, has had some erosion problems. Bedrock crops out in the beach in several places, protecting some houses. These smooth-textured and unbedded rocks, which are dark green to brownish in color, are shown on the geological map as *undivided Mesozoic volcanic and metavolcanic (metamorphosed) rocks*. Where they occur, these rocks appear to be resistant to erosion, as one might expect from hard, volcanic rocks. The fine-grained sand on the beach is dark colored, with a lot of rock fragments in it, presumably reflecting its Franciscan-mélange source in the watershed of Cayucos Creek.

A bit more secluded beach can be found at Morro Strand State Beach (North), located at the southeast end of Cayucos. This beach includes a 350 yard sand spit that closes off the mouth of the stream that drains Whale

Rock Reservoir.

The entrance to an even more secluded beach, Morro Strand State Beach (South), is located about 2.4 miles to the south of Cayucos (see image in Figure 164). This State Beach is 1.7 miles long, reaching south to the mouth of Morro Creek. As you walk out to this beach, you will pass through a relatively wide zone of vegetated sand dunes. There seems to be an abundance of sand at this particular location, in contrast to the beach at Cayucos. Once you walk onto the intertidal beach itself, you will be struck by the starkness of Morro Rock, which seems to project straight up out of the fine-grained sand (see photograph in Figure 166). This beach has large numbers of shorebirds, including long-billed curlews (Figure 167B). If you go there at low tide, you will be treated to a feast of physical structures along the beach surface, including antidunes (Figure 167A) and rhomboid ripples of the type illustrated in Figures 58 and 59D. Looking to the north, the beach is long and straight, and it frequently has large breaking waves, which bring on the accompanying horde of surfers. This flat, fine-grained sand beach is different from most of the beaches you see on the Central California Coast, which typically are steeper and have coarser-grained sand and at least a little fine gravel.

After leaving the South Beach, we recommend that you drive out to the base of Morro Rock, the preserved neck of a volcano that erupted last about 22 million years ago (during the Miocene epoch; Table 1). It now stands 576 feet above sea level. The remaining rock, pictured up close in Figure 168A, is said by some to be **rhyolite** and others **dacite**. Rhyolite is a fine-grained volcanic rock, which means it cooled rapidly when it formed (as it was extruded toward the Earth's surface). It is silica rich, being composed of quartz, K-feldspar, and plagioclase, or, in other words, it is the extrusive equivalent of a granite. Dacite is similar in origin to rhyolite, but contains more plagioclase feldspar. In rocks with this type of origin, they sometimes had crystals, usually of feldspar, already forming in the molten rock before it was extruded. During the quick rise, the groundmass of the rock crystallized rapidly into very small crystals, creating a rock called a **porphyry** (defined earlier and illustrated in Figure 17A). The rock that composes Morro Rock had such an origin and, therefore, is a volcanic porphyritic rhyolite (or dacite?), as the photograph in Figure 168B shows. Earlier, we discussed another porphyritic rock unit, the Porphyritic Granodiorite of Monterey, and showed a photograph of it in Figure 136B. When you get to the rock, you will find lots of blocks of this porphyry that have been used to make a border for the parking area. In earlier times, the main rock was blasted to

280

FIGURE 166. Morro Rock from the north looking across the wide, fine-grained sand beach [Morro Strand State Beach (South); San Luis Obispo County; Zone A, Figure 163]. Photograph courtesy of Patrick Smith Photography.

FIGURE 167. Morro Strand State Beach (South) (San Luis Obispo County; Zone A, Figure 163) at low tide on 6 December 2008. (A) Antidunes formed by the retreating swash. Scale is ~ 6 inches. Compare these with the images in Figure 60. (B) Long-billed curlews feeding near the swash line amongst the antidune traces.

provide riprap for the construction of breakwaters, but that practice has been discontinued. Ed Clifton (pers. comm.) pointed out that *it is illegal (and dangerous) to climb on Morro Rock, unless you are a Chumash Indian,* *with the rare exemption to climb the rock for the annual solstice ceremony.*

In the 22 million years or so since its eruption, all of the material that composed the original volcano, except the resistant rhyolite of its original neck, eventually was eroded away. Morro Rock is one of a line of a number of these remnant volcanic necks that extend in an almost straight line from offshore (one is under water) to the southeast of San Luis Obispo. Alt and Hyndman (2000) hypothesized that they originally were extruded along the same, straight fault line.

Next, we recommend that you go into town and follow the signs to the marina and the state park. Drive along the shoreline of the Bay, and just before you get to the marina, you will come to the Natural History Museum on the right. This Museum is in a great location, on the side of what is probably an old raised sea stack. From the Museum you will have an excellent view of Morro Rock, as well as of the marsh along the margin of Morro Bay. Documents in the museum recommend a walk on the Black Hill Trail, which also has some great views, no doubt. Black Hill is one of the four westernmost of the morros (ancient volcanic necks), that include Morro Rock, Black Hill, Cerro Cabrillo, and Hollister Peak (the tallest at 1,404 feet).

After you leave the Museum, we recommend that you drive toward the east along the margin of the delta built into the bay by Chorro Creek (Figures 164 and 165). This road goes right along the margin of the delta from which you have several views of the marsh and the tidal channels come right up to the road in some places. This is a good place to visit if you want to get a close-up look at the marsh, but the channels are few, and there are no trails. However, there are a few pullovers from which you can look for birds. In addition, this is a place where you have a chance to view a delta up close, because you can drive right along side of its flat top, which is covered by *Salicornia* marsh. This is one of the largest deltas on the Central California Coast, with the obvious exception of the San Joaquin/Sacramento Delta in San Francisco Bay. Studies have shown that 25% of Morro Bay has been filled in with sediments during historic times, with much of that sediment no doubt provided by Chorro Creek. Los Osos Creek has also contributed. This is a good naturalist drive to see a salt marsh and a river delta up close, which isn't a common opportunity along the outer Central California Coast. Also, some of the peaks of the volcanic necks are visible in the background to the east along this drive.

To continue the tour, turn right where this road meets South Bay Boulevard and cross the bridge over Chorro Creek. Continue on South Bay Boulevard for 3.2 miles at which you will meet Los Osos Valley Road, where you should turn right and head back toward the shore. At around 1.5 miles from there, you will come to a big curve where the road turns toward the southwest.

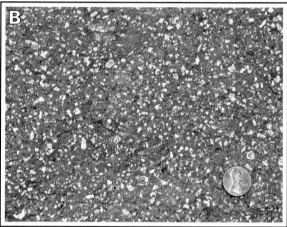

FIGURE 168. Morro Rock (San Luis Obispo County; Zone A, Figure 163). (A) North side of the rock, which is the residual neck of a volcano that was active about 22 million years ago, during the Miocene epoch. (B) Up close view of this volcanic porphyritic rhyolite (or dacite?), which has plagioclase feldspar phenocrysts set in a gray groundmass. Both photographs taken on 6 December 2008.

Somewhere in there the road's name changes to Pecha Valley Road. Within another half mile or so, you will be passing some pullovers from where you can look back and see Morro Rock and the big sand spit that has built to the north, closing off the bay. This spit has some fantastic linear parabolic sand dunes on top of it (see image in Figure 164). There is a trail called Sand Spit Trail along which you can walk down to the beach and get a close-up view of these outstanding, vegetated coastal sand dunes. There is also a paved road down to a parking area near the dunes a little further along.

If you continue the drive along Pecha Valley Road, you will cross the border of Montaña de Oro State Park, which we will discuss in detail in the treatment of the next Zone to the south, that starts at Islay Creek inside the Park. However, before you get to the Creek, at 1.5 miles south of the Park entrance, on the west side of Pecha Valley Road, there is a 1/4 mile long path, called Hazard Canyon Trail, that leads through eucalyptus groves used by monarch butterflies for nesting October through March (California Coastal Commission, 2003). In season, this trail would definitely be worth a visit.

MONTAÑA DE ORO HEADLAND (ISLAY CREEK TO POINT SAN LUIS)

Introduction

This area (Zone B, Figure 163), which occupies the west and southwest border of the northwest-projecting ridge called Irish Hills, or the San Luis Range, is about 12 miles long (see Figure 169). The Diablo Canyon Nuclear Power Plant is located 5.3 miles to the southeast of the mouth of Islay Creek, the northern border of this Zone. Therefore, except for the fabulous Montaña de Oro State

Park, which no one should miss, the rest of the shoreline of this Zone is not readily accessible.

Geological Framework

The layered rocks that compose this headland are complexly folded, with anticlines and synclines composing the northwest/southeast trending ridges of the Irish Hills. In the part of the State Park near Islay Creek (Spooner's Cove), the bedrock is thin-bedded, light-tan-colored, sedimentary rocks that compose the Miguelito Member (basal member) of the Pismo Formation. In this area, this formation consists mainly of *thin- and well-bedded porcelanite, chert, siliceous shale, and mudstone* (Keller and Ba, 1993). This means that the sediments that eventually became these rocks were originally deposited on an ancient sea floor in a somewhat isolated depression that was within the aforementioned Santa Maria Basin. These fine-grained sediments, consisting of **diatoms** and **silicoflagellate** assemblages, rained down out of suspension onto the sea

FIGURE 169. Image of the area from Montaña de Oro State Park to Pismo Beach (San Luis Obispo County; Zones B and C, Figure 163). NAIP 2005 imagery courtesy of the California Spatial Information Library (CaSIL).

283

floor. They were deposited from about 10 to 6 million years ago, late in Miocene time. These rocks are similar to and are thought to be the same age as the rocks of the upper Monterey Formation in nearby basins.

[NOTE: Silicoflagellates are a variety of unicellular algae, found in marine environments, that bear a star-shaped, **silica** skeleton. Hence, the resulting sediments composed of those skeletons contain abundant silica. Diatoms, which were described in some detail earlier, also have silica-bearing skeletons.]

Within the State Park there are abundant world-class geomorphological features, including wave-cut rock platforms (see Figure 25B), sea stacks, and sea arches. The thin-bedded rocks are tilted and folded, being completely vertical in some places, where the eroded cliffs take on rectangular shapes that align with the bedding planes of the dipping layers. From the standpoint of the geomorphology of rocky coasts, this Park cannot be beat.

One way to enjoy the part of the Park near Spooner's Cove is to obtain a copy (or view on the internet) of George Mason's 12-page booklet entitled a *Self-Guided Geology Walk to Valencia Peak (Montaña de Oro State Park, CA)* (Mason, 2006). The walk goes from Spooner's Cove at the mouth of Islay Creek up the side of Valencia Peak (elevation 1,347 feet). Among other interesting things, the text of this booklet describes the five raised marine terraces that are present from Spooner's Cove up the side of the mountain. Orme (2005) observed that *marine terrace and fault evidence indicate that the local mountains are rising at a rate of up to 0.75 feet per 1,000 years*; therefore, more such terraces are due to appear a few tens of thousand years into the future.

A little over half way from Islay Creek to the nuclear plant, the geology changes from the Miocene sedimentary rocks to Paleocene/Neogene volcanic flows mixed with minor pyroclastic deposits (fragments of rock blown out of the volcano into the air, the bulk of which is usually fine-grained volcanic ash). Further around the headland toward Point San Luis, the bedrock changes to the Franciscan Complex, with its sandstones, shale, mélange, and so on.

Some of the most clearly defined **raised marine terraces** on the Central California Coast occur along this shoreline, with the one at Point Buchon being cited frequently in the geological literature. Point Buchon is located about 1.5 miles south of Spooner's Cove. This raised marine platform between Islay Creek and the nuclear plant, which averages 300-400 yards in width, has a very irregular outer margin with abundant sea stacks

offshore. Figure 47B is an oblique aerial view of this raised terrace. The interesting phenomenon of preserved sea stacks on top of the raised marine terrace occurs to the south of the nuclear plant, where the terrace is over 600 yards wide in a few places. As explained in Section I, this lower raised marine terrace was carved around 80,000-125,000 years ago. The four higher terraces up the side of Valencia Peak were obviously carved during earlier highstands of the sea, probably mostly within the Pleistocene epoch.

Beaches and Beach Erosion

There are a few sand beaches north of the nuclear plant in the Miocene sedimentary rocks, but bare wave-cut rock platforms are more common. Pure gravel beaches are sparse. The beach at the mouth of Islay Creek is located at the head of a 300-yard-long bedrock framed embayment. This coarse-grained sand and granule beach is illustrated by the beach profiles in Figure 53A, the field sketch in Figure 53B, and the photograph in Figure 170. As noted in the discussion in Section I of these beach profiles, there was little apparent change between the two surveys at this site. Gravel beaches are much more common further to the south of the nuclear plant, particularly in the area where the Franciscan rocks are present.

With respect to beach erosion, Orme (2005) ranked the entire shoreline as *Stable: low risk*. The ragged edge of the raised marine terrace is evidence that it is eroding, but apparently no data are available on the rate.

FIGURE 170. Spooner's Beach in Montaña de Oro State Park on 5 December 2008 (San Luis Obispo County; Zone B, Figure 163). Two topographic surveys and a field sketch of this beach are given in Figure 53.

General Ecology

This isolated coast is home to ten seabird nesting colonies (black oystercatcher, pelagic and Brandt's cormorant, pigeon guillemot, and western gull) and twelve seal and sea lion haulouts. There can be over 4,200 California sea lions along this stretch of coastline. The kelp beds are very extensive and support nearly 250 sea otters.

Places to Visit

Montaña de Oro State Park is the only place that can be visited conveniently in this area. On our trip there in December 2008, co-author Hayes was heard to exclaim that, if for some reason, he were restricted to going to only one place during a visit to the Central California Coast, "this would be the place!" The coastal geomorphology is in a class by itself. This is a big Park, with over 8,000 acres and three miles of coastline. There are 50 miles of trails; many run along the edge of the wave-cut rock cliff from which you can observe the steeply dipping beds of the thin-bedded, siliceous siltstones and mudstones, the spectacular sea arches, and countless numbers of sea stacks. The sea stacks contain numerous cormorants and other sea birds. Other specialties such as pigeon guillemots are also in abundance. There's no way you would want to miss this place on your next visit to San Luis Obispo County.

SAN LUIS OBISPO BAY (POINT SAN LUIS TO POINT SAL)

Introduction

This area is a nineteen mile long, north/south trending crenulate bay (Zone C, Figure 163). The northern border, which extends in an east-to-west direction for about six miles between Point San Luis on the west and Pismo Beach (Figure 169), constitutes the southern border of the Irish Hills headland. This shoreline contains fine-grained sand beaches and a fairly complex assemblage of rocks, including Paleogene/Neogene volcanic ash deposits. At the village of Shell Beach and the northern end of Pismo Beach, some of the most intricate and beautiful rocky shoreline on the Central California Coast contains an abundance of sea stacks, sea caves, wave-cut cliffs, and rock platforms (Figure 171). The photograph on the cover of this book was taken along this rocky coast a short distance from the town of Avila Beach. Pismo Beach occupies the northern end of a long trend of sand beaches that have

FIGURE 171. Wave-cut rock cliffs and sea stacks viewed from the top of the bluff just beyond the north end of Pismo Beach (San Luis Obispo County; Zone C, Figure 163). Photograph taken on 5 December 2008.

large lobes of coastal dunes extending to the southeast away from them (Figure 172).

Geological Framework

The six mile stretch along the northern arm of the crenulate bay contains a variety of rock types, including miscellaneous Franciscan Complex rocks, Mesozoic volcanic rocks, sedimentary rocks of both Pliocene and Miocene age, and Paleogene/Neogene volcanic deposits, to go along with minor faults and intrusive igneous-rock dikes. Near the north end of Pismo Beach, however, most of the shoreline is covered with Quaternary sediments, mostly in modern sand dunes. Bedrock crops out at two other areas to the south, Mussel Point (Miocene sedimentary rocks) and Point Sal (a famous ophiolite exposure).

Two major features, the northwest/southeast trending peninsula of the Irish Hills on the northern side, and the wide, flat Santa Maria Valley along the eastern side, control the general framework of this area. However, the huge dune fields illustrated in Figures 77, 172 and 173 are the most conspicuous geomorphological highlights in the area.

Beaches and Dunes

Sand beaches compose most of the shoreline in this area. There are a few pocket gravel and sand and gravel beaches along the northern shore. The beaches in the north, most notably at Avila Beach, are flat, fine-grained sand beaches with the sand presumably being supplied by

285

FIGURE 173. Oblique aerial view of parabolic sand dunes along the beach in the Guadalupe Dunes area. The north end (left side) of image is located 2.8 miles south of Oso Flaco Lake, located on Figure 172. Part of the Guadalupe Oil Field is visible in the background. Photograph taken on 28 October 2005, courtesy of the California Coastal Records Project, Kenneth and Gabrielle Adelman.

San Luis Obispo Creek. Further to the south, the beaches are considerably coarser grained, even containing some coarse-grained sand and a few granules. This coarser sand was presumably originally supplied by Arroyo Grande Creek and the Santa Maria River, thought historically to have been significant suppliers of sand to these beaches. It should be pointed out, however, that although the Santa Maria River drains a 1,880 square mile watershed, the outflow of this river has been controlled since 1958 by the Twitchell Reservoir, cutting off much of the sand supply to the beach. During unusually rainy years, the river does breach the bar that normally closes off its mouth (Orme, 2005).

As the image in Figure 172 readily shows, there is a band of active sand dunes off the Pismo Beach/Oceano area. It is 4.6 miles long and has a maximum width of 1.7 miles. This dune field is intercepted on its southern end by the outlet of Oso Flaco Creek. The largest dune field of them all, the Guadalupe-Nipomo dunes, which is four miles long, with a maximum width of 3.4 miles, is located immediately to the south of this Creek. These dunes are mostly vegetated, except along the shore, where the line of classic parabolic dunes shown in the photograph in Figure 173 occurs. This dune field is discussed in some detail in Section I. Another line of dunes, mostly active dunes in this case, extends for a little over three miles to the south of the mouth of the Santa Maria River.

Beach Erosion

Most of the area between Avila Beach and Pismo Beach was classified as *Hazard: high risk* by Orme

FIGURE 172. Image of the Pismo Beach/Guadalupe Dunes area showing the extensive modern and older (vegetated) dune fields (San Luis Obispo County; Zone C, Figure 163). NAIP 2005 imagery courtesy of the California Spatial Information Library (CaSIL).

(2005), with the cliffs around Shell Beach eroding an average of around seven inches per year. Some seawalls and riprap have been installed in the Avila Beach area. The northern end of Pismo Beach is also classified as a high hazard area, and seawalls and riprap are common along that part of the beach, continuing down to Grover Beach. Orme (2005) reported that riprap is used at the mouth of Arroyo Grande Creek to protect dwellings where *storm-wave impacts damage property and erode backshore dunes.* The rest of the shoreline down to Point Sal is classified for the most part as *Caution: moderate risk*, but no man-made structures have been threatened in that undeveloped area.

General Ecology

There are five seabird nesting colonies from Point San Luis to the north end of Pismo Beach, with pigeon guillemot the most common nesting species (February-August) by far. Sooty shearwaters can be present in high numbers (in the thousands) in San Luis Obispo Bay in June-September. Predictably, shorebirds, wading birds, and waterfowl are common along the beaches and wetlands in the southern part of this area. Pismo Beach is an important area for snowy plovers, with 81-200 nesting birds and 154-381 wintering birds observed over the period 2000-2005. The abundant kelp beds to the north generate large amounts of beach wrack full of insects, amphipods, beetles, etc. that are important food items for the shorebirds.

Places to Visit

He wandered by the sea from the border north as far as San Luis Obispo, and he learned to pilfer the tide pools for abalones and eels and mussels and perch, to dig the sandbars for clams, and to trap a rabbit in the dunes with a noose of fishline. And he lay in the sun-warmed sand, counting the waves.

John Steinbeck – EAST OF EDEN

If you visit this area traveling from north to south, the most northerly site to visit is Port San Luis, which is located in the lee of a long breakwater that extends offshore from Point San Luis. To get there, take Highway 101 south from San Luis Obispo, and as you near the shoreline, turn right on Avila Road and keep driving west until you get to the Port. There is a long pier at the Port, which offers various tourist activities, such as boat rides for whale watching and deep-sea fishing. From there, it is about a mile back to the east along a shore road to the town of Avila Beach and Avila Beach State Park. A fine-grained sand beach, called Old Port Beach, extends along the side of this road, which is protected by rip-rap.

Just before you get to the Park, you will cross a bridge over San Luis Obispo Creek, which supplies the fine-grained sand that constitutes Avila Beach. The fact that the watershed of this creek is located mostly within the relatively flat topography of the Los Osos Valley may account for the lack of coarser material in the sediment delivered by it. This is a very pleasant wide beach, but it is backed in places by seawalls and riprap. A long pier within the State Park provides the opportunity for fishing and observing long views along the shore.

A walk of about 650 yards to the east from the pier would provide you with an opportunity to inspect up close one of the complex suites of rocks that typify the stretch of coast between Point San Luis and Pismo Beach. These rocks include Mesozoic volcanic rocks and an extension of the Miocene sedimentary rocks that are so well displayed at Montaña de Oro State Park around the corner to the north. A retired geologist named Al Stevens has reported that spectacular **pillow basalts** occur at the far northwest end of the parking lot for the Avila Beach Pier. He noted that these basalts, which erupted near the equator, are 180 million years old (Jurassic; the oldest rocks in San Luis Obispo County) (Ed Clifton, pers. comm.).

To review, the original sediments that make up the Miocene rocks in this area were deposited in shallow marine conditions as a result of the settling of silica-bearing skeletons of microorganisms, as well as tiny particles of rock and other marine organisms through the water column. This formation, called the San Miguelito Formation, is prevalent throughout the Avila Beach area. Some Paleogene/Neogene pyroclastics (e.g., volcanic ash) and volcanic mudflow deposits are also of special interest in this location. If you do walk over there, be sure to go at low tide to allow for maximum exposure of the rocks.

From Avila Beach, we recommend that you go back to Highway 101, drive south, and take the exit at the very north end of Pismo Beach. Once off the main highway, drive north along side of Highways 1 and 101 to where there is a beach access sign. This access provides a parking area where you can walk around the top of the bluff. This is one of the better spots along the coast to view wave-cut rock cliffs and sea stacks, as the photograph in Figure 171 confirms.

If you continue to the northwest along this road, you will eventually come to the village of Shell Beach, where there are two adjacent parking areas at the top of the wave-cut rock scarp. The best views are from Margo Dodd Park. This site is even better than Santa Cruz for viewing an eroding rocky cliff. Sea stacks are abundant, with one exceptionally large flat-topped stack that has been eroded by waves forming two huge sea arches. Large numbers of brown pelicans and cormorants roost on the surfaces of the sea stacks. Gulls are also numerous. This is really a great spot to visit because of its excellent access. You can park your car in a spacious parking area or in a pullover alongside the road and walk right around the edge of the bluff.

Another place with public access to the shore, Ocean City Park, is located about a half mile west of Margo Dodd Park. Ocean City Park has two stairways down to the sand and gravel beach. A bonus there is that, if you go down one of the stairways, you will be able to inspect up close the strongly dipping beds of what is probably the Miocene Miguelito Formation, a silica-bearing fine-grained siltstone and mudstone. These steeply dipping layers are very similar to the ones you will have seen, or will see, at Montaña de Oro State Park.

As you keep going west after leaving the Ocean City Park, there are several other vista points available. If you drive to the end of El Portal Drive, you can walk out onto a bluff-top trail where you will pass by a number of the Malibu-style houses. This trail is not necessarily recommended for naturalists but, if you are feeling adventuresome, it goes most of the way to Avila Beach, being paved in some parts and passing through forests in others.

The next logical place to visit is the popular swimming beach at Pismo Beach State Park, which has a public fishing pier with excellent views to the south along the coast. The beach is a typical wide, relatively fine-grained sand beach that derives its sediment from San Luis Obispo Creek and other streams to the north.

An interesting sideline to your visit to Pismo Beach would be to take the self-guided geology walk along the beach provided by Jamie Foster (1998). This walk takes place at the northernmost part of Pismo Beach. To get there, return to the point where you turned into Shell Beach. Follow this same road you came in on for 1.2 miles to the southeast and turn right on Wilmar Street, drive to the end of it, and take the stairs down to the beach. On this walk, you will see some yellow and white **tuff** (a somewhat consolidated rock composed of volcanic ash), a fault, and a cross-cutting layer of a fine-grained, dark-colored igneous rock called a **dike** (formed when a band of molten material cuts across pre-existing sediments or rocks, the layers of solidified volcanic ash in this case). One of the layers of ash, or tuff, which is white in color,

is soft enough to crumble in your hands. The volcanism that produced the ash deposits took place during the Paleogene/Neogene periods. Whether or not their origin was in any way associated with the volcanic neck at Morro Rock, which was active during the Neogene period (Miocene epoch), is unclear to us.

The next place we recommend that you visit is the Oso Flaco Lake and dunes. To get there from Pismo Beach, drive south on Highway 1 for approximately 13 miles to where it intersects with Oso Flaco Lake Road. Turn right on this road and head toward the coast. After about another three miles, you will be at the parking area for the start point for a trail that leads across Oso Flaco Lake, the dunes, and finally down to the beach. If you are interested in doing this, leave enough time, because it takes a while to traverse the 0.74 mile to the beach and back. How long it takes depends on which trail is open, as there is more than one option available, and prevailing conditions dictate which one can be used at any given time.

The first part of the walk is about 400 yards along a dirt road through some California live oaks. Then you take a board walk, about 300 yards long, across the Lake. When we were there in early December 2008, two flocks of ruddy ducks were on the Lake, plus 24 white pelicans. After leaving the boardwalk, you walk through vegetated dunes containing a wide range of plant types (Figure 174A). To the south of the trail, there is a large field of barren barchan dunes (see vertical image in Figure 172 and ground photograph in Figure 174B). From the high dunes near the beach, on a clear day you can see the Transverse Range to the south and the Santa Lucia Range to the north – great views. The beach is medium-grained sand, somewhat coarser than the sand at Avila Beach.

To see more of the beach and the dunes of this area, we recommend that you go back out to Highway 1 and head south. At 2.4 miles south after you get back on Highway 1, you will cross the bridge over the Santa Maria River. During this part of the drive, you will be on the flat topography of the flood plain of the Santa Maria River, a very rich agricultural area. If you look to the west as you drive along, you will see the landward side of the Guadalupe-Nipomo dune field (pictured in Figure 172), the site of a major Unocal oil field where our group at RPI responded to a chronic underground spill of diluent in 1994 (discussed briefly in Section I).

After crossing the bridge, go to the south end of the town of Guadalupe (1.4 miles) and turn right on West Main Street. Follow this road west for 4.8 miles, at which time you will have arrived at a paved parking area in the dunes a short distance landward of the beach scarp. The mouth of the Santa Maria River is located about 500

FIGURE 174. Dunes near Oso Flaco Lake (Figure 172). (A) Vegetated dunes near the trail out to the beach. Photograph taken on 5 December 2008. (B) Active barchan dunes south of the lake (see illustration of this type of dune in Figure 76). Photograph taken on 6 November 1992.

yards to the north of the parking area (Figure 175A).

This is a beautiful, medium- to coarse-grained sand beach which typically contains **rhythmic topography** (see example in Figure 175B). When we were there in December 2008, the morphology of this beach conformed exactly to the morphology illustrated in the cross-section in Figure 51. That is, it contained two well-defined berms, a **high berm** and a **lower active berm**. The photographs in Figure 175 show both northerly and southerly views along the beach. This is definitely a spot you don't want to miss if you like beaches and dunes and phenomenal views along the beach. For a broader view, you can look in the distance to the north and see the Santa Lucia Mountain Range, and look to south to see the hills north of Mussel Point covered with sand dunes.

Mussel Point is a rock headland located 3.2 miles south of the mouth of the Santa Maria River. A zone of active sand dunes, approximately three quarters

of a mile wide, is located back of the beach the entire distance between these two points (Figure 172). South of Mussel Point, it is 1.9 miles to Point Sal, the southern boundary of this Zone. Mussel Point is a classic case of a natural groin draped with wind blown sand. However, the shoreline between Mussel Point and Point Sal has no presently active dunes. Cooper (1967) pointed out that, in that stretch of coastline, there is *a strip of stabilized dunes perched on a shelf, doubtless a marine terrace, above a steep bluff rising 100 meters* (330 feet) *from the beach.* Cooper hypothesized that these dunes were formed at a time when more abundant sand was coming around Mussel Point than at the time when he did his study, suggesting that the rock was more efficiently deflecting the sand offshore in the 1960s than it was when the dunes first formed. Another possibility is that we are running short on sand supply in more modern times, so the sand that is now being supplied from the rivers to the shoreline is "used up" to the north with little left over to go around Mussel Point, a probability suggested by Orme (2005).

SAN ANTONIO VALLEY (POINT SAL TO PURISIMA POINT)

Introduction

This area (Zone D, Figure 163) is similar in outline and content to the one to the north. It is also a modified version of a crenulate bay, with its northern flank being along side a bedrock headland, Point Sal headland in this case, and the eastern flank mostly a relatively straight, continuous sand beach with major sand dune fields behind it (Figure 176). It is considerably smaller than its northern neighbor, measuring only about 10 miles from north to south, in contrast to 19 miles for the San Luis Obispo Bay system. The rocky northern shoreline, which is 4.4 miles long in a straight line, is made up mostly of bedrock, but it does contain two sand beaches, each about half a mile long. Some striking raised marine terraces also occur in the northern third of the area.

Geological Framework

The bedrock of the Point Sal headland and the associated high hills to the southeast (Casmalia Hills) is a complex assemblage of rock types – Oligocene sandstone and shale, Miocene fine-grained siltstone and mudstone, Mesozoic volcanic rocks and, most importantly, the famous **Point Sal Ophiolite**. This Ophiolite Sequence was described succinctly by Harden (2004) in her book on *California Geology* as follows: *The*

FIGURE 175. Guadalupe Beach (Santa Barbara County; Zone C, Figure 163). (A) Northerly view of this beautiful, medium- to coarse-grained sand beach from near the parking area. The Santa Lucia Mountains are visible in the background. The arrow points to the closed-off mouth of the Santa Maria River. (B) Southerly view showing rhythmic topography and the hills north of Mussel Point, which are covered with sand dunes. Both photographs taken on 5 December 2008.

ophiolite at Point Sal contains many of the rocks typical of an ophiolite suite, including submarine basalt flows, sheeted dikes, gabbro, and serpentinized ultramafic rocks. The Point Sal Ophiolite remnant has been displaced from the main Coast Range Ophiolite by Cenozoic faulting. In that same book she presents a photograph of the Point Sal Ophiolite, which shows several steeply dipping basalt dikes cutting through gabbro, a coarse-grained, intrusive igneous rock composed mainly of calcic plagioclase and pyroxene. Again, the key attributes of an ophiolite sequence are: 1) their rocks came from the upper mantle beneath the oceanic crust; 2) these rocks are ultramafic, plutonic rocks rich in iron and magnesium and low in

silica; 3) the ultramafic rocks include dunite, composed entirely of the mineral olivine, and peridotite, composed mostly of olivine with some plagioclase feldspar and pyroxene; and 4) as pointed out by Harden, *after they have been displaced from the mantle into the crust, most ultramafic rocks are altered by hot (hydrothermal) fluids to* **serpentinite**, *California's state rock.*

Beaches and Dunes

The southern two thirds of this shoreline is mostly a sand beach, backed by the large tract of sand dunes illustrated by the image in Figure 176. As the image shows, two very wide lobes of vegetated older modern (Holocene) dunes are located on both sides of San Antonio Creek. The lobe to the north has a maximum width of around three miles. The seaward flank of both of these lobes contains a relatively narrow band of active, younger modern (Holocene) dunes. Late Pleistocene dunes extend almost nine miles to the southeast from the present shoreline (Orme, 2005; see Figures 77 and 176). The modern sand supply must run out to some extent along the southern 1.3 miles of this stretch, where the beach is composed primarily of gravel.

Beach Erosion

With regard to beach erosion, Orme (2005) ranked the northern third of this Zone as *Hazard: high risk*, and the southern two thirds as *Caution: moderate risk*. He made special reference to the area south of Point Sal, which he noted is *prone to mass movement and seacliff erosion at rates of 5 to 12 inches per year*. He also observed that in the 1930s, *one landslide pushed far into the surf zone*. The major landslide illustrated in Figure 84B, which is located south of Point Sal, could possibly be the 1930s slide that Orme was referring to. Strangely, the landslide mass pictured in Figure 84B has a flat top, so possibly it is the remnant of a raised marine terrace. As that photograph shows, the region south of Point Sal is certainly an area prone to major landslides.

General Ecology

Point Sal and the adjacent large offshore rock (Lion Rock) provide vital roosting habitat for hundreds of endangered California brown pelicans, as well as two seabird nesting colonies for pelagic cormorant, pigeon guillemot, rhinoceros auklet, and western gull. As many as 2,300 California sea lions haul out on the southeastern tip of Point Sal, along with northern elephant seal, northern fur seal, harbor seal, and the occasional Stellar

FIGURE 176. Image of the area from Point Sal to Purisima Point (Santa Barbara County; Zone D, Figure 163). NAIP 2005 imagery courtesy of the California Spatial Information Library (CaSIL).

sea lion. California sea otters forage in nearshore kelp beds. Both Point Sal and Purisima Point are important foraging areas for seabirds, with high or moderate numbers of sooty shearwater, Pacific loon, western grebe, Clark's grebe, Cassin's auklet, common murre, cormorants, and scoters.

The largest known nesting population of snowy plovers occurs on Vandenberg Air Force Base, along the coast from Shuman Creek to Purisima Point (beach access may be restricted when the birds are nesting and fledging). These beaches also provide roosting and feeding habitat for hundreds of Heermann's gulls and shorebirds. The dunes north of Purisima Point provide important nesting habitat for approximately 160 endangered California least terns. They feed in nearshore waters, diving for northern anchovies, silversides, surfperch, and juvenile rockfish. Nesting least terns readily abandon their nests if disturbed, so this long stretch of beach with very limited access is important habitat for this species that is listed as endangered by both federal and state authorities.

Places to Visit

There is very limited access to the shoreline within this Zone. Most of the coastal area is part of Vandenberg Air Force Base. Point Sal State Park is the least visited State Park in California. As of May 2008, pedestrians may access Point Sal State Beach for day use via Brown Road and Point Sal Road. Motor vehicle and bicycle access and camping are not permitted. Vandenberg Air Force Base officials may close or limit access to Point Sal State Beach and clear the area during missile launches, for public safety or base security reasons at any time.

SANTA YNEZ VALLEY (PURISIMA POINT TO POINT PEDERNALES)

Introduction

Like the two Zones to the immediate north, this area (Zone E, Figure 163) has a modified crenulate bay shape, with a northwest/southeast limb 3.7 miles long that parallels the main runway of the Vandenberg Air Base, and a northeast/southwest straight limb that faces directly into the prevailing waves that approach the shore from the northwest (Figure 177).

Geological Framework

Geomorphologically, this area is a wide, flat river valley sandwiched between a subtle headland in the north at Purisima Point, and the striking, high headland

of the Santa Ynez Mountains to the south. The northerly limb of the subtle crenulate bay that composes this Zone has a low, wave-cut scarp mostly in Quaternary materials with wave-cut rock platforms in some areas. Further to the south, sedimentary rocks make up the lower part of the wave-cut scarp. This appears to be true also for the wave-cut rock platform along the whole area. Though not mapped as such on the state geological map, these rocks appear to be Miocene, fine-grained sedimentary rocks. Gravel beaches occur along the northern shore, and sand beaches are more common in the south. Modern (Holocene) sand dunes occupy the top of the scarp for much of the northern two thirds of this segment (Figure 177). A very conspicuous 300 yard wide band of barren, transverse dunes stretches for a mile away from the main beach just north of Purisima Point (Figure 176). This band of dunes extends to the edge of the wave-cut scarp in places and generally runs parallel to it.

The southern, northwest-facing shoreline starts at its north end with about a mile of sand beach that has a tightly-spaced, partly vegetated band of parabolic sand dunes about a hundred yards wide landward of the beach. The next segment is the area where the mouth of the Santa Ynez River opens during floods, which is about 300 yards wide. Like most of the other rivers and creeks on the outer coast of Central California, it is closed by a sand bar much of the time. On the south side of the river, a wide, medium-grained sand beach extends for about 3.5 miles to the south. The area behind the beach gradually becomes higher in a southerly direction, with the high area being covered with vegetated dunes. From there on, the backbeach becomes a wave-cut bedrock scarp and the sand beach peters out. Some parts of this scarp for the remaining 2.8 miles to Point Pedernales (aka Honda Point), the southern boundary of this Zone, are composed of a fragmented raised marine terrace. The bedrock in the scarp in the first couple of miles is light-colored, sedimentary rocks, probably of the Miocene Monterey Formation, and Paleogene/Neogene volcanic materials are present in the scarp in the vicinity of Point Pedernales.

The hills to the south of the Santa Ynez River Valley are composed of thin-bedded, light-colored sedimentary rocks of the Monterey Formation of Miocene age (19-6 million years ago). Although famous for its overall siliceous composition, the Monterey Formation is quite heterogeneous. In this particular area, however, significant portions of the Monterey Formation are a siliceous **hemipelagic** mudstone deposited in a marine environment. A productive quarry for **diatomite**, a classic case of hemipelagic sedimentation, in the Monterey Formation is located near the town of Lompoc (Eichhubl and Behl, 1998a). Uses of diatomite include

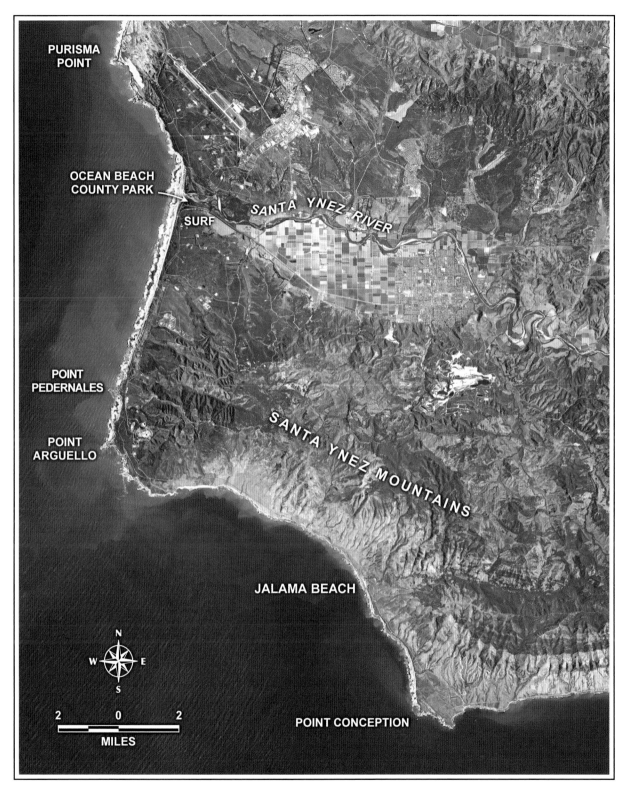

FIGURE 177. Image of the area from Purisima Point to Point Conception (Santa Barbara County; Zones E and F, Figure 163). Much of the shoreline in this area is not readily accessible. NAIP 2005 imagery courtesy of the California Spatial Information Library (CaSIL).

for light abrasives (e.g., toothpaste), for filtration, as filler for paints and other substances, and for miscellaneous other uses of value to mankind.

[NOTE: Hemipelagic sediments are formed in marine waters close to continental margins by settling out of fine particles. Biogenic material is an important constituent of these muddy sediments.]

Beaches, Dunes, and Beach Erosion

The beach at the Santa Ynez River mouth is one of the most pristine and isolated sand beaches along this part of the coast, and its big waves, rhythmic topography, and unspoiled character make it a very desirable place to visit.

No doubt the Santa Ynez River was an important source for the sand on these beaches originally. The river channel is full of sand bars, which indicates that, during floods, sand is moved along the channel floor. However, in 1953, Bradbury Dam was built across the stream about 40 miles upstream from the shore, forming the picturesque Lake Cachuma. This, of course, has had an effect on the flow patterns (hydrograph) and delivery of sand to the coast by this stream.

This beach system is a classic **littoral cell** (Figure 66), with the sand coming into the cell from the stream, with some perhaps coming from the erosion of the Quaternary sediments in the low wave-cut scarps to the north. Sand is lost from the system in the sand dune fields behind the beach. On the other hand, no major submarine canyon is located off this shoreline.

With regard to beach erosion, Orme (2005) classified most of this shoreline as *Caution: moderate risk*, with no significant beach erosion problems having as yet been encountered on this unpopulated shoreline. The relatively low wave-cut scarps behind the beach are not the proper setting for major landslides. The high hills at the southern end of the area are a more likely site for slides but, for the most part, the steep slopes on these hills are shielded from the shore by a flat, raised marine terrace.

General Ecology

The beaches at the Santa Ynez River mouth support a large snowy plover nesting population, with 49-315 nesting birds (March-September) and a wintering population of 113-224 recorded annually over the period 2000-2005. California least terns also nest and roost there. The estuary supports a rich avifauna, with raptors (including peregrine falcon), shorebirds, terns, waterfowl, and marsh birds, such as least bittern.

The Santa Ynez River flows westerly approximately 90 miles to the Pacific Ocean, draining a nearly 900 square mile watershed in southern Santa Barbara County (Figure 178). The river provides about one-third of the water demand in the watershed; the rest is provided by groundwater wells and the importation of State Water Project water from the Sacramento-San Joaquin Delta. Cachuma Reservoir provides about 45% of the total water demand of the southern coastal area of Santa Barbara County.

Until the 1940s, the Santa Ynez River had the largest run of steelhead in Southern California, with estimates of 13,000 to 25,000 fish. High flows during winter storms breached the bar at the river mouth, allowing winter runs of steelhead from December to mid-April. With the completion of Bradbury Dam in 1952 by the Bureau of Reclamation (Figure 178), steelhead are no longer able to access the best spawning and rearing habitat above the dam. It is estimated that steelhead populations have been reduced to less than one percent of their former population size in Southern California (Stoecker, 2004). There are restoration plans under review to provide fish passage around dams, maintain stream flows, and improve spawning and rearing habitat for this highly endangered population of southern steelhead. As elsewhere in California, water allocation issues are very important for the Santa Ynez River, with competing demands on a limited water supply for drinking, industrial, agricultural, and fisheries conservation uses.

Places to Visit

Because most of this shoreline is located either within the Vandenberg Air Force Base, or the Point Arguello U.S. Navy Missile Facility (south of Surf Beach), access to the shore is limited. However, there is one beach stop you can make and shouldn't miss, at Ocean Beach County Park. From the north, take Highway 1 to the beautiful town of Lompoc. As you approach Lompoc (from the north), if you look ahead toward the south, the towering Transverse Range (Santa Ynez Mountains) is a striking reminder that you are approaching the southern end of the Central California Coast.

A short distance north of town, you will cross the Santa Ynez River, a braided channel that contains abundant sand and fine gravel. Continue for about two miles south through town and turn right (west) on Ocean Avenue. As you drive west, you will be near the southern boundary of the Santa Ynez River valley, a rich agricultural area with artichokes and many other crops. This very flat valley is floored by rich-looking, brown soil. Keep an eye out for raptors along this drive, such as

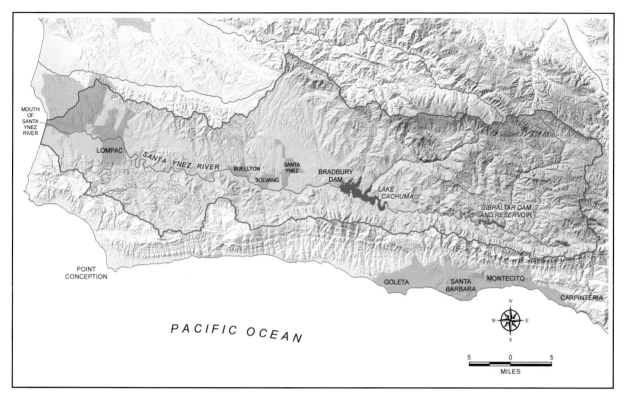

FIGURE 178. Watershed of Santa Ynez River (Santa Barbara County). Shaded relief image and watershed boundary courtesy of the California Spatial Information Library (CaSIL).

the common Coopers hawk. The Santa Ynez Mountains to the south seem to jump up off this valley floor.

At 8.3 miles from the turn off Highway 1, you will reach an exit road that leads down to the parking area for Ocean Beach County Park. You can also reach the beach by driving straight ahead to the Surf Beach railroad station, but access to the beach is a little more difficult there because you have to walk across the railroad track. This is especially a problem if a train has stopped at the station.

Once you make the turn to the beach, you will experience a beautiful, one mile drive along side a *Salicornia* salt marsh. Coots and ducks are numerous in the side channels, and coyotes can be seen. From the parking area, you get to the beach by walking under the railroad bridge along the bank of the Santa Ynez River, which is typically ponded by a sand spit across the river's mouth. As we walked along this trail to the beach in December 2008, we saw snowy egrets, a great blue heron, shovelors, grebes, brown pelicans, coots, and a loon.

Large plunging waves were breaking along the beach on that particular day. A big flock of gulls were resting on the sand bar closing off the mouth of the Santa Ynez River (see Figure 179, a view looking north). Even with the huge waves, there was a bit of a depositional berm present, and well-developed rhythmic topography was in

evidence to the south. A conspicuous line of foredunes with high slip faces on their landward sides were visible to the north of the river mouth.

An inspection with a hand lens showed the beach sediment to be medium-grained sand with a lot of dark-colored rock fragments. It was about 60% quartz with quite a few opaque orange grains, possibly some chert, and goodly number of black heavy minerals. Feldspar was sparse. This is definitely not a pure quartz sand beach, like some of the ones in the Salinian Block to the north (e.g., on the Monterey Peninsula).

In short, this is a beautiful, remote and pristine place, with a very extensive marsh considering its location on the outer Central California Coast. Therefore, this would be a great place for you to end (or start) your visit to the Santa Ynez Valley area.

SANTA YNEZ HEADLAND (POINT PEDERNALES TO POINT CONCEPTION)

Introduction

This area (Zone F, Figure 163) begins with a north/south segment between two small rocky protrusions, Point Pedernales (aka Honda Point) and Point Arguello (see image of the area in Figure 177). A wave-cut rock

295

FIGURE 179. Beach at the mouth of the Santa Ynez River on 5 December 2008 (Santa Barbara County; Zone E, Figure 163). The mouth of the river was closed off by a northerly projecting sand spit.

scarp of moderate height composed mostly of Miocene sedimentary rocks is located between the two points. There is also a raised marine terrace present that is somewhat more elevated than the ones to the north and south of this Zone. Wave-cut rock scarps, with associated wave-cut rock platforms, compose this irregular shore, which has several high scallops into the Miocene sedimentary rocks. Further to the east, there is a six mile stretch of shoreline that has one of the longest, most spectacular **raised marine terraces** on the whole Central California Coast. At its widest point, the terrace is half a mile wide, but it usually only measures a few hundred yards across. The surface of the terrace undulates a little in places, but it is a remarkable terrace.

A small stream, Jalama Creek, which is closed off by a sand bar most of the time (Figure 180), cuts across the beach just north of Jalama Beach County Park, a 28-acre facility with a campground on a slightly raised flat area immediately south of the stream channel.

A raised marine terrace, with an average width of around 150 yards, extends for a mile south of the Park. Beyond the raised terrace to the south, some moderately high wave-cut scarps in the bedrock upland extends in a southeasterly direction for another mile and a half. Gravel beaches are common along this segment. From there it is about two miles to Point Conception, which is on a hillock composed of strongly dipping Miocene sedimentary rocks. The northern two-thirds of this section contains a sand beach, but gravel beaches take over in the near vicinity of the Point. This hillock is flanked on its northeast side by a broad flat area almost

a mile wide. Presumably, this area is a raised marine terrace, even though it isn't completely flat. Also, there is a band of low-elevation, vegetated modern (Holocene) sand dunes with a maximum length of 1,200 yards and maximum width of 350 yards located to the northeast of Point Conception. As expected, these dunes are oriented northwest/southeast.

[NOTE: The authors of this book have flown the coastline of Central California Coast several times, taking numerous photographs and making shoreline habitat maps. However, to see this remote shoreline in more detail than you can by driving the roads, we suggest you "fly" the coastline by looking at the overlapping oblique aerial photographs provided by Kenneth & Gabrielle Adelman (2009).]

Geological Framework

This area is located on the exposed (to wave action) western end of the Santa Ynez Mountains, a part of the east/west trending Transverse Range. The rocks in this part of the mountains are mostly complexly folded and faulted Paleogene/Neogene sedimentary rocks, but some Cretaceous rocks and volcanic rocks of Miocene age are at the surface in a few localities. Along the shoreline, tilted fine-grained sedimentary rocks are the dominant rock type.

The most prominent geomorphological features in the area, in addition to the high mountains, are some of the most extensive and clearly defined **raised marine terraces** along the Central California Coast.

FIGURE 180. View looking north along the beach at low tide at Jalama Beach County Park (Santa Barbara County; Zone F, Figure 163). The mouth of Jalama Creek was closed by a north projecting sand spit on the date this photograph was taken - 6 May 1992.

Beaches and Beach Erosion

Two stretches of beach that are predominantly sand, with gravel commonly mixed in with it, occur: 1) north and south of the mouth of Jalama Creek (2.3 miles long) (see Figure 180); and 2) north of Point Conception (1.4 miles long). Jalama Creek obviously delivers some sand to these beaches but, apparently, another major source for the sand is the Quaternary sediments that make up at least the upper portions of some of the raised marine terraces, as well as most of the wave-cut scarp in some areas (most notably north of the mouth of Jalama Creek). Gravel beaches occur in several places where waves are severely eroding wave-cut scarps. All of the beaches, including sand, sand and gravel, and pure gravel beaches, have wave-cut rock platforms in the Miocene sedimentary rocks in the mid to lower intertidal zones.

With regard to shore erosion, Orme (2005) ranked all three of the major headlands in this area, Point Pedernales, Point Arguello, and Point Conception, as *Hazard: high risk*, because of their exposure to intense wave action, and the fact that they are composed of Miocene sedimentary rocks that erode more readily than many other rock types. The area of the eroding high scarps between 1.0 and 2.5 miles south of Jalama Creek is ranked as a high hazard area also, because these Miocene shales are *subject to frequent rockfalls, occasional landsides, and marine erosion at rates of 12 to 20 inches per year*. It has been necessary to install seawalls at the base of some of the high scarps where the railroad is located close to the shore.

General Ecology

The remote headlands around Point Pedernales and Point Arguello support the farthest-south site on the mainland for breeding seabirds (mainly pelagic cormorant and pigeon guillemot, but also a handful of rhinoceros auklet). This is an important roosting area for California brown pelicans that migrate north from breeding grounds in Mexico. Over 1,000 California sea lions (pupping May-July) and up to 350 harbor seals (pupping March-May) haul out on the rocks.

Jalama Beach hosts large numbers of shorebirds and waterfowl, and harbor seals are often seen hauled out on the sand near the creek mouth. Point Conception supports three large harbor seal haulouts, as well as a northern elephant seal haulout. From March through September, grunion come ashore to spawn on the sand beaches in this area, though they are more common in Southern California. These small fish (5-6 inches in length) come completely out of the water to spawn on

3-4 nights for only a 1-3 hour period after spring high tides. The California Department of Fish and Game (2009) described the spawning process as follows:

Spawning runs typically begin with single fish (usually males) swimming in with a wave and occasionally stranding themselves on the beach. Gradually, more and more fish come in with the waves and by swimming against the outflowing wave strand themselves until the beach is covered by a blanket of grunion. Spawning normally starts about 20 minutes after the first fish appear on the beach. Typically a run lasts 1 to 3 hours, but the number of fish on the beach at any given moment can vary from none, to thousands. Peak activity is reached about an hour after the start of the run and lasts from 30 to 60 minutes. Finally, when the tide has dropped a foot or more, the run slackens and then stops as suddenly as it started. No more fish will be seen that night, and they will not appear again until the next night or the next series of runs.

The eggs incubate while buried in the sand for approximately 2 weeks, until the next spring high tides reach them.

Fishing for grunion is one of the more famous sports in Southern California. On some beaches, there will be more people than grunion. There is a website, Grunion (2009), with up-to-date information on where and when the grunion runs occur. You can only "fish" for grunion by hand, so people go out at night with a flashlight and gunny sack (and a fishing license), running to pick them up as the wave recedes.

Places to Visit

This is a very remote and unpopulated area and much of it is within the Vandenberg Air Base; therefore, ready access to the beach is limited to just one locality, Jalama Beach County Park, another excellent place to experience the shore. To get there, go south from Lompoc on Highway 1 for 4.4 miles to where you turn right on Jalama Road. From there, it is approximately 13 miles to the County Park along this same road.

There are a variety of things to see at this beach, especially if you go at low tide. The mouth of the stream that exits on the north side of the campground is usually closed by a sand bar building to the north, as is shown by the photograph in Figure 180.

Our team surveyed two beach profiles at this location on two separate occasions, on 6 May 1992 and 2 March 1993. About 500 yards south of the stream mouth, the beach is backed by a wave-cut scarp in the seaward edge of a raised marine terrace. As can be seen by the field sketch in Figure 181A, this terrace is underlain by Miocene sedimentary rocks that dip steeply to the west. The berm of this beach is sand or sand and gravel and it has an associated wave-cut rock platform. There was an active accretionary berm over 30 yards wide present at the time of the May 1992 survey. There were also well-developed cusps along the beachface and a minor low-tide terrace that extended roughly 66 feet seaward from the toe of the beachface. The results of the March 1993 survey indicated that the berm had been eroded and the beachface was flattened, undergoing approximately two feet of vertical erosion (Figure 181B). Only a one-foot thick sand veneer remained on top of the rock platform at the time of that survey. It is possible that during major storm events, all of the sand covering the rock platform on this beach could be removed.

A beach survey was also carried out directly off the campground. A 20-yard wide berm along this medium-to fine-grained sand beach was eroding at the time of the May 1992 survey. The low-tide terrace was over 75 yards wide. At the time of the March 1993 survey, the beachface had retreated landward over 25 yards and there was no depositional berm present. The percentage of gravel on the beachface had increased. Storm waves had effectively mined the majority of the sand component from the beach and deposited it offshore, leaving behind a gravel lag deposit on the beach. The gravel made up the horns of a series of beach cusps.

In addition to a chance to observe the classic beach morphology, this location gives you the opportunity to inspect up close a raised marine terrace. North of the stream, the upper part of the terrace is composed of Quaternary sediments which, no doubt, are an important source of sand to the beach when the scarp erodes. If you walk far enough to the north, you will be able to see landslides of this loose sediment that have cascaded down the face of the wave-cut scarp.

If you walk south a few hundred yards, you can examine the steeply dipping Miocene sedimentary rocks up close. As usual, the birding is excellent along this beach.

Just like the **beach sediments** of Compartments 1-3, illustrated in Figure 117, those in these last three Compartments (5, 6, and 7) have also shown some remarkable differences. In December 2008, we collected sediment samples at nine of the beaches in these southernmost Compartments, shown in Figure 182, which illustrates these differences:

1) Sunset Beach – Medium-grained sand (gray) with many black grains.

2) Marina State Beach – Medium-grained sand (light brown) with mixed light-colored grains.

3) Carmel City Beach – Fine-grained sand (white) dominated by quartz.

4) Monastery Beach – Granules with feldspar and granitic rock fragments.

5) Shamel County Park Beach – Pebbles of mixed composition.

6) Cayucos Beach – Fine-grained sand (almost black) with Franciscan source.

7) Avila Beach – Fine-grained sand (gray) with Franciscan source.

8) Guadalupe Beach and Dunes – Medium/coarse grained (beach) and fine-grained (dune) of brownish color. Mixed source rock types.

9) Ocean Beach County Park – Medium/fine-grained (gray to brownish) with mixed source rock types.

There is one last thing you might want to observe in this, the most southern Zone on the Central California Coast. As you are leaving it on Highway 101 and heading toward Santa Barbara or somewhere else to the south, you will pass through the heart of the Santa Ynez Mountains. A little way before you come into Gaviota, there will be some highly tilted sedimentary rocks on both sides of the road that display a **flat-iron** configuration. This means that the rock layers, which are leaning at a high angle against the side of the mountain, have been eroded into parallel triangular shapes along their tops. The resulting individual rock slabs resemble the bottoms of some old-fashioned flat irons. Flat irons occur around the world in mountainous areas with sedimentary rocks. Some of the more famous ones are located along the mountain slopes west of Boulder, Colorado.

Some of the sedimentary rock layers that make up the flat irons in the Santa Ynez Mountains are pretty thick, but mostly not more than a few feet. Eichhubl and Behl (1998b) observed that *the foothills of the Santa Ynez Mountains expose reddish sand- and siltstone of the Oligocene Sespe Formation, underlain by massive sandstones of the Gaviota-Coldwater and the Matilija Formations of Eocene age that form dip slopes facing south toward the basin.* They noted that these particular rocks were exhumed about 1 million years ago, a very young uplift. Another point of interest they made was that Highway 101 crosses the northeast-southwest trending south branch of the Santa Ynez fault about 0.3 miles north of the tunnel at Gaviota Pass, but don't try to see it as you drive by at 65 mph. Pull over somewhere to look for it.

In any event, we are sure you will hate to leave Compartment 7, where you had a chance to see some of the most spectacular coastal geomorphology and experience some of the best birding in the whole world.

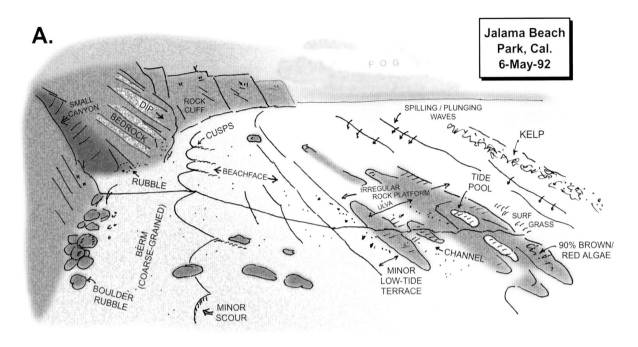

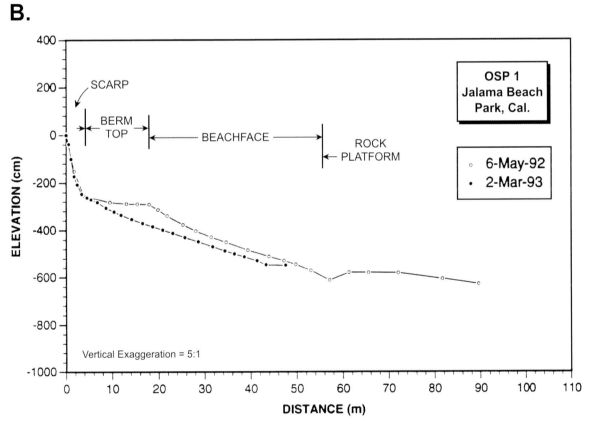

FIGURE 181. The beach south of the mouth of Jalama Creek at Jalama Beach County Park (Santa Barbara County; Zone F, Figure 163). (A) Field sketch of the beach on 6 May 1992. The line surveyed was about 100 yards south of the end of the wave-cut rock cliff (southern edge of Jalama Creek Valley). (B) Beach profiles surveyed for this beach on 6 May 1992 and 2 March 1993. At the time of the March 1993 survey, there was no sandy beach berm present (such as the one shown in the beach sketch in A).

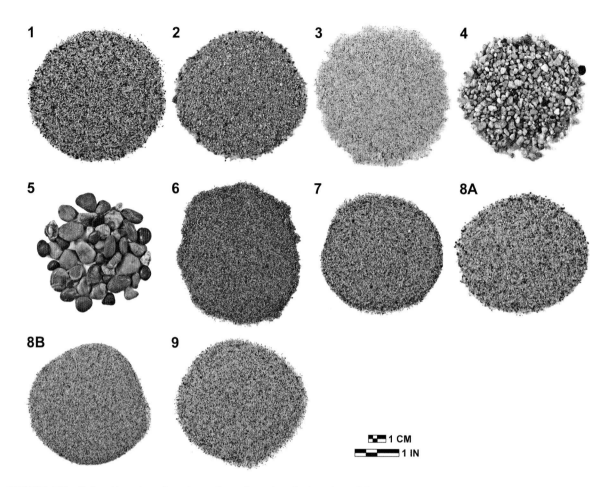

FIGURE 182. Suite of beach sediment samples collected on the beaches of Compartments 5-7 (Figures 1, 130, 140, and 163): 1) Sunset State Beach; 2) Marina State Beach; 3) Carmel City Beach; 4) Monastery Beach; 5) Shamel County Park Beach; 6) Cayucos Beach; 7) Avila Beach; 8) Guadalupe Beach (A – beachface; B – dunes behind beach); and 9) Ocean Beach County Park Beach.

20 A LOOK TO THE FUTURE

It is 0230 on 28 February 2009 and it is time for us to check out of the Hotel California; that is, we have come to the end of the writing of the rough draft of this manuscript. So now maybe it is time for us to close the book on the present Central California Coast and take a brief look into its future. There is much excitement in store for the Central California Coast in the future, that's for sure. With the Pacific Plate moving north at two inches per year (Alt and Hyndman, 2000), this piece of coastline will be off Oregon and Washington in about 15 million years, if the rest of the Farallon Plate has been swallowed up in the subduction zone to the north by then and there is room for the San Andreas Fault to sneak by. Add another 30 million years, or so, and two or our most favorite towns in the whole world, Cambria, California and Cordova, Alaska, will be neighbors.

Before then, most likely, humans will have heard from intelligent beings on one of the numerous planets circling the 200-400 billion stars in the Milky Way Galaxy or from those circling one or more of the even more billions of stars in the Andromeda Galaxy, telling the Earth's residents about their own far away studies of colliding plates and disappearing sea birds, maybe some birds sort of like our own disappearing Xantus' murrelets. And so it goes.

Of more immediate concern for the readers of this book, we would assume, is the prospect of sea level rising 2-5 feet in the next century if something isn't done about the excess of CO_2 in the atmosphere. Such a rise would be a problem for many residents of coastal towns of the California coast, because of increased rates of shore erosion and flooding of the nearshore zone, with Morro Rock becoming an isolated island, etc. However, the impact of such a rise would be devastating on the East and Gulf Coasts, where parts of cities, such as Charleston, South Carolina and New Orleans, Louisiana, would be constantly threatened by inundation, erosion, and loss of habitat.

Climate changes triggered by the increase of CO_2 bode very serious impacts to the Central California shoreline and marine resources, primarily in the form of increased storms, changes in upwelling and coastal productivity, and the like.

In the short term, however, presumably the major concern along this coastline is faulting activity, causing earthquakes, landslides, and so on. We hope you escape most of that and wish you the best of luck on your journeys. We also hope that you will take every opportunity to enjoy spending as much time as possible to explore this very special coast.

And around them the evening crept in as delicately as music. The quail called each other down to the water. The trout jumped in the pool. And the moths came down and fluttered about the pool as the daylight mixed into the darkness.

John Steinbeck – CANNERY ROW

RECOMMENDED ADDITIONAL READING

Alt, D. and D.W. Hyndman. 2000. Roadside Geology of Northern and Central California. Mountain Press Pub. Co., Missoula, MT. 369 pp.

Great general introduction to the geology of the state, with detailed descriptions of rocks and sediments encountered on drives along the roads in the coastal zone. One in a series of these outstanding books about the different states.

California Coastal Commission. 2003. California Coastal Access Guide, 5th Edition. Univ. Cal. Press, Berkeley, CA. 304 pp.

Excellent guide to the scenic and recreational facilities of the California Coast, including directions for finding the different locations and informative discussions on them as well as on a variety of other topics. Every naturalist not already familiar with the Central California Coast should have a copy of this book.

California Coastal Commission. 2007. Beaches and Parks from Monterey to Ventura. Univ. Cal. Press, Berkeley, CA. 318 pp.

With regard to the Central California Coast (as defined in our book), this book covers the shorelines of Monterey County, San Luis Obispo County, and the relevant parts of Santa Barbara County. As stated in the introduction, *this book tells you where to find over 300 beaches, parks, and other recreational facilities along the Central Coast*. Beautifully illustrated with color photographs and maps, it adds supplementary information to go along with the Coastal Access Guide (previous citation).

Collier, M. 1999. A Land in Motion: California's San Andreas Fault. Univ. Cal. Press, Berkeley, CA. 118 pp.

Thorough discussion of the San Andreas Fault, a topic of special concern on the Central California Coast. Excellent maps and photographs to go with a very interesting text.

Cooper, W.S. 1967. Coastal Dunes of California. Memoir 104 of the Geological Society of America. 131 pp.

A golden oldie, but a book anyone interested in the coastal dunes of California should at least review. Contains detailed maps of all the major coastal dune fields in the state, with accompanying engaging discussions and a number of revealing historical photographs.

Davis, R.A., Jr. and D.M. FitzGerald. 2004. Beaches and Coasts. Blackwell Publishing, Malden, MA. 419 pp.

The best modern textbook for giving the novice an introduction to the physical processes of the coastal zone.

Emory, J. 1999. The Monterey Bay Shoreline Guide. Univ. Cal. Press, Berkeley, CA. 307 pp.

Detailed discussion of places to visit between Año Nuevo State Preserve and Point Sur. Some in-depth notes are presented on the individual biological species of interest. A range of perspectives presented on the different preserves, state beaches, and other noteworthy sites. Beautiful photographs.

Griggs, G., K. Patsch and L. Savoy. 2005. Living with the Changing California Coast. Univ. Cal. Press, Berkeley, CA. 540 pp.

If you are into horror stories, put away your Stephen King collection and buy this book, which has no less than 85 photographs of beach-erosion nightmares all along the California Coast. In addition to a clear exposition on the science of dealing with beach-erosion issues, there are detailed maps showing hazard levels, erosion rates, and so on for the entire California Coast, including details on the area covered in this book. Anyone owning property along the shoreline of California should have a copy.

Harden, D.R. 2004. California Geology (2nd edition). Prentice Hall, Upper Saddle River, NJ. 552 pp.
Probably the leading source book for all aspects of the geology of the state as a whole. Excellent illustrations.

Henson, P. and D.J. Usner. 1993. The Natural History of Big Sur. Univ. Cal. Press, Berkeley, CA. 416 pp.
Covers a range of topics, from the resident fauna to fire ecology, including a thorough treatment of the plant communities. Chapter 1, on Big Sur geology, is perhaps the most succinct treatment of this topic available. Beautiful line drawings.

Howard, A.D. 1979. Geologic History of Middle California: California Natural History Guides 43. Univ. Cal. Press, Berkeley, CA. 173 pp.
A somewhat dated, but still useful, summary of the geologic history of Central California.

Hyndman, D. and D. Hyndman. 2006. Natural Hazards and Disasters. Thomson Brooks/Cole, Belmont, CA. 490 pp.
Superbly illustrated treatment of plate tectonics, earthquakes, volcanoes, and other natural hazards with numerous examples from the state of California. Possibly the best place for a non-geologist to begin to learn about these processes.

Lage, J. 2004. Point Reyes: The Complete Guide to the National Seashore & Surrounding Area. Wilderness Press, Berkeley, CA. 249 pp.
Good discussion of many aspects of the National Seashore. The detailed descriptions of the hiking trails are especially valuable for anyone wanting to gain a complete naturalist's experience during a visit to the Point Reyes area.

McConnaughey, B.H. and E.M. McConnaughey. 1985. Pacific Coast: The Audubon Society Nature Guide. Alfred Knopf, New York, NY. 633 pp.
A useful field guide for identifying birds, marine mammals, fish, invertebrates, and seaweeds of the Pacific Coast. Many excellent colored illustrations.

Ricketts, E.F and J.C. Calvin. 1939. Between Pacific Tides. Rickets, An earlier revision by Ricketts, E.F, J. Calvin, and J. W. Hedgepeth. Revised later by D.W. Phillips in1985. 5th/Revised edition, Stanford Univ. Press, Stanford, CA. 652 pp.
Still the "Bible" for marine invertebrates on the U.S. West Coast. It has gone through numerous editions since its first publication with several revisions.

Rigsby, M., ed. 1999. Natural History of the Monterey Bay Marine Sanctuary. Monterey Bay Aquarium, Monterey, CA. 299 pp.
An excellent summary of the oceanography, geology, habitats, and flora and fauna of a marine Sanctuary that extends from north of San Francisco to south of the Big Sur Coast.

Roberson, D. 2002. Monterey Birds (Revised 2nd Edition). Monterey Peninsula Audubon Society, Carmel, CA. 636 pp.
Detailed account of the birds that frequent Monterey County, California. Statistics on occurrence (with maps), life history, and other information.

Schwartz, M.L. ed. 2005. Encyclopedia of Coastal Science. Springer, The Netherlands. 1,211 pp.
The answer to any question you might have on this subject can be found in this phenomenal tome.

Sloan, D. 2006. Geology of the San Francisco Bay Region. Univ. Cal. Press, Berkeley, CA. 335 pp.
Another beautifully illustrated book. The best source for detailed information on the faulting in the Bay area. Also excellent treatment of the geology of the shoreline in the immediate vicinity of the Bay, including the Point Reyes area and the outer shoreline just north and south of the Golden Gate Bridge.

Sloan, D. and D.L Wagner. 1991. Geologic Excursions to Northern California, San Francisco to the Sierra Nevada. California Division of Mines and Geology Special Publication 109. 130 pp.
Includes papers on the geologic setting of the San Francisco Bay area, Point Reyes, and the Merced Formation in the sea cliffs south of San Francisco, as well as other areas in Northern California.

GLOSSARY

Amphibole – An important group of generally dark-colored rock-forming inosilicate minerals, composed of double chain SiO_4 tetrahedra, linked at the vertices and generally containing ions of iron and/or magnesium in their structures (Wikipedia Encyclopedia). Hornblende is a common example of this mineral class.

Amphidromic system – Separate, more-or-less circular, gyrating tidal systems (standing waves) with a nodal point in the middle. They are a large rotary tidal system created by the behavior of tides as giant low waves that are affected by the shape of the world's oceans and the rotation of the Earth. The height of the tide depends on its location within the system. Tides at the axis of the system (nodal point) are nil, but they increase outward with distance from the axis. In the Northern Hemisphere, the tides rotate counterclockwise about the axis; in the Southern Hemisphere, the rotation is clockwise. Both diurnal (one complete rotation of the tides daily) and semi-diurnal (two complete rotations of the tides daily) can coexist in the same ocean basin. The Pacific Ocean contains two diurnal systems and eight partial or complete semidiurnal systems. (Ed Clifton, pers. comm.) See examples in Figure 30.

Anadromous Fish - Fish that live in salt water but migrate to freshwater habitats to spawn, such as salmon, steelhead (a unique form of rainbow trout), and Pacific lamprey.

Angular unconformity – A nearly horizontal, planar surface carved by the erosion of a series of tilted strata. This surface was later covered by layers of younger horizontal strata. Such an erosional surface indicates that a significant amount of geological time is missing from the rock record.

Anticline – The condition where layers of rocks are folded into a form similar to what you would get if you pushed one edge of a rug along the floor, folding it into parallel bands. Where the folded layers bend up, the fold is called an anticline.

Antidunes – Linear and parallel, trochoid-shaped features generated by rapid flow in the upper flow regime. In the process, a relatively thin layer of rapidly moving water flows as a standing wave across the top of the antidune surfaces. The standing waves in the water are in phase with the sediment surface. The form itself moves up current. but the individual sand grains move down current in a series of stops and starts. They have a typical spacing of about 15-20 inches on beaches, where they are commonly associated with flat beds. See Figures 60 and 167.

Armoring of gravel beaches – A process whereby intermediate-sized gravel, or fine gravel not shielded by larger particles, are removed from the surface of a segment of the gravel beach at velocities too low to transport the largest clasts, which remain behind as a moderately well-sorted layer of coarse gravel over the surface of the beach. This residual, coarse-grained armor is quite stable, being moved only during the largest of storms. See Figures 65, 150, and 154B.

Asthenosphere – Part of the Earth's mantle below the lithosphere that behaves in a plastic manner. The rigid and more brittle lithosphere rides over it (Hyndman and Hyndman, 2006).

Astronomical tide – Periodic changes in water levels caused primarily by the gravitational attraction of the moon and sun on the world oceans. See Figure 28.

Backwash of a wave – That portion of the water that rushes up the beachface under the impetus of a breaking wave, doesn't percolate into the beachface, and returns back down the slope of the beachface under the influence of gravity.

Barchan sand dune – A wind-formed, crescent-shaped sand ridge, which has two "horns" that face downwind. A steep slip face is present on the convex side of the dune, which dips at the angle of repose of sand in air (30+ degrees). See Figure 76.

Barrier island – Elongate, shore-parallel accumulations of unconsolidated sediment (usually sand), some parts of which are situated above the high-tide line (supratidal) most of the time, except during major storms. They are separated from the mainland by bays, lagoons, estuaries, or wetland complexes and are typically intersected by deep tidal channels called tidal inlets. They do not occur on shorelines with a tidal range greater than 12 feet (macrotidal shores).

Barycenter – The center of mass of the combined Earth-moon system. The Earth and the moon revolve together around the barycenter, which is displaced a distance from the center of the Earth toward the moon, and is always located on the side of the Earth turned momentarily toward the moon, along a line connecting the individual centers of mass of the Earth and moon (Wood, 1982). See Figure 29A.

Basalt – A dark-colored, very fine-grained igneous rock, the most common type of solidified lava. It is composed mostly of the minerals plagioclase feldspar, pyroxene, and olivine. It is a very prevalent component of the sea floor near mid-ocean ridges. See Figure 19.

Base level – The lowest level to which a river flows.

Batholith – A large emplacement of igneous intrusive (also called plutonic) rocks that form from cooled magma deep within the Earth's crust. They are almost always composed of feldspar-rich rocks such as granite, quartz monzonite, or diorite (Wikipedia Encyclopedia). See Figures 14 and 17.

Beach cusps –The scalloped seaward margin of a beach berm consisting of a regularly spaced sequence of indented bays and protruding horns of a relatively small scale (rarely more than a few tens of feet between horns). They are thought to be primarily depositional features (i.e., on the seaward flank of a prograding depositional berm) that form as a result of the interaction of shore-normal standing waves (edge waves) with shore-parallel, reflective waves, the same general process that produces rip currents (but at a smaller scale) (Guza and Inman, 1975). See Figures 38 and 51.

Beach drift – The transport of sediment along the beachface by the combined processes of uprush and backwash. The uprush from a wave that approaches the beachface obliquely will move sand grains (or pebbles) at an angle across the beachface. Once the approaching uprush loses its momentum, the sand grains (or pebbles) are transported straight down the beachface by the backwash under the influence of gravity. The ultimate result is that the sediment particles move along the beachface in a sawtooth pattern away from the direction of wave approach. See Figure 37 (Lower Left).

Beach nourishment – Artificially adding sediment to a beach to counter beach erosion. The sand can come from a variety of sources, with offshore sand bars being a common one.

Beachface – The zone of wave uprush and backwash during mid to high tide. It usually slopes seaward at an angle of 5-10 degrees and is commonly the seaward face of the berm. See Figures 50 and 51.

Beachrock – A solid rock usually formed recently by rapid solidification of beach sand through the precipitation of calcium carbonate from sea water into the pore spaces between the sand grains, a process more likely to occur where the water is warm.

Berm on a sand beach – A wedge-shaped sediment mass built up along the shoreline by depositional wave action in between storms. It typically has a relatively steep seaward face (5-10 degrees; the beachface) and a gently-sloping landward surface (2-3 degrees; the berm top). See Figures 50 and 51.

Boulder - A sediment particle that is greater than 256 millimeters (~ 10 inches) in size. See Figure 150.

Brackish marsh – A low-lying wetland containing grassy vegetation that is regularly flooded by brackish waters during mid to high tides. Salinities average less than 15 parts of salts per thousand parts of water (ppt).

Breakwater – See offshore breakwater.

Chert – A fine-grained, smooth-textured rock composed of SiO_2 and formed as a sedimentary deposit. The radiolarian chert deposits of the Franciscan Complex are a notable example. See Figure 20.

Chiton – A type of mollusc with only one shell that lives attached to rocks. Its shell is made up of several plates like a suit of armor. They are also called coat-of-mail shells (National Museum Wales Glossary).

Closure point – A term defined by coastal engineers as the most seaward distance for sand to move into the nearshore zone (off the intertidal beach in an offshore direction) during storms. In effect, it is the seaward limit for sand to move during the normal beach cycle that takes place in response to the passage of storms and intervening periods of relative quiescence. Many coastal geologists and some engineers dispute the validity of this concept. See Figure 49.

Coastal geomorphology – A scientific discipline focused on the understanding of how the different coastlines of the world originated, as well as their resulting forms and organization.

Coast Range Ophiolite – Oceanic crustal material upon which much of the sedimentary rocks of the Great Valley Sequence were deposited. It has a complex composition that includes ultramafic and gabbroic rocks and so on (Hopson, Mattinson, and Pessagno, 1981). See Figures 14 and 158B.

Cobble – A sediment particle that is between 64-256 millimeters (2.5-10 inches) in size. See Figure 152B.

Collision coast – Shorelines defined by Inman and Nordstrom (1971) as those that occur along the leading edges of continental plates, such as the ones along much of the west coasts of North and South America. They usually are mountainous and contain rocky shores with short, steep rivers emptying into the sea.

Competency of flow – The ability of moving water to transport sediment.

Concretion – Local concentration of chemical compounds, such as calcium carbonate or iron oxide, within a sedimentary rock layer that usually take on the form of a spherical nodule.

Conglomerate – Sedimentary rock made up mostly of gravel-sized particles or clasts. See Figure 142.

Continental drift – A theory proposed by Wegener (1912) that states that the present continents on the Earth's surface were joined together and later moved apart, based to some extent on the almost perfect fit of the east coast of South America with the west coast of Africa, plus other comprehensive evidence. Lacking in his theory was a provable mechanism to move the continents apart. See Figure 13B.

Continental shelf – The gently sloping seabed between the shoreline and the shelf break, where there is an abrupt increase in slope, at depths measured in hundreds of feet. See Figure 81.

Continental slope – A steeply descending planar surface that adjoins the outer margin of the continental shelf. See Figure 81.

Core of the Earth – The central, interior portion of the Earth that is composed mostly of iron and makes up ~30% of the Earth's volume. It consists of two layers, an outer extremely hot zone, which is molten, and an inner layer, also very hot, that is solid because of the intense pressure at those depths within the Earth. See Figure 6.

Coriolis effect – An effect whereby a body of air, water, etc. moving in a rotating frame of reference experiences a force that acts perpendicular to the direction of motion and to the axis of rotation. On the rotating Earth, this effect deflects moving gases and liquids to the right in the northern hemisphere and to the left in the southern hemisphere. Named after French engineer G.G. Coriolis (1792-1843).

Crenulate bay – An embayment along the shore that is usually separated by two rocky headlands. They typically have a fishhook, shape with the shank of the hook pointing in the direction of sediment transport. Three of the most prominent ones on the Central California Coast are Bodega Bay, Drakes Bay, and Half Moon Bay. See Figure 67.

Crust of the Earth – The solid upper layer of the Earth composed of a variety of igneous, metamorphic, and sedimentary rocks. The crust under the oceans, which averages about 6 miles thick, is composed mostly of rocks that contain an abundance of iron and magnesium-bearing minerals (e.g., basalt), whereas the crust on the continents, which averages 20-30 miles thick, is composed of rocks made up of lighter minerals (e.g., granite) (Wikipedia Encyclopedia).

Cuspate spit – A triangular-shaped bulge of sand or sand and gravel beaches that projects directly offshore. They are most common on shorelines where the wind blows in two opposite directions (parallel to shore). They form where there is a relative equality between the two longshore sediment transport volumes generated by the counterbalanced winds. See Figure 98.

Debris flow – Rapid movement of a partially consolidated mass of sediment and broken up rock fragments, some up to boulders in size, down a steep slope, usually in a marine setting. Debris-flow deposits are commonly associated with turbidites, which are deposited by more fluid plumes of sediment-laden water (Figure 21A).

Depositional coast – Shoreline type defined by Hayes (1965) as one characterized by: 1) coastal zone made up chiefly of broad coastal and deltaic plains; 2) bedrock composed generally of Tertiary and Quaternary sediments (young geologically speaking); 3) tectonically subsiding areas; and 4) many large and long rivers emptying into the sea. They most commonly occur on the trailing edges of continental plates or along the shorelines of epicontinental seas, such as the Black and Caspian Seas.

Diatom – One-celled marine algae that are roughly circular in shape and secrete intricate skeletons of silica. The Santa Cruz Mudstone, which is a prominent cliff-forming rock formation along the Central California Coast at Santa Cruz and for many miles to the north of there, contains an abundance of siliceous diatom skeletons, as does the Monterey Shale.

Dike – A relatively narrow rock band (or zone) of distinctly different composition (usually) that has been intruded into and cuts across an older rock.

Diorite – A rock similar in texture to granite made up largely of white to light gray plagioclase feldspar (ranging from $NaAlSi_3O_8$ to $CaAl_2Si_2O_8$) and a certain amount of hornblende (a range of compositions including $CaNa(Mg,Fe)_4(Al,Fe,Ti)_3Si_6O_{22}(OH,F)_2$) with <5% quartz. See Figure 17B.

Diurnal inequality of the tides – A condition where, although the tides are roughly semi-diurnal, four significantly different levels (rather than two) occur during one tidal day (referred to as higher-high, lower-high, higher-low and lower-low tides). One possible cause of this is the tilt of the Earth's axis, because the depth of the water under the tidal bulge will be different on opposite sides of the Earth when the moon is not directly over the equator (Figure 29C). Another probable cause is a combination of tides from separate diurnal and semidiurnal amphidromic systems, which appears to the case on the Central California Coast (Ed Clifton, pers. comm.). See Figure 31.

Downdrift – The direction in which individual sediment grains are moving along the shore within the longshore sediment transport system. See Figure 37.

Ebb-tidal current – Current generated during the falling tide.

Ebb-tidal delta – A lobe of sediment, usually composed of sand, much of which has been transported from the scoured-out throat of a tidal inlet by ebb-tidal currents. The sand is deposited on the seaward side of the inlet throat as the ebb currents slow down upon entering the open ocean. The morphological components of ebb-tidal deltas include a main ebb channel (with slightly stronger ebb currents than flood currents) flanked by linear sand bars on both sides and a terminal lobe at the seaward end. The main ebb channel is bordered by a platform of sand (swash platform) dominated by swash bars (intertidal bars built up by wave action). The swash platform is separated from adjacent beaches by marginal flood channels (which are dominated by flood currents). See Figure 69.

Echinoderm – Any of numerous radially symmetrical marine invertebrates of the phylum Echinodermata that have an internal calcareous skeleton and are often covered with spines. Examples include sea stars, sea urchins, and sea cucumbers (Answers.com).

Eelgrass – Common name for a group or genus of plants called *Zostera* that grow under water in estuaries and in shallow coastal areas. It is neither a grass nor a seaweed. It is an angiosperm, or flowering plant, that can live for many years (i.e., it is a perennial) (Eelgrass, 2005).

El Niño **event** – A globally coupled, poorly understood ocean-atmosphere phenomena that results in: 1) relatively low pressure over the Pacific Ocean; 2) a weakening of the trade winds (during normal conditions, these strong winds generate a very strong westerly flow of the surface waters to the west in the equatorial region) that contributes to an increase in water level on the California coast (as much as 8-12 inches); and 3) the thermal expansion of the warmer water because of a decrease in upwelling along the coast. During these events, unusually heavy rains typically accompany an increased number of winter storms. Leads to significant coastal inundation and sea cliff erosion during the associated coastal storms (Inman and Jenkins, 2005b).

Ensoulment – A concept that suggests that humans may participate with the earth as if it were a living being (Cajete, 1995).

Entrainment – The selective transport of sediment particles by water or air.

Environmental Sensitivity Index (ESI) – A system for mapping coastal environments, the first version of which was proposed for the coast of Lower Cook Inlet, Alaska by Hayes, Michel, and Brown in 1977. It ranks shorelines on a scale of 1-10 based upon their sensitivity to oil-spill impacts. Exposed rocky shores are ranked 1 (least sensitive) and sheltered salt marshes are ranked 10 (most sensitive). ESI maps, which are used for oil-spill response and other types of environmental assessments on a world-wide basis, also include detailed biological and human-use information. See examples in Figures 85 and 165.

Epicenter of an earthquake – The point on the surface of the Earth that lies directly above the hypocenter, the place at depth under the Earth's surface where the fracture that causes the earthquake is located.

Estuary – A coastal water body that, according to Pritchard (1967), has three defining qualities: 1) a flooded river valley that was formed during the lowstand of sea level that terminated about 12,000 years ago; 2) a water body with a substantial freshwater influx; and 3) a water body subject to tidal fluctuations. See Figures 78 and 79.

Evaporites – A class of sedimentary rocks usually, but not always, formed as the result of evaporation of sea water, typically under high surface temperatures. Evaporite deposits are most commonly composed of halite (NaCl) and gypsum $[CaSO_4 \cdot (2H_2O)]$.

Fault – A fracture along which there has been slipping of the contiguous rock or sediment masses against one another. Points formerly together have been dislocated or displaced along the fracture (Lahee, 1952). See Figures 9 and 10.

Fetch – The length of water surface over which the wind blows to form waves. See Figure 33B.

Fire-obligate pine trees – Pine trees that cannot reproduce without a fire to open up the pine cones and free the seeds.

Flandrian time – The segment of geological time that has occurred since the end of the last glaciation about 12,000 years ago (also called the present interglacial or Holocene time).

Flocculation – A process that takes place in estuaries when colliding clay particles tend to adhere together under the influence of the increased concentration of cations, such as Na+, Ca++ and Mg++, derived from the sea water. As a result, the clay particles, which may have diameters as little as one micron, cluster into groups (**flocs**) that may have diameters measured in hundreds of microns. Consequently, many of the flocs sink to the bottom at slack tide. See Figure 79B.

Floodplain of a river – Many river channels run through valleys of different widths that were carved during the last lowstand of sea level during the last major glaciation, which terminated about 12,000 years ago. During floods, sediment rich (muddy) water spills out of the channel and builds up a flat plain between the channel and the sides of the valley. These flat-lying deposits occupy a zone known as the floodplain.

Flood-tidal current – Current generated by the rising tide.

Flood-tidal delta – A lobe of sediment, usually composed of sand, most of which has been transported from the scoured-out throat of a tidal inlet by flood-tidal currents. The sand is deposited on the landward side of the inlet throat as the flood currents slow down upon entering the more open bay on the landward side of the inlet. See Figure 68.

Flow regime – A concept worked out by hydraulic engineers to distinguish among features that develop in sediment (most appropriate for sand) along the bed of a stream (or beach, etc.) under the influence of flowing water. When the water is flowing slowly (or is relatively deep), the bed changes from small ripples to larger megaripples and sand waves (asymmetric features that move in the same direction as the current is flowing) under increasing flow strength. This first part of the sequence, during which the value of the Froude number (the ratio of water velocity to the square root of the gravitational constant times water depth) is <1.0, is called the lower flow regime. As the Froude number increases to values >1.0, by either increasing the velocity of flow or decreasing the water depth, the sand bed goes into the upper flow regime, where either flat beds or antidunes are formed. See Figures 56, 57, and 60.

Forearc basin – A depression in the sea floor located between a subduction zone and a volcanic arc. One was located in the area of the present Central Valley of California during Late Cretaceous and Early Paleocene time. See Figure 14.

Foredunes – Wind-blown sand dunes that form immediately adjacent to the backbeach. They are most commonly vegetated.

Franciscan Complex – A sequence of rocks composed of materials that accumulated in the subducting Franciscan trench from 130-70 million years ago. Pillow basalts, radiolarian cherts, turbidite sandstones, and mélange are included in the complex, which crops out along the Central California Coast north of where the San Andreas Fault line trends offshore (south of San Francisco) and as the major component of the Nacimiento Block south of Monterey. See Figures 16 and 18.

Franciscan trench – A major oceanic trench formed as the oceanic Farallon Plate was subducting beneath the continental North American Plate between 130-70 million years ago. It was located approximately 60 miles west of the present Sierra Nevada. See Figures 12 and 14.

Garnet – A member of a complex suite of silicate minerals that are commonly reddish in color, one example being glossular garnet $[Ca_3Al_2(SiO_4)_3]$. Common constituent of metamorphic rocks, being present in both schists and gneisses. See Figures 147B and 148.

Gastropod – A class of molluscs typically having a one-piece coiled shell, flattened muscular foot, and head bearing stalked eyes (WordNet Search).

Geomorphology – A scientific discipline devoted to understanding the origin and three-dimensional shape of the landforms of the Earth.

Glacial rebound coast – Shoreline which continues to rise even to this day as a result of the removal of the vast ice sheets associated with the last Pleistocene glaciation.

Gondwana – A super continent that existed between 300-400 million years ago. It consisted of the present areas of South America, Africa, India, Australia, and Antarctica. It was separated from another supercontinent, Laurasia, by the Rheic Ocean. Figure 13A.

Graded bed – A sediment layer that is coarse-grained at the bottom and fine-grained at the top. A common way for them to form is to be dropped out of suspension from the sediment-laden plume of a turbidity current. See Figure 21.

Granite – A light colored, coarsely crystalline igneous rock that is composed of 20-60% quartz, with most of the remaining component being between 10 and 65% of a mixture of alkali feldspar and plagioclase. It may also contain a few dark-colored minerals, such as hornblende and biotite. See Figure 17.

Granodiorite – A light colored, coarsely crystalline igneous rock that is composed of 20-60% quartz, with the remaining mostly feldspar component being between 65 and 90% plagioclase. It may also contain a few dark-colored minerals, such as hornblende and biotite. See Figure 17B.

Granule – A sediment particle that is 2-4 millimeters in size. See Figure 116B.

Gravel – Sediment with a median diameter greater than 2 millimeters. Gravel is defined according to the Wentworth (1922) scale, which includes four classes – *granules* (2-4 millimeters), *pebbles* (4-64 millimeters), *cobbles* (64-256 millimeters), and *boulders* (>256 millimeters).

Graywacke – A type of sandstone that is usually dark-colored, made that way by either containing a lot of clay matrix between the sand grains or having abundant sand particles composed of dark-colored rock fragments from a variety of rock sources, such as metamorphic slates or schists. A common component of turbidite deposits.

Groin – A linear, relatively short, shore-protection structure built perpendicular to the shoreline for the purpose of widening the beach by trapping a portion of the sand in the littoral drift. Commonly composed of riprap, but other materials, such as sand bags and wooden timbers, are sometimes used. See Figure 82B.

Half-life – The amount of time it takes half of a radioactive parent element to decay to a stable daughter element. For example, it takes 1.25 billion years for half of Potassium-40 (the radioactive parent) to decay to Argon-40 (its stable daughter).

Halophytes – Plants that are adapted to grow in salty soil.

Heavy minerals – Minerals with specific gravities significantly greater than quartz, which has a specific gravity of 2.65 (the specific gravity of water is 1.00; i.e., a cubic centimeter of water has a mass of 1 gram). Examples include magnetite [(Fe_3O_4); specific gravity of 5.1)], garnet, and hornblende.

Highstand – The period when sea level was high for a significant amount of time, measured in thousands of years, during intervals between the four major glaciations of the Pleistocene epoch, called interglacials. The last highstand before the present one occurred around 125,000 years ago, and it is thought to have been considerably higher that the one at present (several feet in areas with no tectonic uplift).

Holocene time – The segment of geological time that has occurred since the end of the last glaciation, starting about 12,000 years ago.

Hornblende – A dark-colored silicate mineral of the ferromagnesium class that is a common constituent of both igneous and metamorphic rocks with a range in compositions, including $Ca_2(Mg,Fe,Al)_5(Al,Si)_8O_{22}(OH)_2$ and $CaNa(Mg,Fe)_4$ $(Al,Fe,Ti)_3Si_6O_{22}(OH,F)_2$.

Hydraulic equivalence – A size–density relationship that governs the deposition of mineral particles from flowing (or standing) water. Two particles of different sizes and densities are said to be hydraulically equivalent if they are deposited at the same time under a given set of conditions (Encyclopedia Britannica; e.g., if dropped into a cylinder of water, the different particles would settle to the bottom at the same speed despite their differences in size).

Hydrodynamic regime – For purposes of interpreting the morphology of coastal features, the hydrodynamic regime is commonly expressed as the ratio of wave energy to tidal energy (i.e., how big the waves are versus how large the tidal range is). It exerts a major control on the geomorphological makeup of depositional coasts.

Hypocenter – The focus, or location of the fracture, where the movement starts during an earthquake. Usually at some depth below the Earth's surface.

Ice plant (*Carpobrotus edulis*) – A highly invasive, creeping mat-forming succulent plant. Since its introduction into California from South Africa in the early 1900s to stabilize soil along railroad tracks and for other purposes, this plant has become a serious ecological problem, covering vast areas, lowering biodiversity, and competing directly with several threatened or endangered plant species (Wikipedia Encyclopedia).

Igneous rock – Rock that crystallized from an extremely hot molten mass of material (magma). One class of igneous rock, called granitic rock (e.g., granite and diorite), develops by slow cooling of the molten mass at great depths within the Earth's crust. As a result of the slow cooling, large crystals of individual minerals form that can be as much as an inch or so in diameter. A second class of igneous rock, called volcanic rock (e.g., basalt and rhyolite), crystallize rapidly as a result of molten lava being extruded suddenly onto the Earth's surface. The crystals of these minerals that form so rapidly are, for the most part, too small to be seen by the naked eye. See Figure 17.

Imbrication of gravel clasts – A process whereby the maximum projection areas of the clasts face (dip toward) the strongest force acting upon them, usually at angles of 20-25 degrees. It is best developed by platy or disc-shaped clasts. See Figure 63.

Invertebrates – Animals that do not have an internal skeleton made of bone, such as anemones, sea stars, jellyfish, clams, snails, crabs, shrimp, and worms.

Island arc – An arc-shaped zone of volcanic islands built above a subduction zone.

Jetties – Structures built at a river mouth, a tidal inlet, or even an artificially dug harbor, to stabilize the channel, prevent its shoaling by sediments, and protect its entrance from waves. They are commonly long, parallel, and composed of large pieces of rock of boulder size called riprap. See Figure 82A.

Joints in the rocks – Planar surfaces along which the rocks have split apart, usually during one or more of the ever present tectonic movements along this coast.

Kelp – A marine algae that consists of three components: a holdfast that attaches to the rocky seafloor; a stipe that is similar to a plant stalk; and fronds that attach to the stipe. Kelp forests require wave energy and currents to bring in nutrients, clear water to allow photosynthesis, and cool water temperatures. Kelp forests also require an appropriate water depth (typically shallower than 60-80 feet unless the water is unusually clear) and a hard substrate for attachment.

Krill - Shrimp-like planktonic crustaceans, a major food source for some whale species.

La Niña **event** – A globally coupled, poorly understood ocean-atmosphere phenomena that generally shows effects opposite to those of the *El Niño*. For example, a decrease in water levels along the California coast is possible. Rainfall is greatly diminished, and the trade winds are usually strengthened, resulting in the intensification of upwelling along the coast. A generally high-pressure anomaly over the eastern North Pacific that hugs fairly close to the North American shoreline

typically accompanies these events. This forces storm tracks to come ashore further north than those that occur during *El Niño* events (Inman and Jenkins, 2005b).

Landslide – Various processes whereby loosened soil and rock materials on a hill/slope are moved from their original site of formation to lower lying areas under the influence of gravity. Earthquakes and water saturation in the soil are two of the primary causes of landslides. They can occur both above or below the sea. See Figure 84.

Laurasia – A supercontinent that existed between 300-400 million years ago. It consisted of the present areas of North America, Greenland, Europe, and part of Asia. It was separated from another supercontinent, Gondwana, by the Rheic Ocean. See Figure 13A.

Limestone – Sedimentary rock composed of calcium carbonate ($CaCO_3$), which usually develop in marine waters by a combination of chemical precipitation and accumulation of the hard parts of marine organisms, such as sea shells.

Limpet – A type of gastropod mollusc which has a characteristic conical shell and a suckerlike foot. Limpets (like chitons) are well adapted for life on rocky surfaces exposed to wave action, and alternating tidal submergence and exposure to air (Sci-tech Encyclopedia).

Lithosphere – The rigid outer rind of the Earth approximately 40-60 miles thick. It composes the tectonic plates (Hyndman and Hyndman, 2006).

Littoral cell – A segment of the shoreline with updrift and downdrift boundaries that do not allow sediment in the longshore sediment transport system to pass into or out of the segment in a longshore direction. On the California coast, these boundaries are rock headlands. Sediment is added to the California cells from rivers and bluff erosion, and sediment is lost from the cell primarily down submarine canyons and into onshore dune fields. See Figure 66.

Littoral drift – Volume of sand moving along the shore in the longshore sediment transport system.

Longshore current – Current formed when an obliquely approaching wave piles up water in the surf zone that flows away from the direction of the approaching wave. The resulting current runs parallel to the shore. See Figure 37.

Longshore sediment transport – The movement of sediment along the beachface and within the nearshore zone when waves approach the shoreline at an oblique angle. This movement of sediment is usually produced by the combined processes of beach drift and longshore currents. See Figure 37.

Lowstand – Term that refers to the period when sea level was low for a significant amount of time, measured in thousands of years, during the four major glaciations of the Pleistocene epoch. The lowstand during the last glaciation, the Wisconsin, which occurred from 20,000 to 12,000 years ago, was about 350-400 feet lower than it is today.

Low-tide terrace – A relatively flat surface that is located seaward of the toe of the beachface and slopes offshore at a small angle (2-3 degrees). The landward margin is usually near mean sea level and the seaward margin near mean low water. See Figures 50 and 51.

Macrotidal – Term originally used by Hayes (1976) to define shoreline types. Davies (1973) defined macrotidal coasts as those with a tidal range (vertical distance between high and low tide) of >4 meters (approximately 12 feet), whereas Hayes (1979) proposed the same >12 feet boundary, with low-macrotidal being 12-16 feet and macrotidal being >16 feet.

Magma – A molten mass of material that usually forms deep within the Earth's crust. See Figure 14.

Magnitude-Moment (M) scale – A logarithmic scale (base 10) used to determine the amount of energy released by an earthquake. It is based on several factors, including: a) how far the fault slipped (i.e., how far a fence line, railroad line, and so on was offset); b) length of the fault line that was affected; c) depth at which movement was initiated; and d) the "stickiness" of the surfaces being faulted (rock rigidity). This is a more complex way to measure earthquake magnitude than the Richter scale, which is based on the amplitude of shear waves as measured by a seismograph. The Loma Prieta earthquake on 17 October 1989, which resulted in severe damage in San Francisco, measured 6.9 on the Magnitude-Moment scale and 7.1 on the Richter scale.

Mantle – The thick layer of material below the relatively thin Earth's crust. It contains more of the heavier elements, such as iron and magnesium, than the crust, is about 1,800 miles thick, and makes up approximately 70% of the Earth's total volume. It is composed mostly of the igneous rock peridotite. See Figure 6.

Marble – A metamorphic rock formed when limestones are subjected to high heat and/or pressure. They are usually made up of relatively large, tightly melded crystals of calcium carbonate ($CaCO_3$).

Mass wasting – A mechanism whereby weathered materials are moved from their original site of formation to lower lying areas under the influence of gravity.

Meandering channel – Channel with a highly winding path, like the loops one throws into a rope. Such channels tend to have: 1) relatively flat slopes; 2) a high ratio of suspended to bedload sediments; and 3) a steady discharge. Meandering tidal channels are by far the most common channel type in the coastal water bodies of Central California. See Figure 80.

Megaripples in sand – Asymmetric, linear, or cuspate features moved along a sandy sediment surface by flowing water with a spacing between their crests of 2-18 feet. On the shoreline, they are seen more commonly in tidal channels, where the water is deeper and the currents are stronger. They can also be produced by wave surge beneath or just seaward from the breaker zone (Ed Clifton, pers. comm.). See Figures 56B, 56C, and 57.

Mélange – A prevalent rock assemblage in the Franciscan Complex formed under intense tectonic forces at great depths within the Franciscan trench. It consists of a soft crushed shale or sepentinite matrix with blocks of other rocks "floating" in it (Sloan, 2006). These foreign blocks have a variety of compositions and sizes.

Mesotidal – Term originally used by Hayes (1976) to define shoreline types. Davies (1973) defined mesotidal coasts as those with a tidal range (vertical distance between high and low tide) of 2-4 meters (approximately 6-12 feet), whereas Hayes (1979) proposed 3-12 feet, with low-mesotidal being 3-6 feet.

Metamorphic rocks – Rocks that result from dramatic changes in igneous and sedimentary rocks affected by heat, pressure, and water that usually creates a more compact and more crystalline condition. These changes usually take place at significant depths below the Earth's surface.

Microtidal – Term originally used by Hayes (1976) to define shoreline types. Davies (1973) defined microtidal coasts as those with a tidal range (vertical distance between high and low tide) of <2 meters (approximately 6 feet), whereas Hayes (1979) proposed the boundary of <3 feet.

Mid-ocean ridge – A long ridge with a deep rift along its crest that usually runs the length of, as well as along the middle of, the major oceans (with the Pacific Ocean being an exception; Alt and Hyndman, 2000). These ridges mark the lines where tectonic plates pull away from each other. As new lava flows spill out onto the ocean floor along fissures in these ridges, new oceanic crust is created. See Figures 2 and 8.

Mixed-energy coast – Term used by Hayes (1976) to define shoreline types on depositional coasts. Most mixed-energy depositional coasts have short, drumstick-shaped barrier islands with complex tidal flats and coastal wetlands on their landward flanks. They usually have tides in the mesotidal range (3-12 feet). See Figure 3B.

Mollusc – Shellfish; invertebrate having a soft unsegmented body usually enclosed in a shell (WordNet Search).

Mud – Sediment with a median diameter finer than 0.0625 millimeters (mm). It includes two classes: silt (median diameter 0.00391-0.0625 millimeters) and clay (median diameter <0.00391 millimeters). Silt rubbed between your teeth feels gritty, whereas clay feels smooth, a practice recommended only for pristine areas.

Mudflat – Tidal flat composed primarily of muddy sediments. They occur mostly in the more sheltered (from wave action) parts of coastal water bodies, such as the head of Morro Bay. See Figure 27A.

Mudstone – Sedimentary rock made up mostly of silt and/or clay.

Nacimiento Block – Located in the Southern Coast Ranges of California and consists mainly of Franciscan Complex accretionary prism rocks. The Block is bounded on the west by the San Gregorio-Hosgri fault zone, and to the east by the Nacimiento Fault. The Nacimiento Fault separates the Nacimiento Block from the Salinian Block, a displaced piece of the Sierra Nevada batholith (Trasko, 2007). See Figure 18.

Natural groin – A natural, usually rocky protrusion that sticks out into the longshore sediment transport system that can serve the same function as a man-built groin, that is, trap sand and form a relatively permanent sand beach.

Neap tides – The minimum tides of the month that exist when the gravitational attraction of the moon and sun on the water bodies of the world ocean are working in opposition, during the first and third quarter of the moon (condition called quadrative). See Figure 28.

Nonconformity – An unconformable erosional surface along which sedimentary rocks, or semiconsolidated sediments, overlie igneous or metamorphic rocks that are massive in character (i.e., not bedded). This surface represents a loss of a significant amount of time from the rock record. See Figure 144.

Normal, or gravity, fault – The type of fault that occurs when a rock or sediment mass located on the upper side of a sloping planar surface (a fracture in the rocks) slides down the fault relative to the lower side under the influence of gravity (and/or tension). See Figure 10A.

Nudibranch – Sea slug; any of various marine gastropods of the order *Nudibranchia* having a shell-less and often beautifully colored body (WordNet Search).

Offshore breakwaters – Offshore segments of riprap or heavy concrete blocks of miscellaneous shapes several tens of feet long that are detached from and usually oriented parallel to the beach (but not always). They reduce wave energy on their landward sides, which causes sand moving along shore in the longshore sediment transport system to accumulate in their lee, commonly forming tombolos.

Orthoclase feldspar – A silicate mineral with an abundance of aluminum and potassium (e.g., $KAlSi_3O_8$).

Pangaea – A single, large mass consisting of all the land area on the Earth's surface that formed about 300 million years ago, during the Carboniferous period of the Paleozoic era. It was created when two supercontinents, Gondwana and Laurasia, collided and welded together. See Figure 13A.

Parabolic sand dune – A wind-formed, scoop-shaped sand ridge with the plan-view shape of a parabola. Two linear arms that trail along side of the "scoop" point upwind. The most common type of sand dunes on the Central California Coast, which tend to be at least partially vegetated. See Figures 76 and 173.

Pebble – A sediment particle that is 4-64 millimeters in size. See Figure 111.

Pegmatite dike – A band of extremely coarse, crystalline granitic rock that cuts across adjacent rocks.

Peridotite – A coarsely crystalline, dark-colored igneous rock that consists mainly of olivine and other ferromagnesian minerals. It is the principal rock type in the mantle.

Period of a wave – The amount of time it takes two succeeding wave crests to pass a single point.

Permafrost – Permanently frozen subsoil.

Permeability of sediments – A measure of the ease with which a fluid flows through open pore spaces (e.g., between sand grains or gravel clasts).

Phytoplankton – Tiny plants that are the basis of the food chain in the ocean.

Pillow basalt – A rock type that typically forms when basaltic lava flows onto the sea floor in deep water, as is common along the mid-ocean ridges. The cold water in those depths quenches the surface of the flow and the lava commonly takes on a pillow shape. Common component of the basal units of the Franciscan Complex. See Figure 19.

Pinniped – A group of carnivorous marine mammals with fin-like limbs. Some common species occurring on the Central California Coast include California sea lion (*Zalophus californianus)*, harbor seal (*Phoca vitulina*), and northern elephant seal (*Mirounga angustirostri).*

Pivotability – The property of a sediment particle determined by how much it projects above the sediment bed it is lying on relative to its neighbors. One that projects relatively high will be more apt to be turned over and moved away by the current. This is a key factor determining whether a sediment particle will be moved by a water current, particularly if it is subjected to shallow sheet flow. See Figure 63B.

Placer deposits – Layers of the heavier grains of sediment (mostly sand; typically called heavy minerals) that were deposited primarily as the result of the selective sorting out and transport to elsewhere of most of the coarser-grained, light-weight minerals in the original sediment source. This may take place by a variety of processes, with increased pivotability and transportability of the larger, lighter grains in the shallow, sheet-flow of the wave swash being the primary process creating placers on beachface or berm top surfaces (see example in Figure 148). The accumulation of individual sediment grains composed of gold within sand bars in a stream bed is a notable type of placer deposits.

Plagioclase feldspar – A white to light gray silicate mineral ranging in composition from $NaAlSi_3O_8$ to $CaAl_2Si_2O_8$. Common constituent of coarsely crystalline igneous rocks such as diorite, tonalite, and granodiorite. See Figure 17B.

Plate tectonics – The theory named by J. Tuzo Wilson which states that individual blocks of the lithosphere, called plates, compose the outer portion of the Earth's surface. The individual plates either move apart (as along mid-ocean ridges), into each other (as in subduction zones), or parallel to each other in a few places (as along the San Andreas Fault). The plates move slowly (2-4 inches per year) above (riding on) the portion of the Earth's mantle called the asthenosphere, which behaves in a plastic manner. See Figure 7.

Plateau-shield coast – Coastline defined by Hayes (1965) as one characterized by plateaus and relatively narrow continental shelves. The bedrock along the shoreline is typically composed of ancient basement complexes of granite and gneiss (in shield areas) and Paleozoic and Mesozoic sedimentary rocks (in plateau areas). They are tectonically stable and moderately long rivers may be present, depending upon the climate.

Plunging waves – Waves that take on a cylindrical shape (Hawaii Five-O type) when they break, falling abruptly down (collapsing) with considerable force. See Figure 35A.

Pocket beach – Usually a small beach, between two headlands. In an idealized setting, there is very little or no exchange of sediment between the pocket beach and the adjacent shorelines (Wikipedia Encyclopedia).

Porosity of sediments – The ratio of the open pore spaces between the individual sediment grains or gravel clasts to the space taken up by the sand or gravel clasts themselves (when they are stacked together).

Porphyritic igneous rock – An igneous rock that has exceptionally large, separate, and distinct individual crystals imbedded within a finer-grained matrix. The larger crystals formed first and "floated around" within the molten magma as they slowly grew larger. The matrix crystallized more abruptly when a cooler temperature of the molten magma was reached that promoted the more rapid crystallization of the matrix minerals. See Figure 17A, 136B, and 168B.

Precocial chicks – The young of shorebirds, such as the snowy plover, that hatch with their eyes open, covered with down, and feed themselves.

Pyroxene – A silicate mineral that occurs in a variety of igneous and metamorphic rock types and is an important constituent of the upper mantle. Pyroxenes have the general formula $XY(Si,Al)_2O_6$ (where X represents calcium, sodium, iron^{+2} and magnesium and more rarely zinc, manganese and lithium, and Y represents ions of smaller size, such as chromium and aluminum) (Wikipedia Encyclopedia).

Quadrative – The times when the gravitational forces of the moon and sun work in opposition to each other with regard to generating tides in the world ocean (first and third quarter of the moon). This condition produces the smallest tides of the month (neap tides). See Figure 28.

Radioactive decay – A process whereby naturally occurring radioactive materials break down into other materials at known rates, or half lives.

Radiolaria – Microscopic-sized, unicellular marine protozoans. [Protozoans are a large group of single-celled eukaryotic organisms (i.e., their cells have a nucleus that contains its chromosomes).] Radiolaria have silica skeletons that accumulate on the seafloor in phenomenal numbers to form a "radiolarian ooze." See Figure 20B.

Radiolarian chert – A sedimentary rock composed of the siliceous skeletons of millions of radiolaria. The chemical composition is SiO_2, which is the same as quartz, but the individual crystals are so small the rock has a smooth texture. This type of rock, sometimes called ribbon chert because it is interlayered with thin shale layers, is a prominent member of the Franciscan Complex that crops out along the shoreline immediately north of the Golden Gate. See Figure 20A.

Raised marine terrace – A flat, or slightly seaward dipping, wave-cut rock terrace usually at least a few hundred yards wide that is flanked on its landward side by a somewhat degraded former wave-cut rock cliff and on its seaward side by a high, more recently active wave-cut rock cliff. It was originally formed during the last highstand of sea level between 80-125,000 years ago (for the youngest ones); then it was raised well above the present high-tide level by continuing tectonic uplift. There may be a series of them up the sides of the mountains that face the shore. Where such multiple terraces exist, the highest ones are the oldest and they get progressively younger toward the sea. See Figures 47 and 48.

Revetment – A facing built to protect a scarp or shoreline from erosion. It can be composed of a variety of materials including concrete, riprap, wood, and sand bags.

Rhythmic topography – Bulges of sand on the beach that move down the shore like the sinusoidal loops one might throw along a rope. They are created by waves that approach the beach at an oblique angle and are typically spaced 10s to 100s of feet apart. See Figure 37 (Upper Right).

Rhyolite – A fine-grained, extrusive volcanic rock, similar to granite in composition.

Richter scale – A scale that reflects the magnitude of earthquakes based on the calculation of the maximum amplitudes of the shear waves as recorded on a seismograph. The scale is logarithmic (base 10). For example, a magnitude (M) 5 earthquake on the Richter scale would result in ten times the level of ground shaking as an M 4 earthquake (and 32 times as much energy would be released). Quakes with Richter scale values of 3.5 to 5.4 are often felt, but rarely cause damage. Those above 7.0 on the scale are classified as major earthquakes and can cause widespread damage (Wikipedia Encyclopedia).

Right-lateral fault – A type of strike-slip fault, of which the San Andreas Fault is a prime example. If you are standing on one side of the fault line looking across it, the *terra firma* on the other side of the fault line has moved to the right during an earthquake. The same thing is true if you step across the fault line and look back at the side you were standing on (i.e., that side also has moved to the right in a relative sense).

Rip currents – Wave-generated currents that flow straight off the beach and tend to be very regularly spaced. They form most commonly during conditions when the waves come straight on shore (not at an angle usually, but there are some exceptions to this general rule). Also, there is typically some degree of reflection of the waves off the beachface when they are forming. Most rip currents owe their formation to the occurrence of edge waves – standing waves with their crests normal to the shoreline, according to Bowen and Inman (1969) and Komar (1976). See Figure 39.

Ripples in sand – Asymmetric, linear or cuspate features mostly composed of sand that are moved along the sediment surface by flowing water (or wind) with a spacing between their crests of less than 2 feet. On the beach, they usually occur in intertidal troughs on the landward side of intertidal bars, where water depths are greater than elsewhere on the beach. See Figures 56A and 59.

Riprap – Large pieces of rock of boulder size used in shore-protection structures, such as groins, revetments, and jetties.

River delta – A protruding mass of sediment that builds into a standing body of water off a river mouth. See Figure 97B.

Salinian Block – A geologic province that lies west of the trace of the San Andreas Fault (system) and east of the Nacimiento Fault. It crops out along the Central California Coast from Bodega Head to the Big Sur Country. Its granitic core apparently shares its origins with the Sierra Nevada, being a southern extension of the same granitic batholith (Wikipedia Encyclopedia).

Salt marsh – A low-lying wetland containing halophyte vegetation that is regularly flooded by marine waters during mid to high tides. Salinities range from 15-36 parts of salts per thousand parts of water (ppt).

Salt pruning – An occurrence along many shorelines where the trees or shrubs back of the shore bend away from a persistent prevailing wind, with growth only occurring on the side of the tree or shrub that is protected from the salt spray. Thus, the limbs extend downwind like flags blowing in the wind. This presumably has something to do with the dehydrating effects of the salt spray, but there may be other explanations.

Saltation – A mode of sand transport by the wind, whereby the sand grains bounce off the sand bed and "fly" for some distance at heights usually not more than a couple of feet. See Figure 75.

San Andreas Fault – A geological transform fault that runs a length of roughly 800 miles through Western California. Its motion is right lateral strike-slip (horizontal motion). It forms the tectonic boundary between the Pacific Plate and the North American Plate (Wikipedia Encyclopedia). See Figure 15.

Sand – Sediment with a median diameter between 0.0625 and 2 millimeters. Individual grains smaller than this are difficult to see with the naked eye.

Sand bypassing system – Mechanical systems, such as land-based dredging plants and pumping systems, which move the sand from the updrift to the downdrift sides of jetties, or other features, such as man-made harbors, that block the natural longshore transport of sand along the shore. The sand by-passing system at the Santa Barbara Harbor, California is a classic example.

Sand waves – Asymmetric, linear features (ripple shaped) moved along a sandy sediment surface by flowing water with a spacing between their crests of >18 feet. On the shoreline, they are mostly found in deep tidal channels or tidal inlets. See Figure 71.

Sandstones – Sedimentary rocks made up mostly of sand-sized particles of minerals, most notably quartz and rock fragments. See Figure 124.

Sea arch – A bedrock arch usually located offshore of a rocky coast in relatively shallow water. Many form where a narrow peninsula composed of layered rocks with a more resistant top layer projects offshore. Waves refracting around the headland focus the wave erosion such that two caves eroded into the opposite flanks of the peninsula meet, forming the arch. See Figure 45A.

Sea cave – A natural, arcuate horizontal chamber in a zone of relatively weak rocks that extends into the face of a wave-cut rock cliff. The cave is quarried out by wave action. See Figure 44.

Sea stacks – Isolated pinnacles of rock that are typically taller than they are wide with tops at a somewhat lower elevation than the top of the adjacent rock cliff (Davis and FitzGerald, 2004). The stacks are usually composed of rocks that are significantly more resistant to wave erosion than their neighboring rocks in the

formerly existing wave-cut rock cliff. See Figure 42.

Seafloor spreading – A process first described by Hess (1962) after he observed that the floor of the Pacific Ocean was in motion and "expanding." This is happening in response to additional volcanic flows being added to the sea floor along mid-ocean ridges at the same time as similar volcanic materials are carried to great depths in subduction zones along the outer margin of the oceans. *A dramatic proof of sea-floor spreading was discovered in the mid 1960s when data revealed alternating stripes of magnetic orientation on the sea floor, parallel to the mid-ocean ridges and symmetric across them – that is, a thick or thin stripe on one side of the ridge is always matched by a similar stripe at a similar distance on the other side. This mirror-image magnetic orientation pattern is created by steady sea-floor spreading combined with recurrent reversals of Earth's magnetic field. Iron atoms in liquid rock welling up along a mid-ocean ridge align with Earth's magnetic field* (Enotes, 2009). See Figure 8.

Seas – Waves in the area where the wind that forms them is blowing. Under sea conditions, the water surface is choppy and the waves are asymmetrical. Whitecaps are common. See Figure 33B.

Seawall – Vertical, hard structures commonly built along the shoreline to protect man-made structures. See Figure 83.

Sedimentary rock – Rock most commonly formed by weathering and erosion of pre-existing rocks on the Earth's surface. This process creates sediment that can be transported by water, or wind in some cases, to be deposited in masses of significant size. Once deposited, these sediments may become buried where they are consolidated by chemical cementation and other processes to form solid rock. See Figures 20 and 124.

Seismic wave – Waves that pass through the Earth caused by the sudden release of energy into the surrounding Earth from the spot where an earthquake originates (the hypocenter or focus of the quake).

Seismic P wave – The primary seismic wave that moves through the Earth in a fashion similar to sound waves due to compression of the rock particles, with each particle moving backward and forward in a straight line from the origin of disturbance. They are faster moving than their associated seismic S waves (Collier, 1999).

Seismic S wave – The secondary seismic waves that are due to shearing or transverse motion that moves rock particles not forward and backward, but side to side or up and down the way ripples radiate out when a rock breaks the surface of a calm pond. They are the seismic waves that produce the rolling motion of the ground during an earthquake and are, therefore, more damaging to structures than P waves (Collier, 1999).

Seismograph – An instrument used to measure seismic waves, which runs continuously, recording all incoming waves. Because of their different speeds, the seismic P waves from a specific earthquake will arrive ahead of the seismic S waves. Although their rates of speed may vary significantly, depending upon what they are moving through, the ratio of the difference between the speeds of the two waves is quite constant; therefore, this difference can be used to determine how far away the hypocenter of the earthquake is from the seismograph. The closer together the arrival times are, the closer the hypocenter.

Serpentinite – A metamorphic rock composed of complex, usually hydrated, ferromagnesium minerals. It usually forms at the crests of oceanic ridges, where seawater sinks into fractures that open as the two plates separate. The tremendous pressure of the depths forces the water down into the extremely hot mantle rocks below the ridge crest. The water reacts with hot peridotite to make serpentinite, the California state rock (Alt and Hyndman, 2000). See Figure 156.

Setback line – A "soft" solution for dealing with potential beach erosion in which a line is established behind the shoreline, seaward of which building is prohibited. They work best and are easiest to implement in areas that have not been developed as yet. The criteria used to establish such lines include historical analysis of shoreline trends, preservation of the line of foredunes, and defining areas of flooding and storm wave uprush.

Shale – Sedimentary rock made up mostly of silt and/or clay that is fissile (i.e., composed of very fine, multiple layers).

Shelf break – Point where the outer edge of the continental shelf joins a steeply descending planar surface called the continental slope. See Figure 81.

Sierran trench – A major oceanic trench that first appeared about 200 million years ago. It was located along the western margin of the North American Plate near the position of the present Sierra Nevada. See Figure 12.

Significant wave height – The average height of the waves that comprise the highest 33% of waves in a given sampling period.

Silica tetrahedron – The basic building block of all silicate minerals, which consists of a single silicon cation surrounded by four oxygen anions. Silicon cations, which have a charge of +4, are less than 1/3 the size of the oxygen anions, which have a charge of −2. As a result of these size differences, one silicon cation fits "snugly" between 4 oxygen anions, forming the tetrahedron.

Silicates – A group of minerals that are the fundamental building block of the Earth's crust, with the atoms of silicon and oxygen combined making up about 75% of it. The basic structure of the silicate minerals is the silica tetrahedron (see definition above). The tetrahedron is perfectly arranged geometrically, but the resultant ionic charge of −4 is far from balanced. To reduce the negative charge, the tetrahedra must bond with other positive ions, such as Mg^{2+} or link up with other tetrahedra in various ways, such as in sheets (e.g., mica;), chains (e.g., pyroxenes and micas), and so on. Because of its particular arrangement of bonding tetrahedra, called framework bonding, quartz (SiO_2) is the most stable of all the silicate minerals, both physically and chemically (Eggar, 2006).

Slip face – A steeply sloping face of a migrating wedge of sand, such as a bedform created by flowing water (e.g., ripples and sand waves or on the downwind side of a sand dune created by blowing wind). The angle of slope of the slip face, which is essentially the angle of repose of the sand (dip angle achieved when a pile of sand is created by simply dumping the sand out of a container, such as a truck bed), averages around 22 degrees in water and a little over 30 degrees in air. See Figures 56C and 57.

Spilling waves – Waves that start disintegrating into foaming lines fairly far offshore and continue to foam their way to shore, gradually decreasing in height as they go. See Figure 35B.

Spit – A linear projection of sediment across an embayment, stream outlet, and so on that builds in the direction that the longshore transport system is moving the sediment. See Figures 72, 94B, and 164.

Spring tides – The maximum tides of the month that exist when the gravitational attraction of the moon and sun on the water bodies of the world ocean are working together, during full and new moons (condition called syzygy). See Figure 28.

Storm berm – A linear, pyramidal-shaped mound of cobbles and pebbles that typically occurs on the most landward portion of exposed gravel beaches. It is only activated during major storms and, unless the storm is unusually severe, it is built up even higher during the storm event and not eroded away. See Figure 61.

Storm profile – A term suggested by Paul Komar, who was working on the Oregon coast, to indicate a flat beach profile eroded by storm waves. Hayes and Boothroyd (1969) used the term post-storm beach. See Figure 50B.

Strike-slip fault – A geologic fault in which the blocks of rock on either side of the fault slide horizontally past each other without any significant up or down motion. The strike of a bedding surface is represented by a horizontal line oriented perpendicular to the dip of the strata. This kind of fault moves exactly along that line. See Figure 10.

Subduction zone – A region where a portion of one of the Earth's tectonic plates dives beneath a portion of another plate, in some instances resulting in the formation of an oceanic trench. Along ocean margins, an oceanic plate dives under a continental plate.

Submarine canyon – A deep canyon in the continental shelf and slope offshore and extending about perpendicular to the shore. See Figure 81.

Surf zone – The nearshore zone off the beachface where the waves usually break. See Figure 34B.

Surging waves – Waves that approach shore and wash directly up the beachface without breaking.

Swash zone – Area where the water released by a broken wave washes across the portion of the beach known as the beachface. The released water at first rushes up the beachface until it loses its momentum, at which point it stops momentarily before returning back down the beachface under the influence of gravity. Usually, some of this water is lost through percolation into the beachface sediments.

Swell – Waves that have passed beyond the area where they were formed. Swell waves are typically longer than those in the area of formation. The swell surface is regular and more-or-less symmetrical (whitecaps are rare). See Figure 33B.

Swell profile – A term suggested by Paul Komar, who was working on the Oregon coast, to indicate a beach with a well-developed depositional berm (formed between storms). Hayes and Boothroyd (1969) used the terms late accretional or mature profile. See Figure 50B.

Syncline – The condition where layers of rocks are folded into a form similar to what you would get if you pushed one edge of a rug along the floor, folding it into parallel bands. Where the folded layers bend down, the fold is called a syncline.

Syzygy – The times when the gravitational forces of the moon and sun work together with regard to generating tides in the world ocean (full and new moon). The largest tides of the month (spring tides) occur during syzygy. See Figure 28.

Talus – Debris piles of fragments of rocks up to boulders in size (and soil in some cases) at the base of cliffs.

Tarball –Weathered oil that is less than 6 inches in size.

Tectonic (lithospheric) plates – A dozen or so segments of the lithosphere that covers the Earth's outer part (Hyndman and Hyndman, 2006). The plates that comprise the continents (e.g., the North American Plate) are made up of relatively light material composed of silica, oxygen, and so on. The plates that constitute the ocean floors are composed of relatively heavy material that contains abundant iron, magnesium, and so on. See Figure 7.

Tectonism – The forces that produce deformation of the Earth's crust, such as folding or faulting of the rocks, with the motion of tectonic plates commonly being the driving mechanism for this deformation.

Thrust or reverse, fault – The type of fault that occurs when a rock or sediment mass located on the upper side of a sloping planar surface (a fracture in the rocks) slides up the fault relative to the lower side under the influence of compression. See Figure 10B.

Tidal flat – A flat intertidal surface covered and uncovered as the tides rise and fall. It is usually a depositional surface with the sediments supplied by tidal currents. In more exposed areas, where the waves are somewhat active and the tidal currents are strong, the flats are composed of sand. In more sheltered areas, they are composed mostly of mud.

Tidal inlet – Usually defined as major tidal channels that intersect barrier islands, or separate two barrier islands, usually to depths of tens of feet (Hayes, 2009). During our work developing protection strategies to be used during oil spills on the California Coast, we broadened the definition to include any channel-like opening that connects the sea with the mainland. Of the 170 tidal inlets (thus defined) that are present on the California Coast, the most common inlets are the ones where a relatively small stream mouth, and its associated valley, meets the sea. See Figures 68 and 73.

Tidal range – The vertical distance between high and low tide. On the Central California Coast, the average spring tidal range increases slightly from south to north, from an average of 5.2 feet at Point Arguello to 5.8 feet at Point Reyes.

Tide-dominated coast – Term used by Hayes (1976) to define shoreline types on **depositional coasts**.

They are shorelines characterized by such features as open mouthed, multi-lobate river deltas, abundant large estuarine complexes, and extensive tidal flats and salt marshes. They usually occur on coasts with a combination of relatively small waves and large tides (>12 feet).

Tide pool – Depressions in wave-cut rock platforms that hold water during low tide, noted for the wide diversity of plant and animal life that inhabit them.

Tombolo – A triangle-shaped body of sediment that forms in the lee of an offshore structure, such as a small island or ship wreck. Waves approaching shore bend around this offshore structure, with the result that two opposing longshore sediment transport directions are created that converge in the shelter of the offshore structure, resulting in the formation of the tombolo. See Figure 36.

Tonalite – A light colored, coarsely crystalline igneous rock that is composed of 20-60% quartz, with the remaining mostly feldspar component being >90% plagioclase. It may also contain a few dark-colored minerals, such as hornblende and biotite. See Figure 17B.

Trailing edge coast – Shorelines defined by Inman and Nordstrom (1971) as those that occur on the trailing edges of continental plates. They are typically characterized by wide coastal plains, numerous river deltas, and barrier islands. Rocky coasts are extremely rare.

Transform fault – A form of a strike-slip fault that occurs between oceanic plates or within oceanic plates along the length of mid-ocean ridges in which the rocks on either side of the fault simply slide past each other in a horizontal plane without colliding or pulling apart. Defined and described by J. Tuzo Wilson in 1965, they provided an understandable mechanism whereby the new lava flows moving away from the mid-ocean ridges could fracture and fit onto the spherical surface of the Earth. See Figures 8 and 9.

Transverse sand dune – A wind-formed, linear sand ridge that has a steep slip face on its downwind side. The slip face dips at the angle of repose of sand in air (30+ degrees). See Figure 76.

Tsunami – A Japanese word translated as *harbor wave*. It is a large ocean wave usually generated by a submarine earthquake, although it may also be generated by volcanic eruptions, landslides (both those initiated on shore and submarine), and even by a comet or asteroid impact. *Seismic seawaves* is another term that scientists commonly use for this phenomenon. These are typically very long waves (up to several hundred miles) with relatively small wave heights (commonly less than 3 feet), and long periods (over an hour in some cases) that have sufficient energy to travel across entire oceans. They move very fast in the open water at speeds measured in hundreds of miles per hour, but they slow down before breaking in shallow water.

Turbidite – A coarse-grained sediment deposit formed by a swiftly moving, bottom-flowing current along the ocean floor (usually, they can also form in lakes) called a turbidity current. See Figures 21C and 124.

Turbidity current – A rapidly flowing, bottom-hugging current that is composed of a dense turbulent plume that contains an unusually large amount of suspended sediment. This type of flow is commonly generated by a mechanism such as a landslide on a steep slope of the sea floor (there are several other possibilities). See Figure 21A.

Unconformity – The general term for an erosion surface, usually near horizontal when it was formed, that represents a significant loss of time from the rock record. Therefore, it separates two rock masses, or strata, of significantly different ages. See Figure 144.

Updrift – The direction from which the individual sand grains moving along the shore within the longshore transport system have come. See Figure 37.

Upwelling – An oceanographic process by which warm, less-dense surface water is drawn away from a shore by offshore currents and replaced by colder, denser, and more nutrient-rich water brought up from the depths. An example of this occurs on the California coast where southward flowing currents generated by northerly winds are deflected offshore because of the Coriolis effect (also called the Ekman spiral). This offshore current that moves the surface water offshore generates the upwelling.

Washover fan – A feature found in California on sand spit surfaces where waves wash all the way across the top of the spit during storms or during very high spring tides. This process creates a flattened-off surface along which sand is transported across the spit's surface and into the standing water behind it. The resulting sand deposit commonly has a fan-like shape.

Wave-cut rock platform – A flat, usually intertidal, rock platform left behind as an eroding, wave-cut rock cliff retreats. Some geologists refer to these features as shore platforms. See Figure 25B.

Wave-dominated coast – Term used by Hayes (1976) to define shoreline types on **depositional coasts** (not leading edge coasts like the Pacific coast of North

America). They are characterized by such features as deltas with smooth outer margins and abundant long barrier islands. They are usually found on coasts with a combination of relatively large waves and small tides (<6 feet).

Wave diffraction – The phenomenon by which energy is transmitted laterally along a wave crest. When part of a wave is interrupted by a barrier, the effect of diffraction is manifested by propagation of waves into the sheltered region within the barrier's geometric shadow. A good example would be waves passing between two offshore breakwaters and into the sheltered areas landward of the breakwaters.

Wave refraction – Any change in the direction of a wave resulting from the bottom contours. For example, a submerged hump would cause the two ends of the wave crest located away from the hump to bend toward each other. Such differences in the offshore topography can cause the energy of the waves to be focused in some areas along the shore and less energetic in others.

Wave uprush – The water that moves up the beachface under the impetus of a breaking wave.

Young mountain range coast – Defined by Hayes (1965) as coastlines that abut high mountains created by relatively recent orogenic (mountain building) activity. In most places, the land is continuing to rise. The bedrock is usually dominated by relatively young sedimentary and volcanic rocks. Mostly short, steep rivers empty into the sea along such coasts. They are most common on the leading edges of continental plates.

Zone of the turbidity maximum – A zone in the middle reaches of an estuary that contains an abundance of fine-grained sediment in suspension, because of the mixing of salt water with fresh water containing a profusion of suspended clay particles, which tend to flocculate. See Figure 79B.

Zooplankton – Animals that drift in the water column without the ability to effectively swim against currents, including small animals that are the next level in the food chain above phytoplankton, such as krill and the larvae of many fish and shellfish.

REFERENCES CITED

Adams, J., J.T. Harvey, K.D. Hyrenbach, C.L. Baduini and H.M. Nevins. 2008. Tracking the movements and trans-Pacific migration of sooty shearwaters (*Puffinis griseus*) captured off California. Available at: http://birdmam. mlml.calstate.edu/jadams_sooty.

Adelman, K. and G. Adelman. 2009. California Coastal Records Project. Available at: http://www.californiacoastline. org.

Allen, J.R.L. 1968. Current ripples – their patterns in relation to water and sediment motion. North-Holland Pub. Co., Amsterdam, Holland. 433 pp.

Allen, J. and P. Komar. 2000. Are ocean wave heights increasing in the eastern North Pacific. EOS 81:561-567.

Alt, D. and D.W. Hyndman. 1989. Roadside Geology of Idaho. Mountain Press Pub. Co., Missoula, MT. 394 pp.

Alt, D. and D.W. Hyndman. 2000. Roadside Geology of Northern and Central California. Mountain Press Pub. Co., Missoula, MT. 369 pp.

American Geological Institute. 2002. Earthquake activities. Available at: http://earthsci.org/education/investigations/ agi/Earthquakes/Earthquakes.htm.

Anderson, R.S. 2005. Contrasting Vegetation and Fire Histories on the Point Reyes Peninsula During the Pre-Settlement and Settlement Periods: 15,000 Years of Change. Center for Environmental Sciences & Education, & Quaternary Sciences Program, Northern Arizona University, Flagstaff, AZ. 43 pp.

Anima, R. 1990. Pollution studies of Drakes Estero, and Abbotts Lagoon, Point Reyes National Seashore. U.S. Geological Survey. 233 pp.

Anima, R.J., J.L. Bick and H.E. Clifton. 1988. Sedimentologic consequences of the storm in Tomales Bay. In: Ellen, S.D. and G.F. Wieczorek, eds. Landslides, Floods, and Marine Effects of the Storm of January 3-5, 1982, in the San Francisco Bay Region, California. U.S. Geological Survey Professional Paper 1434. pp. 283-310.

Atwater, B.F., M. Satoko, S. Kenji, T. Yoshinobu, U. Kazue and D.K. Yamaguchi. 2005. The Orphan Tsunami of 1700. U.S. Geological Survey Professional Paper 1707. 144 pp.

Bagnold, R.A. 1940. Beach formation by waves; some model experiments in a wave tank. Journal of the Institute of Civil Engineering 15:27-52.

Bailey, E.H., W.P. Irwin and D.L Jones. 1964. Franciscan and Related Rocks, and Their Significance in the Geology of Western California. Bulletin 183, California Division of Mines and Geology. pp. 1-177.

Bailey, E.H., M.C. Blake, Jr. and D.L. Jones. 1970. On-land Mesozoic Oceanic Crust in California Coast Ranges. U.S. Geological Survey Professional Paper 700-C. pp. C70-C81.

Barnard, P.L., D.M. Hanes, D.M. Rubin and R.G. Kvitek. 2006. Giant sand waves at the mouth of San Francisco Bay. EOS Transactions 87:285-286.

Basan, P.B. and R.W. Frey. 1977. Actual-palaeontology and neoichnology of salt marshes near Sapelo Island, Georgia. In: Crimes, T.P. and J.C. Harber, eds. Trace Fossils 2. Geology Jour., Spec. Issue 9. pp. 41-70.

Bascom, W. 1951. The relation between sand size and beachface slope. Amer. Geophys. Union, Trans. 32:866-874.

Bascom, W. 1954. Characteristics of natural beaches. In: Proc. 4th Conf. on Coastal Engineering. pp. 163-180.

BeachCalifornia.com. 2009. Rip currents at Huntington Beach. Available at: http://www.beachcalifornia.com/beach/ rip-current-rescue.html.

Benson, S.R., K.A. Forney, J. Harvey, J.V. Carretta and P.H. Dutton. 2007. Abundance, distribution, and habitat of leatherback turtles (*Dermachelys coriacea*) off California, 1990-2003. Fisheries Bulletin 105:337-347.

Berlin, M. 2005. Tectonic wedging in the California Coast Ranges: Useful model or an unreasonable hypothesis.

Available at: http://www.colorado.edu/GeolSci/Resources/WUSTectonics/CACoastRanges/websiteberlin.html.

Beyene, A. and J.H. Wilson. 2006. Comparison of wave energy flux for northern, central, and southern coasts of California based on long-term statistical wave data. Energy 31:1856-1869.

Biggs, R.B. 1978. Coastal bays. In: Davis, R.A., Jr., ed. Coastal Sedimentary Environments. Springer Verlag, New York. pp. 69-99.

Billings, M.P. 1954. Structural Geology. 2nd edition, Prentice Hall, Englewood Cliffs, NJ. 514 pp.

Bird, E.C.F. 1996. Beach Management. John Wiley and Sons, Chichester, NY. 292 pp.

Bluck, B.J. 1967. Sedimentation of beach gravels: examples from South Wales. Journal of Sedimentary Petrology 37:128-156.

Bogdanov, N.A. and N.L. Dobretsov. 1987. The ophiolites of California and Oregon. Geotectonics 12:472-479.

Boothroyd, J.C. 1969. Hydraulic conditions controlling the formation of estuarine bedforms. In: Hayes, M.O., ed. Coastal Environments – NE Massachusetts and New Hampshire: Field Trip Guidebook for Eastern Section of SEPM, May 9-11, 1969. pp. 417-427.

Bottin, R.R., Jr. and E.F. Thompson. 2002. Comparisons of Physical and Numerical Model Wave Predictions with Prototype Data at Morro Bay Harbor Entrance, California. U.S. Army Corps of Engineers Rept., ERDC/CHL CHETN-1-65. 15 pp.

Bowen, O.E. 1965. Stratigraphy, structure, and oil possibilities in Monterey and Salinas Quadrangles, California (Abstract). Amer. Assoc. Petroleum Geol. Bull. v. 49.

Bowen, A.J. and D.L. Inman. 1966. Budget of Littoral Sands in the Vicinity of Point Arguello, California. Tech. Memo. 19. Coastal Engineering Research Center, Washington, DC. 56 pp.

Bowen, A.J. and D.L. Inman. 1969. Rip currents: laboratory and field observations. Journal of Geophysical Research 74:5479-5490.

Bradley, W.C. and G.B. Griggs. 1976. Form, genesis, and deformation of central California wave-cut platforms. Geological Society of American Bulletin 87:433-449.

Brian Atwater. 2009. Wikipedia free encyclopedia. Available at: http://en.wikipedia.org/wiki/Brian_Atwater.

Brown, D. 1970. Bury My Heart at Wounded Knee: An Indian History of the American West. H. Holt and Co., New York. 487 pp.

Browning, R.M. 1972. The natural resources of Elkhorn Slough, their present and future use. Coastal Wetland Series No. 4, California Dept. of Fish and Game, Sacramento, CA. 105 pp.

Bues, S.S. and M. Morales. 1990. Grand Canyon Geology. Oxford Univ. Press, New York. 9 pp.

Burnham, K. 2005. Point Lobos to Point Reyes: Evidence of ± 180 km offset on the San Gregorio and northern San Andreas Faults. In: Stevens, C. and J. Cooper, eds. Cenozoic Deformation in the Central Coast Ranges, California. Field Trip Guidebook, Pacific Section SEPM (Society for Sedimentary Geology). pp. 1-29.

Cajete, G.A. 1995. Ensoulment of nature. In: Hirschfelder, A.B., ed. Native Heritage: Personal Accounts by American Indians, 1790 to the Present. Macmillan Pub. Co., New York. 298 pp.

California Beaches.com. 2009. Available at: http://www.beachcalifornia.com/

California Coastal Commission. 2003. California Coastal Access Guide. 5th Edition, Univ. Cal. Press, Berkeley, CA. 304 pp.

California Coastal Commission. 2007. Beaches and Parks from Monterey to Ventura. Univ. Cal. Press, Berkeley, CA. 318 pp.

California Department of Conservation. 2001. Landslides in the Highway 1 Corridor: Geology and Slope Stability along the Big Sur Coast. Div. of Mines & Geology, prep. for Caltrans District 5. 32 pp.

California Department of Fish and Game. 2009. The Amazing Grunion. Available at: http://www.dfg.ca.gov/marine/grnindx3.asp.

Cardwell, R. and M. Detterman. 2003. Geology of the Point Reyes Area, California. Available at: http://ncgeolsoc.org/FieldTripInfo/Pt%20Reyes%20Area%20FT.htm.

Carter, R.W.G. and C.D. Woodroffe, eds. 1994. Coastal Evolution. Cambridge Univ. Press, Cambridge. 517 pp.

Cartwright, D.E. 1969. Deep sea tides. Science Journal 5:60-67.

CBNMS. 2009. Cordell Bank National Marine Sanctuary. Available at: http://cordellbank.noaa.gov/.

Chappell, J.M. 1983. A revised sea-level record for the last 300,000 years from Papua, New Guinea. Jour. of Geophysical Research 14:99-101.

Cherry, J.A. 1965. Sand Movement along a Portion of the Northern California Coast. Tech. Memo No. 14. Coastal

Eng. Res. Center, U.S. Army Corps of Engineers, Washington, DC.

Cherry, J.A. 1966. Sand movement along equilibrium beaches north of San Francisco. Jour. of Sedimentary Petrology 36:341-357.

Chin, J.L., H.E. Clifton and H.T. Mullins. 1988. Seismic stratigraphy and Late Quaternary shelf history, south-central Monterey Bay, California. Marine Geology 81:137-157.

Clark, J.C. and E.E. Brabb. 1997. Geology of Pt. Reyes National Seashore: A Digital Database. U.S. Geological Survey Open-File Report. pp. 97-456. Available at: http://pubs.usgs.gov/of/1997/of97-456/.

Clark, J.C., E.E. Brabb and W.O. Addicott. 1979. Tertiary Paleontology and Stratigraphy of the Central Santa Cruz Mountains, California Coast Ranges. Geological Society of America, Cordilleran Section, Field Trip Guidebook. 23 pp.

Clifton, H.E. 1981. Submarine canyon deposits, Point Lobos, California. In: Frizzel, V., ed. Upper Cretaceous and Paleocene Turbidites, Central California Coast. Field Trip Guidebook, Field Trip no. 6. pp. 79-92.

Clifton, H.E. 2007. Conglomeratic Submarine Fill, Point Lobos State Reserve. Pacific SEPM Book 105. 77 pp. Available at; http://www.sci.sdsu.edu/pacsepm/publications/105/105cover.pdf.

Clifton, H.E. and E.L. Leithold. 1991. Quaternary coastal and shallow marine facies sequences, northern California and the Pacific Northwest. In: Morrison, R.B., ed. Quaternary Nonglacial Geology, Conterminous U.S. Geological Society of America DNAG v. K-2. pp. 143-156.

Cole, R.B. and A.R. Basu. 1995. Nd-Sr isotopic geochemistry and tectonics of ridge subduction and middle Cenozoic volcanism in western California. Geological Society of America Bulletin 107:167-179.

Coleman, J.M. and L.D. Wright. 1975. Modern river deltas: Variability of processes and sand bodies. In: Broussard, M.L., ed. Deltas. 2nd ed. Houston Geological Society, Houston, TX. pp. 99-150.

Collier, M. 1999. A Land in Motion: California's San Andreas Fault. Univ. Cal. Press, Berkeley, CA. 118 pp.

Comet, P.A. 1996. Geological reasoning: Geology as an interpretive and historical science. Discussion. Geological Society of America Bulletin 108:1508-1510.

Concerned Coastal Geologists. 1981. Saving the American Beach. Skidaway Inst. Ocean. Conf., Savannah, GA, March 1981. 12 pp.

Cooper, W.S. 1967. Coastal Dunes of California. Memoir 104 of the Geological Society of America. 131 pp.

Covault, A. 2004. Uplifted, Quaternary marine terraces along the margin of the North American Plate: North Central California between Alder Creek and Mendocino. Proc. 17th Keck Symposium in Geology, pp. 11-14. Available at: http://keckgeology.org/files/pdf/symvol/17th/california/covault.pdf.

Cox, A. and D.C. Engebretsen. 1985. Change in motion of the Pacific Plate at 5 Ma. Nature 313:472-474.

Cummings, J.C., R.M. Touring and E.E. Brabb. 1962. Geology of the northern Santa Cruz Mountains, California. In: Bowen, O.E., Jr., ed. Geologic Guide to the Gas and Oil Fields of Northern California. California Division of Mines and Geology Bulletin 181:179-220.

Davies, J.L. 1973. Geographical Variation in Coastal Development. Hafner Pub. Co., New York. 204 pp.

Davis, M. 2001. Late Victorian Holocausts: *El Niño* Famines and the Making of the Third World. Verso, London. 271 pp.

Davis, R.A., Jr. and D.M. FitzGerald. 2004. Beaches and Coasts. Blackwell Publishing, Malden, MA. 419 pp.

Dean, R.G. and T.C. Walton. 1975. Sediment transport processes in the vicinity of inlets with special reference to sand trapping. In: Cronin, I.E., ed. Estuarine Research, Volume 11, Geological and Engineering. Academic Press, New York. pp. 129-150.

Dickinson, W.R. 1983. Cretaceous sinistral strike slip along Nacimiento Fault in Coastal California. American Association of Petroleum Geologists Bulletin 67:624-645.

Dietz, R.S. and J.C. Holden. 1970. Reconstruction of Pangaea: Breakup and dispersion of continents, Permian to Present. Journal Geophysical Research 75:4939-4956.

Dingler, J.R. 1981. Stability of a very coarse-grained beach at Carmel, California. Marine Geology 44:241-252.

Dingler, J.R. and R.J. Anima. 1989. Subaqueous grain flows at the head of Carmel Submarine Canyon, California. Journal of Sedimentary Petrology 59:280-286.

Dingler, J.R. and T.E. Reiss. 2002. Changes to Monterey Bay beaches from the end of the 1982-83 *El Niño* through the 1997-98 *El Niño*. Marine Geology 181:249-263.

Domurat, G.W., D.M. Pirie and J.F. Sustar. 1979. Beach erosion control study Ocean Beach, San Francisco, CA. Shore & Beach 47:20-32.

Dietrich, G. 1963. General Oceanography. Wiley-Interscience, New York. 588 pp.

Dupré, W.R. 1990. Quaternary geology of the Monterey Bay region, California. In: Garrison, R.E. et al., eds. Geology and Tectonics of the Central California Coast Range, San Francisco to Monterey – Volume and Guidebook. Pacific Section, AAPG. pp. 185-193.

Eaton, M. 2009. Ocean Beach and Fort Funston – Birds. Available at: http://sffo1.markeaton.org/SFBirding/Ocean%20Beach.htm.

Eelgrass. 2005. Eelgrass Habitat. Available at: http://www.botos.com/marine/egrass01.html.

Eggar, A.E. 2006. Minerals III: The Silicates. Available at: http://www.visionlearning.com/library/module_viewer.php?mid=140.

Eichhubl, P. and R.J. Behl. 1998a. Diagenesis, deformation, and fluid flow in the Miocene Monterey Formation of Coastal California. In: Eichhubl, P., ed. Diagenesis, Deformation, and Fluid Flow in the Miocene Monterey Formation of Coastal California. SEPM Pacific Section Special Publication 83:5-13.

Eichhubl, P. and R.J. Behl. 1998b. Field guide to diagenesis, deformation, and fluid flow in the Miocene Monterey Formation—Ventura-Santa Barbara-Jalama Beach-Grefco Quarry/Lompoc. In: Eichhubl, P., ed. 1998. Diagenesis, Deformation, and Fluid Flow in the Miocene Monterey Formation of Coastal California. SEPM Pacific Section Special Publication 83:85-98.

Elder, W.P. 2001. Geology of the Golden Gate Headlands. In: Stoffer, P.W. and L.C. Gordon, eds. Geology and Natural History of the San Francisco Bay Area: A Field-Trip Guidebook. U.S. Geological Survey Bulletin 2188. pp. 61-86. Available at: nps.gov/goga/forteachers/upload/Geology%20of%20the%20Golden%20Gate%20Headlands%.

Emory, J. 1999. The Monterey Bay Shoreline Guide. Univ. Cal. Press, Berkeley. 307 pp.

Enotes. 2009. Sea-floor spreading. Available at: http//enotes.com/earth-science/sea-floorspreading.

Evgeni Sergeev. 2009. Evgeni Sergeev/Artwork photo galleries on the web. Available at: http://en.wikipedia.org/wiki/User:Evgeni_Sergeev/Artwork_photo_galleries_o.

Federov, A.V. and S.G. Philander. 2000. Is *El Niño* changing? Science 288:1997-2002.

Finkelstein, K. 1977. Morphologic variations and sediment transport in crenulate bay beaches, Kodiak Island, Alaska. M.S. Thesis, Dept. Geol., Univ. of South Carolina, Columbia. 96 pp.

FitzGerald, D.M. 1977. Hydraulics, morphology, and sediment transport at Price Inlet, South Carolina. Ph.D. Dissertation, Dept. Geol., University of South Carolina, Columbia. 84 pp.

FitzGerald, D.M., D. Nummedal and T.W. Kana. 1976. Sand circulation patterns of Price Inlet, South Carolina. In: Proc. 15th Coastal Engineering Conf., American Society Civil Eng., Honolulu, HI. pp. 1868-1880.

Folk, R.L. 1955. Student operator error in determination of roundness, sphericity, and grain size. Journal of Sedimentary Petrology 25:297-301.

Foster, J. 1998. Pismo Beach, CA., Beach Geology Walk. Available at: http://www.jf2.com/geowalk/geowalk.html.

Femmer, R. 2005. National Biological Information Infrastructure. Available at: http://life.nbii.gov.

Frey, R.W. and J.D. Howard. 1969. A profile of biogenic sedimentary structures in a Holocene barrier island-salt marsh complex. Trans. Gulf Coast Assoc. Geol. Societies 19:427-444.

Geology News Blog. 2008. Weekend adventures – Capitola fossils. Available at: http://geology.rockbandit.net/tag/santa-cruz/.

GFNMS. 2009. Gulf of Farallones National Marine Sanctuary. Available at: http://farallones.noaa.gov/.

Glen, W. 1975. Continental Drift and Plate Tectonics. C.E. Merrill Pub. Co., Columbus, OH. 188 pp.

Gore, A. 2006. An Inconvenient Truth. Rodale Pub., Emmaus, PA. 325 pp.

Gore, P.J.W. 1999. Radiometric dating. Available at: http://facstaff.gpc.edu/~pgore/geology/geo102/radio.htm.

Graham, N.E. and H.F Diaz. 2001. Evidence for intensification of North Pacific winter cyclones since 1948. Bulletin of the American Meteorological Society 82:1869-1893.

Griggs, G. 1986. Littoral cells and harbor dredging along the California Coast. Environmental Geology 10:7-20.

Griggs, G. 1998. California needs a coastal hazards policy. California Coast & Ocean Magazine, Autumn. 7 pp.

Griggs, G.B. and A.S. Trenhaile. 1994. Coastal cliffs and platforms. In: Carter, R.W.G. and C.D. Woodroffe, eds. Coastal Evolution, Cambridge Univ. Press, Cambridge. pp. 425- 450.

Griggs, G.B. and K. Brown. 1998. Erosion and shoreline damage along the Central California Coast: A comparison between the 1997-98 and 1982-83 winters. Shore and Beach 66:3:18-23.

Griggs, G.B. and K. Patsch 2002. Littoral cells and sand budgets along the coast of California. Proposal to the California coastal Sediment Management Working Group and California Department of Boating and

Waterways. 7 pp. Available at: http://dbw.ca.gov/csmw/PDF/GriggslSedBudgetTechProposal061704.

Griggs, G.B. and K. Patsch. 2004. Cliff erosion and bluff retreat along the California Coast. Sea Technology 45:36-40.

Griggs, G.B. and K. Patsch. 2005. Chapter Fourteen: Año Nuevo to the Monterey Peninsula. In: Griggs, G.B., K. Patsch and L. Savoy, eds. Living with the Changing California Coast. Univ. Cal. Press, Berkeley, CA. pp. 270-310.

Griggs, G.B. and K. Patsch. 2006. Littoral Cells, Sand Budgets, and Beaches: Understanding California's Shoreline. Institute of Marine Sciences, Univ. Cal. Santa Cruz, Cal. Dept. of Boating and Waterways, and Cal. Coastal Sed. Management Work Group. 39 pp. Available at: http://dbw.ca.gov/csmw/PDF/LittoralDrift.pdf.

Griggs, G.B. and D.S. Ross. 2006. Santa Cruz Coast. Arcadia Publishing, San Francisco, CA. 95 pp.

Griggs, G.B., K. Patsch and L. Savoy. 2005. Living with the Changing California Coast. Univ. Cal. Press, Berkeley, CA. 540 pp.

Griggs, G.B., K. Fulton-Bennett and L. Savoy. 2005. Chapter Twelve: The San Francisco Coastline. In: Griggs, G.B., K. Patsch and L. Savoy, eds. Living with the Changing California Coast. Univ. Cal. Press, Berkeley, CA. pp. 222-227.

Griggs, G.B., J. Weber, D.R. Lajoie and S. Mathieson. 2005. Chapter Thirteen: San Francisco to Año Nuevo. In: Griggs, G.B., K. Patsch and L. Savoy, eds. Living with the Changing California Coast. Univ. Cal. Press, Berkeley, CA. pp. 228-269.

Grunion. 2009. The timing of grunion runs. Available at: http://www.Grunion.org.

Guadalupe Restoration Project – Dunes Center Manual. 2008. Manual by Chevron Corporation. Available at: http://guaddunes.com/duneManual.html.

Gulf of Farallones National Marine Sanctuary. 2008. Geology. Available at: http://sanctuarysimon.org/farallones/sections/geology/overview.php?sec=g.

Guza, R.T. and D.L Inman. 1975. Edge waves and beach cusps. Journal of Geophysical Research 80:1285-1291.

Hall, C.A., D.L Jones and S.A. Brooks. 1959. Pigeon Point Formation of Late Cretaceous age. AAPG Bulletin 43:12:2855-2859.

Hand, B.M., J.M. Wessel and M.O. Hayes. 1969. Antidunes in the Mt. Toby Conglomerate (Triassic), Massachusetts. Journal of Sedimentary Research 39:4:1310-1316.

Hands, E.B. 1977. Implications of submergence for coastal engineers. Proc. Coastal Sediments '77, Amer. Soc. Civil Eng., New York. pp. 149-166.

Hanson, K.L., W.R. Lettis, M.K. McLaren, W.U. Savage and N.T. Hall. 1995. Style and Rate of Quaternary Deformation of the Hosgri Fault Zone, Offshore South-Central California. U.S. Geological Survey Bulletin 1995-BB. 33 pp.

Hapke, C., D. Reid, B.M. Richmond, P. Ruggiero and J. List. 2006. National Assessment of Shoreline Change Part 3: Historical Shoreline Change and Associated Coastal Land Loss along Sandy Shorelines of the California Coast. U.S. Geological Survey, Open File Report 2006-1229. 79 pp. Available at: http://pubs.usgs.gov/of/2006/1219/.

Hapke, C. 2005. Chapter Fifteen: The Monterey Peninsula to Morro Bay. In: Griggs, G.B., K. Patsch and L. Savoy, eds. Living with the Changing California Coast. Univ. Cal. Press, Berkeley, CA. pp. 311-333.

Harden, D.R. 2004. California Geology (2nd edition). Prentice Hall, Upper Saddle River, NJ. 552 pp.

Hartness, N.B. 2001. The 1964 Good Friday earthquake tsunami. Available at: http://www.geo.arizona.edu/~nhartnes/alaska/tsunam.html.

Hayes, M.O. 1964. Lognormal distribution of inner continental shelf widths and slopes. Deep-Sea Research 11:53-78.

Hayes, M.O. 1965. Sedimentation on a semiarid, wave-dominated coast (south Texas); with emphasis on hurricane effects. Ph.D. Dissertation, Dept. of Geol., University of Texas, Austin. 350 pp.

Hayes, M.O. 1967a. Relationship between coastal climate and bottom sediment type on the inner continental shelf. Marine Geology 5:111-132.

Hayes, M.O. 1967b. Hurricanes as Geological Agents: Case Studies of Hurricanes Carla (1961) and Cindy (1963). University of Texas, Austin, Bureau of Economic Geology. Rep. 61. 56 pp.

Hayes, M.O., ed. 1969. Coastal Environments – NE, Massachusetts and New Hampshire: Field Trip Guidebook for Eastern Section of SEPM, May 9-11. 462 pp.

Hayes, M.O. 1976. Lecture Notes. In: Hayes, M.O. and T.W. Kana, eds. Terrigenous Clastic Depositional Environments. Tech Rep 11-CRD, Geol. Dept., Univ. of South Carolina, Columbia. pp. I-1-131.

Hayes, M.O. 1979. Barrier island morphology as a function of tidal and wave regime. In: Leatherman, S., ed. Barrier Islands, from the Gulf of St. Lawrence to the Gulf of Mexico, Academic Press, New York. pp. 1-27.

Hayes, M.O. 1980. General morphology and sediment patterns in tidal inlets. Sedimentary Geology 26:139-156.

Hayes, M.O. 1985. Beach erosion. In: Clark, J.R., ed. Coastal Resources Management: Development Case Studies. National Park Service and U.S. Agency for International Development. pp. 67-201.

Hayes, M.O. 1999. Black Tides. Univ. of Texas Press, Austin. 287 pp.

Hayes, M.O. 2005. Fan Deltas as Exploration Targets. Report by RPI Louisiana, Inc., New Orleans, Louisiana for Devon Energy Corporation. 97 pp.

Hayes, M.O. 2009. Barrier islands. In: Gillespie, R. and D. Clague, eds. Encyclopedia of Islands. Univ. Cal. Press, Berkeley, CA. pp. 82-88.

Hayes, M.O. and W.F. Baird. 1993. Shoreline erosional/depositional patterns in Oman. Coastal Engineering Considerations in Coastal Zone Management, American Society of Civil Engineers, New York. pp. 144-158.

Hayes, M.O. and J.C. Boothroyd. 1969. Storms as modifying agents in the coastal environment. In: M.O. Hayes, ed. Coastal Environments – NE Massachusetts and New Hampshire: Field Trip Guidebook for Eastern Section of SEPM, May 9-11. pp. 290-315.

Hayes, M.O. and T.W. Kana, eds.1976. Terrigenous Clastic Depositional Environments. Tech Rep 11-CRD, Geol, Dept., Univ. of South Carolina, Columbia. 315 pp.

Hayes, M.O. and J. Michel. 1982. Shoreline sedimentation within a forearc embayment, lower Cook Inlet, Alaska. Journal of Sedimentary Petrology 52:251-263.

Hayes, M.O. and T.M Montello. 1993a. Coastal Inlet Protection Strategies for Oil-spill Response. Vol. I. San Diego, Orange, Los Angeles, and Santa Barbara Channel Planning Areas. Rept. by RPI for Marine Spill Response Corporation and California Department of Fish and Game Office of Oil Spill Prevention and Response.

Hayes, M.O. and T.M Montello. 1993b. Coastal Inlet Protection Strategies for Oil-spill Response. Vol. II. Central Coast and San Francisco Bay/Delta Planning Areas. Rept. by RPI for Marine Spill Response Corporation and California Department of Fish and Game Office of Oil Spill Prevention and Response.

Hayes, M.O. and C.H. Ruby. 1994 Chapter 10: Pacific Coast of Alaska. In: Davis, R.A., Jr., ed. Barrier Islands. Springer Verlag, New York. pp. 395-433.

Hayes, M.O. and J. Michel. 1994. Options for replacing sand excavated during cleanup at the UNOCAL Guadalupe oil field. Memorandum to J. Morris, NOAA Scientific Support Coordinator, 18 July 1994. 10 pp.

Hayes, M.O. and J. Michel. 1999. Factors determining the long-term persistence of *Exxon Valdez* oil in gravel beaches. Marine Pollution Bulletin 38:92-101.

Hayes, M.O. and J. Michel. 2008. A Coast for All Seasons: A Naturalist's Guide to the Coast of South Carolina. Pandion Books, Columbia, SC. 285 pp.

Hayes, M.O., C.H Ruby, M.F. Stephen and S.J Wilson. 1976a. Geomorphology of the southern coast of Alaska. In: Proc. Fifteenth Conference on Coastal Engineering, Honolulu, HI. pp. 1992-2008.

Hayes, M.O., D.M. FitzGerald, L.J. Hulmes and S.J. Wilson. 1976b. Geomorphology of Kiawah Island, South Carolina. In: Hayes, M.O. and T.W. Kana, eds. Terrigenous Clastic Depositional Environments. Tech. Rep. 11-CRD, Geology Dept., Univ. of South Carolina, Columbia, SC. pp. II-80-100.

Hayes, M.O., J. Michel and F.J. Brown. 1977. Vulnerability of coastal environments of Lower Cook Inlet, Alaska to oil spill impact. In: Proc. 4th International Conference on Port and Ocean Engineering Under Arctic Conditions, 26-30 September 1977, Memorial Univ. of Newfoundland. 12 pp.

Hayes, M.O., E.R. Gundlach and C.D. Getter. 1980. Sensitivity ranking of energy port shorelines. In: Specialty Conference Proceedings, American Society of Civil Engineers, New York. pp. 697-709.

Hayes, M.O., J. Michel and B. Fichaut. 1990. Oiled gravel beaches: a special problem. Proc. Conference on Oil Spills: Management and Legislative Implications, 15-18 May 1990, Newport, RI. pp. 444-457.

Hayes, M.O., W.J. Sexton and K.N. Sipple. 1994. Fluid-bearing capacity of strandline deposits – Implication for hydrocarbon exploration (Abstract). American Assoc. Petrol. Geologists Bulletin 68:485.

Henson, P. and D.J. Usner. 1993. The Natural History of Big Sur. Univ. Cal. Press, Berkeley, CA. 416 pp.

Hess, H.H. 1962. History of ocean basins. In: Engel., A.E.J., H.L. James and B.F. Leonard, eds. Petrologic Studies: A Volume in Honor of A.F. Buddington. Geology Society of America. pp. 599-620.

Highway Research Board. 1958. Landslides and Engineering Practice. Special Report 29. Highway Research Board, Washington, DC. 544 pp.

Hopson, C.A., J.M. Mattinson and E.A Pessagno, Jr. 1981. Coast Range Ophiolite, western California. In: Ernst, W.G., ed. The Geotectonic Development of California. Prentice-Hall, Englewood Cliffs, NJ. pp. 418-510.

Howard, J.D. and J. Dorjes. 1972. Animal-sediment relationships in two beach-related tidal flats, Sapelo Island,

Georgia. Journal of Sedimentary Petrology 42:608-623.

Huber, N.K. and C. Wahrhaftig. 1987. The Geologic Story of Yosemite National Park. U.S. Geological Survey Bulletin 1595. 64 pp.

Hutton, J. 1785. Theory of the Earth with Proof and Illustrations. Royal Society of Edinburgh.

Hyndman, D. and D. Hyndman. 2006. Natural Hazards and Disasters. Thomson Brooks/Cole, Belmont, CA. 490 pp.

Inman, D.L. 2005. Littoral cells. In: Schwartz, M.L., ed. Encyclopedia of Coastal Science. Springer, The Netherlands. pp. 594-599.

Inman, D.L. and T.K Chamberlain. 1960. Littoral sand budget along the southern California coast. In: Volume of Abstracts. Report of the 21st International Geological Congress, Copenhagen, Denmark. pp. 245-246.

Inman, D.L. and C.E. Nordstrom. 1971. On the tectonic and morphologic classification of coasts. Journal of Geology 79:1-21.

I. Masters. 1991. Budget of sediment and prediction of future state of the coast. In: State of t, San Diego Region, Coast of California Storm and Tidal Waves Study. U.S. Army Corps of p.

I. Masters. 1994. Status of research on the nearshore. Shore & Beach 62:11-20.

. Jenkins. 2005a. Energy and sediment budgets of the global coastal zone. In: Schwartz, M.L., ed. Coastal Science. Springer, The Netherlands. pp. 408-415.

. Jenkins. 2005b. Climate patterns in the coastal zone. In: Schwartz, M.L., ed. Encyclopedia of . Springer, The Netherlands, pp. 243-246.

ission on Stratigraphy. 2008. Revised geological time scale. Available at: http://www.stratigraphy. hart2008.pdf.

ke, Jr. and R.N. Brother. 1986. Blue schist metamorphism of the eastern Franciscan belt, northern California. Geological Society Amer. Memoir No. 164. pp. 107-123.

Jessey, D.R., R.E. Burns, L.L Michalka and D.M. Wall. 2003. Spring 2003 Field Trip, Big Sur. Cal Poly-Pomona, Geology Club. 32 pp. Available at: http://geology.csupomona.edu/docs/BigSur.pdf.

Kana, T.W. 1977. Suspended sediment transport of Price Inlet, South Carolina. Proc. Coastal Sediments '77, American Society of Civil Engineers. pp. 366-382.

Kana, T.W. 1979. Suspended sediments in breaking waves. Ph.D. Dissertation, Dept. of Geology, Univ. of South Carolina, Columbia. 153 pp.

Kana, T.W. and P. McKee. 2003. Relocation of Captain Sam's Inlet – 20 years later. Proc. Coastal Sediments '03, American Society Civil Engineers, New York. 12 pp.

Kana, T.W., T.E. White and P. McKee. 2004. Management and engineering guidelines for groin rehabilitation. Journal of Coastal Research 33:57-82.

Karl, S.M. 1984. Sedimentologic, diagenetic, and geochemical analysis of Upper Mesozoic ribbon cherts from the Franciscan assemblage at the Marin Headlands, California. In: Blake, M.C., Jr., ed. Franciscan Geology of Northern California. Pacific Section SEPM 43:71-88.

Karnovsky, N.J., L.B. Spear, H.R. Carter, D.G. Ainley, K.D. Amey, L.T. Ballance, K.T. Briggs, R.G. Ford, G.L. Hunt, Jr., J.W. Keiper, K.H Morgan, R.L Pitman and C.T. Tynan. 2005. At-sea distribution, abundance and habitat affinities of Xantus's Murrelets. Marine Ornithology 33:89-104.

Keller, M. and J.A. Ba. 1993. Re-evaluation of the Miguelito Member of the Pismo Formation of Montaña de Oro State Park, California, including new diatom age data (Abstract). Amer. Assoc. of Petroleum Geologists Bulletin, vol. 77, Pacific Section Mtg.

Kelly, J.P. 1998. Status of waterbirds on Tomales Bay, California. In: Thomas, T., ed. Symposium on Current Research in Golden Gate National Recreation Area. San Francisco, CA, September 10, 1998.

Kelly, J.P. 2001. Distribution and abundance of winter shorebirds on Tomales Bay, California: Implications for conservation. Western Birds 32:145-166.

Kistler, R.W. and D.E. Champion. 2001. Rb-Sr whole-rock and mineral ages, K-Ar, 40Ar/39Ar, and U-Pb mineral ages, and strontium, lead, neodymium, and oxygen isotopic compositions for granitic rocks from the Salinian composite terrane, California. USGS Open File Report 01-453. 85 pp.

Komar, P.D. 1976. Beach Processes and Sedimentation. Prentice-Hall, Inc., Englewood Cliffs, NJ. 429 pp.

Komar, P.D. and C. Wang. 1984. Processes of selective grain transport and the formation of placers on beaches. Journal of Geology 92:637-655.

Komar, P.D. and Z. Li. 1988. Pivoting analyses of the selective entrainment of sediments by shape and size with application to gravel threshold. Sedimentology 33:425-436.

Krumbein, W.C. 1944. Shore Processes and Beach Characteristics. Beach Erosion Board, Tech. Memo. No. 3, U.S. Army Corps of Engineers, Washington, DC.

Lage, J. 2004. Point Reyes: The Complete Guide to the National Seashore & Surrounding Area. Wilderness Press, Berkeley, CA. 249 pp.

Lahee, F.H. 1952. Field Geology. McGraw-Hill Book Co., New York. 883 pp.

Lambeck, K. and J. Chappell. 2001. Sea-level change through the last glacial cycle. Science 292:679-686.

Leech, M. 2001. Field Trip to the Marin Headlands and Half Moon Bay Coast. Geol. 116, Continuing Education Studies, Stanford University, 21 May 2001. Available at: http://online.sfsu.edu/~leech/geo116/coastalgeology.pdf.

Lorang, M.S. 1991. An artificial perched-gravel beach as a shore protection structure. Proc. Coastal Sediments '91, American Society of Civil Engineers, New York. pp. 1916-1925.

Lorang, M.S. 2000. Predicting threshold entrainment mass for a boulder beach. Journal of Coastal Research 16:432-445.

Lowry, M.S., J.V. Carretta and K.A. Forney. 2005. Pacific Harbor Seal, *Phoco vitulina richardsi*, Census in California during May-July 2004. National Marine Fisheries Service, Southwest Fisheries Science Center Administrative Report LJ-05-06, La Jolla, CA. 38 pp.

Lyday, S., J. Roletto, D. Devlin, J. Hall, N. Bubert, R. Stallcup, J. Ferlin, D. Klotz, F Higgason, K. Gulland and D. Greig. 2008. Beach Watch Fifteen Year Report: 1993-2008. Unpublished report, National Oceanic and Atmospheric Administration, Office of National Marine Sanctuaries, Gulf of the Farallones National Marine Sanctuary, San Francisco, CA. 16 pp.

Manuwal, D.A., H.R. Carter, T.S Zimmerman and D.L. Orthmeyer, eds. 2001. Biology and Conservation of the Common Murre in California, Oregon, Washington, and British Columbia. Volume 1: Natural History and Population Trends. U.S. Geological Survey, Biological Resources Division, Information and Technology Report USGS/BRD/ITR–2000-0012, Washington, DC. 132 pp.

Marine Mammal Center. 2008. Northern elephant seal. Available at: http://www.marinemammalcenter.org/pdfs/library/Northern_Elephant_Seal.pdf.

Martin, J.W. 2009. The Great Valley Sequence. Available at: http://www.johnmartin.com/earthquakes/eqsafs/safs_335.htm.

Mason, G. 2006. Self-guided walk to Valencia Peak. Available at: http://morro-bay.com/docents/geo-mason/.

Mavericks. 2007. Mavericks surf contest. Available at: http://www.maverickssurf.com/Pulse/archive/2007/December/carbon_neutrality.asp.

MBNMS. 2009. Monterey Bay National Marine Sanctuary list of intertidal rocky communities. Available at: http://montereybay.noaa.gov/sitechar/rock1.html.

MBNMS-Maps. 2009. Monterey Bay National Marine Sanctuary – Sanctuary Maps. Available at: http://montereybay.noaa.gov/Intro/maps.html.

McChesney, G.J. and 13 others. 2006. Restoration of Common Murre Colonies in Central California. Annual Report 2005. U.S. Fish and Wildlife Service, Newark, CA. 78 pp.

McWilliams, M.O. and D.G. Howell. 1982. Exotic terranes of Western California. Nature 5863:215-217.

Mehring, P. and M. Cloos. 2008. Petrology of the San Simeon Coast Range Ophiolite, San Simeon, California. Abstracts with Programs, Geol. Soc. of America 40:530.

Michel, J. and M.O. Hayes. 1991. Geomorphological Controls on the Persistence of Shoreline Contamination from the *Exxon Valdez* Oil Spill. Report HMRAD 91-2. Hazardous Materials Response and Assessment Division, National Oceanic and Atmospheric Administration, Seattle, WA. 306 pp.

Michel, J. and M.O. Hayes. 1999. Weathering patterns of oil residues eight years after the *Exxon Valdez* oil spill. Marine Pollution Bulletin 38:855-863.

Monroe, J.S. and R. Wicander. 1998. Physical Geology (3rd edition). Wadsworth Publishing. Company, Belmont, CA. 646 pp.

Montello, T.M., M.O. Hayes and J. Michel. 1993. Results of Beach-monitoring Surveys, Northern and Central California. RPI report R-93-7. Office of Spill Prevention and Response. California Department of Fish and Game, Sacramento, CA. 53 pp.

Monterey Bay National Marine Sanctuary. 2008. Geology. Available at: http://sanctuarysimon.org/monterey/sections/

geology/overview.php?sec=g.

Moore, L.J. and G.B. Griggs. 2002. Long-term cliff retreat and erosion hotspots along the central shores of Monterey Bay National Marine Sanctuary. Marine Geology 181:265-283.

Moores, E.M. and R.J. Twiss. 1995. Tectonics. W.H. Freeman, New York. 415 pp.

Moores, E.M., and J.E. Moores. 2001. Geology of Puta-Cache: The Great Valley Sequence. Available at: http://bioregion.ucdavis.edu/book/06_Monticello_Dam/06_03_moores_geo_gvs.html.

MTYcounty.com. 2002. Gray whale. Available at: http://www.mtycounty.com/pgs-animals/g-whale.html.

My Highlife. 2004. Definition of tsunami. Available at: http://highlife.proweblog.com/print.php?p=2.

Nairn, R. and M.O. Hayes. 1997. Large-scale coastal evolution in the vicinity of Keta Lagoon, Ghana. Proc. International Conference on Coast Research through Large-Scale Experiments, University of Plymouth.

National Data Buoy Center. 2009. Wave observations, California Coast. Available at: http://www.ndbc.noaa.gov/.

National Seashore. 2009. Point Reyes National Seashore – Birds. Available at: nps.gov/pore/naturescience/birds.htm.

National Ocean Survey. 1973. Tidal current charts, San Francisco Bay, 6th Edition. U.S. Dept. of Commerce, National Oceanic and Atmospheric Administration, Rockville, MD.

Nevins, H.M. and H.R. Carter. 2003. Age and sex of Common Murres *Uria aalge* recovered during the 1997-98 Point Reyes tarball incidents in central California. Marine Ornithology 31:51-58.

NMFS. 1997. Draft Environmental Assessment: Use of Acoustic Pingers as a Management Measure in Commercial Fisheries to Reduce Marine Mammal Bycatch. NMFS, Office Of Protected Resources, Silver Spring, MD.

NGCD. 2009. Age of the ocean floor. NGDC Data Announcement Number: 96-MGG-04NGDC, Report MGG-12. Available at: http://www.ngdc.noaa.gov/mgg/image/images/g01167-pos-a0001.pdf.

NOAA. 2008. Northern elephant seal. Available at: http://www.afsc.noaa.gov/nmml/species/species_ele.php.

NOAA. 2009. Map of California pinniped rookeries and haulout sites. Available at: http://swfscdata.nmfs.noaa.gov/pinniped/viewer.htm.

NOAA/RPI. 1994. Environmental Sensitivity Index (ESI) Maps of the Shoreline of Northern California. By Research Planning, Inc. (RPI) under contract to the National Oceanic and Atmospheric Administration, NOS Data Explorer. 51 maps.

NOAA/RPI. 1996a. Environmental Sensitivity Index (ESI) Maps of the Shoreline of South Carolina. By Research Planning, Inc. (RPI) under contract to the National Oceanic and Atmospheric Administration, NOS Data Explorer. 63 maps.

NOAA/RPI. 1996b. Environmental Sensitivity Index (ESI) Maps of the Shoreline of North Carolina (3 volumes). By Research Planning, Inc. (RPI) under contract to the National Oceanic and Atmospheric Administration, NOS Data Explorer. 135 maps.

NOAA/RPI. 1993-2001. Environmental Sensitivity Index (ESI) Maps of the Shoreline of Southeast Alaska (four volumes). By Research Planning, Inc. (RPI) under contract to the National Oceanic and Atmospheric Administration, NOS Data Explorer. 237 maps.

NOAA/RPI. 1999. Environmental Sensitivity Index (ESI) Maps of the Shoreline of Massachusetts. By Research Planning, Inc. (RPI) under contract to the National Oceanic and Atmospheric Administration, NOS Data Explorer. 55 maps.

NOAA/RPI. 2006. Environmental Sensitivity Index (ESI) Maps of the Shoreline of Central California. By Research Planning, Inc. (RPI) under contract to the National Oceanic and Atmospheric Administration, NOS Data Explorer. 41 maps.

NRC (National Research Council). 2009. Shellfish Mariculture in Drakes Estero, Point Reyes National Seashore, California. National Academies Press, Washington, DC. 138 pp

O'Brien, M.P. 1931. Estuary tidal prisms related to entrance areas. Jour. Civil Engineers 1(8):738-793.

Orford, J.D. 1975. Discrimination of particle zonation on a pebble beach. Sedimentology 22:441-463.

Orme, A.R. 2005. Chapter Sixteen: Morro Bay to Point Conception. In: Griggs, G.B., K. Patsch and L. Savoy, eds. Living with the Changing California Coast. Univ. Cal. Press, Berkeley, CA. pp. 334-358.

Orme, A.R. and V.P. Tchakerian. 1989. Quaternary dunes of the Pacific Coast of the Californias. In: Nickling, W.G., ed. Aeolian Geomorphology. Allen and Unwin, Boston, MA. pp. 149-175.

Packwood, A.R. 1983. The influence of beach porosity on wave uprush and backwash. Coastal Engineering 7:29-40.

Page, B.M. 1970. Sur-Nacimiento Fault Zone of California – Continental margin tectonics. Geol. Soc. Amer. Bulletin 83:957-972.

Page, B.M. 1989. Coast Range uplifts and structural valleys. In: Wahrhaftig, C. and D. Sloan, eds. Geology of San Francisco and vicinity. 28th International Geological Congress Field Trip Guidebook T105. pp. 30-32.

Page, B.M., G.A. Thompson and R.G. Coleman. 1998. Late Cenozoic tectonics of the central and southern Coast Ranges in California. Geol. Soc. Amer. Bulletin 110(7): 846-876.

Pararas-Carayannis, G. 2009a. The tsunami page. The March 27, 1964 Great Alaska Tsunami. Available at: http://www.drgeorgepc.com/Tsunami1964GreatGulf.html.

Pararas-Carayannis, G. 2009b. The tsunami page. Historical tsunamis in California, Available at: http://www.drgeorgepc.com/TsunamiCalifornia.html.

Patsch, K. and G. Griggs. 2006. Littoral Cells, Sand Budgets, and Sand Beaches: Understanding California's Shoreline. Inst. of Marine Sciences, U.C. Santa Cruz, Cal. Sed. Management Workshop. 40 pp.

Peery, M.Z., L.A. Hall and J.T. Harvey. 2008. Abundance and Productivity of Marbled Murrelets off Central California during the 2008 Breeding Season. Report submitted to California State Parks, Half Moon Bay, CA.

Pierce, D.W., T.P Barnett and M. Latif. 2000. Connections between the Pacific ocean tropics and midlatitudes on decadal time scales. Jour. of Climate 13:1173-1194.

Pilkey, O.H. and M.E. Fraser. 2003. A Celebration of the World's Barrier Islands. Columbia Univ. Press, New York. 309 pp.

Pilkey, O.H. and L. Pilkey-Jarvis. 2006. Useless Arithmetic: Why Environmental Scientists Can't Predict the Future. Columbia Univ. Press, New York. 230 pp.

Plafker, G. 1969. Tectonics of the 27 March 1964 Alaska Earthquake. U.S. Geological Survey Prof. Paper 543-1. 74 pp.

Plafker, G. 1971. Possible future petroleum resources of Pacific Margin Tertiary Basin, Alaska. Am. Assoc. Petroleum Geol. Mem. 15. pp. 120-135.

Playfair, J. 1802. Illustrations of the Huttonian Theory of the Earth. Edinburgh. 528 pp.

Point Sur. 2009. Point Sur State Historic Park & Lighthouse. Available at: http://www.pointsur.org/

Postma, H. 1967. Sediment transport and sedimentation in the estuarine environment. Estuaries 83:158-179.

PRBO. 2009. PRBO Conservation Science. Bolinas Lagoon. Available at: http://www.prbo.org/cms/366

Pritchard, D.W. 1967. What is an estuary?: Physical viewpoint. In: Lauff, G.W., ed. Estuaries, Am. Assoc. Advancement of Science 83:3-5.

Reeg, H. 1999. Structural variations in deformed serpentinite, Monterey County, California. Extract from M.S. thesis with a slightly different title, Pomono College Geology Department, Claremont, California. 6 pp. Available at: http://keckgeology.org/files/pdf/symvol/12th/California/reeg.pdf.

Reimnitz, E. 1966. Late Quaternary history and sedimentation of the Copper River Delta and vicinity, Alaska. Ph.D. Dissertation, Univ. Cal., San Diego. 160 pp.

Reinen, L.A. and C. Davidson. 1999. Paleoseismicity of serpentinized shear zones in the Coast Range. Available at: http://keckgeology.org/files/pdf/symvol/12th/California/reinen_davidson.pdf.

Ricketts, E.F. and J. Calvin. 1939. Between Pacific Tides. Ricketts, E.F, J. Calvin and J. W. Hedgepeth. Revised by D.W. Phillips, 1985. 5th/Revised edition, Stanford Univ. Press, Stanford. 652 pp.

Roletto, J., J. Mortenson, I. Harrald, J. Hall and L. Grella. 2003. Beached bird surveys and chronic oil pollution in Central California. Marine Ornithology 31:21-28.

Rust, B.R. 1978. Depositional models for braided alluvium. In: Miall, A.D., ed. Fluvial Sedimentology. Can. Soc. Petrol. Geol. Memoir 5. pp. 605-625.

Savoy, L., D. Merritts, K. Grove and R. Walker. 2005. Chapter Eleven: Point Arena to San Francisco. In: Griggs, G.B., K. Patsch and L. Savoy, eds. Living with the Changing California Coast. Univ. Cal. Press, Berkeley, CA. pp. 204-221.

SCBC. 2004. Coastal Shore Stewardship: A Guide for Planners, Builders, and Developers. Published by the Stewardship Centre of British Columbia. Available at: http://www.stewardshipcentre.bc.ca/cdirs/st_series/index.php/17.

Seaturtle.org. 2009. Satellite tracking. Available at: http://www.seaturtle.org/tracking.

Sexton, W.J. and M.O. Hayes. 1996. Holocene deposits of reservoir-quality sand along the central South Carolina coastline. Amer. Assoc. Petrol. Geologists Bulletin 80:831-855.

Shepard, F.P. 1950a. Longshore Bars and Longshore Troughs. Tech. Memo 15, Beach Erosion Board. U.S. Army Corps of Engineers, Washington, DC. 38 pp.

Shepard, F.P. 1950b. Beach cycles in Southern California. Tech. Memo 20, Beach Erosion Board. U.S. Army Corps. of

Engineers, Washington, DC. 26 pp.

Shepard, F.P. and R.F. Dill. 1966. Submarine Canyons and Other Sea Valleys. Rand McNally and Co., Chicago. 381 pp.

Shepard, F.P. and H.R. Wanless. 1971. Our Changing Coastlines. McGraw-Hill, New York. 579 pp.

Shervais, J.W. 1990. Island arc and oceanic crust ophiolites: Contrasts in the petrology, geochemistry, and tectonic style of ophiolite assemblages in the California Coast Ranges. In: Malpas, J. et al., eds. Ophiolite, Oceanic Crust Analogues; Nicosia, Cyprus. Cyprus Geological Survey Department. pp. 507-520.

Shervais, J.W. 2003. The Coast Range Ophiolite (CRO) California and the Jurassic tectonic evolution of the Western Cordillera in North America (Abstract). Geol. Soc. Amer., Cordilleran Section, 99th annual Meeting, 1-3 April 2003.

Silberstein, M. and E. Campbell. 1989. Elkhorn Slough. Monterey Bay Aquarium, Monterey, CA

Silvester, R. 1974. Coastal Engineering, Vol. II. Elsevier Pub. Co. 338 pp.

Silvester, R. 1977. The role of wave reflection in coastal processes. Proc. Coastal Sediments '77, Amer. Soc. Civil Engineers, New York. pp. 639-654.

Simons, D.B. and E.V. Richardson. 1962. Resistance to flow in alluvial channels. Am. Soc. Civil Eng. Transactions 127:927-953.

Slingerland, R.L. 1977. The effects of entrainment on the hydraulic equivalence relationships of light and heavy minerals in sands. Journal of Sedimentary Petrology 47:753-770.

Sloan, D. 2006. Geology of the San Francisco Bay Region. Univ. Cal. Press, Berkeley, CA. 335 pp.

Snyder, M.A., L.C. Sloan, N.S. Diffenbaugh and J.L. Bell. 2003. Future climate change and upwelling in the California Current. Geophysical Research Letters 30:CLM8.1-CLM8.4.

Sorensen, S., G.E. Harlow and D. Rumble. 2006. The origin of jadeitite-forming subduction-zone fluids: CL-guided SIMS oxygen-isotope and trace-element evidence. American Mineralogist 91:979-996.

Stamski, R. 2005. The Impacts of Coastal Protection Structures in California's Monterey National Marine Sanctuary. Report, Marine Sanctuaries Div., NOAA, Silver Springs, MD. 18 pp.

Steinbeck, J. 1933. To a God Unknown. Penguin Books, New York. 188 pp.

Steinbeck, J. 1935. Tortilla Flat. Penguin Books, New York. 174 pp.

Steinbeck, J. 1945. Cannery Row. Penguin Books, New York. 185 pp.

Steinbeck, J. 1952. East of Eden. Penguin Books, New York. 601 pp.

Steinbeck, J. 1954. Sweet Thursday. Penguin Books, New York. 260 pp.

Stephenson, W. and R. Kirk. 2005. Shore platforms. In: Schwartz, M.L., ed. Encyclopedia of Coastal Science. Springer, New Zealand. pp. 873-875.

Stoecker, M.W. 2004. Steelhead Migration Barrier Inventory and Recovery Opportunities for the Santa Ynez River, Ca. Community Environmental Council, Santa Barbara, CA. 238 pp.

Storlazzi, C.D. and G.B. Griggs. 1998. The 1997-1998 *El Niño* and erosion processes along the Central Coast of California. Shore and Beach 66:3:12-17.

Sunamura, T. 1992. Geomorphology of Rocky Coasts. Wiley, New York. 314 pp.

Thornton, E.B., A. Sallenger, J. Conforto Sesto, L. Egley, T. McGee and R. Parsons. 2006. Sand mining impacts on long-term dune erosion in southern Monterey Bay. Marine Geology 229:45-58.

Trask, P.D. 1959. Beaches near San Francisco, California, 1956-57. Tech. Memo. No. 110, Beach Erosion Board. U.S. Army Corps of Engineers, Washington, DC.

Trasko, K.P. 2007. Faulting in the Lopez Point-Ragged Point segment of Highway 1. Southern Big Sur Area, California: Implication for the Nacimiento Fault (Abstract). 103rd Ann. Mtg., Cordilleran Sect., Geol. Soc. America.

Trenhaile, A.S. 1987. The Geomorphology of Rock Coasts. Oxford Univ. Press, New York. 393 pp.

UCal-Berkeley. 2009. San Francisco History. Available at: http://seismo.berkeley.edu/seismo/faq/1957_0.html.

Underwood, M.B., M.W. Laughland, K.L. Shelton and R.L. Sedlock. 1995. Thermal-maturity trends within Franciscan rocks near Big Sur, California: Implications for offset along the San Gregorio-San Simeon-Hosgri fault zone. Geology 23:839-842.

Univ. Nevada-Reno. 2009. Available at: http://quake.unr.edu/ftp/pub/louie/class/100/magnitude.html.

U.S. Army Corps of Engineers. 1971. National Shoreline Study. Washington, DC.

U.S. Army Corps of Engineers. 2002. Coastal Engineering Manual. Engineer Manual 1110-2-1100, U.S. Army Corps of Engineers, Washington, DC. Available at: http://bigfoot.wes.army.mil/cem026.html.

USGS. 1997. Geology of Point Reyes National Seashore and Vicinity, California: A Digital Database. Open-File Report 97-456. Available at: http://pubs.usgs.gov/of/1997/of97-456/.

USGS. 2005. Diagram of the interior of the earth. Available at: http://pubs.usgs.gov/gip/dynamic/inside.html.

USGS/Seal Cove. 2009. The San Andreas and San Gregorio Fault Systems in San Mateo County. Available at: http://pubs.usgs.gov/of/2005/1127/chapter8.pdf.

USGS-earthquake. 2009. Quaternary fault and fold database of the U.S. Available at: http://earthquake.usgs.gov/regional/qfaults/.

Van Dyke, E. and K. Wasson. 2005. Historical ecology of a Central California estuary: 150 years of habitat change. Estuaries 28 (2):173-189.

Van Straaten, L.M.J.U. 1950. Environment of formation and facies of the Wadden Sea sediments. Koninkl.Ned. Aardrijkskde Genoot. 67:94-108.

Wahrhaftig, C. and B. Murchey. 1987. Marin Headlands, California: 100-million-year Record of Sea Floor Transport and Accretion. Geological Society of America Centennial Field Guide, Volume 1 – Cordilleran Section. pp. 263-268.

Wallace, R., ed. 1990. The San Andreas Fault System, California. U.S. Geological Survey Professional Paper 1515. 283 pp.

Wang, P., B.A. Ebersole and E.R. Smith. 2002. Longshore sand transport – initial results from large-scale sediment transport facility. ERDC/CHL CHETN-II-46. pp. 1-14.

Weber, G.E. and A.O. Allwardt. 2001. Geology from Santa Cruz to Point Año Nuevo—The San Gregorio Fault Zone and Pleistocene Marine Terraces. NAGT Field Trip Guidebook. Available at: http://pubs.usgs.gov/bul/b2188ch1.pdf.

Wegener, A. 1912. "Die Entstehung der Kontinente." Peterm. Mitt. 185–195, 253–256, 305–309.

Wentworth, C.K. 1922. A scale of grade and class terms for clastic sediments. J. of Geology 30:377-39.

White, W.R. and T.J. Day. 1982. Transport of graded gravel bed material. In: Hey, R.D., J.D. Bathurst and C.R. Thorne, eds. Gravel-bed Rivers. John Wiley and Sons, New York. pp. 181-223.

Wiegel, R.L. 1964. Oceanographical Engineering. Prentice-Hall, Englewood Cliffs, NJ. 532 pp.

Wilcock, P.R. and B.T. DeTemple. 2005. Persistence of armor layers in gravel-bed streams. Geophysical Research Letters 32, L08402, doi:10.1029/2004GL021772.

Wilson, J.T. 1965. A new class of faults and their bearing on continental drift. Nature 207:343-347.

Winter, J.D. 2001. An Introduction to Igneous and Metamorphic Rocks. Prentice Hall, Upper Saddle River, NJ. 697 pp.

Wood, F.J. 1982. Tides. In: Schwartz, M.L., ed. The Encyclopedia of Beaches and Coastal Environments. Hutchinson Ross Pub. Co., Stroudsburg, PA. pp. 826-837.

Woodroffe, C.D. 2002. Coasts. Cambridge University Press, Cambridge. 623 pp.

Woolfolk, A. 2005. Evolution of Elkhorn Slough and Associated Wetlands 20,000 years Before Present (ybp) to 1880 A.D. Elkhorn Slough Tidal Wetland Plan, 9/7/05. 7 pp.

Wright, T. 1996. The Geology of Bodega Head: The Salinian Terrane West of the San Andreas Fault. Available at: sonoma.edu/geology/wright/Bhead.html.

Zenkovitch, V.P. 1967. Processes of Coastal Development. Interscience, New York. 738 pp.

Zingg, T.H. 1935. Beitrag zur Schotternalyse. Schweiz. Min. u. Pet. Mitt. 15:39-140.

INDEX

ABOUT THE AUTHORS

Miles O. Hayes

Dr. Miles O. Hayes is a coastal geomorphologist with over 50 years of research experience. He has authored over 250 articles and reports and three books on numerous topics relating to tidal hydraulics, river morphology and processes, beach erosion, barrier island morphology, oil pollution, and petroleum exploration. Hayes' teaching experience includes both undergraduate and graduate courses while a Professor at the Universities of Massachusetts and South Carolina. Seventy-two graduate students received their degrees under his supervision, most of whom are now leaders in their respective academic, government and industry positions. He is considered to be the "Father of Coastal Geology." A review of his 1999 book – Black Tides, said

"A skilled raconteur, Hayes tells engrossing stories of responding to most of the recent, headline-grabbing oil spills, including the Gulf War spills, the *Exxon Valdez*, the *Amoco Cadiz* spill in France, and the *Ixtoc I* blowout in Mexico. Interspersed among them are personal events and adventures, such as his survival of a plane crash while mapping a remote part of Alaska." He is Chairman of the Board of Research Planning, Inc. (RPI), a science technology company located in Columbia, S. C.

Jacqueline Michel

Dr. Jacqueline Michel is an internationally recognized expert in oil and hazardous materials spill response and assessment with a primary focus in the areas of oil fates and effects, non-floating oils, shoreline cleanup, alternative response technologies, and natural resource damage assessment. As of this time, she has participated in research projects in 33 countries. Much of her expertise is derived from her role, since 1982, as part of the Scientific Support Team to the U.S. Coast Guard provided by the National Oceanic and Atmospheric Administration. Under this role, she is on 24-hour call and provides technical support for an average of 50 spill events per year. She leads shoreline assessment teams and assists in selecting cleanup methods to minimize the environmental impacts of the spill. She has written over 150 manuals, reports, and scientific papers on coastal resource impacts, mapping, and protection. As a member of the Ocean Studies Board at the National Academy of Sciences for four years, she served on four National Research Council committees (chairing two), and is a Lifetime Associate of the National Academies. One of the original founders of RPI, which started in 1977, she now serves as the company President.